DISCRETE WAVELET TRANSFORMATIONS

BICENTENNIAL
1807
WILEY
2007
BICENTENNIAL

THE WILEY BICENTENNIAL–KNOWLEDGE FOR GENERATIONS

*E*ach generation has its unique needs and aspirations. When Charles Wiley first opened his small printing shop in lower Manhattan in 1807, it was a generation of boundless potential searching for an identity. And we were there, helping to define a new American literary tradition. Over half a century later, in the midst of the Second Industrial Revolution, it was a generation focused on building the future. Once again, we were there, supplying the critical scientific, technical, and engineering knowledge that helped frame the world. Throughout the 20th Century, and into the new millennium, nations began to reach out beyond their own borders and a new international community was born. Wiley was there, expanding its operations around the world to enable a global exchange of ideas, opinions, and know-how.

For 200 years, Wiley has been an integral part of each generation's journey, enabling the flow of information and understanding necessary to meet their needs and fulfill their aspirations. Today, bold new technologies are changing the way we live and learn. Wiley will be there, providing you the must-have knowledge you need to imagine new worlds, new possibilities, and new opportunities.

Generations come and go, but you can always count on Wiley to provide you the knowledge you need, when and where you need it!

WILLIAM J. PESCE
PRESIDENT AND CHIEF EXECUTIVE OFFICER

PETER BOOTH WILEY
CHAIRMAN OF THE BOARD

DISCRETE WAVELET TRANSFORMATIONS
An Elementary Approach with Applications

PATRICK J. VAN FLEET
University of St. Thomas

WILEY-
INTERSCIENCE

A JOHN WILEY & SONS, INC., PUBLICATION

Library of Congress Cataloging-in-Publication Data:

Van Fleet, Patrick J., 1962–
  Discrete wavelet transformations : an elementary approach with applications
  Patrick J. Van Fleet
    p. cm.
  Includes bibliographical references and index.
  ISBN 978-0-470-18311-3 (cloth : acid-free paper)
  1. Wavelets (Mathematics) 2. Transformations (Mathematics) 3. Digital images—Mathematics. I. Title.
  QA403.3.V36 2008
  515'.2433—dc22   2007027174

Printed in the United States of America.

10 9 8 7 6 5 4 3 2 1

For Andy

CONTENTS

Preface   xiii

Acknowledgments   xxiii

1   Introduction: Why Wavelets?   **1**

2   Vectors and Matrices   **15**

   2.1   Vectors, Inner Products, and Norms   16
          Problems   20
          Computer Lab   22
   2.2   Basic Matrix Theory   22
          Problems   36
          Computer Lab   38
   2.3   Block Matrix Arithmetic   38
          Problems   45
          Computer Lab   48

3	**An Introduction to Digital Images**		**49**
	3.1	The Basics of Grayscale Digital Images	50
		Problems	66
		Computer Labs	70
	3.2	Color Images and Color Spaces	70
		Problems	75
		Computer Labs	78
	3.3	Qualitative and Quantitative Measures	78
		Problems	84
		Computer Labs	88
	3.4	Huffman Encoding	88
		Problems	94
		Computer Labs	95
4	**Complex Numbers and Fourier Series**		**97**
	4.1	The Complex Plane and Arithmetic	98
		Problems	103
		Computer Lab	104
	4.2	Complex Exponential Functions	104
		Problems	109
	4.3	Fourier Series	110
		Problems	121
		Computer Lab	126
5	**Convolution and Filters**		**127**
	5.1	Convolution	128
		Problems	135
		Computer Lab	137
	5.2	Filters	138
		Problems	147
		Computer Lab	150
	5.3	Convolution as a Matrix Product	150
		Problems	155
6	**The Haar Wavelet Transformation**		**157**
	6.1	Constructing the Haar Wavelet Transformation	158

		Problems	171
		Computer Labs	173
	6.2	Iterating the Process	173
		Problems	182
		Computer Labs	183
	6.3	The Two-Dimensional Haar Wavelet Transformation	183
		Problems	198
		Computer Labs	199
	6.4	Applications: Image Compression and Edge Detection	199
		Problems	216
		Computer Labs	221

7   Daubechies Wavelet Transformations   **223**

	7.1	Daubechies Filters of Length 4 and 6	224
		Problems	243
		Computer Labs	252
	7.2	Daubechies Filters of Even Length	253
		Problems	263
		Computer Labs	265
	7.3	Algorithms for Daubechies Wavelet Transformations	265
		Problems	278
		Computer Labs	279

8   Orthogonality and Fourier Series   **281**

	8.1	Fourier Series and Lowpass Filters	282
		Problems	288
	8.2	Building $G(\omega)$ from $H(\omega)$	288
		Problems	297
	8.3	Coiflet Filters	298
		Problems	312
		Computer Labs	316

9   Wavelet Shrinkage: An Application to Denoising   **317**

	9.1	An Overview of Wavelet Shrinkage	318
		Problems	323
		Computer Labs	324

	9.2	VisuShrink	325
		Problems	333
		Computer Labs	334
	9.3	SureShrink	335
		Problems	346
		Computer Labs	350

10	**Biorthogonal Filters**		**351**
	10.1	Constructing Biorthogonal Filters	353
		Problems	368
	10.2	Biorthogonal Spline Filters	371
		Problems	391
		Computer Lab	393
	10.3	The Cohen–Daubechies–Feauveau 9/7 Filter	394
		Problems	403
		Computer Lab	406

11	**Computing Biorthogonal Wavelet Transformations**		**407**
	11.1	Computing the Biorthogonal Wavelet Transformation	408
		Problems	416
		Computer Labs	416
	11.2	Computing the Inverse Biorthogonal Wavelet Transformation	417
		Problems	434
		Computer Labs	434
	11.3	Symmetry and Boundary Effects	435
		Problems	453
		Computer Labs	457

12	**The JPEG2000 Image Compression Standard**		**459**
	12.1	An Overview of JPEG	460
		Problems	467
		Computer Lab	467
	12.2	The Basic JPEG2000 Algorithm	468
		Problems	475
	12.3	Lifting and Lossless Compression	476
		Problems	484

	Computer Lab	486
12.4	Examples	486
	Computer Lab	492

Appendix A: Basic Statistics   **493**

A.1	Descriptive Statistics	493
	Problems	495
A.2	Sample Spaces, Probability, and Random Variables	496
	Problems	499
A.3	Continuous Distributions	499
	Problems	505
A.4	Expectation	506
	Problems	511
A.5	Two Special Distributions	512
	Problems	515

| References | 519 |

| Index | 525 |

PREFACE

Why This Book?

How do you apply wavelets to images? This question was asked of me by a bright undergraduate student while I was a professor in the mid-1990s at Sam Houston State University. I was part of a research group there and we had written papers in the area of multiwavelets, obtained external funding to support our research, and hosted an international conference on multiwavelets. So I fancied myself as somewhat knowledgeable on the topic. But this student wanted to know how they were actually *used* in the applications mentioned in articles she had read. It was quite humbling to admit to her that I could not exactly answer her question. Like most mathematicians, I had a cursory understanding of the applications, but I had never written code that would apply a wavelet transformation to a digital image for the purposes of processing it in some way. Together, we worked out the details of applying a discrete Haar wavelet transform to a digital image, learned how to use the output to identify the edges in the image (much like what is done in Section 6.4), and wrote software to implement our work.

My first year at the University of St. Thomas was 1998-1999 and I was scheduled to teach Applied Mathematical Modeling II during the spring semester. I wanted to select a current topic that students could immediately grasp and use in concrete applications. I kept returning to my positive experience working with the undergraduate student at Sam Houston State University on the edge detection problem. I was surprised by the number of concepts from calculus and linear algebra that we had reviewed in the process of coding and applying the Haar wavelet transform. I was also impressed with the way the student embraced the coding portion of the work and connected to it ideas from linear algebra. In December 1998, I attended a wavelet workshop organized by Gilbert Strang and Truong Nguyen. They had just authored the book *Wavelets and Filter Banks* [73], and their presentation of the material focused a bit more on an engineering perspective than a mathematical one. As a result, they developed wavelet filters by using ideas from convolution theory and Fourier series.

I decided that the class I would prepare would adopt the approach of Strang and Nguyen and I planned accordingly. I would attempt to provide enough detail and background material to make the ideas accessible to undergraduates with backgrounds in calculus and linear algebra. I would concentrate only on the development of the *discrete* wavelet transform. The course would take an "applications first" approach. With minimal background, students would be immersed in applications and provide detailed solutions. Moreover, the students would make heavy use of the computer by writing their own code to apply wavelet transforms to digital audio or image files. Only after the students had a good understanding of the basic ideas and uses of discrete wavelet transforms would we frame general filter development using classical ideas from Fourier series. Finally, wherever possible, I would provide a discussion of *how* and *why* a result was obtained versus a statement of the result followed by a concise proof and example. The first course was enough of a success to try again. To date, I have taught the course seven times and developed course materials (lecture notes, software, and computer labs) to the point where colleagues can use them to offer the course at their home institutions.

As is often the case, this book evolved out of several years' worth of lecture notes prepared for the course. The goal of the text is to present a topic that is useful in several of today's applications involving digital data in such a way that it is accessible to students who have taken calculus and linear algebra. The ideas are motivated through applications — students learn the ideas behind discrete wavelet transforms and their applications by using them in image compression, image edge detection, and signal denoising. I have done my best to provide many of the details for these applications that my SHSU student and I had to discover on our own. In so doing, I found that the material strongly reinforces ideas learned in calculus and linear algebra, provides a natural arena for an introduction of complex numbers, convolution, and Fourier series, offers motivation for student enrollment in higher-level undergraduate courses such as real analysis or complex analysis, and establishes the computer as a legitimate learning tool. The book also introduces students to late-twentieth century mathematics. Students who have grown up in the digital age learn how mathematics is utilized to solve problems they understand and to which they can easily relate. And although students who read this book may not be ready to perform high-level mathematical research, they will be at a point where they can identify problems and open questions studied by researchers today.

To the Student

Many of us have learned a foreign language. Do you remember how you formulated an answer when your instructor asked you a question in the foreign language? If you were like me, you mentally translated the question to English, formulated an answer to the question, and then translated the answer back to the foreign language. Ultimately, the goal is to omit the translation steps from the process, but the language analogy perfectly describes an important mathematical technique.

Mathematicians are often faced with a problem that is difficult to solve. In many instances, mathematicians will *transform* the problem to a different setting, solve the problem in the new setting, and then transform the answer back to the original domain. This is exactly the approach you use in calculus when you learn about u-substitutions or integration by parts. What you might not realize is that for applications involving discrete data (lists or tables of numbers), *matrix multiplication* is often used to transform the data to a setting more conducive to solving the problem. The key, of course, is to choose the correct matrix for the task.

In this book, you will learn about discrete wavelet transforms and their applications. For now, think of the transform as a matrix that we multiply with a vector (audio) or another matrix (image). The resulting product is much better suited than the original image for performing tasks such as compression, denoising, or edge detection. A wavelet filter is simply a list of numbers that is used to construct the wavelet matrix. Of course, this matrix is very special, and as you might guess, some thought must go into its construction. What you will learn is that the ideas used to construct wavelet filters and wavelet transform matrices draw largely from calculus and linear algebra!

You will need to learn about convolution and Fourier series, but soon you will be building wavelet filters and transformations. It will surprise you that the construction is a bit ad hoc. On the other hand, problem solving often utilizes ideas from different areas. Questions you are often asked in mathematics courses start with phrases such as "solve this equation," "differentiate/integrate this function,", or "invert this matrix." In this book, you will see *why* you need to perform these tasks since the questions you will be asked often start with "denoise this signal," "compress this image," or "build this filter." At first you might find it difficult to solve problems without clear-cut instructions, but understand that this is exactly the approach used to solve problems in mathematical research or industry.

At some point, we will need to develop some mathematical theory in order to build more sophisticated wavelet filters. The hope is that by this point, you will understand the applications sufficiently to be highly motivated to learn the theory.

Although you will need to work to master the material in this book, I encourage you to work particularly hard on the material on Fourier series in Section 4.3. A firm grasp on the material in this section is paramount to the development of a deep understanding of wavelet filter construction. Finally, if your instructor asks you to write software to implement wavelet transforms and their inverses, understand that learning to write good mathematical programs takes time. In many cases, you can simply translate the pseudocode from the book to the programming language you are using. Resist this temptation and take the extra time necessary to deeply understand how the algorithm works. You will develop good programming skills and you will be surprised at the amount of mathematics you can learn in the process!

To the Instructor

The technique of solving problems in the transform domain is common in applied mathematics as used in research and industry, but we do not devote as much time to it as we should in the undergraduate curriculum. It is my hope that faculty can use this book to create a course that can be offered early in the curriculum and fill this void.

I have found that it is entirely tractable to offer this course to students who have completed calculus I and II, a computer programming course, and sophomore linear algebra. I view the course as a post-sophomore capstone course that strengthens student knowledge in the prerequisite courses and provides some rationale and motivation for the mathematics they will see in courses such as real analysis.

The aim is to make the presentation as elementary as possible. Toward this end, explanations are not quite as terse as they could be, applications are showcased, rigor is sometimes sacrificed for ease of presentation, and problem sets and labs often elicit subjective answers. It is difficult as a mathematician to minimize the attention given to rigor and detail (convergence of Fourier series, for example) and it is often irresistible to omit ancillary topics (the FFT and its uses, other filtering techniques, wavelet packets) that almost beg to be included. It is important to remind yourself that you are preparing students for more rigorous mathematics and that any additional topics you introduce takes time away from a schedule that is already quite full.

I have prepared the book and software so that instructors have several options when offering the course. When I teach the course, I typically ask students to write modules (subroutines) for constructing wavelet filters or computing wavelet transforms and their inverses. Once the modules are working, students are given instructions for including them in a software package. This package is used for work on inquiry-type labs that are included in the text. For those faculty members who wish to offer a less programming-intensive course and concentrate on the material in the labs, the complete software package is available for download.

In lieu of a final exam, I ask my students to work on final projects. I usually allow four or five class periods at the end of the semester for student work on the projects. Project topics are either theoretical in nature, address a topic in the book we did not cover, or introduce a topic or application involving the wavelet transform that is not included in the book. In many cases, the projects are totally experimental in nature.

Projects are designed for students to work either on their own or in small groups. To make time for the projects, I usually have to omit some chapters from the book (typically Chapters 8 and 9). Faculty members who do not require the programming described above and do not assign final projects can probably complete the book in a single semester. Detailed course offerings are suggested on the text Web site.

Text Topics

The book begins with a short chapter (essay) entitled *Why Wavelets?* The purpose of this chapter is to give the reader information about the mathematical history of wavelets, how they are applied, and why we would want to learn about them. In Chapter 2, we review some basic ideas from linear algebra. Vectors and matrices are central tools in our development. We review vector inner products and norms, matrix algebra, and block matrix arithmetic. The material in Sections 2.1 and 2.2 is included to make the text self-contained and can be omitted in most cases. Students rarely see a detailed treatment of block matrices in an introductory linear algebra course, and because the text makes heavy use of block matrices, it is important to cover Section 2.3.

Since most of our applications deal with digital images, we cover basic ideas from this topic in Chapter 3. We show how to write images as matrices, discuss some elementary digital processing tools, and define popular color spaces. We will want to make decisions about the effectiveness of our wavelet transforms in applications. Toward this end, we discuss the qualitative/quantitative measures *entropy*, *cumulative energy*, and *peak signal-to-noise ratio*. The chapter concludes with an introduction to *Huffman coding*. In Chapter 4 we provide an introduction to *complex numbers* and arithmetic and *Fourier series*. We construct Fourier series, learn rules for computing one Fourier series from another, and discuss how Fourier series are used and interpreted by engineers. In Chapter 5 we introduce *convolution* as a fundamental tool for processing a digital signal or image. Convolution is a binary operation involving two bi-infinite sequences. We designate one sequence as a *filter* and learn how to identify *lowpass* and *highpass filters*. We conclude the chapter by writing convolution as the product of a matrix and vector. In this context we will learn that it is typically impossible to "de-convolve" and recover our signal.

The second part of the book starts with a chapter on the *discrete Haar wavelet transformation*. We learn how to combine lowpass and highpass filters to form an *orthogonal wavelet transform matrix* and then *downsample* to produce an invertible transform. We develop software for applying the transform and then extend the results and code so that we can transform two-dimensional data (images). Data compression and image edge detection are studied in detail using the Haar transformation. In Chapter 7 we introduce the orthogonal lowpass filters of Ingrid Daubechies [26]. Filters of length 4 and 6 are constructed and the procedure is generalized to produce orthogonal even-length filters. Algorithms for implementing these filters in wavelet transforms are presented in Section 7.3. The algorithm development in Chapter 6 and Section 7.3 is optional and intended for those instructors who ask their students to write programs for computing wavelet transforms and their inverses.

The development in Chapters 6 and 7 uses ideas from Fourier series to develop conditions that lowpass wavelet filters must satisfy. We write constraints on the lowpass filters so that the matrix we build from them is orthogonal. In the process we show how to construct the highpass wavelet filter for the transform. The construction is a bit ad hoc but practical. Chapter 8 offers a first glimpse of abstraction as we see how to use Fourier series to completely characterize orthogonal wavelet filters. Fourier series are also used to construct highpass wavelet filters from lowpass filters. We use the general results of this chapter to construct Coiflet filters. Chapter 9 is devoted to the problem of signal or image denoising using a technique called *wavelet shrinkage*. Noisy signals or images are processed by a wavelet transform and the resulting output is quantized. The quantized transform is inverted and the hope is that the approximation to the original signal or image has been denoised. We present the basic algorithm and a rationale for why it works as well as two methods for performing the quantization. Appendix A is provided for students who need a review of some elementary statistical concepts.

The final part of the book deals with *biorthogonal wavelet filters*. Chapter 10 generalizes the techniques presented in Chapter 8 for creating wavelet filters using Fourier series. We remove the requirement that the transform matrix is orthogonal. Consequently, we learn that we need *two* lowpass wavelet filters to build a nonorthogonal transform and its inverse. The good news is that we can replace the orthogonality condition with the practical insistence that our lowpass filters are *symmetric*. We learn that symmetric biorthogonal filter pairs are desirable in applications involving image compression and then construct some filter pairs (biorthogonal spline filter pairs and the CDF97 filter pair) that are used in application. In Chapter 11 we develop algorithms for applying biorthogonal wavelet transforms and their inverses to signals and images. Sections 11.1 and 11.2 are optional and can be omitted if the instructor does not require the students to write code that implements the algorithms. In Section 11.3 we see how to exploit the symmetry of the biorthogonal filter pair so that our transform does a better job of handling signal and image boundaries. In the last chapter of the book we describe the JPEG2000 image compression standard. The original JPEG standard utilizes the discrete cosine transform to compress images. The JPEG2000 standard uses the biorthogonal wavelet transform to compress images. The new standard attempts to address some of the limitations of the JPEG standard. One such limitation is the inability of the original JPEG standard to perform *lossless compression*. The JPEG2000 standard performs lossless compression through a technique due to Wim Sweldens [75] called *lifting*. We conclude the chapter with examples of JPEG2000 compression and compare the results to those obtained via JPEG compression.

Problem Sets, Software Package, Computer Labs, and Live Examples

Problem sets and computers labs are given at the end of each section. There are almost 400 problems in the book. Some of the problems can be viewed as basic drill work or an opportunity to step through an algorithm "by hand." There are basic problem-solving exercises designed to increase student understanding of the material, problems where the student is asked to either provide or complete the proof of a result from the section, and problems that allow students to furnish details for ancillary topics that we do not completely cover in the text. To help students develop good problem-solving skills, many problems contain hints, and more difficult problems are often broken into several steps. The results of problems marked with a ★ are used later in the text.

We have prepared a software package called DiscreteWavelets that contains three types of modules. One set of modules can be classified as *computational modules*. These modules can be used to create orthogonal and biorthogonal wavelet filters; apply one- and two-dimensional wavelet transforms and their inverses; compute cumulative energy, entropy, or peak signal-to-noise ratios; perform Huffman coding or signal/image denoising; and much more. *Visualization modules* allow you to display transformed signals or images, hear transformed audio signals, or view Huffman trees. *Pedagogical modules* are available for building generic wavelet transform matrices, creating and manipulating Fourier series, or animating convolution products. The purpose of the pedagogical modules is to allow you to identify patterns that will help you write code, analyze a filter, or better understand how convolution works.

The package comes in two forms. One form is complete as described above while the second form contains complete visualization and pedagogical modules but only instructions for creating the computational modules. The second form is designed for those instructors who wish to include more basic programming in their course.

There are two types of computer labs at the end of each section. Over 20 software development labs (marked with ♦) are provided for those instructors who ask their students to write modules that perform tasks such as the construction of wavelet filters or the computation of a wavelet transform. Approximately 40 more labs allow students to experiment with the ideas discussed in the text. Of particular note are the labs that involve digital audio — it is much easier to use software to "hear" how wavelets can be used to denoise audio rather than to attempt to describe the results.

Textbook authors use examples to help the reader understand concepts. In this book, there are over 100 examples and many of these examples utilize images, audio, or real data sets. Other examples illustrate how various algorithms or techniques work. The software that has been used to prepare the examples in this book are available on the Web site. You can simply download the software for the example of interest and experiment with it yourself. You can change parameters, try a different image, *hear* the output rather than only view it, reproduce a result (often at better resolution than in the book), or modify software to solve your own problem. All software is available in Mathematica, Matlab, and Maple.

Course Outlines

The book contains more material than can be covered in a one-semester course. The extra material allows some flexibility when teaching the course. I have used two different outlines when teaching the course. For both approaches, I assign Chapter 1 as reading and then cover portions of Chapter 2 (depending on the linear algebra background of the students in the course). I then cover the material in Chapters 3 to 6 and Sections 7.1 and 7.2. Some of the elementary material may either be omitted or covered in a cursory manner if the students have a basic knowledge of complex numbers and arithmetic. I usually ask students to do the software development labs in Sections 3.2, 3.3, 6.1 to 6.3, and 7.1. On some occasions, I ask advanced students to complete the software development labs from Sections 3.4 and 7.2. If I have a group of students who are strong programmers, I cover Section 7.3. I also make sure to assign several computer labs in each chapter and dedicate class time for students to work on the labs in Section 6.4.

One way to proceed is to cover the material in Chapters 8 and 9 and Appendix A. This approach introduces students to the idea of signal/image denoising via wavelet shrinkage. To adequately cover the material in Section 9.3, students must have some basic statistical background. Students are asked to complete all the labs (including software development labs) in Chapter 9. If I follow this approach, I typically end the course about three weeks early (in a 15-week semester) and allow the students to use the remainder of the time to work on final projects.

A second approach emphasizes image compression. Here, I omit Chapters 8 and 9 and cover Chapter 10, Section 11.3, and Chapter 12. I ask the students to complete all labs from this material. If the students in the class are strong programmers, I also cover the first two sections of Chapter 11 and assign the software development labs. This approach leaves about two weeks of time for students to work on final projects. Several ideas for final projects are given on the text Web site. I also use material from the text that I did not cover during the semester for final project topics.

Text Web Site

The best way to understand the ideas presented in this book is to experiment with them! We have prepared a Web site that is replete with material to help you do just that. The URL is

```
http://www.stthomas.edu/wavelets
```

At this Web site you will find computer labs, the complete software package and a template for creating it, live examples, an audio and image repository, full-size versions of all text images, and final projects. For classroom presentation, lecture slides can be downloaded. The site also contains several course syllabi (suggested by the author or others who have taught the course), teaching tips, and an errata list.

What You *Won't* Find in This Book

There are two points to be made in this section. The first laments the fact that several pet topics were omitted from the text and the second provides some rationale for the presentation of the material.

For those readers with a background in wavelets, it might pain you to know that except for a short discussion in Chapter 1 and an aside in Section 10.2, there is no mention of scaling functions, wavelet functions, dilation equations, or multiresolution analyses. In other words, this presentation does not follow the classical approach where wavelet filters are constructed in $L^2(\mathbb{R})$. The approach here is entirely discrete — the point of the presentation is to draw as much as possible from courses early in a mathematics curriculum, augment it with the necessary amount of Fourier series, and then use applications to motivate more general filter constructions. For those students who might desire a more classical approach to the topic, I highly recommend the texts by Boggess and Narcowich [7], Frazier [35], and Walnut [83].

Scientists who have used nonwavelet methods (Fourier or otherwise) to process signals and images will not find them discussed in this book. There are many methods for processing signals and images. Some work better than wavelet-based methods, some do not work as well. We should view wavelets as simply another tool that we can use in these applications. Of course, it is important to compare other methods used in signal and image processing to those presented in this book. If those methods were presented here, it would not be possible to cover all the material in a single semester.

Some mathematicians might be disappointed at the lack of rigor in several parts of the text. This is the conundrum that comes with writing a book with modest prerequisites. If we get too bogged down in the details, we lose sight of the applications. On the other hand, if we concern ourselves only with applications, then we never develop a deep understanding of how the mathematics works. I have skewed my presentation to favor more applications and fewer technical details. Where appropriate, I have tried to point out arguments that are incomplete and provided suggestions (or references) that can make the argument rigorous.

Finally, despite the best efforts of friends, colleagues, and students, you will not find an error-free presentation of the material in this book. For this, I am entirely to blame. Any corrections, suggestions, or criticisms will be greatly appreciated!

P. J. VAN FLEET

St. Paul, Minnesota USA
June 2007

ACKNOWLEDGMENTS

There are many people I wish to thank for their help on this project. This book evolved from a series of lecture notes for a course on wavelets and their applications and as a result, several students were exposed to less refined versions of the text. I appreciate their patience, suggestions for improvements, and willingness to test algorithms and point out errors in the manuscript. In particular, I am indebted to Jessica Klaers for her thorough review of the first six chapters of the book. Her interpretations of my mistakes were informative as well as entertaining. University of St. Thomas colleagues Jeffrey Jalkio, Arkady Shemyakin, and Heekyung Youn read various chapters of the book and provided valuable feedback. My department chair, John Kemper, was a source of constant support and also attended a workshop where he worked through and commented on many of the computer labs that appear in the book. David Ruch at Metropolitan State College of Denver has read much of the text and provided invaluable suggestions and encouragement. Dan Flath at Macalester College taught the course during the spring 2007 semester and suggested many improvements for the book; I am particularly grateful for his careful reading of Chapter 10. I wish to thank Colm Mulcahy from Spelman College for finding numerous typographical errors in one of the final versions of the manuscript. I owe a large debt of gratitude to my friend and mentor David Kammler at Southern Illinois

University – Carbondale. David read much of the first seven chapters and provided many words of wisdom. I am also grateful to David for helping me through the rigors of writing a book — his enthusiasm for and love of mathematics has greatly influenced me during my career.

My thanks go to the National Science Foundation for their support (DUE-0442684) for the development of the book, computer packages and computer labs. I wish to recognize Michael Pearson and Silja Sistok-Katz of the Mathematical Association of America for helping me organize two workshops (with NSF support) that allowed me to present much of the material in this book. I am indebted to the workshop participants who provided feedback and suggestions for the book, computer labs, and course structure. Thanks also go to my editor, Susanne Stietz-Filler, for her help in completing the project and to Amy Hendrickson of TeXnology, Inc. for answering my LaTeX questions.

What would a book be without pictures? In this case, quite incomplete! I am fortunate that my good friend, former colleague, and wavelets expert Radka (Turcajov á) Tezaur is an excellent photographer. I am even more fortunate that she generously donated several images for use in the book. All images that you see in the book are photographs taken by Radka.

Finally, I would like to express my profound gratitude to my wife, Verena, and our three children Sam, Matt, and Rachel. Verena was a pillar of support throughout the entire project, provided endless encouragement, and allowed me large blocks of time to work on the book. Her unselfish sacrifice of time was a principal reason that the project could be transformed from an idea to a finished book. Sam, Matt, and Rachel were equally supportive — they allowed me time to work on the book and played with me when I needed to get away.

Thanks in advance go to all readers who suggest improvements in the presentation, create new computer labs, or develop new problems. Please contact me with your ideas!

P.V.F.

CHAPTER 1

INTRODUCTION: WHY WAVELETS?

Why wavelets? This is certainly a fair question for those interested in this book and also one that can be posed in multiple contexts. Students wonder why they need to learn about wavelets. Researchers ask why wavelets work so well in certain applications. Scientists want to know why they should consider using wavelets in applications instead of other popular tools. Finally, there is curiosity about the word itself — why *wavelets*? What are wavelets? Where did they come from?

In this short introduction to the book, we answer these questions and in the process, provide some information about what to expect from the chapters that follow.

Image Compression

In keeping with the true spirit of the book, let's start with an application in image compression.

Suppose that you wish to use the Internet to send a friend the digital image file plotted in Figure 1.1. Since the file is quite large, you first decide to *compress* it. Suppose that it is permissible to sacrifice image resolution in order to minimize

Discrete Wavelet Transformations: An Elementary Approach With Applications. By P. J. Van Fleet
Copyright © 2008 John Wiley & Sons, Inc.

transmission time. The dimensions of the image, in pixels,[1] are 512 rows by 512 columns. The total number of elements that comprise the image is $512 \times 512 = 262{,}144$.

Figure 1.1 A digital image.

What does it even mean to compress the image? How do you suppose that we compress the image? How do we measure the effectiveness of the compression method?

As you will learn in Chapter 3, each pixel value is an integer from 0 (black) to 255 (white) and each of these integers is stored on a computer using 8 *bits* (the value of a bit is either 0 or 1). Thus we need $262{,}144 \times 8 = 2{,}097{,}152$ bits to represent the image. To compress an image, we simply wish to reduce the number of bits needed to store it. Most compression methods follow the basic algorithm given in Figure 1.2.

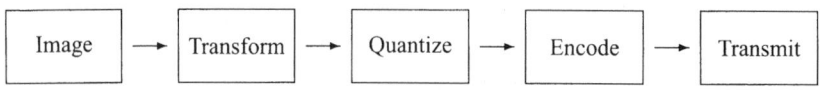

Figure 1.2 A basic algorithm for compressing a digital image.

[1] You can think of a pixel as a small rectangle on your computer screen or paper that is rendered at some gray level between black and white.

Note that the first step is to *transform* the image. The goal here is to map the integers that comprise the image to a new set of numbers. In this new setting we alter (*quantize*) some or all of the values so that we can write (*encode*) the modified values using fewer bits. In Figure 1.3, we have plotted one such transformation of the image from Figure 1.1. The figure also contains a plot of the quantized transform. The transformation in Figure 1.3 is a *discrete wavelet transform*. In fact, it is the same transformation as that used in the JPEG2000 image compression standard.[2]

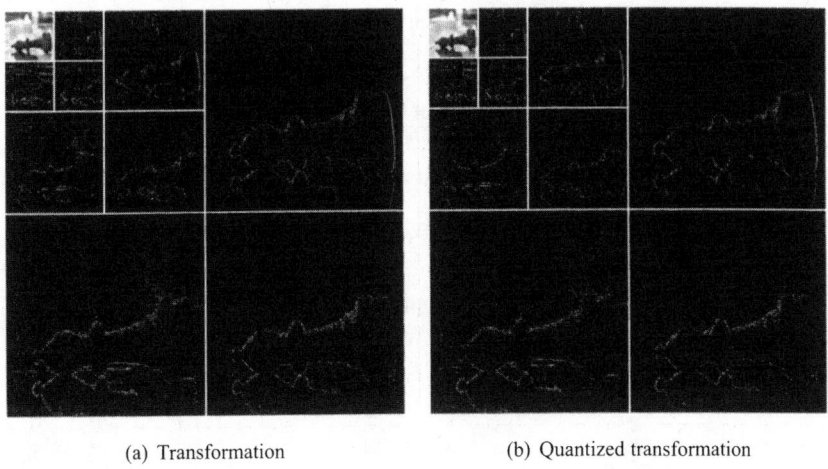

(a) Transformation   (b) Quantized transformation

Figure 1.3   A transformation and the quantized transformation of the image in Figure 1.1. The white lines that appear in both images are superimposed to separate the different parts of the transform.

In mathematical terms, if the image is represented by the square matrix A, then the transformation is performed by constructing special matrices W and $\tilde{W}$[3] and then computing $WA\tilde{W}^T$. Of course, the key to the process is understanding the construction of $W, \tilde{W}$ and how $WA\tilde{W}^T$ concentrates most of the nonzero values (nonblack pixels) in the upper left-hand corner of the transformation. You might already begin to see why the wavelet transformation is useful for reducing the number of bits. The transformation has created large regions that are black or nearly black (i.e., regions where most of values are either zero or near zero). It is natural to believe that you would need less information to store these values than what is required for the original image.

The next step in the process is to quantize the transform. It turns out that zero is the only integer value in the transformed output. Thus each value must be converted

---

[2] You will learn about JPEG2000 in Chapter 12.
[3] In some cases, we choose $\tilde{W} = W$.

(rounded, for example) to an integer. The quantizer also makes decisions about certain values and may reduce them in size or even convert them to zero. Values that are reduced or converted to zero are those that the quantizer believes will not adversely affect the resolution of the compressed image. After quantization, it is impossible to exactly retrieve the original image[4]. It may be hard to see in Figure 1.3, but nearly all the values of the quantized transformation are different than the corresponding values in the transformation.

The final step before transmission is to *encode* the quantized transformation. That is, instead of using 8 bits to store each integer, we will try to group together like integers and possibly use a smaller number of bits to store those values that occur with greater frequency. Since the quantized wavelet transformation contains a large number of zeros (black pixels), we expect the encoded transform to require fewer bits. Indeed, using a very naive encoding scheme (see Section 3.4), we find that only 548,502 bits are needed to store the quantized wavelet transform. This is about 26% of the number of bits required to store the original!

To view the compressed image, the receiver decodes the transmitted file and then applies the inverse wavelet transform. The compressed image appears in Figure 1.4. It is very difficult to tell the images in Figures 1.1 and 1.4 apart!

Other Applications That Use Wavelet Transformations

Wavelets, or more precisely, wavelet transformations, abound in image-processing applications. The Federal Bureau of Investigation uses wavelets in their Wavelet/Scalar Quantization Specification to compress digital images of fingerprints [8]. In this case there is an objective measure of the effectiveness of the compression method — the uncompressed fingerprint file must be uniquely matched with the correct person. In many instances it is desirable to identify edges of regions that appear in digital images. Wavelets have proved to be an effective way to perform edge detection in images [56, 55]. Other image-processing applications where wavelets are used are image morphing [43] and digital watermarking [86, 32]. Image morphing is a visualization technique that transforms one image into another. You have probably seen image morphing as part of a special effect in a movie. Digital watermarking is the process by which information is added (either visible or invisible) to an image for the purpose of establishing the authenticity of the image.

Wavelets are used in applications outside image processing. For example, wavelet-based methods have proved effective in the area of signal denoising [28, 85, 81]. Cipra [19] recounts an interesting story of Yale University mathematician Ronald Coifman and his collaborators using wavelets to denoise an old cylinder recording (made of a radio broadcast) of Brahms playing his *First Hungarian Dance* [4]. It is impossible to identify the piano in the original but a restored version (using wavelets)

---

[4]For those readers with some background in linear algebra, $W^{-1}, \tilde{W}^{-1}$ both exist so it is possible to completely recover A before the quantization step.

Figure 1.4   The decompressed digital image. Whereas each pixel in the original image is stored using 8 bits, the pixels in this image require 2.09 bits storage on average.

clearly portrays the piano melody. Stanford University music professor and Coifman collaborator Jonathan Berger maintains a very interesting Web site on this project [5]. Wavelet-based denoising (wavelet shrinkage) is also used in applications in finance and economics [38]. We study wavelet shrinkage in Chapter 9.

Wavelets have been used to examine electroencephalogram data in order to detect epileptic seizures [3], model distant galaxies [6], and analyze seismic data [45, 46]. Finally, in more mathematical applications, wavelets are used to estimate densities and to model time series in statistics [81] and in numerical methods for solving partial differential equations [22].

Wavelet Transforms Are Local Transforms

We have by no means exhausted the list of applications that utilize wavelets. Understand as well that wavelets are not always the best tool to use for a particular application. There are many applications in audio and signal processing for which Fourier methods are superior, and applications in image processing where filtering methods other than wavelet-based methods are preferred.

There are several reasons that wavelets work well in many applications. Probably the most important is the fact that a discrete wavelet transform is a *local transform*. To understand what is meant by this term, consider the 100-term signal in Figure 1.5(a). Four of the signal values are altered and the result is plotted in Figure 1.5(b).

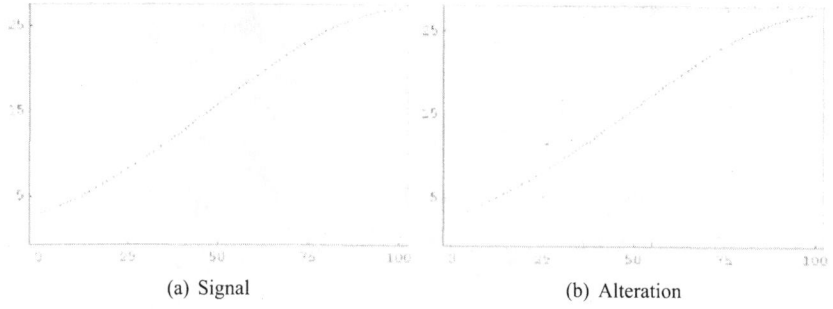

(a) Signal   (b) Alteration

Figure 1.5   A signal and alteration of the signal.

In Figure 1.6 we have plotted a discrete wavelet transformation of each signal from Figure 1.5. The wavelet transform in Figure 1.6(a) is again simply a matrix we use to multiply with the input vector. The output consists of two parts. The first 50 components of the transform serve as an approximation of the original signal. The last 50 elements basically describe the differences between the original signal and the approximation. Since the original signal is relatively smooth, it is not surprising that the differences between the original signal and the approximation are quite small.

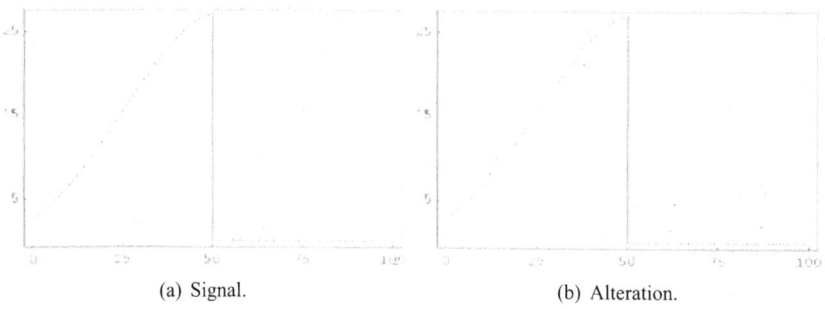

(a) Signal.   (b) Alteration.

Figure 1.6   Discrete wavelet transformations of each of the signals in Figure 1.5.

Figure 1.7 gives a better view of the difference portion of the wavelet transform in Figure 1.6(a). Now look at the transformation in Figure 1.6(b). In both portions of the transform, we see two small regions that are affected by the changes we made to produce the altered signal in Figure 1.5(b). The important point to note is that the regions where the transform was affected by the altered data are relatively small. Thus,

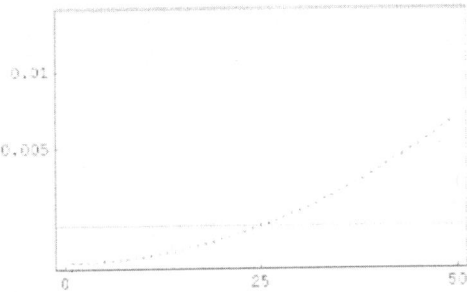

Figure 1.7   The difference portion of the discrete wavelet transformation from Figure 1.6(a).

small changes in the input data result in small changes in the wavelet-transformed data.

One of the most popular and useful tools for signal-processing applications is the *discrete Fourier transformation* (DFT). It is built using sampled values of cosine and sine. Although we will not delve into the specifics of the construction of the transform and how it is used in applications, it is worthwhile to view the DFT of the signals in Figure 1.5. We have plotted the modulus (see Section 4.1) of these DFTs in Figure 1.8.

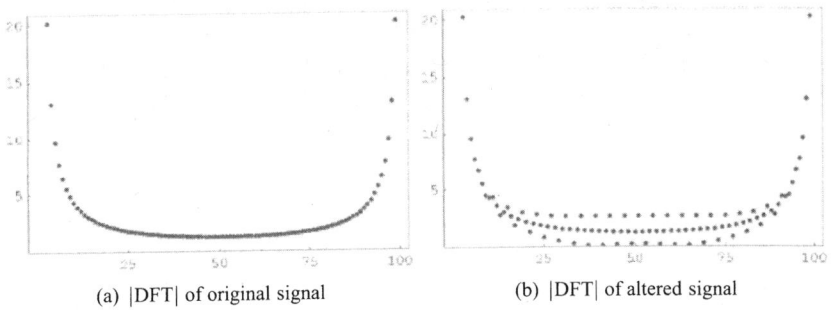

(a) |DFT| of original signal   (b) |DFT| of altered signal

Figure 1.8   The moduli of the discrete Fourier transforms of each of the signals in Figure 1.5.

Look at what happened to the DFT of the altered data. Although we changed only four values, the effect of these changes on the transform is *global*[5]. The DFT is constructed from samples of sine and cosine functions.[6] These functions oscillate

---

between -1 and 1 for all time and never decay to 0. Thus the effects of any minor change in the input data will be felt throughout the entire transformed output.

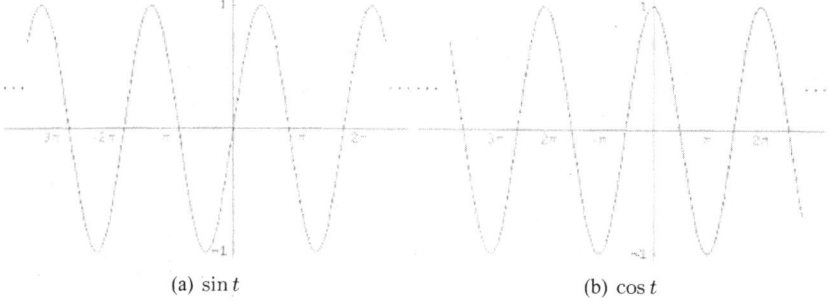

(a) $\sin t$ (b) $\cos t$

Figure 1.9   Sine and cosine. Both functions oscillate between -1 and 1 and never decay to zero.

Classical Wavelet Theory in a Nutshell

The classical approach to wavelet theory utilizes the setting of square integrable functions on $\mathbb{R}$ and seeks to elicit the elements of the discrete transform from oscillatory functions (e. g., sine and cosine; Figure 1.9) that are required to decay to zero. That is, the underlying functions are tiny waves or *wavelets*. A couple of popular wavelet functions are plotted in Figure 1.10.

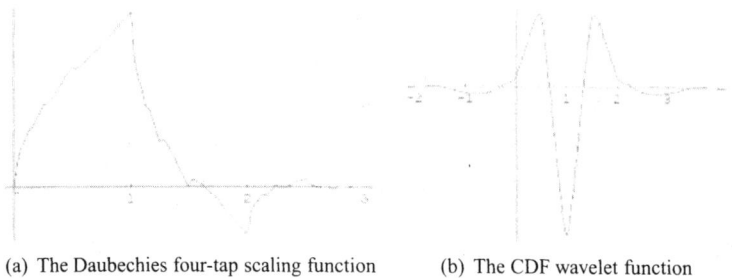

(a) The Daubechies four-tap scaling function   (b) The CDF wavelet function

Figure 1.10   Some functions from classical wavelet theory.

Note that the functions in Figure 1.10 are oscillatory and have value zero outside a finite interval. The function in Figure 1.10(a) is probably one of the most famous in all of wavelet theory. It is Ingrid Daubechies' *four-tap scaling function* [7] [26]. It gives rise

---

[7] A scaling function is called a *father wavelet function* by some researchers.

to the four numbers we use to build a wavelet transformation matrix in Section 7.1. The function in 1.10(b) is a wavelet function constructed by Albert Cohen, Ingrid Daubechies, and Jean-Christophe Feauveau [20]. It provides the elements of the wavelet transform matrix used in the JPEG2000 image compression standard.

If these functions really do lead to the discrete wavelet transformations used in so many applications, several questions naturally arise: Where do these functions come from? How are they constructed? Once we have built them, how do we extract the elements that are used in the discrete wavelet transforms?

Although the mathematics of wavelets can be traced back to the early twentieth century,[8] most researchers mark the beginning of modern research in wavelet theory by the 1984 work [41] of French physicists Jean Morlet and Alexander Grossmann. Yves Meyer [59] (in 1992) and Stéphane Mallat [57] (in 1989) produced foundational work on the so-called *multiresolution analysis* and using this construct, Ingrid Daubechies constructed wavelet functions [23, 24] (in 1988, 1993, respectively) that give rise to essentially all the discrete wavelet transforms found in this book.

To (over)simplify the central idea of Daubechies, we seek a function $\phi(t)$ (use the function in Figure 1.10(a) as a reference) that satisfies a number of properties. In particular, $\phi(t)$ should be zero outside a finite interval and $\phi(t)$ and its integer translates $\phi(t - k), k = 0, \pm 1, \pm 2, \pm 3, \ldots$, (see Figure 1.11) should form a *basis* for a particular space. The function $\phi(t)$ should be suitably smooth[9] and the functions should also satisfy

$$\int_{\mathbb{R}} \phi(t)\phi(t - k)\, dt = 0, \qquad k = \pm 1, \pm 2, \ldots \tag{1.1}$$

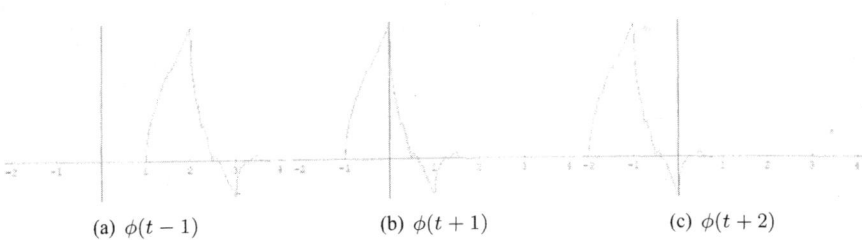

(a) $\phi(t - 1)$ (b) $\phi(t + 1)$ (c) $\phi(t + 2)$

Figure 1.11   Some integer translates of the function $\phi(t)$ from Figure 1.10(a).

Finally, we should be able to write $\phi(t)$ as a combination of dilations (contractions, actually) and translations of itself. For example, the function $\phi(t)$ in Figure 1.10(a) satisfies the *dilation equation*

---

[8]Yves Meyer [59] gives a nice discussion of the origins of wavelet theory in the twentieth century, and a wonderful exposition on the history of wavelet theory, its origins, and uses in applications can be found in a 1998 book by Barbara Burke Hubbard [47].

[9]Believe it or not, in a certain sense, the function in Figure 1.10(a) and its integer translates can be used to represent linear polynomials!

$$\phi(t) = h_0\phi(2t) + h_1\phi(2t - 1) + h_2\phi(2t - 2) + h_3\phi(2t - 3) \qquad (1.2)$$

The function $\phi(t)$ in Figure 1.10(a) is nonzero only on the interval $[0,3]$. The function $\phi(2t)$ is a contraction — it is nonzero only on $[0, \frac{3}{2}]$. For $k = 1, \phi(2t-1) = \phi(2(t - \frac{1}{2}))$, so we see that $\phi(2t - 1)$ is simply $\phi(2t)$ translated $1/2$ unit right. In a similar way, we see that $\phi(2t - 2)$ and $\phi(2t - 3)$ are obtained by translating $\phi(2t)$ 1 and $3/2$ units right, respectively. The functions on the right side of (1.2) are plotted in Figure 1.12.

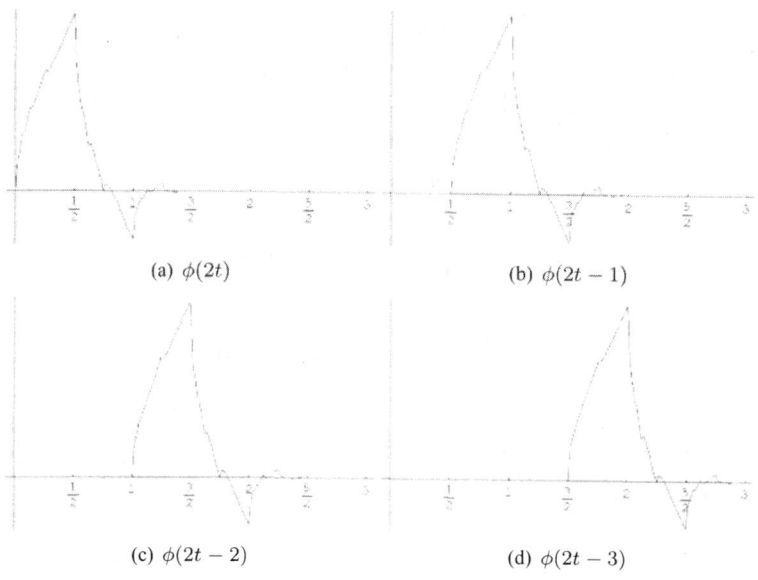

(a) $\phi(2t)$

(b) $\phi(2t - 1)$

(c) $\phi(2t - 2)$

(d) $\phi(2t - 3)$

Figure 1.12   The functions that comprise the right-hand side of (1.2).

It turns out that the four numbers $h_0, \ldots, h_3$ we need to use in combination with the functions in Figure 1.12 to build $\phi(t)$ are

$$h_0 = \frac{1 + \sqrt{3}}{4\sqrt{2}} \approx 0.482963 \qquad\qquad h_1 = \frac{3 + \sqrt{3}}{4\sqrt{2}} \approx 0.836516$$

$$h_2 = \frac{3 - \sqrt{3}}{4\sqrt{2}} \approx 0.224144 \qquad\qquad h_3 = \frac{1 - \sqrt{3}}{4\sqrt{2}} \approx -0.129410$$

We derive these numbers in Section 7.1. Each of the properties satisfied by $\phi(t)$ affect the discrete wavelet transform. The integral condition (1.1) ensures that the discrete wavelet transform matrix W is not only invertible, but *orthogonal*. As we will see in Section 2.2, orthogonal matrices satisfy the simple inverse formula $W^{-1} = W^T$. Insisting that $\phi(t)$ is suitably smooth allows us to form accurate approximations of smooth data. Requiring $\phi(t)$ to be zero outside a finite interval results in a finite list of numbers $h_0, \ldots, h_N$ that are used in the discrete wavelet transform. Finally, the dilation equation allows us to build a transform that can "zoom" in on complicated segments of a signal or image. This zoom-in property is not entirely obvious — we talk more about it in Chapters 6 and 7.

The Approach in This Book

Using the classical approach of Daubechies [23] to derive the numbers h_0, h_1, h_2, and h_3 for the function in Figure 1.10(a) is a journey through some beautiful mathematics, but it is difficult and requires more mathematical background than most undergraduate students possess.[10]

On the other hand, the matrix used to produce the output in Figure 1.6 is quite simple in structure. It is 100 rows by 100 columns and looks as follows:

$$
W = \left[
\begin{array}{ccccccccccccc}
\frac{3}{4} & \frac{1}{2} & -\frac{1}{4} & 0 & 0 & 0 & 0 & 0 & 0 & 0 & \cdots & 0 \\
-\frac{1}{8} & \frac{1}{4} & \frac{3}{4} & \frac{1}{4} & -\frac{1}{8} & 0 & 0 & 0 & 0 & 0 & \cdots & 0 \\
0 & 0 & -\frac{1}{8} & \frac{1}{4} & \frac{3}{4} & \frac{1}{4} & -\frac{1}{8} & 0 & 0 & 0 & \cdots & 0 \\
0 & 0 & 0 & 0 & -\frac{1}{8} & \frac{1}{4} & \frac{3}{4} & \frac{1}{4} & -\frac{1}{8} & 0 & \cdots & 0 \\
\vdots & & & & & \ddots & & & & & & \vdots \\
0 & 0 & 0 & 0 & \cdots & 0 & -\frac{1}{8} & \frac{1}{4} & \frac{3}{4} & \frac{1}{4} & -\frac{1}{8} & 0 \\
0 & 0 & 0 & 0 & \cdots & 0 & 0 & 0 & -\frac{1}{8} & \frac{1}{4} & \frac{5}{8} & \frac{1}{4} \\
\hline
-\frac{1}{2} & 1 & -\frac{1}{2} & 0 & 0 & 0 & 0 & 0 & 0 & 0 & \cdots & 0 \\
0 & 0 & -\frac{1}{2} & 1 & -\frac{1}{2} & 0 & 0 & 0 & 0 & 0 & \cdots & 0 \\
0 & 0 & 0 & 0 & -\frac{1}{2} & 1 & -\frac{1}{2} & 0 & 0 & 0 & \cdots & 0 \\
0 & 0 & 0 & 0 & 0 & -\frac{1}{2} & 1 & -\frac{1}{2} & 0 & & \cdots & 0 \\
\vdots & & & & & \ddots & & & & & & \vdots \\
0 & 0 & 0 & 0 & 0 & 0 & \cdots & 0 & -\frac{1}{2} & 1 & -\frac{1}{2} & 0 \\
0 & 0 & 0 & 0 & 0 & 0 & \cdots & 0 & 0 & 0 & -1 & 1
\end{array}
\right]
\qquad (1.3)
$$

Note that we have placed a line between the first 50 rows and the last 50 rows. There are a couple of reasons for partitioning the matrix in this way. First, you might notice that rows 2 through 49 are built using the same sequence of five nonzero values

[10]For those students interested in a more classical approach to wavelet theory, I strongly recommend the books by David Walnut [83], Albert Boggess and Francis Narcowich [7], or Michael Frazier [35].

$(-\frac{1}{8}, \frac{1}{4}, \frac{3}{4}, \frac{1}{4}, -\frac{1}{8})$, and rows 51 through 99 are built using the same sequence of three nonzero values $(-\frac{1}{2}, 1, -\frac{1}{2})$.[11] Another reason for the separator is the fact that the product of W with an input signal consisting of 100 elements has to produce an approximation to the input (recall Figure 1.6) that consists of 50 elements. So the first 50 rows of W perform that task. In a similar manner, the last 50 rows of W produce the differences we need to combine with the approximation to recover the original signal.

There is additional structure evident in W. If we consider row 2 as a list of 100 numbers, then row 3 is obtained by cyclically shifting the elements in row 2 to the right 2 units. Row 4 is obtained from row 3 in a similar manner, and the process is continued to row 49. Row 52 can be constructed from row 51 by cyclically shifting the list $(-\frac{1}{2}, 1, -\frac{1}{2}, 0, 0, \ldots, 0, 0)$ two unit right. It is easy to see how to continue this process to construct rows 53 through 99. Rows 1, 50, and 100 are different from the other rows and you will have to wait until Section 11.3 to find out why!

Let's look closer at the list $(-\frac{1}{8}, \frac{1}{4}, \frac{3}{4}, \frac{1}{4}, -\frac{1}{8})$. When we compute the dot product of row 2 say with an input signal $(x_1, \ldots, x_{100})$, we obtain

$$-\frac{1}{8} x_1 + \frac{1}{4} x_2 + \frac{3}{4} x_3 + \frac{1}{4} x_4 - \frac{1}{8} x_5$$

If we add the elements of our list, we obtain

$$-\frac{1}{8} + \frac{1}{4} + \frac{3}{4} + \frac{1}{4} - \frac{1}{8} = 1$$

and see that the list creates a weighted average of the values $(x_1, \ldots, x_5)$. The idea here is no different from what your instructor uses to compute your final course grade. For example, if three hourly exams are worth 100 points each and the final exam is worth 200 points, then your grade is computed as

$$\text{grade} = \frac{1}{5} \cdot \text{exam 1} + \frac{1}{5} \cdot \text{exam 2} + \frac{1}{5} \cdot \text{exam 3} + \frac{2}{5} \cdot \text{final}$$

Of course, most instructors do not use negative weights like we did in W! The point here is that by computing weighted averages, we are creating an approximation of the original input signal.

Taking the dot product of row 51 say with the input signal produces

$$-\frac{1}{2} x_1 + 1 \cdot x_2 - \frac{1}{2} x_3$$

This value will be near zero if x_1, x_2, and x_3 are close in size. If there is a large change between x_1 and x_2 or x_2 and x_3 then the output will be a relatively large

---

[11]We mentioned earlier in this chapter that we seek orthogonal matrices for use as wavelet transforms. The reader with a background in linear algebra will undoubtedly realize that W in (1.3) is not orthogonal. As we will see late in the book, it is desirable to relinquish orthogonality so that our wavelet matrices can be built with lists of numbers that are symmetric.

(absolute) value. For applications, we may want to convert small changes to zero (compression) or identify the large values (edge detection).

The analysis of W and its entries leads to several questions: How do we determine the structure of W? How do we build the lists of numbers that populate it? How are the rows at the top and bottom of the different portions of W formed? How do we design W for use in different applications? How do we make decisions about the output of the transformation? The matrix W consists of a large number of zeros; can we write a program that exploits this fact and computes the product of W and the input signal faster than by using conventional matrix and vector multiplication?

We answer all of these questions and many more in the chapters that follow. But we will consider the problem in the discrete setting. We will see that our lists of numbers are called *filters* by electrical engineers and other scientists. We first learn about the structure of W by understanding how signals are processed by filters. We then learn how to create filters that perform tasks such as approximating signals or creating the differences portion of the transform. These are the ideas that are covered in Chapters 4 and 5.

We further refine W in Chapter 6 to build the *discrete Haar wavelet transformation* and then use this transform to detect edges in images and to perform naive data compression. More sophisticated transforms are constructed in Chapter 7. Unfortunately, as we ask more of the transformations, the method we use to construct them increases in complexity. But at this point, you should have a good understanding of the discrete wavelet transform, how to use it in applications, how to write software for its implementation, and what limitations we still face. With this understanding comes an appreciation of the need to step back and model our construct in a more general mathematical setting. In the second half of the book we take a more theoretical tack toward transform design. The payoff is the ability to design wavelet transforms like that given in (1.3) and use them in current applications such as JPEG2000 or signal and image denoising.

Thus, in the pages that follow, you will discover the answers to the questions asked at the beginning of the chapter. It is my sincere hope after working through the material, problem sets, software, and computer labs that your response to these questions is: *Why wavelets, indeed!*

CHAPTER 2

VECTORS AND MATRICES

In this chapter we cover some basic concepts from linear algebra necessary to understand the ideas covered later in the book. Signals and digital audio are typically represented as vectors, and digital images are usually described using matrices. Therefore, it is imperative that before proceeding, you have a solid understanding of some elementary concepts associated with vectors and matrices.

If you have completed a sophomore linear algebra class, the first two sections will undoubtedly be a review for you. Make sure though, that you take time to understand concepts such as orthogonal vectors and orthogonal matrices. We also discuss the matrix as a way of "processing" another matrix or a vector via matrix multiplication. Although you may have mastered multiplication with matrices, it is worthwhile to think about the concepts related to matrix multiplication presented in Section 2.2.

The final section of the chapter deals with block (partitioned) matrix arithmetic. This material is typically not covered in detail in a sophomore linear algebra class but is very useful for analyzing and designing the wavelet transformations that appear in later chapters.

Discrete Wavelet Transformations: An Elementary Approach With Applications. By P. J. Van Fleet   **15**
Copyright © 2008 John Wiley & Sons, Inc.

2.1 VECTORS, INNER PRODUCTS, AND NORMS

We remember vectors from a linear algebra course. In $\mathbb{R}^2$, vectors take the form

$$\mathbf{v} = \begin{bmatrix} v_1 \\ v_2 \end{bmatrix}$$

where v_1 and v_2 are any real numbers. You may recall the idea of vectors in $\mathbb{R}^n$:

$$\mathbf{v} = \begin{bmatrix} v_1 \\ v_2 \\ \vdots \\ v_n \end{bmatrix}$$

where $v_1, \ldots, v_n$ are any real numbers. When we think of vectors we generally think of the vectors described above. It is worth mentioning, however, that in a general sense, a vector is best thought of as some element of a given space. For example, if the space we are working with is the space of all quadratic polynomials, then we'll consider $f(x) = x^2 + 3x + 5$ a vector in that space. If the space we're working with is all 2×2 matrices, then certainly

$$A = \begin{bmatrix} 3 & 2 \\ 5 & 1 \end{bmatrix}$$

is a vector in that space.

But in many cases we'll think of vectors as elements of $\mathbb{R}^n$. You should be familiar with how to add and subtract vectors. Figure 2.1 is a plot of the sum and difference of $\mathbf{u} = \begin{bmatrix} 3 \\ 2 \end{bmatrix}$ and $\mathbf{v} = \begin{bmatrix} 1 \\ 5 \end{bmatrix}$:

$$\begin{bmatrix} 3 \\ 2 \end{bmatrix} + \begin{bmatrix} 1 \\ 5 \end{bmatrix} = \begin{bmatrix} 3+1 \\ 2+5 \end{bmatrix} = \begin{bmatrix} 4 \\ 7 \end{bmatrix}$$

and

$$\begin{bmatrix} 3 \\ 2 \end{bmatrix} - \begin{bmatrix} 1 \\ 5 \end{bmatrix} = \begin{bmatrix} 3-1 \\ 2-5 \end{bmatrix} = \begin{bmatrix} 2 \\ -3 \end{bmatrix}$$

Two important concepts associated with vectors are the *inner (dot) product* and *norm* (length).

Inner Products

We now define the inner product for vectors in $\mathbb{R}^n$.

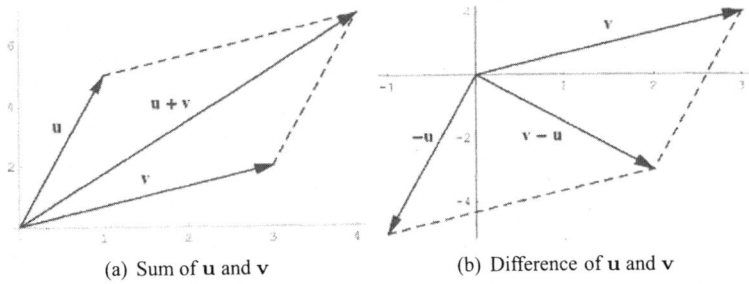

(a) Sum of **u** and **v**   (b) Difference of **u** and **v**

Figure 2.1 Plots of the sum and difference of two vectors.

Definition 2.1 (Inner Product). *Let* $\mathbf{v} = \begin{bmatrix} v_1 \\ v_2 \\ \vdots \\ v_n \end{bmatrix}$ *and* $\mathbf{w} = \begin{bmatrix} w_1 \\ w_2 \\ \vdots \\ w_n \end{bmatrix}$ *be two vectors*

in $\mathbb{R}^n$*. We define the* inner product $\mathbf{v} \cdot \mathbf{w}$ *as*

$$\mathbf{v} \cdot \mathbf{w} = \sum_{k=1}^{n} v_k w_k \tag{2.1}$$

□

You might recall the *transpose* of a matrix from your linear algebra class. We cover the matrix transpose in Section 2.2, but since we are discussing inner products, perhaps it will be useful, to talk about transposing a vector **v**.

Definition 2.2 (Vector Transpose). *We define the* transpose *of vector* $\mathbf{v} \in \mathbb{R}^n$ *as*

$$\mathbf{v}^T = [v_1, v_2, \ldots, v_n] \tag{2.2}$$

□

We will also find it convenient to write vectors as $\mathbf{v} = [v_1, v_2, \ldots, v_n]^T$:

$$\mathbf{v}^T = \begin{bmatrix} 1 \\ 2 \\ 3 \end{bmatrix}^T = [1, 2, 3]$$

The transpose of a vector is often used to express the inner product of two vectors. Indeed, we write

$$\mathbf{v} \cdot \mathbf{w} = \mathbf{v}^T \mathbf{w} \tag{2.3}$$

Next, we look at some examples of inner products.

Example 2.1 (Inner Products). *Compute the following inner products:*

(a) $\mathbf{v} = \begin{bmatrix} 1 \\ 4 \\ -1 \end{bmatrix}$ *and* $\mathbf{w} = \begin{bmatrix} 3 \\ 2 \\ -2 \end{bmatrix}$.

(b) $\mathbf{v} = \begin{bmatrix} -1 \\ 3 \end{bmatrix}$ *and* $\mathbf{w} = \begin{bmatrix} 6 \\ 2 \end{bmatrix}$.

Solution. *We have*

$$\begin{bmatrix} 1 \\ 4 \\ -1 \end{bmatrix} \cdot \begin{bmatrix} 3 \\ 2 \\ -2 \end{bmatrix} = [1, 4, -1] \begin{bmatrix} 3 \\ 2 \\ -2 \end{bmatrix} = 1 \cdot 3 + 4 \cdot 2 + (-1) \cdot (-2) = 13$$

and

$$\begin{bmatrix} -1 \\ 3 \end{bmatrix} \cdot \begin{bmatrix} 6 \\ 2 \end{bmatrix} = [-1, 3] \begin{bmatrix} 6 \\ 2 \end{bmatrix} = (-1) \cdot 6 + 3 \cdot 2 = 0$$

□

The inner product value of zero that we computed in part (b) of Example 2.1 has geometric significance. We plot these vectors in Figure 2.2.

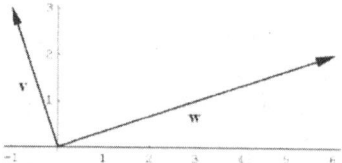

Figure 2.2   The vectors $\mathbf{v}$ and $\mathbf{w}$ from Example 2.1(b).

Notice that these vectors are perpendicular to each other. Using a term more common in linear algebra, we say that $\mathbf{v}$ and $\mathbf{w}$ are *orthogonal*. We can extend this notion to vectors of length n and make the following definition.

Definition 2.3 (Orthogonal Vectors). *Let* $\mathbf{v}, \mathbf{w} \in \mathbb{R}^n$. *Then we say that* $\mathbf{v}$ *and* $\mathbf{w}$ *are* orthogonal *if*

$$\mathbf{v} \cdot \mathbf{w} = \mathbf{v}^T \mathbf{w} = 0$$

We say thay a set of vectors is orthogonal or forms an orthogonal set *if all distinct vectors in the set are orthogonal to each other.*   □

As we shall see, there are many advantages to working with orthogonal vectors.

Vector Norms

It is important to be able to measure the length or *norm* of a vector. Typically, the norm of a vector $\mathbf{v}$ is written as $\|\mathbf{v}\|$. To motivate our definition, let's assume that $\mathbf{v} \in \mathbb{R}^2$. Consider Figure 2.3:

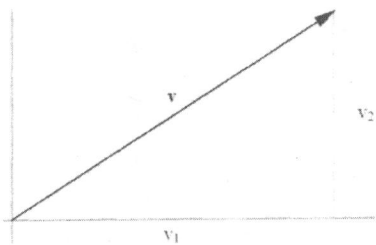

Figure 2.3   Vector $\mathbf{v} = [v_1, v_2]^T$.

In addition to graphing $\mathbf{v}$ in Figure 2.3, we have added the horizontal and vertical line segments with lengths v_1 and v_2, respectively. We see that the length of $\mathbf{v}$ might be measured by considering $\mathbf{v}$ as the hypotenuse of a right triangle and using the Pythagorean theorem to write the length as

$$\|\mathbf{v}\| = \sqrt{v_1^2 + v_2^2}$$

This idea can be formalized for vectors of length n. We give the following definition for $\|\mathbf{v}\|$.

Definition 2.4 (Vector Norm). *Let $\mathbf{v} \in \mathbb{R}^n$. We define the* norm *of $\mathbf{v}$ by*

$$\|\mathbf{v}\| = \sqrt{\sum_{k=1}^{n} v_k^2} \tag{2.4}$$

$\square$

You should notice a connection between $\|\mathbf{v}\|$ and the definition of the inner product (2.1). In Problem 2.4 you will be asked more about this relationship.

Example 2.2 (Vector Norms). *Let $\mathbf{v} = \begin{bmatrix} 2 \\ -5 \\ 4 \end{bmatrix}$ and $\mathbf{w} = [\frac{1}{2}, \frac{1}{2}, \frac{1}{2}, \frac{1}{2}]^T$. Compute*
$\|\mathbf{v}\|$ *and* $\|\mathbf{w}\|$.
Solution.

$$\|\mathbf{v}\| = \sqrt{2^2 + (-5)^2 + 4^2} = 3\sqrt{5} \qquad and \qquad \|\mathbf{w}\| = \sqrt{\frac{1}{4} + \frac{1}{4} + \frac{1}{4} + \frac{1}{4}} = 1$$

☐

PROBLEMS

2.1   For each of the pairs $\mathbf{v}$ and $\mathbf{w}$ below, compute the inner product $\mathbf{v} \cdot \mathbf{w}$. (*Hint:* To simplify the inner products in parts (c) – (e), you can use well-known summation formulas for $\sum\limits_{k=1}^{n} k$ and $\sum\limits_{k=1}^{n} k^2$. See, e. g., Stewart [69], Chap. 5.)

(a) $\mathbf{v} = [2, 5, -3, 1]^T$ and $\mathbf{w} = [-1, 0, -1, 0]^T$.

(b) $\mathbf{v} = [1, 0, 1]^T$ and $\mathbf{w} = [0, 50, 0]^T$.

(c) $\mathbf{v} = [1, 2, 3, \ldots, n]^T$ and $\mathbf{w} = [1, 1, 1, \ldots, 1]^T$.

(d) $\mathbf{v} = [1, 1, 2, 4, 3, 9, 4, 16, \ldots, n, n^2]^T$ and $\mathbf{w} = [1, 0, 1, 0, \ldots, 1, 0]^T$.

(e) $\mathbf{v} = [1, 1, 2, 4, 3, 9, 4, 16, \ldots, n, n^2]^T$ and $\mathbf{w} = [0, 1, 0, 1, \ldots, 0, 1]^T$.

2.2   Compute $\|\mathbf{v}\|$ and $\|\mathbf{w}\|$ for each of the vectors in Problem 2.1(a) – (c).

2.3   Find $\|\mathbf{v}\|$ when $\mathbf{v} = [v_1, \ldots, v_n]^T$ with

(a) $v_k = c, k = 1, \ldots, n$, where c is any real number.

(b) $v_k = k, k = 1, \ldots, n$.

(c) $v_k = \sqrt{k}, k = 1, \ldots, n$.

(*Hint:* The summation formulas referred to in Problem 2.1(c) – (e) will be useful for parts (b) and (c).)

2.4   Let $\mathbf{v} \in \mathbb{R}^N$ and show that $\|\mathbf{v}\|^2 = \mathbf{v} \cdot \mathbf{v}$.

2.5   Show that for vectors $\mathbf{u}, \mathbf{v}, \mathbf{w} \in \mathbb{R}^n$ and $c \in \mathbb{R}$, we have

(a) $\mathbf{u} \cdot \mathbf{v} = \mathbf{v} \cdot \mathbf{u}$.

(b) $(c\mathbf{u}) \cdot \mathbf{v} = c(\mathbf{u} \cdot \mathbf{v})$.

(c) $(\mathbf{u} + c\mathbf{v}) \cdot \mathbf{w} = \mathbf{u} \cdot \mathbf{w} + c\mathbf{v} \cdot \mathbf{w}$.

2.6   Let c be any real number. Show that for any vector $\mathbf{v} \in \mathbb{R}^n$, $\|c\mathbf{v}\| = |c| \cdot \|\mathbf{v}\|$.

2.7   Let $\mathbf{u}, \mathbf{v} \in \mathbb{R}^N$. Show that $\mathbf{u} \cdot \mathbf{v} = \frac{1}{4}\|\mathbf{u} + \mathbf{v}\|^2 - \frac{1}{4}\|\mathbf{u} - \mathbf{v}\|^2$.

2.8   Prove the *Cauchy – Schwarz inequality.* That is, show that for vectors $\mathbf{u}, \mathbf{v} \in \mathbb{R}^n$, we have

$$|\mathbf{u} \cdot \mathbf{v}|^2 \leq \|\mathbf{u}\|^2 \|\mathbf{v}\|^2 \tag{2.5}$$

(*Hint:* Consider the nonnegative expression $0 \leq (\mathbf{u} - t\mathbf{v}) \cdot (\mathbf{u} - t\mathbf{v})$. Expand this expression and observe that it is a quadratic polynomial in t that is nonnegative. What do we know, then, about the roots of this quadratic polynomial? Write this condition using the discriminant from the quadratic formula.)

2.9 Show that in Problem 2.8, equality holds if and only if $\mathbf{u} = c\mathbf{v}$ for some $c \in \mathbb{R}$.

2.10 There are other possible norms for $\mathbb{R}^n$. The *taxicab norm* $\|\mathbf{v}\|_1$ is defined to be

$$\|\mathbf{v}\|_1 = \sum_{k=1}^{n} |v_k|.$$

In the case of 2-vectors, the taxicab norm is simply $\|\mathbf{v}\|_1 = |v_1| + |v_2|$, the sum of the horizontal and vertical distances of the vector from the origin. The *max norm* is defined as

$$\|\mathbf{v}\|_\infty = \max(|v_1|, |v_2|, \dots, |v_n|)$$

A function f that maps $\mathbf{v} \in \mathbb{R}^n$ into the nonnegative real numbers is a *norm* if for $\mathbf{v} \in \mathbb{R}^n$ and $c \in \mathbb{R}$ we have

(i) $f(\mathbf{v}) \geq 0$ for all $\mathbf{v} \in \mathbb{R}^n$ with equality if and only if $\mathbf{v}$ is the zero vector.

(ii) $f(c\mathbf{v}) = |c| f(\mathbf{v})$.

(iii) $f(\mathbf{u} + \mathbf{v}) \leq f(\mathbf{u}) + f(\mathbf{v})$.

Item (iii) is referred to as the *triangle inequality*.

(a) Show that $\| \cdot \|_1$ is a norm.

(b) Show that $\| \cdot \|_\infty$ is a norm.

2.11 Use the definition of norm in Problem 2.9 to show that $\| \cdot \|$ is a norm. (*Hint:* Condition (ii) of the definition is Problem 2.6. The triangle inequality condition (iii) can be proved by first expanding $\|\mathbf{u} + \mathbf{v}\|^2$ and next applying the Cauchy – Schwarz inequality from Problem 2.8.)

2.12 In geometry, we learn that a circle is the set of all points of equal distance from a center point. In this problem you will see that the shape of the circle is completely determined by the norm we use. Sketch a graph of the "circle" centered at $[0,0]^T$ with radius 1 that results if we measure distance in $\mathbb{R}^2$ using norms. Do your sketches look like conventional circles?

(a) $\| \cdot \|$.

(b) $\| \cdot \|_1$.

(c) $\| \cdot \|_\infty$.

⋆**2.13**   In this problem you will see how to construct a vector that is orthogonal to a given vector. We refer to this problem again in the Problems for Section 5.2.

(a) Let $\mathbf{u} = [1, 2, 3, 4]^T$. Set $\mathbf{v} = [4, -3, 2, -1]^T$. Show that $\mathbf{u}$ and $\mathbf{v}$ are orthogonal and that $\|\mathbf{v}\| = \|\mathbf{u}\|$. Do you see how $\mathbf{v}$ was constructed from $\mathbf{u}$?

(b) If you have guessed what the construction is, try it on your own examples in $\mathbb{R}^n$. Make sure to choose vectors that are of even length and vectors that are of odd length. What did you find?

(c) Let n be even and let $\mathbf{u} = [u_1, u_2, \ldots, u_n]^T$. Define $\mathbf{v} = [v_1, v_2, \ldots, v_n]^T$ by the rule
$$v_k = (-1)^k u_{n-k+1}, \qquad k = 1, 2, \ldots, n$$

Show that $\mathbf{u}$ and $\mathbf{v}$ are orthogonal and that $\|\mathbf{v}\| = \|\mathbf{u}\|$.

⋆**2.14**   Often, it is desirable to work with vectors of length 1. Suppose that $\mathbf{v} \in \mathbb{R}^n$ with $\mathbf{v} \neq \mathbf{0}$. Let $\mathbf{u} = \frac{1}{\|\mathbf{v}\|}\mathbf{v}$. Show that $\|\mathbf{u}\| = 1$. This process is known as *normalizing* the vector $\mathbf{v}$.

2.15   Use Problem 2.14 to normalize the following vectors:

(a) $\mathbf{v} = [5, 0, 12]^T$.

(b) $\mathbf{v} = [2, 6, 3, 1]^T$.

(c) $\mathbf{v} \in \mathbb{R}^n$ with $v_k = -2, k = 1, 2, \ldots, n$.

Computer Lab

2.1   **Inner Products and Norms.** From the text Web site, access the lab vectors. This lab will familiarize you with how your **CAS** defines vectors and computes inner products and norms.

2.2 BASIC MATRIX THEORY

One of the most fundamental tools to come from linear algebra is the *matrix*. It would be difficult to write down all the applications that involve matrices. For our purposes we will learn that matrices can be used to represent the transformations that we apply to digital images and audio data for the purpose of image enhancement, data compression, or denoising.

A matrix can be thought of as simply a table of numbers. More generally, we call these numbers elements since in some instances we want matrices comprised of functions or even other matrices. Unless stated otherwise, we will denote matrices by uppercase letters. Here are three examples of matrices:

$$A = \begin{bmatrix} 2 & 3 \\ 6 & -5 \end{bmatrix} \qquad B = \begin{bmatrix} b_{11} & b_{12} \\ b_{21} & b_{22} \\ b_{31} & b_{32} \end{bmatrix} \qquad C = \begin{bmatrix} 3 & 5 & 0 \\ 2 & \frac{1}{2} & 6 \end{bmatrix}$$

We say the first matrix is a 2×2 matrix: two rows (horizontal lists) and two columns (vertical lists). The matrix A has four elements: The element in row 1, column 1 is 2; the element in row 1, column 2 is 3; and so on.

We give a generic 3×2 matrix for B. Here we use the common notation b_{ij}, $i = 1, 2, 3$, and $j = 1, 2$. The notation b_{ij} denotes the value in B located at row i, column j.

Matrix C is a 2×3 matrix. If we were to assign the variables c_{ij}, $i = 1, 2$ and $j = 1, 2, 3$ to the elements of C, we would have, for example, $c_{12} = 5$, $c_{21} = 2$, and $c_{23} = 6$.

We now give a formal definition of the dimension of a matrix.

Definition 2.5 (Dimension of a Matrix). *Let A be a matrix consisting of m rows and n columns. We say that the dimension of A is $m \times n$.* ☐

Note that the vectors introduced in Section 2.1 are actually $n \times 1$ matrices. The transpose of $\mathbf{v} = (v_1, v_2, \ldots, v_n)^T$ is also a matrix with dimension $1 \times n$.

Matrix Arithmetic

Some of the most common operations we perform on matrices are addition, subtraction, and scalar multiplication. To add two matrices or subtract one matrix from another, they must both have the same dimensions. If A is a $m \times n$ matrix, then the only matrices that we can add to it or subtract from it must also be $m \times n$. If A and B are of the same dimension then addition and subtraction is straightforward. We simply add or subtract the corresponding elements.

Definition 2.6 (Addition and Subtraction of Matrices). *Let A and B be the $m \times n$ matrices*

$$A = \begin{bmatrix} a_{11} & a_{12} & \cdots & a_{1n} \\ a_{21} & a_{22} & \cdots & a_{2n} \\ \vdots & \vdots & \ddots & \vdots \\ a_{m1} & a_{m2} & \cdots & a_{mn} \end{bmatrix} \qquad B = \begin{bmatrix} b_{11} & b_{12} & \cdots & b_{1n} \\ b_{21} & b_{22} & \cdots & b_{2n} \\ \vdots & \vdots & \ddots & \vdots \\ b_{m1} & b_{m2} & \cdots & b_{mn} \end{bmatrix}$$

The sum S of A and B is the $m \times n$ matrix given by

$$S = A + B$$

$$\begin{bmatrix} s_{11} & s_{12} & \cdots & s_{1n} \\ s_{21} & s_{22} & \cdots & s_{2n} \\ \vdots & \vdots & \ddots & \vdots \\ s_{m1} & s_{m2} & \cdots & s_{mn} \end{bmatrix} = \begin{bmatrix} a_{11}+b_{11} & a_{12}+b_{12} & \cdots & a_{1n}+b_{1n} \\ a_{21}+b_{21} & a_{22}+b_{22} & \cdots & a_{2n}+b_{2n} \\ \vdots & \vdots & \ddots & \vdots \\ a_{m1}+b_{m1} & a_{m2}+b_{m2} & \cdots & a_{mn}+b_{mn} \end{bmatrix}$$

and the difference D of A and B is the $m \times n$ matrix given by

$$D = A - B$$

$$
\begin{bmatrix}
d_{11} & d_{12} & \cdots & d_{1n} \\
d_{21} & d_{22} & \cdots & d_{2n} \\
\vdots & \vdots & \ddots & \vdots \\
d_{m1} & d_{m2} & \cdots & d_{mn}
\end{bmatrix}
=
\begin{bmatrix}
a_{11}-b_{11} & a_{12}-b_{12} & \cdots & a_{1n}-b_{1n} \\
a_{21}-b_{21} & a_{22}-b_{22} & \cdots & a_{2n}-b_{2n} \\
\vdots & \vdots & \ddots & \vdots \\
a_{m1}-b_{m1} & a_{m2}-b_{m2} & \cdots & a_{mn}-b_{mn}
\end{bmatrix}
$$

□

Another basic operation we perform on matrices is scalar multiplication. This operation basically consists of multiplying every element in a matrix by the same number. Here is a formal definition.

Definition 2.7 (Matrices and Scalar Multiplication). *Let A be the $m \times n$ matrix*

$$
A =
\begin{bmatrix}
a_{11} & a_{12} & \cdots & a_{1n} \\
a_{21} & a_{22} & \cdots & a_{2n} \\
\vdots & \vdots & \ddots & \vdots \\
a_{m1} & a_{m2} & \cdots & a_{mn}
\end{bmatrix}
$$

and let c be any real number. Then we define the scalar product cA *as follows:*

$$
cA = c
\begin{bmatrix}
a_{11} & a_{12} & \cdots & a_{1n} \\
a_{21} & a_{22} & \cdots & a_{2n} \\
\vdots & \vdots & \ddots & \vdots \\
a_{m1} & a_{m2} & \cdots & a_{mn}
\end{bmatrix}
=
\begin{bmatrix}
ca_{11} & ca_{12} & \cdots & ca_{1n} \\
ca_{21} & ca_{22} & \cdots & ca_{2n} \\
\vdots & \vdots & \ddots & \vdots \\
ca_{m1} & ca_{m2} & \cdots & ca_{mn}
\end{bmatrix}
\quad (2.6)
$$

□

Matrices have all kinds of practical uses. The example that follows illustrates how matrix addition, subtraction, and multiplication by a scalar applies to images.

Example 2.3 (Matrix Arithmetic). *Let $a = \frac{1}{5}$, $b = 2$, and consider the following 4×4 matrices:*

$$
A =
\begin{bmatrix}
100 & 50 & 50 & 25 \\
0 & 25 & 25 & 125 \\
0 & 100 & 75 & 50 \\
125 & 50 & 0 & 25
\end{bmatrix}
\qquad
B =
\begin{bmatrix}
50 & 50 & 50 & 50 \\
50 & 50 & 50 & 50 \\
50 & 50 & 50 & 50 \\
50 & 50 & 50 & 50
\end{bmatrix}
$$

and

$$
C =
\begin{bmatrix}
255 & 255 & 255 & 255 \\
255 & 255 & 255 & 255 \\
255 & 255 & 255 & 255 \\
255 & 255 & 255 & 255
\end{bmatrix}
$$

We can compute the sum and difference

$$A + B = \begin{bmatrix} 150 & 100 & 100 & 75 \\ 50 & 75 & 75 & 175 \\ 50 & 150 & 125 & 100 \\ 175 & 100 & 50 & 75 \end{bmatrix} \qquad C - A = \begin{bmatrix} 155 & 205 & 205 & 230 \\ 255 & 230 & 230 & 130 \\ 255 & 155 & 180 & 205 \\ 130 & 205 & 255 & 230 \end{bmatrix}$$

and the scalar products

$$aA = \begin{bmatrix} 20 & 10 & 10 & 5 \\ 0 & 5 & 5 & 25 \\ 0 & 20 & 15 & 10 \\ 25 & 10 & 0 & 5 \end{bmatrix} \qquad bA = \begin{bmatrix} 200 & 100 & 100 & 50 \\ 0 & 50 & 50 & 250 \\ 0 & 200 & 150 & 100 \\ 250 & 100 & 0 & 50 \end{bmatrix}$$

easily enough, but when we view these results as images, we can see the effects of addition, subtraction, and scalar multiplication.

In Section 3.1 we will learn that pixels in grayscale (monochrome) images are often represented as integer values ranging from 0 (black) to 255 (white). In Figure 2.4 we display the matrices A, B, and C and various arithmetic operations by shading their entries with the gray level that the value represents.

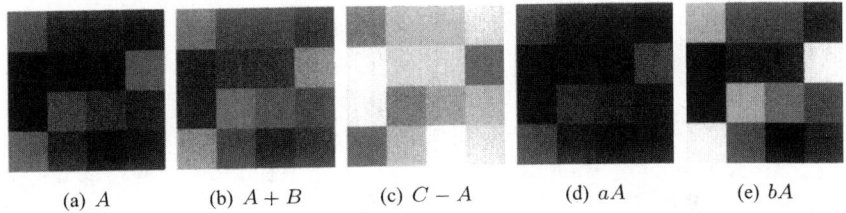

(a) A (b) $A + B$ (c) $C - A$ (d) aA (e) bA

Figure 2.4   Matrices A, B, and C and various arithmetic operations with entries shaded to the appropriate gray level.

In Figure 2.4 you can see that the sum $A + B$ is about 20% lighter than A. The difference $C - A$ is more relevant. Subtracting each of the entries of A from 255 has the effect of reversing the gray intensities. The scalar multiplication by $\frac{1}{5}$ has the effect of darkening each pixel of A by about 20%, and 2A has the effect of lightening each pixel in A by a factor of 2.   □

Matrix Multiplication

We are now ready to review matrix multiplication. It is not an elementwise operation as is the case for addition, subtraction, and scalar multiplication. Let's motivate the idea with an example. Let $\mathbf{v} = [150, 200, 250, 150]^T$ and assume that $\mathbf{v}$ represents the grayscale values in a column of some image. Suppose that we want to compute not only the average of the components of $\mathbf{v}$, but also some weighted averages.

If we want to average the components of $\mathbf{v}$, then we could dot $\mathbf{v}$ with the vector $\mathbf{a} = \left[\frac{1}{4}, \frac{1}{4}, \frac{1}{4}, \frac{1}{4}\right]^T$. Suppose that we want the middle two components weighted twice as much as the first and last. Then we could dot $\mathbf{v}$ with the vector $\mathbf{b} = \left[\frac{1}{6}, \frac{1}{3}, \frac{1}{3}, \frac{1}{6}\right]^T$. And finally, suppose that we want to average just the first and last components of $\mathbf{v}$. Then we would dot $\mathbf{v}$ with the vector $\mathbf{c} = \left[\frac{1}{2}, 0, 0, \frac{1}{2}\right]^T$.

So we would compute three inner products: $\mathbf{a}^T\mathbf{v}$, $\mathbf{b}^T\mathbf{v}$, and $\mathbf{c}^T\mathbf{v}$. We store these row vectors in a matrix A and then consider the product $A\mathbf{v}$ as a 3-vector containing our three dot products.

$$
A\mathbf{v} = \begin{bmatrix} \mathbf{a}^T \\ \mathbf{b}^T \\ \mathbf{c}^T \end{bmatrix} \mathbf{v} = \begin{bmatrix} \frac{1}{4} & \frac{1}{4} & \frac{1}{4} & \frac{1}{4} \\ \frac{1}{6} & \frac{1}{3} & \frac{1}{3} & \frac{1}{6} \\ \frac{1}{2} & 0 & 0 & \frac{1}{2} \end{bmatrix} \begin{bmatrix} 100 \\ 200 \\ 250 \\ 150 \end{bmatrix}
$$

$$
= \begin{bmatrix} \frac{1}{4}(100) + \frac{1}{4}(200) + \frac{1}{4}(250) + \frac{1}{4}(150) \\ \frac{1}{6}100) + \frac{1}{3}(200) + \frac{1}{3}(250) + \frac{1}{6}(150) \\ \frac{1}{2}(100) + 0 \cdot 200 + 0 \cdot 250 + \frac{1}{2}(150) \end{bmatrix} = \begin{bmatrix} 175 \\ 191\frac{2}{3} \\ 125 \end{bmatrix}
$$

Of course, we might want to apply our weighted averages to other columns in addition to $\mathbf{v}$. We would simply store these in a matrix V (with $\mathbf{v}$ as one of the columns) and dot each new column with the rows of A and store the results in the appropriate place. Suppose that we create the matrix

$$
V = \begin{bmatrix} 100 & 150 & 150 \\ 200 & 100 & 200 \\ 250 & 50 & 250 \\ 150 & 200 & 50 \end{bmatrix}
$$

and we wish to compute the product AV. Then we would simply dot each row of A with each column of V. There would be nine total dot products. We would have three dot products for each of the three columns of V, so it would make sense to store the result in a 3×3 matrix Y with the element y_{ij} in row i, column j of Y obtained by dotting row i of A with column j of V. This would correspond to applying the ith weighted average to the jth vector in V. We have

$$
AV = \begin{bmatrix} \frac{1}{4} & \frac{1}{4} & \frac{1}{4} & \frac{1}{4} \\ \frac{1}{6} & \frac{1}{3} & \frac{1}{3} & \frac{1}{6} \\ \frac{1}{2} & 0 & 0 & \frac{1}{2} \end{bmatrix} \begin{bmatrix} 100 & 150 & 150 \\ 200 & 100 & 200 \\ 250 & 50 & 250 \\ 150 & 200 & 50 \end{bmatrix}
$$

$$
= \begin{bmatrix} 175\frac{1}{2} & 125 & 162\frac{1}{2} \\ 191\frac{2}{3} & 108\frac{1}{3} & 183\frac{1}{3} \\ 125 & 175 & 100 \end{bmatrix}
$$

We are now ready to define matrix multiplication. To compute AB we dot the columns of A with the rows of B. Thus, for the inner products to make sense, the number of elements in each column of A must be the same as the number of elements in each row of B. Also note that while we can compute AB, the product BA might not make sense. For example, if A is 3×4 and B is 4×2, then AB will consist of six inner products stored in a 3×2 matrix. But the product BA makes no sense because the rows of B consist of two elements, whereas the columns of A consist of four elements and we can't perform the necessary inner products.

Definition 2.8 (Matrix Multiplication). *Suppose that A is an $m \times p$ matrix and B is an $p \times n$ matrix. Then the product AB is the $m \times n$ matrix C whose entries c_{ij} are obtained by dotting row i of A with column j of B where $i = 1, 2, \ldots, m$ and $j = 1, 2, \ldots, n$. That is,*

$$
c_{ij} = \sum_{k=1}^{p} a_{ik} b_{kj}
$$

□

Example 2.4 (Matrix Multiplication). *Compute the following products.*

(a) $\begin{bmatrix} 1 & 5 & 2 \\ -2 & 6 & 9 \end{bmatrix} \cdot \begin{bmatrix} 6 & 0 & 3 & 7 \\ 1 & 2 & 5 & 4 \\ 0 & -1 & 3 & -2 \end{bmatrix}.$

(b) $\begin{bmatrix} 1 & 1 & 1 \\ 2 & 2 & 2 \\ 3 & 3 & 3 \end{bmatrix} \cdot \begin{bmatrix} 1 & 3 & -2 \\ 4 & 2 & -5 \\ -1 & 3 & 0 \end{bmatrix}.$

(c) $\begin{bmatrix} 6 & 11 & 5 \\ 1 & 0 & -1 \end{bmatrix} \cdot \begin{bmatrix} 3 & -2 \\ 8 & -3 \end{bmatrix}.$

(d) $\begin{bmatrix} 2 & 5 \\ 1 & 3 \end{bmatrix} \cdot \begin{bmatrix} 3 & -5 \\ -1 & 2 \end{bmatrix}.$

Solution. *Since we are multiplying a* 2 × 3 *matrix by a* 3 × 4 *matrix, the product in part (a) will be a* 2 × 4 *matrix. We compute the eight inner products to obtain*

$$\begin{bmatrix} 1 & 5 & 2 \\ -2 & 6 & 9 \end{bmatrix} \cdot \begin{bmatrix} 6 & 0 & 3 & 7 \\ 1 & 2 & 5 & 4 \\ 0 & -1 & 3 & -2 \end{bmatrix} = \begin{bmatrix} 11 & 8 & 34 & 23 \\ -6 & 3 & 51 & -8 \end{bmatrix}$$

In part (b) we are multiplying two 3 × 3 *matrices, so the result will be a* 3 × 3 *matrix. We compute the nine inner products to obtain*

$$\begin{bmatrix} 1 & 1 & 1 \\ 2 & 2 & 2 \\ 3 & 3 & 3 \end{bmatrix} \cdot \begin{bmatrix} 1 & 3 & -2 \\ 4 & 2 & -5 \\ -1 & 3 & 0 \end{bmatrix} = \begin{bmatrix} 4 & 8 & -7 \\ 8 & 16 & -14 \\ 12 & 24 & -21 \end{bmatrix}$$

For the matrices in part (c), the product is undefined. We are unable to multiply a 2 × 3 *matrix and a* 2 × 2 *matrix since the number of columns in the first matrix is not the same as the number of rows in the second matrix.*

In part (d) we multiply two 2 × 2 *matrices so that the result is a* 2 × 2 *matrix. We compute the four inner products to obtain*

$$\begin{bmatrix} 2 & 5 \\ 1 & 3 \end{bmatrix} \cdot \begin{bmatrix} 3 & -5 \\ -1 & 2 \end{bmatrix} = \begin{bmatrix} 1 & 0 \\ 0 & 1 \end{bmatrix}$$

□

Part (b) of Example 2.4 gives us a chance to describe what a matrix multiplication actually does. First note that dotting each column of the second matrix by $(1, 1, 1)$ (the top row of the first matrix) has the effect of summing the components of the column. Then it is easy to see that multiplying each column of the second matrix by the row $(2, 2, 2)$ has the effect of summing the components of each column of the second matrix and doubling the result and multiplying each column of the second matrix by the row $(3, 3, 3)$ has the effect of summing the components of each column of the second matrix and tripling the result.

Although the product in part (c) is undefined, it is true that we can compute

$$\begin{bmatrix} 3 & -2 \\ 8 & -3 \end{bmatrix} \cdot \begin{bmatrix} 6 & 11 & 5 \\ 1 & 0 & -1 \end{bmatrix} = \begin{bmatrix} 16 & 33 & 17 \\ 46 & 88 & 42 \end{bmatrix}$$

We immediately see that matrix multiplication is not a commutative operation. It is a common error for students to perform the computation above and give it as the answer for the product requested in part (c) of Example 2.4 — do not make this mistake! The product in Example 2.4(c) is undefined.

The Inverse of a Matrix

The product in Example 2.4(d) is a very special matrix.

Definition 2.9 (Identity Matrix). *Let I_n be the $n \times n$ matrix whose kth column is given by the* standard basis vector

$$\mathbf{e}^k = \left[0, 0, \ldots, \underbrace{1}_{k\text{th}}, 0, \ldots, 0 \right]^T \tag{2.7}$$

Then I_n is called the identity matrix of order n. □

We write out three identity matrices:

$$I_2 = \begin{bmatrix} 1 & 0 \\ 0 & 1 \end{bmatrix} \qquad I_3 = \begin{bmatrix} 1 & 0 & 0 \\ 0 & 1 & 0 \\ 0 & 0 & 1 \end{bmatrix} \qquad I_4 = \begin{bmatrix} 1 & 0 & 0 & 0 \\ 0 & 1 & 0 & 0 \\ 0 & 0 & 1 & 0 \\ 0 & 0 & 0 & 1 \end{bmatrix}$$

The special property about identity matrices is that $I_n A = A$ and $B I_n = B$. That is I_n is to matrix multiplication what the number 1 is to ordinary multiplication. Consider the product

$$I_m A = \begin{bmatrix} 1 & 0 & 0 & \cdots & 0 \\ 0 & 1 & 0 & \cdots & 0 \\ 0 & 0 & 1 & \cdots & 0 \\ \vdots & \vdots & \vdots & \ddots & \vdots \\ 0 & 0 & 0 & \cdots & 1 \end{bmatrix} \begin{bmatrix} a_{11} & a_{12} & \cdots & a_{1n} \\ a_{21} & a_{22} & \cdots & a_{2n} \\ \vdots & \vdots & \ddots & \vdots \\ a_{m1} & a_{m2} & \cdots & a_{mn} \end{bmatrix}$$

The first row of I_n is $(1, 0, 0, \ldots, 0)$ and when we dot it with each column of A, the result is simply 1 times the first element in the column. So $(1, 0, 0, \ldots, 0) A$ gives the first row of A. In a similar manner we see that the kth row of I_n picks out the kth element of each column of A and thus returns the kth row of A.

We now turn our attention to the important class of nonsingular or invertible matrices. We start with a definition.

Definition 2.10 (Nonsingular Matrix). *Let A be an $n \times n$ matrix. We say that A is* nonsingular *or* invertible *if there exists an $n \times n$ matrix B such that*

$$AB = BA = I_n$$

In this case, B is called the inverse of A *and is denoted by $B = A^{-1}$. If no such matrix B exists, then A is said to be* singular. □

Let's take a closer look at $BA = I_n$ in the case where $B = A^{-1}$. We have

$$
\begin{bmatrix}
b_{11} & b_{12} & \cdots & b_{1n} \\
b_{21} & b_{22} & \cdots & b_{2n} \\
\vdots & \vdots & \ddots & \vdots \\
b_{n1} & b_{n2} & \cdots & b_{nn}
\end{bmatrix}
\begin{bmatrix}
a_{11} & a_{12} & \cdots & a_{1n} \\
a_{21} & a_{22} & \cdots & a_{2n} \\
\vdots & \vdots & \ddots & \vdots \\
a_{n1} & a_{n2} & \cdots & a_{nn}
\end{bmatrix}
=
\begin{bmatrix}
1 & 0 & 0 & \cdots & 0 \\
0 & 1 & 0 & \cdots & 0 \\
0 & 0 & 1 & \cdots & 0 \\
\vdots & \vdots & \vdots & \ddots & \vdots \\
0 & 0 & 0 & \cdots & 1
\end{bmatrix}
$$

In particular, we see that the first row of B must be dotted with the first column of A and produce 1 and the first row of B dotted with any other column of A results in an answer of zero. We can generalize this remark and say that when we dot row j of B with column k of A, we get 1 when $j = k$ and 0 when $j \neq k$.

You probably learned how to use elementary row operations or Gaussian elimination in a linear algebra class to find A^{-1} given A. We do not review that technique in this book. If you need to review Gaussian elimination, you should consult your linear algebra book or see, for example Strang, [71].

Some inverses are really easy to compute. Consider the following example.

Example 2.5 (Inverse of a Diagonal Matrix). *Let D be the $n \times n$ diagonal matrix*

$$
D =
\begin{bmatrix}
d_1 & 0 & 0 & \cdots & 0 \\
0 & d_2 & 0 & \cdots & 0 \\
\vdots & \vdots & \vdots & \ddots & \vdots \\
0 & 0 & 0 & \cdots & d_n
\end{bmatrix}
$$

where $d_k \neq 0$, $k = 1, 2, \ldots, n$.

Suppose that D is nonsingular and let's think about how to compute D^{-1}. If we consider the product $DD^{-1} = I_n$, we see that the first row of D, $(d_1, 0, 0, \ldots, 0)$ must dot with the first column of D^{-1} and produce 1. An obvious selection for this first column of D^{-1} is to make the first element $\frac{1}{d_1}$ and set the rest of the elements in the column equal to zero. That is, the first column of the inverse is $\begin{bmatrix} \frac{1}{d_1} & 0 & \cdots & 0 & 0 \end{bmatrix}^T$.

Note also that if we take any other row of D and dot it with this column, we obtain zero. Thus, we have the first column of D^{-1}.

It should be clear how to choose the remaining columns of D^{-1}. The kth column of D^{-1} should have 0's in all positions except the kth position where we would want $\frac{1}{d_k}$. In this way we see that

$$
D^{-1} =
\begin{bmatrix}
\frac{1}{d_1} & 0 & 0 & \cdots & 0 \\
0 & \frac{1}{d_2} & 0 & \cdots & 0 \\
\vdots & \vdots & \vdots & \ddots & \vdots \\
0 & 0 & 0 & \cdots & \frac{1}{d_n}
\end{bmatrix}
$$

$\square$

Here are some basic properties obeyed by invertible matrices.

Proposition 2.1 (Properties of Nonsingular Matrices). *Suppose that A and B are $n \times n$ nonsingular matrices and c is any nonzero real number.*

(a) A^{-1} is unique. That is, there is only one matrix A^{-1} such that

$$A^{-1}A = AA^{-1} = I_n$$

(b) A^{-1} is nonsingular and $(A^{-1})^{-1} = A$.

(c) AB is nonsingular with $(AB)^{-1} = B^{-1}A^{-1}$.

(d) cA is nonsingular with $(cA)^{-1} = \frac{1}{c}A^{-1}$.

□

Proof of Proposition 2.1. The proof is left as Problem 2.20.

□

Why are we interested in the class of nonsingular matrices? Suppose that an $m \times m$ matrix A represents some procedure that we want to apply to an $m \times n$ image matrix M. For example, suppose that AM is used to compress our image so that we can send it over the Internet. The recipient would have only AM. But if A is nonsingular, the recipient can compute

$$A^{-1}(AM) = (A^{-1}A)M = I_m M = M$$

and recover the original image matrix M.

In effect, if A is nonsingular, we can solve the matrix equation $AX = B$ for unknown matrix X by computing $X = A^{-1}B$ or the vector equation $Ax = b$ for unknown vector x by computing $x = A^{-1}b$. This is completely analogous to solving $ax = b$, where a, b are real numbers with $a \neq 0$. To solve this equation, we multiply both sides by a^{-1}, the multiplicative inverse of a, to obtain $x = a^{-1}b$.

So why can't we always solve $Ax = b$ when A is singular? Let's look at an example.

Example 2.6 (Solving $Ax = b$). *Solve the equation*

$$\begin{bmatrix} 1 & 1 & 1 \\ 2 & 2 & 2 \\ 3 & 3 & 3 \end{bmatrix} \begin{bmatrix} x_1 \\ x_2 \\ x_3 \end{bmatrix} = \begin{bmatrix} 0 \\ 0 \\ 0 \end{bmatrix}$$

Solution. *Let's take an alternative approach to solving this problem. Rather than using the elementary row operation techniques that you learned in linear algebra, we'll first note that*

$$\begin{bmatrix} 1 & 1 & 1 \\ 2 & 2 & 2 \\ 3 & 3 & 3 \end{bmatrix} = \begin{bmatrix} 1 & 0 & 0 \\ 0 & 2 & 0 \\ 0 & 0 & 3 \end{bmatrix} \begin{bmatrix} 1 & 1 & 1 \\ 1 & 1 & 1 \\ 1 & 1 & 1 \end{bmatrix}$$

and use the results of Example 2.5 to note that

$$
\begin{bmatrix} 1 & 0 & 0 \\ 0 & 2 & 0 \\ 0 & 0 & 3 \end{bmatrix}^{-1} = \begin{bmatrix} 1 & 0 & 0 \\ 0 & \frac{1}{2} & 0 \\ 0 & 0 & \frac{1}{3} \end{bmatrix} \tag{2.8}
$$

So we have

$$
\begin{bmatrix} 1 & 1 & 1 \\ 2 & 2 & 2 \\ 3 & 3 & 3 \end{bmatrix} \begin{bmatrix} x_1 \\ x_2 \\ x_3 \end{bmatrix} = \begin{bmatrix} 0 \\ 0 \\ 0 \end{bmatrix}
$$

$$
\begin{bmatrix} 1 & 0 & 0 \\ 0 & 2 & 0 \\ 0 & 0 & 3 \end{bmatrix} \begin{bmatrix} 1 & 1 & 1 \\ 1 & 1 & 1 \\ 1 & 1 & 1 \end{bmatrix} \begin{bmatrix} x_1 \\ x_2 \\ x_3 \end{bmatrix} = \begin{bmatrix} 0 \\ 0 \\ 0 \end{bmatrix}
$$

Multiplying both sides of this equation by the inverse found in (2.8) gives

$$
\begin{bmatrix} 1 & 1 & 1 \\ 1 & 1 & 1 \\ 1 & 1 & 1 \end{bmatrix} \begin{bmatrix} x_1 \\ x_2 \\ x_3 \end{bmatrix} = \begin{bmatrix} 1 & 0 & 0 \\ 0 & \frac{1}{2} & 0 \\ 0 & 0 & \frac{1}{3} \end{bmatrix} \begin{bmatrix} 0 \\ 0 \\ 0 \end{bmatrix}
$$

$$
\begin{bmatrix} x_1 + x_2 + x_3 \\ x_1 + x_2 + x_3 \\ x_1 + x_2 + x_3 \end{bmatrix} = \begin{bmatrix} 0 \\ 0 \\ 0 \end{bmatrix}
$$

It is easy to see that $x_1 = x_2 = x_3 = 0$ or $x_1 = x_2 = 1$ and $x_3 = -2$ are solutions to this equation. In fact, there are infinitely many solutions to this equation, so if you were sent $[0, 0, 0]^T$ and told it was obtained by computing

$$
\begin{bmatrix} 1 & 1 & 1 \\ 2 & 2 & 2 \\ 3 & 3 & 3 \end{bmatrix} \begin{bmatrix} x_1 \\ x_2 \\ x_3 \end{bmatrix}
$$

you would have no way of knowing exactly what vector **x** *was sent to you.*   □

The Transpose of a Matrix

There is another matrix related to A that is important in our subsequent work. This matrix is called the *transpose* of A and denoted by A^T. Basically, we form A^T by taking the rows of A and making them the columns of A^T. So if A is an $m \times n$ matrix, A^T will be an $n \times m$ matrix. Here is the formal definition:

Definition 2.11 (Transpose of a Matrix). *Let A be an $m \times n$ matrix with elements a_{ij}, $i = 1, 2, \ldots, m$, $j = 1, 2, \ldots, n$. Then the* transpose *of A, denoted A^T, is the $n \times m$ matrix whose elements are given by a_{ji} for $j = 1, 2, \ldots, n$ and $i = 1, 2, \ldots, m$. That is, if*

$$A = \begin{bmatrix} a_{11} & a_{12} & \cdots & a_{1n} \\ a_{21} & a_{22} & \cdots & a_{2n} \\ \vdots & \vdots & \ddots & \vdots \\ a_{m1} & a_{m2} & \cdots & a_{mn} \end{bmatrix}, \quad then \quad A^T = \begin{bmatrix} a_{11} & a_{21} & \cdots & a_{m1} \\ a_{12} & a_{22} & \cdots & a_{m2} \\ \vdots & \vdots & \ddots & \vdots \\ a_{1n} & a_{2n} & \cdots & a_{mn} \end{bmatrix}$$

□

Next we present some elementary examples of transposes of matrices.

Example 2.7 (Matrix Transposes). *Find the transpose of*

$$A = \begin{bmatrix} 1 & 6 & -2 \\ 2 & 3 & 0 \end{bmatrix} \quad B = \begin{bmatrix} 2 & 3 \\ -1 & 1 \end{bmatrix} \quad C = \begin{bmatrix} 2 & 6 & 3 \\ 6 & 4 & -1 \\ 3 & -1 & 1 \end{bmatrix}$$

Solution. *We have*

$$A^T = \begin{bmatrix} 1 & 2 \\ 6 & 3 \\ -2 & 0 \end{bmatrix} \quad B^T = \begin{bmatrix} 2 & -1 \\ 3 & 1 \end{bmatrix} \quad C^T = \begin{bmatrix} 2 & 6 & 3 \\ 6 & 4 & -1 \\ 3 & -1 & 1 \end{bmatrix}$$

□

Note that in Example 2.7 C^T was the same as C. We call such matrices *symmetric*.

Definition 2.12 (Symmetric Matrix). *We say that A is symmetric if $A^T = A$.*   □

Orthogonal Matrices

We leave this section by reviewing one last class of matrices called *orthogonal matrices*. Consider the matrix

$$U = \begin{bmatrix} \frac{1}{2} & \frac{\sqrt{3}}{2} & 0 \\ 0 & 0 & 1 \\ -\frac{\sqrt{3}}{2} & \frac{1}{2} & 0 \end{bmatrix}$$

Notice that if you dot any row with any other row, you get zero. In addition, any row dotted with itself returns 1. So if we were interested in finding U^{-1}, it seems that the rows of U would work very well as the columns of U^{-1}. Indeed, if we compute UU^T, we have

$$
UU^T = \begin{bmatrix} \frac{1}{2} & \frac{\sqrt{3}}{2} & 0 \\ 0 & 0 & 1 \\ -\frac{\sqrt{3}}{2} & \frac{1}{2} & 0 \end{bmatrix} \begin{bmatrix} \frac{1}{2} & 0 & -\frac{\sqrt{3}}{2} \\ \frac{\sqrt{3}}{2} & 0 & \frac{1}{2} \\ 0 & 1 & 0 \end{bmatrix}
$$

$$
= \begin{bmatrix} 1 & 0 & 0 \\ 0 & 1 & 0 \\ 0 & 0 & 1 \end{bmatrix}
$$

The matrix U above is an example of an orthogonal matrix. There are a couple of ways to classify such matrices. A good thing to remember about orthogonal matrices is that their rows (or columns), taken as a set of vectors, forms an orthogonal set. We give the standard definition below.

Definition 2.13 (Orthogonal Matrix). *Suppose that U is an $n \times n$ matrix. We say U is an* orthogonal matrix *if*

$$
U^{-1} = U^T \tag{2.9}
$$

$\square$

Recall that the columns of U^T are the rows of U. So when we say that $UU^T = I$, we are saying that if row j of U is dotted with row k of U (or column k of U^T), then we must either get 1 if $j = k$ or zero otherwise — this is the condition on the rows of U given in the text preceding Definition 2.13.

Orthogonal matrices also possess another important property: They preserve distances.

Theorem 2.1 (Orthogonal Matrices Preserve Distance). *Suppose that U is an $n \times n$ orthogonal matrix and $\mathbf{x}$ is an n-vector. Then*

$$
\|U\mathbf{x}\| = \|\mathbf{x}\| \tag{2.10}
$$

$\square$

Proof of Theorem 2.1. From Problem 2.4 in Section 2.1 we have

$$
\|U\mathbf{x}\|^2 = (U\mathbf{x})^T U\mathbf{x}
$$

Starting with this identity, Problem 2.22 from Section 2.2, and (2.9), we have

$$
\|U\mathbf{x}\|^2 = (U\mathbf{x})^T U\mathbf{x} = \mathbf{x}^T U^T U\mathbf{x} = \mathbf{x}^T U^{-1} U\mathbf{x} = \mathbf{x}^T I_n \mathbf{x} = \mathbf{x}^T \mathbf{x} = \|\mathbf{x}\|^2
$$

$\square$

We use the following example to further describe the geometry involved in the result of Theorem 2.1.

Example 2.8 (Rotations by an Orthogonal Matrix). *Consider the set of vectors*

$$C = \{\mathbf{x} \in \mathbb{R}^2 \mid \|\mathbf{x}\| = 1\}$$

If we plot all these vectors in C, we would have a picture of a circle centered at $(0,0)$ with radius 1. Suppose that U is the orthogonal matrix

$$U = \begin{bmatrix} \frac{\sqrt{2}}{2} & \frac{\sqrt{2}}{2} \\ -\frac{\sqrt{2}}{2} & \frac{\sqrt{2}}{2} \end{bmatrix} = \frac{\sqrt{2}}{2} \begin{bmatrix} 1 & 1 \\ -1 & 1 \end{bmatrix}$$

Then we know from Theorem 2.1 that $\|U\mathbf{x}\| = 1$ for any $\mathbf{x} \in C$. This means that applying U to some vector $\mathbf{x} \in C$ will result in some other vector in C.

In particular, if we apply U to $\mathbf{x} = [\,1,0\,]^T$, $\mathbf{y} = [\,\frac{1}{2}, \frac{\sqrt{3}}{2}\,]^T$, and $\mathbf{z} = [\,-\frac{\sqrt{2}}{2}, \frac{\sqrt{2}}{2}\,]^T$ we obtain the vectors

$$U\mathbf{x} = \begin{bmatrix} \frac{\sqrt{2}}{2} \\ -\frac{\sqrt{2}}{2} \end{bmatrix} \qquad U\mathbf{y} = \begin{bmatrix} \frac{\sqrt{2}+\sqrt{6}}{4} \\ \frac{\sqrt{3}-1}{2\sqrt{2}} \end{bmatrix} \qquad U\mathbf{z} = \begin{bmatrix} 0 \\ 1 \end{bmatrix}$$

The vectors $\mathbf{x}, \mathbf{y}, \mathbf{z}$, and $U\mathbf{x}, U\mathbf{y}, U\mathbf{z}$ are plotted in Figure 2.5. Do you see that U rotated each vector $\frac{\pi}{4}$ radians clockwise? □

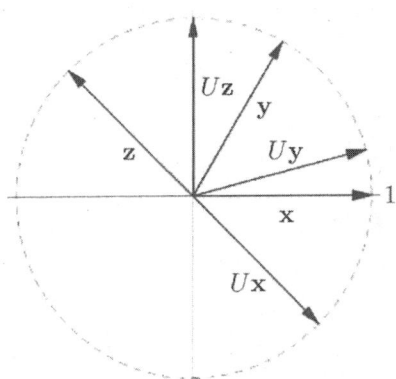

Figure 2.5 The vectors $\mathbf{x}, \mathbf{y}, \mathbf{z}$, and $U\mathbf{x}, U\mathbf{y}, U\mathbf{z}$ on the unit circle.

PROBLEMS

2.16   Properties of matrix multiplication.

(a) Suppose that S is the set of all $n \times n$ matrices. Is matrix multiplication commutative for the elements in S? That is, is it true that for any A and B in S, we have $AB = BA$?

(b) Assume that A has dimension $n \times p$, B has dimension $p \times m$, and C has dimension $m \times \ell$. Show that matrix multiplication is associative. That is, show that $(AB)C = A(BC)$.

2.17   Suppose that D is an $n \times n$ diagonal matrix with diagonal entries d_{ii}, $i = 1, 2, \ldots, n$. Let A be any $n \times n$ matrix.

(a) Describe in words the product DA.

(b) Describe in words the product AD.

2.18   Suppose that A is an $m \times n$ matrix. Explain how to write cA, where c is some real number, as a matrix product CA, where C is an $m \times m$ matrix.

2.19   In a linear algebra class, we learn a variety of methods for determining if a matrix is nonsingular. We can compute its *determinant* or attempt to find the inverse via Gaussian elimination. We have discussed neither of these techniques in this section. In this problem we learn that sometimes we can use basic matrix multiplication to determine if a matrix is singular. Consider the matrix

$$B = \begin{bmatrix} 1 & 1 & 1 \\ 1 & 1 & 1 \\ 1 & 1 & 1 \end{bmatrix}$$

If B is nonsingular, then $BB^{-1} = I$. But what happens if we multiply B with any 3×3 matrix? Use this observation to explain why B is singular.

2.20   Prove Proposition 2.1.

2.21   Let A be an $m \times n$ matrix. Show that AA^T is a symmetric matrix.

★**2.22**   Show that $(A^T)^T = A$ and $(AB)^T = B^T A^T$.

★**2.23**   Explain why $(A^{-1})^T = (A^T)^{-1}$.

2.24   Suppose that A and B are $m \times n$ matrices. Is it true that $A^T + B^T = (A+B)^T$?

2.25   This section requires a knowledge of eigenvalues. We give a basic definition here. Let A be an $n \times n$ matrix. We say that λ is an *eigenvalue* of A with associated eigenvector $\mathbf{v} \neq \mathbf{0}$ if $A\mathbf{v} = \lambda\mathbf{v}$. The eigenvalue may possibly be a complex number.

Suppose that U is an orthogonal matrix. Show that any eigenvalue λ of U satisfies $|\lambda| = 1$.

⋆**2.26** You saw in Example 2.8 that the orthogonal matrix U is of the form

$$U = \begin{bmatrix} a & b \\ -b & a \end{bmatrix}$$

Certainly, the rows (and columns) of U are orthogonal to each other.

(a) What additional condition on a, b must be satisfied so that U is an orthogonal matrix?

(b) Use part (a) and your knowledge of trigonometry to write a, b in terms of an angle θ.

(c) Using part (b), write down the general form of all 2×2 orthogonal matrix for a given angle θ.

(d) Construct 2×2 orthogonal matrix U that rotates the vectors in C from Example 2.8 $\frac{\pi}{3}$ radians counterclockwise.

(e) Construct 2×2 orthogonal matrix U that rotates the vectors in C from Example 2.8 $\frac{\pi}{2}$ radians counterclockwise.

(f) Write down the general form of $U^{-1} = U^T$ for the orthogonal matrix in part (c). Describe in words what U^T does to vectors in C if U rotates the vectors in C by θ radians.

⋆**2.27** Let n be an even positive integer, and let $\mathbf{v} = \begin{bmatrix} v_1 \\ v_2 \\ \vdots \\ v_n \end{bmatrix}$

(a) Find a matrix H so that

$$H\mathbf{v} = \frac{1}{2} \begin{bmatrix} v_1 + v_2 \\ v_3 + v_4 \\ \vdots \\ v_{n-1} + v_n \end{bmatrix}$$

(b) Note that the rows of H are orthogonal to each other. However, any inner product of a row with itself does not result in the value 1. Can you create a matrix $\tilde{H}$ that has the same structure (i. e., the placement of zeros is the same, so that the rows remain orthogonal, and the nonzero numbers are the same but not equal to $\frac{1}{2}$), so that the rows of $\tilde{H}$ have unit length?

Computer Lab

2.2   Matrix Arithmetic. From the text Web site, access the lab matrices. In this lab, you will become familiar with how your **CAS** performs matrix arithmetic.

2.3   BLOCK MATRIX ARITHMETIC

We discussed matrix arithmetic in Section 2.2 and we will learn how matrices can be utilized in image processing in Example 3.5. In many instances it is desirable to process only part of an image. In such a case, knowledge of *block matrices* is invaluable. Block matrices have applications throughout mathematics and sciences, and their manipulation is an important skill to master. Along with image-processing applications, we will see that block matrices are very useful when developing the wavelet transformations that first appear in Chapter 6.

Partitioning Matrices

The idea of block matrices is quite simple. Given an $N \times M$ matrix A, we simply partition it into different blocks. There is no set rule for how the partition is created — we are simply driven by the application. These blocks are called *submatrices* because they are matrices themselves. If you have learned to compute the determinant of a matrix by the cofactor expansion method, then you have already worked with block matrices.

We denote our submatrices with capital letters, and often, we will add subscripts to indicate their position in the original matrix. Let's look at a simple example.

Example 2.9 (Block Matrices). *Consider the matrix*

$$A = \begin{bmatrix} -3 & 2 & 0 & 1 \\ 2 & 6 & 3 & 5 \\ 0 & -1 & -1 & 9 \end{bmatrix}$$

We can partition A in a variety of ways. We present two possibilities.

(a) We partition A into two 3×2 matrices R and L. Thus,

$$A = \begin{bmatrix} -3 & 2 & 0 & 1 \\ 2 & 6 & 3 & 5 \\ 0 & -1 & -1 & 9 \end{bmatrix} = [L \mid R]$$

where

$$L = \begin{bmatrix} -3 & 2 \\ 2 & 6 \\ 0 & -1 \end{bmatrix} \quad and \quad R = \begin{bmatrix} 0 & 1 \\ 3 & 5 \\ -1 & 9 \end{bmatrix}$$

(b) We partition A into two 2 × 2 blocks and two 1 × 2 blocks. That is,

$$A = \left[\begin{array}{cc|cc} -3 & 2 & 0 & 1 \\ 2 & 6 & 3 & 5 \\ \hline 0 & -1 & -1 & 9 \end{array}\right] = \left[\begin{array}{c|c} A_{11} & A_{12} \\ \hline A_{21} & A_{22} \end{array}\right]$$

where

$$A_{11} = \left[\begin{array}{cc} -3 & 2 \\ 2 & 6 \end{array}\right] \qquad A_{12} = \left[\begin{array}{cc} 0 & 1 \\ 3 & 5 \end{array}\right]$$

and

$$A_{21} = \left[\begin{array}{cc} 0 & -1 \end{array}\right] \qquad A_{22} = \left[\begin{array}{cc} -1 & 9 \end{array}\right]$$

The notation should be intuitive — we can view A as a 2 × 2 block matrix. In this case, A_{11} is the submatrix in the first "row," first "column," A_{12} is positioned at row 1, column 2; A_{21} occupies row 2, column one; and A_{22} is positioned at row 2, column 2.

□

Adding and Subtracting Block Matrices

If A and B are matrices whose dimensions are equal, then we can always compute $A + B$ and $A - B$. In many instances it is desirable to partition A and B and then write the sum or difference as sums or differences of the blocks. We can only add and subtract block by block if A and B are *partitioned in exactly the same way.*

Let's consider two 4×4 matrices:

$$A = \left[\begin{array}{ccc|c} a_{11} & a_{12} & a_{13} & a_{14} \\ a_{21} & a_{22} & a_{23} & a_{24} \\ \hline a_{31} & a_{32} & a_{33} & a_{34} \\ a_{41} & a_{42} & a_{43} & a_{44} \end{array}\right] = \left[\begin{array}{c|c} A_{11} & A_{12} \\ \hline A_{21} & A_{22} \end{array}\right]$$

and

$$B = \left[\begin{array}{ccc|c} b_{11} & b_{12} & b_{13} & b_{14} \\ b_{21} & b_{22} & b_{23} & b_{24} \\ \hline b_{31} & b_{32} & b_{33} & b_{34} \\ b_{41} & b_{42} & b_{43} & b_{44} \end{array}\right] = \left[\begin{array}{c|c} B_{11} & B_{12} \\ \hline B_{21} & B_{22} \end{array}\right]$$

Disregarding the partitioning, it is a straightforward task to compute

$$
A + B = \begin{bmatrix} a_{11}+b_{11} & a_{12}+b_{12} & a_{13}+b_{13} & a_{14}+b_{14} \\ a_{21}+b_{21} & a_{22}+b_{22} & a_{23}+b_{23} & a_{24}+b_{24} \\ a_{31}+b_{31} & a_{32}+b_{32} & a_{33}+b_{33} & a_{34}+b_{34} \\ a_{41}+b_{41} & a_{42}+b_{42} & a_{43}+b_{43} & a_{44}+b_{44} \end{bmatrix}
$$
(2.11)

and since A and B are partitioned exactly the same, we can also simply add blocks

$$
A + B = \left[\begin{array}{c|c} A_{11}+B_{11} & A_{12}+B_{12} \\ \hline A_{21}+B_{21} & A_{22}+B_{22} \end{array} \right]
$$

noting that $A_{11}+B_{11}$, $A_{12}+B_{12}$, $A_{21}+B_{21}$, and $A_{22}+B_{22}$ are all valid matrix operations themselves. In particular,

$$
A_{11}+B_{11} = \begin{bmatrix} a_{11} & a_{12} & a_{13} \\ a_{21} & a_{22} & a_{23} \end{bmatrix} + \begin{bmatrix} b_{11} & b_{12} & b_{13} \\ b_{21} & b_{22} & b_{23} \end{bmatrix}
$$

$$
= \begin{bmatrix} a_{11}+b_{11} & a_{12}+b_{12} & a_{13}+b_{13} \\ a_{21}+b_{21} & a_{22}+b_{22} & a_{23}+b_{23} \end{bmatrix}
$$

is exactly the upper 3×2 corner of $A + B$ in (2.11). The other sums can be easily computed and verified as well.

Block Matrix Multiplication

We have learned that two matrices must be identically partitioned in order to add or subtract them. What about multiplication with blocks? Is it possible to partition matrices A and B into blocks so that the product AB can be computed by multiplying the blocks of A and B? Let's consider a couple of matrices as we formulate answers to these questions.

Example 2.10. *Let*

$$
A = \begin{bmatrix} 2 & 1 & -1 \\ 0 & 3 & 1 \end{bmatrix} \quad and \quad B = \begin{bmatrix} 2 & 1 \\ -1 & 0 \\ 5 & 4 \end{bmatrix}
$$

Certainly we can compute

$$
AB = \begin{bmatrix} -2 & -2 \\ 2 & 4 \end{bmatrix}
$$
(2.12)

What if we partition A as

$$
A = \left[\begin{array}{cc|c} 2 & 1 & -1 \\ 0 & 3 & 1 \end{array} \right] = \left[\begin{array}{c|c} A_{11} & A_{12} \end{array} \right]?
$$

The dimension of A, in terms of blocks, is 1×2. If we were to suppose for the moment that we could indeed partition B so that we could compute AB by multiplying blocks, then in block form, B must have two "rows" of blocks. Since A_{11} has two rows, the blocks in the first row of B must have two columns so we need to partition B as

$$B = \begin{bmatrix} 2 & 1 \\ -1 & 0 \\ \hline 5 & 4 \end{bmatrix} = \begin{bmatrix} B_{11} & B_{12} \\ \hline B_{21} & B_{22} \end{bmatrix} \quad or \quad B = \begin{bmatrix} 2 & 1 \\ -1 & 0 \\ \hline 5 & 4 \end{bmatrix} = \begin{bmatrix} B_{11} \\ \hline B_{21} \end{bmatrix} \quad (2.13)$$

In the case of the first partitioning of B in (2.13), we have

$$AB = \begin{bmatrix} A_{11} & A_{12} \end{bmatrix} \cdot \begin{bmatrix} B_{11} & B_{12} \\ \hline B_{21} & B_{22} \end{bmatrix}$$

A is 1×2 blocks and B is 2×2 blocks, so the result should be 1×2 blocks:

$$AB = \begin{bmatrix} A_{11} & A_{12} \end{bmatrix} \cdot \begin{bmatrix} B_{11} & B_{12} \\ \hline B_{21} & B_{22} \end{bmatrix} = \begin{bmatrix} A_{11}B_{11} + A_{12}B_{21} & A_{11}B_{12} + A_{12}B_{22} \end{bmatrix}$$

It is straightforward to verify that the four block matrix products $A_{11}B_{11}$, $A_{12}B_{21}$, $A_{11}B_{12}$, and $A_{12}B_{22}$ are all well defined. Indeed,

$$A_{11}B_{11} + A_{12}B_{21} = \begin{bmatrix} 2 & 1 \\ 0 & 3 \end{bmatrix} \cdot \begin{bmatrix} 2 \\ -1 \end{bmatrix} + \begin{bmatrix} -1 \\ 1 \end{bmatrix} \cdot \begin{bmatrix} 5 \end{bmatrix}$$

$$= \begin{bmatrix} 3 \\ -3 \end{bmatrix} + \begin{bmatrix} -5 \\ 5 \end{bmatrix} = \begin{bmatrix} -2 \\ 2 \end{bmatrix}$$

This is indeed the first column of AB, and if you think about it, the block procedure is simply another way to rewrite the multiplication $\begin{bmatrix} 2 & 1 & -1 \\ 0 & 3 & 1 \end{bmatrix} \begin{bmatrix} 2 \\ -1 \\ 5 \end{bmatrix}$ *that produces the first column of AB.*

In a similar manner we see that

$$A_{11}B_{12} + A_{12}B_{22} = \begin{bmatrix} 2 & 1 \\ 0 & 3 \end{bmatrix} \cdot \begin{bmatrix} 1 \\ 0 \end{bmatrix} + \begin{bmatrix} -1 \\ 1 \end{bmatrix} \cdot \begin{bmatrix} 4 \end{bmatrix}$$

$$= \begin{bmatrix} 2 \\ 0 \end{bmatrix} + \begin{bmatrix} -4 \\ 4 \end{bmatrix} = \begin{bmatrix} -2 \\ 4 \end{bmatrix}$$

and this result is the second column of AB. In Problem 2.28 you will verify that the second method for partitioning B in (2.13) also produces in a block matrix multiplication that results in AB.

What would happen if we partitioned the columns of A as before but partitioned the rows of A so that

$$A = \left[\begin{array}{cc|c} 2 & 1 & -1 \\ \hline 0 & 3 & 1 \end{array} \right] ? \tag{2.14}$$

You will verify in Problem 2.29 that either partition (2.13) will result in well-defined block matrix multiplication products that give AB as the answer.

What happens if we switch the column partition of A? That is, what if we write A as

$$A = \left[\begin{array}{c|cc} 2 & 1 & -1 \\ \hline 0 & 3 & 1 \end{array} \right] = \left[\begin{array}{c|c} A_{11} & A_{12} \end{array} \right] ? \tag{2.15}$$

In this case the partitions (2.13) of B do not work. Since A_{11} is now a 2×1 matrix, we cannot multiply it against either B_{11} in (2.13). To make the block multiplication product work when A is partitioned as in (2.15), we must adjust the row paritions of B. That is, we must use either

$$B = \left[\begin{array}{c|c} 2 & 1 \\ \hline -1 & 0 \\ \hline 5 & 4 \end{array} \right] = \left[\begin{array}{c|c} B_{11} & B_{12} \\ \hline B_{21} & B_{22} \end{array} \right] \quad or \quad B = \left[\begin{array}{c|c} 2 & 1 \\ \hline -1 & 0 \\ 5 & 4 \end{array} \right] = \left[\begin{array}{c} B_{11} \\ \hline B_{21} \end{array} \right] \tag{2.16}$$

In Problem 2.30 you will verify that the partitions of B given in (2.16) along with the partition (2.15) of A will result in block matrix multiplication products that yield AB.   □

The results of Example 2.10 generalize to matrices A and B where the product AB is defined. The main point of Example 2.10 was to illustrate that to successfully partition A and B so that we can use block matrix multiplication to compute AB, we need the *column partitioning* of A to be the same as the *row partitioning* of B. That is, if A has been partitioned into $n_1, n_2, \ldots, n_k$ columns, then in order to make the block multiplications welldefined, B must be partitioned into $n_1, n_2, \ldots, n_k$ rows. The row partitioning of A and the column partitioning of B can be discretionary. Indeed, the row partitioning of A and the column partitioning of B are usually functions of the application at hand.

A Useful Way to Interpret Matrix Multiplication

In practice, we are either given or construct matrix M that we use to process an input matrix A by way of MA to produce some desired effect. In particular, it is often desirable to view the product MA as a way of *processing the columns* of A. Consider the following example.

Example 2.11. *Let M be the 2×6 matrix*

$$M = \begin{bmatrix} \frac{1}{3} & \frac{1}{3} & \frac{1}{3} & 0 & 0 & 0 \\ 0 & 0 & 0 & \frac{1}{3} & \frac{1}{3} & \frac{1}{3} \end{bmatrix}$$

and suppose that A is any matrix that consists of six rows. For the purposes of this example, suppose that

$$A = \begin{bmatrix} a_{11} & a_{12} \\ a_{21} & a_{22} \\ a_{31} & a_{32} \\ a_{41} & a_{42} \\ a_{51} & a_{52} \\ a_{61} & a_{62} \end{bmatrix} = \left[\begin{array}{c|c} a_{11} & a_{12} \\ a_{21} & a_{22} \\ a_{31} & a_{32} \\ a_{41} & a_{42} \\ a_{51} & a_{52} \\ a_{61} & a_{62} \end{array} \right] = \left[\mathbf{a}^1 \mid \mathbf{a}^2 \right]$$

We can easily compute the 2×2 product MA and see that

$$MA = \begin{bmatrix} \frac{a_{11}+a_{21}+a_{31}}{3} & \frac{a_{12}+a_{22}+a_{32}}{3} \\ \frac{a_{41}+a_{51}+a_{61}}{3} & \frac{a_{42}+a_{52}+a_{62}}{3} \end{bmatrix}$$

Let's look at the first column of this product. The first entry, $\frac{a_{11}+a_{21}+a_{31}}{3}$ is simply the average of the first three elements of $\mathbf{a}^1$, and the second entry is the average of the last three elements of $\mathbf{a}^1$. The same holds true for the second column of the product. The first entry of the second column is the average of the first three elements of $\mathbf{a}^2$, and the second entry is the average of the last three elements of $\mathbf{a}^2$.

So we can see that M processes the columns of A by computing averages of elements of each column and we can easily verify that

$$MA = \left[M\mathbf{a}^1 \mid M\mathbf{a}^2 \right]$$

If A consists of p columns, the interpretation is the same — the only difference is that the product would consist of p columns of averages rather than the two that resulted in this example. □

In general, suppose that M is an $m \times p$ matrix and A is a $p \times n$ matrix and we are interested in interpreting the product MA. We begin by partitioning A in terms of its columns. That is, if $\mathbf{a}^1, \mathbf{a}2, \ldots, \mathbf{a}^n$ are the columns of A, then block matrix multiplication allows us to write

$$MA = M\left[\mathbf{a}^1 \mid \mathbf{a}^2 \mid \cdots \mid \mathbf{a}^n \right]$$

and view this product as a 1×1 block (M) times $1 \times n$ blocks. Thus,

$$\begin{aligned} MA &= M\left[\mathbf{a}^1 \mid \mathbf{a}^2 \mid \cdots \mid \mathbf{a}^n \right] \\ &= \left[M\mathbf{a}^1 \mid M\mathbf{a}^2 \mid \cdots \mid M\mathbf{a}^n \right] \end{aligned} \qquad (2.17)$$

and we see that when we compute MA, we are applying M to each column of A. Thus, when we multiply A on the left by M, we can view this product as processing the columns of A. Problem 2.33 allows you to further investigate this idea.

Alternatively, suppose that we are either given or construct some $p \times n$ matrix P and we wish to apply it to input matrix A with dimensions $m \times p$ by way of the product AP. Using block matrix multiplication, we can view this product as a way of *processing the rows* of A with matrix P. In Problem 2.34 you will produce the block matrix multiplication that supports this claim.

Transposes of Block Matrices

As we will see in subsequent chapters, it is important to understand how to transpose block matrices. As with block matrix multiplication, let's start with a generic matrix and its transpose

$$
A = \begin{bmatrix} a_{11} & a_{12} & a_{13} & a_{14} \\ a_{21} & a_{22} & a_{23} & a_{24} \\ a_{31} & a_{32} & a_{33} & a_{34} \\ a_{41} & a_{42} & a_{43} & a_{44} \end{bmatrix} \qquad A^T = \begin{bmatrix} a_{11} & a_{21} & a_{31} & a_{41} \\ a_{12} & a_{22} & a_{32} & a_{42} \\ a_{13} & a_{23} & a_{33} & a_{43} \\ a_{14} & a_{24} & a_{34} & a_{44} \end{bmatrix}
$$

Suppose that we partition A as

$$
A = \left[\begin{array}{cc|cc} a_{11} & a_{12} & a_{13} & a_{14} \\ a_{21} & a_{22} & a_{23} & a_{24} \\ \hline a_{31} & a_{32} & a_{33} & a_{34} \\ a_{41} & a_{42} & a_{43} & a_{44} \end{array} \right] = \left[\begin{array}{c|c} A_{11} & A_{12} \\ \hline A_{21} & A_{22} \end{array} \right]
$$

If we partition A^T as

$$
A^T = \left[\begin{array}{cc|cc} a_{11} & a_{21} & a_{31} & a_{41} \\ a_{12} & a_{22} & a_{32} & a_{42} \\ \hline a_{13} & a_{23} & a_{33} & a_{43} \\ a_{14} & a_{24} & a_{34} & a_{44} \end{array} \right]
$$

then it is easy to see that in block form,

$$
A^T = \left[\begin{array}{c|c} A_{11}^T & A_{21}^T \\ \hline A_{12}^T & A_{22}^T \end{array} \right]
$$

We can try a different partition,

$$
A = \begin{bmatrix} a_{11} & a_{12} & a_{13} & a_{14} \\ a_{21} & a_{22} & a_{23} & a_{24} \\ a_{31} & a_{32} & a_{33} & a_{34} \\ \hline a_{41} & a_{42} & a_{43} & a_{44} \end{bmatrix} = \left[\begin{array}{c|c} A_{11} & A_{12} \\ \hline A_{21} & A_{22} \end{array} \right]
$$

and we can again easily verify that if we partition A^T as

$$
A^T = \begin{bmatrix} a_{11} & a_{21} & a_{31} & a_{41} \\ a_{12} & a_{22} & a_{32} & a_{42} \\ a_{13} & a_{23} & a_{33} & a_{43} \\ \hline a_{14} & a_{24} & a_{34} & a_{44} \end{bmatrix}
$$

then again we have

$$
A^T = \left[\begin{array}{c|c} A_{11}^T & A_{21}^T \\ \hline A_{12}^T & A_{22}^T \end{array} \right]
$$

The results of this simple example hold in general. That is, if we partition A as

$$
A = \begin{bmatrix} A_{11} & A_{12} & \cdots & A_{1q} \\ A_{21} & A_{22} & & A_{2q} \\ \vdots & & \ddots & \vdots \\ A_{p1} & A_{p2} & \cdots & A_{pq} \end{bmatrix}, \quad \text{then} \quad A^T = \begin{bmatrix} A_{11}^T & A_{21}^T & \cdots & A_{p1}^T \\ A_{12}^T & A_{22}^T & & A_{p2}^T \\ \vdots & & \ddots & \vdots \\ A_{1q}^T & A_{2q}^T & \cdots & A_{pq}^T \end{bmatrix}
$$

PROBLEMS

2.28   Confirm that the the second partition of B in (2.13) results in a block matrix multiplication that gives AB (2.12).

2.29   Verify that either partition given in (2.13) will result in well-defined block matrix multiplication products that give AB (2.12) as the answer.

2.30   Verify that the partitions of B given in (2.16) along with the partition (2.15) of A will result in a block matrix multiplication product that yields AB (2.12).

2.31   Let

$$
A = \left[\begin{array}{cc|cc|c}
3 & -2 & 4 & 0 & 1 \\
1 & 0 & 1 & -2 & 5 \\
-2 & 4 & 3 & 0 & 1 \\
\hline
1 & 2 & -2 & 3 & 3
\end{array}\right]
\quad \text{and} \quad
B = \left[\begin{array}{ccc}
2 & -2 & 1 \\
1 & 0 & -1 \\
0 & 0 & 5 \\
-3 & 1 & -2 \\
4 & 1 & 0
\end{array}\right]
$$

Partition B in an appropriate way so that AB can be computed in block matrix form.

2.32   Matrix multiplication as we defined it in Section 2.2 is nothing more than block matrix multiplication. Suppose that A is an $n \times p$ matrix and B is a $p \times m$ matrix. Let $\mathbf{a}^{k^T}$, $k = 1, \ldots, n$, denote the rows of A, and $\mathbf{b}^j$, $j = 1, \ldots, m$, denote the columns of B. Partition A by its rows and B by its columns and then write down the product AB in block matrix form. Do you see why this block matrix form of the product is exactly matrix multiplication as stated in Definition 2.8?

2.33   Let

$$
M = \left[\begin{array}{ccc}
1 & 5 & 2 \\
-2 & 6 & 9
\end{array}\right]
\quad \text{and} \quad
A = \left[\begin{array}{cccc}
6 & 0 & 3 & 7 \\
1 & 2 & 5 & 4 \\
0 & -1 & 3 & -2
\end{array}\right]
$$

Verify (2.17) by showing that

$$
MA = \left[\, M\mathbf{a}^1 \mid M\mathbf{a}^2 \mid M\mathbf{a}^3 \mid M\mathbf{a}^4 \,\right]
$$

$$
= \left[\, M\begin{bmatrix} 6 \\ 1 \\ 0 \end{bmatrix} \,\middle|\, M\begin{bmatrix} 0 \\ 2 \\ -1 \end{bmatrix} \,\middle|\, M\begin{bmatrix} 3 \\ 5 \\ 3 \end{bmatrix} \,\middle|\, M\begin{bmatrix} 7 \\ 4 \\ -2 \end{bmatrix} \,\right]
$$

2.34   In this problem you will use block matrix multiplication to show that we can view the matrix R as processing the rows of matrix A when we compute the product AR.

(a) Suppose that R is a $p \times m$ matrix and A is an $n \times p$ matrix so that the product AR is defined. Suppose that $\mathbf{a}^{1^T}, \mathbf{a}^{2^T}, \ldots, \mathbf{a}^{n^T}$ are the rows of A (remember that each $\mathbf{a}^k$ is a vector consisting of p elements, so we need to transpose it to get a row). Partition A by rows and the use this block version of A to compute AR.

(b) Let

$$
A = \left[\begin{array}{cc}
3 & 1 \\
0 & 5
\end{array}\right]
\quad \text{and} \quad
R = \left[\begin{array}{ccc}
-2 & 3 & 0 \\
1 & 1 & -2
\end{array}\right]
$$

Use part (a) to verify that

$$AR = \begin{bmatrix} [3\ 1]\,R \\ \hline [0\ 5]\,R \end{bmatrix}$$

★**2.35**   Suppose that H and G are $\frac{N}{2} \times N$ matrices and that $\mathbf{v}$ and $\mathbf{w}$ are vectors of length $\frac{N}{2}$. Compute

$$\begin{bmatrix} H \\ G \end{bmatrix}^T \cdot \begin{bmatrix} \mathbf{v} \\ \mathbf{w} \end{bmatrix}$$

★**2.36**   Consider the 8×8 matrix

$$W = \begin{bmatrix} H \\ G \end{bmatrix}$$

where

$$H = \begin{bmatrix} 1 & 1 & 0 & 0 & 0 & 0 & 0 & 0 \\ 0 & 0 & 1 & 1 & 0 & 0 & 0 & 0 \\ 0 & 0 & 0 & 0 & 1 & 1 & 0 & 0 \\ 0 & 0 & 0 & 0 & 0 & 0 & 1 & 1 \end{bmatrix}$$

and

$$G = \begin{bmatrix} 1 & -1 & 0 & 0 & 0 & 0 & 0 & 0 \\ 0 & 0 & 1 & -1 & 0 & 0 & 0 & 0 \\ 0 & 0 & 0 & 0 & 1 & -1 & 0 & 0 \\ 0 & 0 & 0 & 0 & 0 & 0 & 1 & -1 \end{bmatrix}$$

Verify that $W^T = \begin{bmatrix} H^T \mid G^T \end{bmatrix}$ and that

$$W^T W = H^T H + G^T G = 2I_8$$

where I_8 is the 8×8 identity matrix.

2.37   Let

$$A = \begin{bmatrix} A_{11} & A_{12} \\ \hline A_{21} & A_{22} \end{bmatrix}$$

(a) Describe a block matrix L so that LA results in the product

$$LA = \begin{bmatrix} 3A_{11} & 3A_{12} \\ \hline 5A_{21} & 5A_{22} \end{bmatrix}$$

(b) Describe a block matrix R so that AR results in the product

$$
AR = \left[\begin{array}{c|c} 2A_{11} & 7A_{12} \\ \hline 2A_{21} & 7A_{22} \end{array}\right]
$$

(c) What is LAR?

$\star$**2.38**   Suppose that A is an $N \times N$ matrix with N even and further suppose that A has the block structure

$$
A = \left[\begin{array}{c|c} W & 0 \\ \hline 0 & I \end{array}\right]
$$

where 0, I and W are $\frac{N}{2} \times \frac{N}{2}$ matrices. 0 is a zero matrix and I is an identity matrix. Let $\mathbf{v} \in \mathbb{R}^N$ be partitioned as

$$
\mathbf{v} = \left[\begin{array}{c} \mathbf{v}^\ell \\ \hline \mathbf{v}^h \end{array}\right]
$$

where $\mathbf{v}^\ell, \mathbf{v}^h \in \mathbb{R}^{\frac{N}{2}}$. Write $A\mathbf{v}$ in block form.

2.39   Show that

$$
\left[\begin{array}{cc} A & B \\ B^T & C \end{array}\right]
$$

has the inverse

$$
\left[\begin{array}{cc} D & -DBC^{-1} \\ -C^{-1}B^T D & C^{-1} + C^{-1}B^T DBC^{-1} \end{array}\right]
$$

whenever C and $D = (A - BC^{-1}B^T)^{-1}$ are nonsingular.

Computer Lab

2.3   Block Matrix Arithmetic. From the text Web site, access the lab blocks. In this lab you will learn how to use your **CAS** to perform basic operations involving block matrices.

CHAPTER 3

AN INTRODUCTION TO DIGITAL IMAGES

One of the main applications of wavelets is in the area of image processing. Wavelets can be used to denoise images, perform image compression, search for edges in images, and enhance image features. Indeed, since two chapters of this book are dedicated to denoising and compression, and examples involving images are featured throughout the text, we include this chapter on the basics of digital images.

It is quite tempting to dedicate many pages to the various aspects of image processing. In this chapter we discuss the basic material necessary to successfully navigate through the remainder of the text. If you are interested in more thorough treatments on the numerous topics that comprise digital image processing, I might suggest you start by consulting Russ [64], Gonzalez and Woods [39], Gonzalez, Woods, and Eddins [40], or Wayner [84].

This chapter begins with a section that introduces digital grayscale images. We learn how digital images can be interpreted by computer software and how matrices are important tools in basic image-processing applications. We also discuss intensity transformations. These transformations are commonly used to improve the image display. Section 3.2 deals with digital color images as well as some popular color

Discrete Wavelet Transformations: An Elementary Approach With Applications. By P. J. Van Fleet **49**
Copyright © 2008 John Wiley & Sons, Inc.

spaces that are useful for image processing. In the third section of the chapter we introduce some qualitative and quantitative measures that are useful when assessing the effectiveness of image-processing methods. The final section covers Huffman codes. These codes are extremely popular in image compression applications. We use Huffman codes in various examples to illustrate how effective wavelet transformations can be when utilized in image compression problems.

3.1 THE BASICS OF GRAYSCALE DIGITAL IMAGES

In this section we learn how a *grayscale* (monochrome) image can be interpreted by computer software, how we can represent a grayscale image using a matrix, and how matrices can be used to manipulate images. The section concludes with an introduction to some basic intensity transformations.

Bits and Bytes

The fundamental unit on a computer is the *bit*. The value of a bit is either 0 or 1. A *byte* is the standard amount of information needed to store one keyboard character. On most computers, there are 8 bits in a byte. Considering that each bit can assume one of two values, we see that there can be $2^8 = 256$ different bytes. Not coincidentally, there are 256 characters on a standard keyboard and each character is assigned a value from the range 0 to 255.[1] The American Standard Code for Information Interchange (ASCII) is the standard used to assign values to characters. For example, the ASCII standard for the characters Y, y, 9, $?$, and sp are 89, 121, 57, 63, and 32, respectively. Here sp denotes the space key. The ASCII codes for the characters that appear on a common computer keyboard are given in Table 3.1. The table of all 256 ASCII codes can be found in numerous books or on many web sites. One possible reference is the book by Oualline [60].

Since computers work with bits, it is often convenient to write the ASCII codes in base 2. where now the intensities range from $0 = 0000000_2$ to $255 = 11111111_2$. When an intensity is expressed in base 2, it is said to be in *binary form*. The following example recalls how the conversion works.

Example 3.1 (Converting from Base 10 to Base 2). *The character Y has ASCII value 89. Convert this value to base 2.*
Solution.
The largest power of 2 that is less than or equal to 89 is $2^6 = 64$. So we start with the seventh bit. $89 - 64 = 25$ and the largest power of 2 that is less than or equal to

---

[1] If you take a quick glance at your keyboard, it is not clear that there are 256 characters. If you have access to a PC, open a new text file in Notepad or MS Word. You can display the characters by holding down the Alt key and then typing a number between 0 and 255 on the numeric keypad.

Table 3.1   ASCII character codes 32 through 126 in decimal and binary (base 2) format.

Char.	Dec.	Bin.	Char.	Dec.	Bin.	Char.	Dec.	Bin.
sp	32	00100000	@	64	01000000	`	96	01100000
!	33	00100001	A	65	01000001	a	97	01100001
"	34	00100010	B	66	01000010	b	98	01100010
#	35	00100011	C	67	01000011	c	99	01100011
$	36	00100100	D	68	01000100	d	100	01100100
%	37	00100101	E	69	01000101	e	101	01100101
&	38	00100110	F	70	01000110	f	102	01100110
'	39	00100111	G	71	01000111	g	103	01100111
(	40	00101000	H	72	01001000	h	104	01101000
)	41	00101001	I	73	01001001	i	105	01101001
*	42	00101010	J	74	01001010	j	106	01101010
+	43	00101011	K	75	01001011	k	107	01101011
,	44	00101100	L	76	01001100	l	108	01101100
-	45	00101101	M	77	01001101	m	109	01101101
.	46	00101110	N	78	01001110	n	110	01101110
/	47	00101111	O	79	01001111	o	111	01101111
0	48	00110000	P	80	01010000	p	112	01110000
1	49	00110001	Q	81	01010001	q	113	01110001
2	50	00110010	R	82	01010010	r	114	01110010
3	51	00110011	S	83	01010011	s	115	01110011
4	52	00110100	T	84	01010100	t	116	01110100
5	53	00110101	U	85	01010101	u	117	01110101
6	54	00110110	V	86	01010110	v	118	01110110
7	55	00110111	W	87	01010111	w	119	01110111
8	56	00111000	X	88	01011000	x	120	01111000
9	57	00111001	Y	89	01011001	y	121	01111001
:	58	00111010	Z	90	01011010	z	122	01111010
;	59	00111011	[	91	01011011	{	123	01111011
<	60	00111100	\	92	01011100	|	124	01111100
=	61	00111101	]	93	01011101	}	125	01111101
>	62	00111110	^	94	01011110	~	126	01111110
?	63	00111111	_	95	01011111			

25 is $2^4 = 16$. $25 - 16 = 9$, and we can write 9 as $2^3 + 2^0$. Thus, we see that

$$89 = 0 \cdot 2^7 + 1 \cdot 2^6 + 0 \cdot 2^5 + 1 \cdot 2^4 + 1 \cdot 2^3 + 0 \cdot 2^2 + 0 \cdot 2^1 + 1 \cdot 2^0$$
$$= 01011001_2$$

☐

Storing Images on Computers

The basic unit of composition of a digital image is a *pixel*.[2] The *resolution* of a rectangular digital image[3] is the number of rows × the number of columns that comprise the image. In order to work with a grayscale digital image, we must know the location of the pixel (i.e., the row and column of the pixel) and the *intensity* of the pixel. Since the fundamental unit on a computer is a bit, it is natural to measure these intensities using bits. Thus, the intensity scale of a grayscale image is typically 0 to $2^m - 1, m = 0, 1, 2, \ldots$. The value 0 typically corresponds to black, and $2^m - 1$ corresponds to white. Thus, the intensities between 0 and $2^m - 1$ represent the spectrum of gray values between black and white in increments of $1/2^m$. We say that an image is an *m-bit image* when the intensity scale ranges from 0 to $2^m - 1$.

A common value for use on computers is $m = 8$ for an intensity scale of 256 values that range from 0 (black) to 255 (white). This intensity scale is plotted in Figure 3.1.

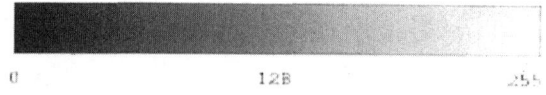

0   128   255

Figure 3.1   The 8-bit grayscale intensity scale.

Certainly, you can see why an 8-bit image is convenient for computer graphics. One byte is 8 bits and we know that we can store the numbers 0 through 255 as single characters from the ASCII table (see Table 3.1 or [60]). For example, suppose that the first row of pixels in a digital image had the intensities 50, 100, 80, and 124. Then we would only need the four characters *2, d, p*, and | to store these intensities. This is 4 bytes, compared to the minimum of 10 bytes needed for the number of digits that comprise the intensities in the row. This is exactly how file formats such as *portable gray maps* (PGMs) store digital images.

If a finer intensity scale is desired, it is typically stored as a 16-bit image. Then there are $2^{16} = 65,536$ possible intensity levels. Now $65,536 = 256^2$, so for storage purposes we only need to use two characters! Let's illustrate this statement with an example.

Example 3.2 (16-Bit Image Intensities as ASCII Characters). *Write the intensity level 31,785 as two ASCII characters.*

---

[2]Pixel is short for *picture* (pix) *element* (el).
[3]We only consider rectangular digital images in this book.

Solution. *We write* $31{,}785$ *in base* 2:

$$31{,}785 = 1 \cdot 2^{14} + 1 \cdot 2^{13} + 1 \cdot 2^{12} + 1 \cdot 2^{11} + 1 \cdot 2^{10} + 1 \cdot 2^{5} + 1 \cdot 2^{3} + 1 \cdot 2^{0}$$
$$= 0111110000101001_2$$
$$= \underline{01111100} \ \underline{00101001} \ _2$$

Now the leftmost 8 *bits can be viewed as a number between* 0 *and* 255 *and indeed,* $01111100_2 = 124$. *Using Table 3.1, we see that ASCII code for* 124 *is* | . *The second* 8 *bits are* 00101001_2, *and in base* 10, *the value is* 41. *The ASCII code for* 41 *is*). *So we could store the five-digit number* 31,785 *as* |).

Converting to base 10 *is quite simple. We can think of the base* 2 *representation as the sum*

$$0111110000101001_2 = \mathbf{0111110000000000}_2 + 000000000\mathbf{00101001}_2$$

Now the second term is simply 41, *and to get the first term we simply shift the bits* $01111100_2 = 124$ 8 *bits to the left. This is nothing more than the multiplication* $124 \cdot 2^8 = 31{,}744$.

So to obtain the original base 10 *intensity level, we find the ASCII codes* 124 *and* 41 *for the characters* | *and*), *respectively, and then compute* $124 \cdot 2^8 + 41$. $\qquad\square$

Note: Unless otherwise noted, all images in this book are 8-bit images.

Grayscale Images as Matrices

A natural tool with which to describe the locations and intensities of pixels in a digital image is a matrix. The dimensions of a matrix correspond directly to the resolution of an image, and the location of a pixel in an image is nothing more than the (i, j) element of a matrix. If we denote the matrix by A, then a_{ij} is the intensity value. Thus, we can store a digital grayscale image in a matrix A whose integer entries range from 0 (black) to 255 (white).

Figure 3.2 illustrates this concept. In the figure, we have plotted a 200×300 image. Although we do not want to display a 200×300 matrix, we choose a particular 10×10 region, zoom in on it, and display those values in a matrix.

Matrices and Basic Image Processing

Basic operations involving matrices are important in image processing. Let's look at a few simple examples.

$$\begin{bmatrix} 180 & 132 & 73 & 68 & 70 & 115 & 156 & 158 & 166 & 172 \\ 191 & 204 & 210 & 68 & 69 & 78 & 79 & 167 & 160 & 168 \\ 193 & 196 & 193 & 210 & 109 & 76 & 78 & 71 & 158 & 163 \\ 198 & 210 & 192 & 196 & 201 & 51 & 72 & 74 & 54 & 156 \\ 200 & 205 & 205 & 197 & 194 & 177 & 69 & 68 & 68 & 167 \\ 206 & 194 & 201 & 199 & 185 & 185 & 116 & 64 & 64 & 54 \\ 200 & 202 & 199 & 199 & 204 & 182 & 201 & 62 & 66 & 62 \\ 205 & 204 & 204 & 202 & 201 & 191 & 191 & 211 & 55 & 60 \\ 201 & 204 & 203 & 198 & 201 & 199 & 191 & 199 & 92 & 63 \\ 198 & 200 & 194 & 192 & 192 & 188 & 197 & 187 & 193 & 62 \end{bmatrix}$$

Figure 3.2   Storing a digital image as a matrix.

Example 3.3 (Image Negation). *One common image-processing tool involves a simple matrix operation. Image negation is the process of inverting values on the intensity scale. For example, 0 maps to 255, 255 maps to 0, 100 maps to 155, and 20 maps to 235. In general, pixel p is mapped to $255 - p$.*

We can easily perform image negation using matrix subtraction. Consider the 160×240 image plotted on the left in Figure 3.3. Suppose that we represent this image in matrix A.

(a) A 160×240 grayscale image   (b) Negation of the image

Figure 3.3   Image negation.

Define S to be the 160×240 matrix where each element of S is set to 255. If we were to plot S as an image, it would be a white square. To produce the negative image, we simply compute $S - A$. The result is plotted on the right in Figure 3.3. □

Example 3.4 (Inner Products and Grayscale Images). *Suppose that we have an image that is 10 pixels wide and that the pixel values in the top row are 100, 102, 90, 94, 91, 95, 250, 252, 220, and 210. We can load these pixel values into a vector $\mathbf{v}$:*

$$\mathbf{v} = [100, 102, 90, 94, 91, 95, 250, 252, 220, 210]^{T}$$

Practically speaking the row starts out a medium dark gray and then the last four pixels are quite close to white. Pictorially, the vector is shown in the figure below.

Figure 3.4   A pictorial representation of **v**.

Suppose that our task is to produce a 5-vector that consists of the averages of two consecutive elements of **v**. *This is a common task in image compression — the average values of some pixels are used to represent a string of values. So we want as output the vector*

$$\mathbf{u} = [101, 92, 93, 251, 215]^T$$

It is easy to compute these averages, but how do we express it mathematically? Let's concentrate on the average of the first two numbers. We want to compute

$$\frac{100 + 102}{2} = \frac{100}{2} + \frac{102}{2}$$

So we need a 10-vector **a** *that performs this task. It should be clear to you that the last eight values of* **a** *should be zero since we don't want to involve these values of* **v**. *Then we need only to choose* $a_1 = a_2 = \frac{1}{2}$ *so that*

$$\mathbf{a}^1 = \left[\frac{1}{2}, \frac{1}{2}, 0, 0, 0, 0, 0, 0, 0, 0\right]^T$$

The inner product $\mathbf{a}^T \mathbf{v}$ *gives us* $u_1 = 101$. *If we construct the vector*

$$\mathbf{a}^2 = \left[0, 0, \frac{1}{2}, \frac{1}{2}, 0, 0, 0, 0, 0, 0\right]^T$$

and dot it with **v**, *we obtain* $u_2 = 92$. *We can continue this process and additionally create the vectors*

$$\mathbf{a}^3 = \left[0, 0, 0, 0, \frac{1}{2}, \frac{1}{2}, 0, 0, 0, 0\right]^T \qquad \mathbf{a}^4 = \left[0, 0, 0, 0, 0, 0, \frac{1}{2}, \frac{1}{2}, 0, 0\right]^T$$

and

$$\mathbf{a}^5 = \left[0, 0, 0, 0, 0, 0, 0, 0, \frac{1}{2}, \frac{1}{2}\right]^T$$

and dot them each in turn with **v** *to obtain* $u_3 = 93$, $u_4 = 251$, *and* $u_5 = 215$. *A pictorial representation of* **u** *appears in Figure 3.5.*

Figure 3.5 A pictorial representation of **u**.

We can combine all of these inner products and perform the task by computing the matrix product

$$
\frac{1}{2}
\begin{bmatrix}
1 & 1 & 0 & 0 & 0 & 0 & 0 & 0 & 0 & 0 \\
0 & 0 & 1 & 1 & 0 & 0 & 0 & 0 & 0 & 0 \\
0 & 0 & 0 & 0 & 1 & 1 & 0 & 0 & 0 & 0 \\
0 & 0 & 0 & 0 & 0 & 0 & 1 & 1 & 0 & 0 \\
0 & 0 & 0 & 0 & 0 & 0 & 0 & 0 & 1 & 1
\end{bmatrix}
\cdot
\begin{bmatrix}
100 \\ 102 \\ 90 \\ 94 \\ 91 \\ 95 \\ 250 \\ 252 \\ 220 \\ 210
\end{bmatrix}
=
\begin{bmatrix}
101 \\ 92 \\ 93 \\ 251 \\ 215
\end{bmatrix}
$$

$\square$

We next look at a practical example where we illustrate what happens to an image matrix M when we compute HM and MH^T.

Example 3.5 (Matrices and Naive Edge Detection). *Let M be the 160×160 matrix that represents the image plotted in Figure 3.6. Then M has integer entries ranging from 0 to 255 and is of the form*

$$
M =
\begin{bmatrix}
m_{11} & m_{12} & \cdots & m_{1,160} \\
m_{21} & m_{22} & \cdots & m_{2,160} \\
\vdots & \vdots & \ddots & \vdots \\
m_{160,1} & m_{160,2} & \cdots & m_{160,160}
\end{bmatrix}
$$

Now consider the 160×160 matrix

$$
H =
\begin{bmatrix}
1 & -1 & 0 & 0 & 0 & \cdots & 0 & 0 & 0 \\
0 & 1 & -1 & 0 & 0 & \cdots & 0 & 0 & 0 \\
0 & 0 & 1 & -1 & 0 & \cdots & 0 & 0 & 0 \\
\vdots & \vdots & \vdots & \vdots & \vdots & \ddots & \vdots & \vdots & \vdots \\
0 & 0 & 0 & 0 & 0 & \cdots & 1 & -1 & 0 \\
0 & 0 & 0 & 0 & 0 & \cdots & 0 & 1 & -1 \\
-1 & 0 & 0 & 0 & 0 & \cdots & 0 & 0 & 1
\end{bmatrix}
$$

Figure 3.6   The image represented by matrix M.

*What happens when we compute HM? It is certainly easy enough to write down
the answer, HM:*

$$
\begin{bmatrix}
1 & -1 & 0 & \cdots & 0 & 0 \\
0 & 1 & -1 & \cdots & 0 & 0 \\
0 & 0 & 1 & \cdots & 0 & 0 \\
\vdots & \vdots & \vdots & \ddots & \vdots & \vdots \\
0 & 0 & 0 & \cdots & -1 & 0 \\
0 & 0 & 0 & \cdots & 1 & -1 \\
-1 & 0 & 0 & \cdots & 0 & 1
\end{bmatrix}
\cdot
\begin{bmatrix}
m_{11} & m_{12} & \cdots & m_{1,160} \\
m_{21} & m_{22} & \cdots & m_{2,160} \\
\vdots & \vdots & \ddots & \vdots \\
m_{160,1} & m_{160,2} & \cdots & m_{160,160}
\end{bmatrix}
$$

$$
=
\begin{bmatrix}
m_{11}-m_{21} & m_{12}-m_{22} & \cdots & m_{1,160}-m_{2,160} \\
m_{21}-m_{31} & m_{22}-m_{32} & \cdots & m_{2,160}-m_{3,160} \\
\vdots & \vdots & \ddots & \vdots \\
m_{160,1}-m_{11} & m_{160,2}-m_{12} & \cdots & m_{160,160}-m_{1,160}
\end{bmatrix}
$$

*Can you describe the result of computing HM? The first row of HM is obtained by
subtracting row 2 of M from row 1 of M. In general, for $j = 1, \ldots, 159$, the jth row
of HM is obtained by subtracting row $j + 1$ from row j of M. Row 160 of HM is
obtained by subtracting the first row of M from the last row of M. In areas where
the pixels in consecutive rows do not change much in value, the components of HM
are pretty close to zero (black). Conversely, whenever there is a boundary (along a
row), the values of HM will be large (white), and the larger they are the whiter the
resulting pixel. So multiplying by H might be thought of as a very naive method for
detecting boundaries along the rows in M. We plot HM in Figure 3.7(a).*

Suppose that we wanted to compute differences along the consecutive columns of M. That is, we want to subtract column $k + 1$ from column k, $k = 1, \ldots, 159$ and then subtract column 1 from column 100. What computation should we perform?

We learned in Chapter 2 that multiplying M on the left by H processes the columns of M. That is, if we write M as a matrix of column vectors, $M = [\mathbf{m}_1, \mathbf{m}_2, \ldots, \mathbf{m}_{160}]$, then $HM = [H\mathbf{m}_1, H\mathbf{m}_2, \ldots, H\mathbf{m}_{160}]$. If we want to process the rows of M, then it is natural to try computing MH. But if we compute MH, we obtain

$$
\begin{bmatrix} m_{11} & m_{12} & \cdots & m_{1,160} \\ m_{21} & m_{22} & \cdots & m_{2,160} \\ \vdots & \vdots & \ddots & \vdots \\ m_{160,1} & m_{160,2} & \cdots & m_{160,160} \end{bmatrix} \cdot \begin{bmatrix} 1 & -1 & 0 & \cdots & 0 & 0 \\ 0 & 1 & -1 & \cdots & 0 & 0 \\ 0 & 0 & 1 & \cdots & 0 & 0 \\ \vdots & \vdots & \vdots & \ddots & \vdots & \vdots \\ 0 & 0 & 0 & \cdots & -1 & 0 \\ 0 & 0 & 0 & \cdots & 1 & -1 \\ -1 & 0 & 0 & \cdots & 0 & 1 \end{bmatrix}
$$

$$
= \begin{bmatrix} m_{11}-m_{1,160} & m_{12}-m_{11} & \cdots & m_{1,160}-m_{1,159} \\ m_{21}-m_{2,160} & m_{22}-m_{21} & \cdots & m_{2,160}-m_{2,159} \\ \vdots & \vdots & \ddots & \vdots \\ m_{160,1}-m_{160,160} & m_{160,2}-m_{160,1} & \cdots & m_{160,160}-m_{160,159} \end{bmatrix}
$$

and we can see that this computation will not process the rows of M in a manner consistent with the column processing in HM. We need the $(1, -1)$ pairs oriented vertically rather than horizontally. So the correct solution is to multiply M by H^T. The resulting image is displayed in Figure 3.7(b).

We can compute MH^T by

$$
\begin{bmatrix} m_{11} & m_{12} & \cdots & m_{1,160} \\ m_{21} & m_{22} & \cdots & m_{2,160} \\ \vdots & \vdots & \ddots & \vdots \\ m_{160,1} & m_{160,2} & \cdots & m_{160,160} \end{bmatrix} \cdot \begin{bmatrix} 1 & 0 & \cdots & 0 & -1 \\ -1 & 1 & \cdots & 0 & 0 \\ 0 & -1 & \cdots & 0 & 0 \\ \vdots & \vdots & \ddots & \vdots & \vdots \\ 0 & 0 & \cdots & 0 & 0 \\ 0 & 0 & \cdots & 1 & 0 \\ 0 & 0 & \cdots & -1 & 1 \end{bmatrix}
$$

$$
= \begin{bmatrix} m_{11}-m_{12} & m_{12}-m_{13} & \cdots & m_{1,160}-m_{11} \\ m_{21}-m_{22} & m_{22}-m_{23} & \cdots & m_{2,160}-m_{21} \\ \vdots & \vdots & \ddots & \vdots \\ m_{160,1}-m_{160,2} & m_{160,2}-m_{160,3} & \cdots & m_{160,160}-m_{160,1} \end{bmatrix}
$$

You can learn more about this example by completing Problem 3.7 and Computer Lab 3.2. □

Block matrix multiplication is also quite useful in image processing.

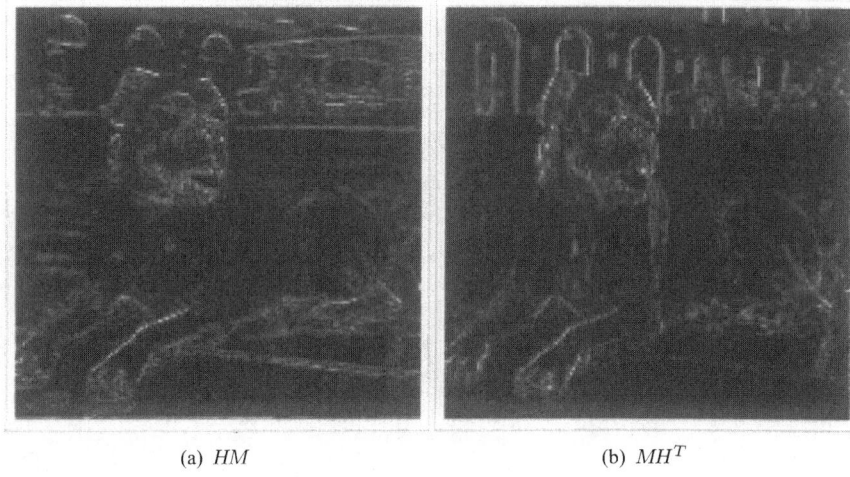

(a) HM (b) MH^T

Figure 3.7   Plots of HM and MH^T.

Example 3.6 (Block Matrices and Image Masking). *Consider the* 450×450 *grayscale image plotted in Figure 3.8. Denote the matrix of grayscale values by* A.

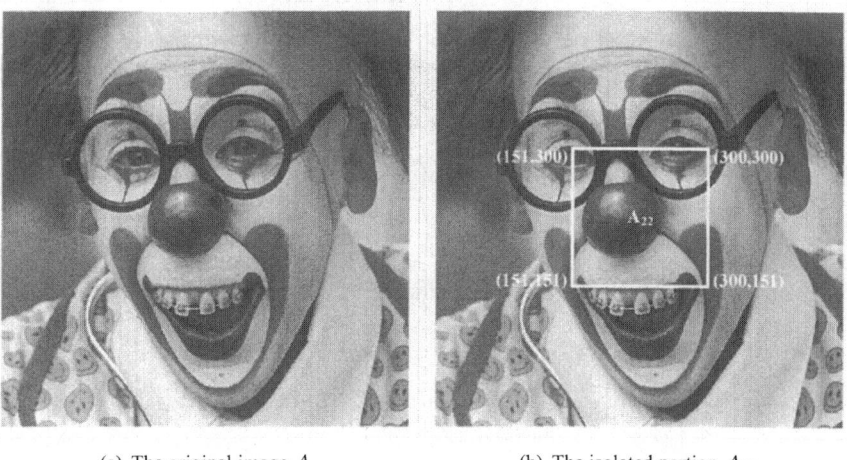

(a) The original image A (b) The isolated portion A_{22}

Figure 3.8   A digital image and marking it for masking.

Suppose that we were only interested in displaying the center of the image. That is, we only wished to display the region enclosed by the square with corners $(151, 151)$, $(300, 151)$, $(300, 300)$, *and* $(151, 300)$ *(see Figure 3.8) and convert the remaining pixels to black. We denote the matrix enclosed by this square as* A_{22}.

We can mask this portion of the image using block matrices. Consider the 450×450 *matrix* M *that has the following block structure:*

$$
M = \left[\begin{array}{c|c|c}
Z_{150} & Z_{150} & Z_{150} \\
\hline
Z_{150} & I_{150} & Z_{150} \\
\hline
Z_{150} & Z_{150} & Z_{150}
\end{array} \right]
$$

where Z_{150} *is a* 150×150 *matrix whose entries are* 0 *and* I_{150} *is the* 150×150 *identity matrix.*

We then partition A into nine blocks each of size 150×150. *We label these blocks as* A_{jk}, $j, k = 1, 2, 3$. *The partitioning is illustrated in Figure 3.9.*

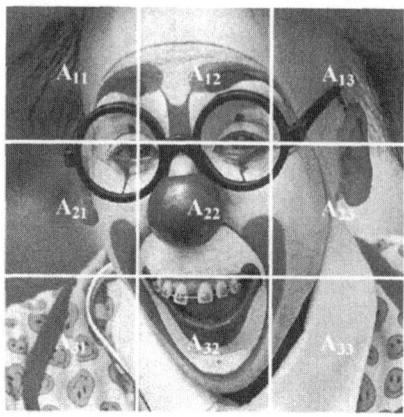

Figure 3.9   Partitioning of A into nine blocks A_{jk} each of size 150×150.

If we compute MA using block matrix multiplication, we obtain

$$
MA = \left[\begin{array}{c|c|c}
Z_{150} & Z_{150} & Z_{150} \\
\hline
Z_{150} & I_{150} & Z_{150} \\
\hline
Z_{150} & Z_{150} & Z_{150}
\end{array} \right] \cdot \left[\begin{array}{c|c|c}
A_{11} & A_{12} & A_{13} \\
\hline
A_{21} & A_{22} & A_{23} \\
\hline
A_{31} & A_{32} & A_{33}
\end{array} \right] = \left[\begin{array}{c|c|c}
Z_{150} & Z_{150} & Z_{150} \\
\hline
A_{21} & A_{22} & A_{23} \\
\hline
Z_{150} & Z_{150} & Z_{150}
\end{array} \right]
$$

Thus, multiplication on the left by M leaves only the middle row of blocks of A as nonzero blocks. The result of MA is plotted in Figure 3.10(a).

We now multiply MA on the right by $M = M^T$ to obtain

$$
MAM = \begin{bmatrix} Z_{150} & Z_{150} & Z_{150} \\ A_{21} & A_{22} & A_{23} \\ Z_{150} & Z_{150} & Z_{150} \end{bmatrix} \cdot \begin{bmatrix} Z_{150} & Z_{150} & Z_{150} \\ Z_{150} & I_{150} & Z_{150} \\ Z_{150} & Z_{150} & Z_{150} \end{bmatrix} = \begin{bmatrix} Z_{150} & Z_{150} & Z_{150} \\ Z_{150} & A_{22} & Z_{150} \\ Z_{150} & Z_{150} & Z_{150} \end{bmatrix}
$$

The product MAM accomplishes our goal of leaving the middle block A_{22} as the only nonzero block. It is plotted in Figure 3.10(b).

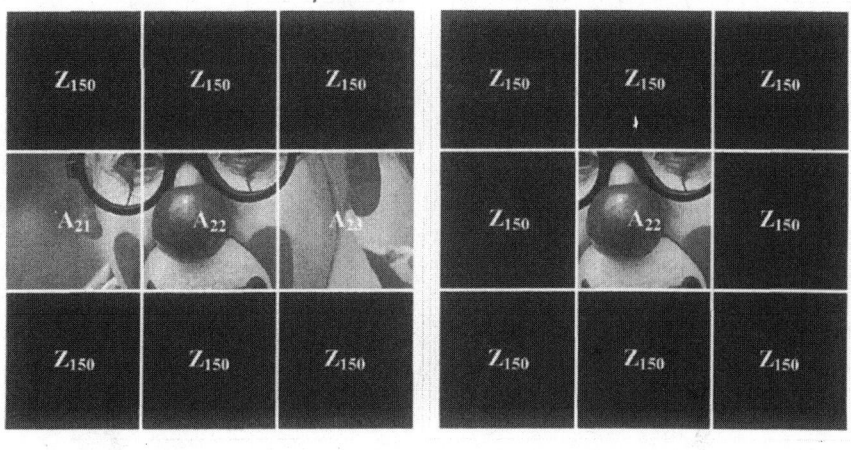

(a) The product MA (b) The product MAM

Figure 3.10   Masking an image.

□

Intensity Transformations

Intensity transformations are useful in several areas. We can alter the intensities in an image to present a more balanced display. If we are given an image whose intensities are disproportionately dark or light, an intensity transformation might give us a better idea of the true image. In this book, we learn how to transform images for the purposes of compressing the information necessary to store them or to remove noise. In some cases it is desirable to view the transformation. It is quite common for the transformation to move the intensities out of the $[0, 255]$ range. In such a case it is useful to have methods available that allow us to produce a view of the transformation that we can analyze.

We discuss two different intensity transformations. Each transformation is quite easy to implement on a computer, and Computer Labs 3.4 and 3.5 are available on the book Web site. The first lab assists you in developing intensity transformations as software modules, and the second lab allows you to experiment with them on various images.

Gamma Correction

Perhaps the simplest intensity transformation is *gamma correction*. The idea is to pick an exponent $\gamma > 0$ and then create a new pixel p_{new} by raising the old pixel to the power of γ. That is

$$p_{new} = p_{orig}^{\gamma}$$

To ensure that the data stay within the $[0, 255]$ range, we first map the pixels in the image to the interval $[0, 1]$. We can perform this map by simply dividing all of our pixels by 255. Figure 3.11 shows gamma correction for various values of γ.

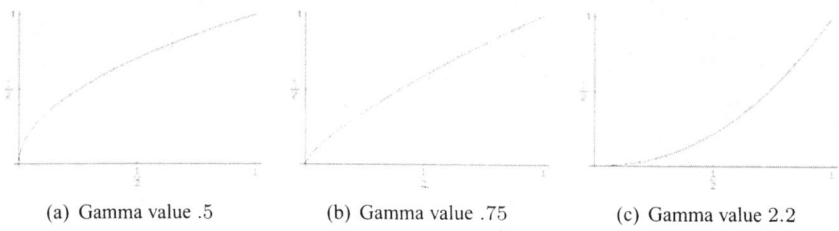

(a) Gamma value .5   (b) Gamma value .75   (c) Gamma value 2.2

Figure 3.11   Gamma correction transformations.

Note that $\gamma = 1$ is a linear map of the intensities. Exponents $\gamma > 1$ create concave-up functions on $(0, 1]$ that tend to compress the lower intensities toward 0, while $\gamma < 1$ result in concave-down functions on $(0, 1]$ that tend to compress higher intensities toward 1.

The transformation is quite simple. The intensities in an image are first divided by 255 so that they reside on the interval $[0, 1]$. Then each pixel is raised to the γ power. We next multiply by 255 and round the products to the nearest integer. Figure 3.12 shows the results of gamma corrections for the values of gamma given in Figure 3.11.

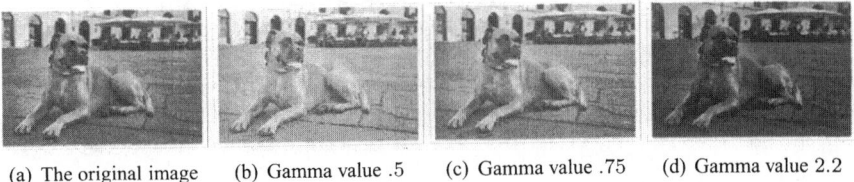

(a) The original image   (b) Gamma value .5   (c) Gamma value .75   (d) Gamma value 2.2

Figure 3.12   Gamma correction transformations applied to a digital image.

Histogram Equalization

We can learn much about the contrast of intensities if we plot a histogram of the frequencies at which the intensities occur. In the top row of Figure 3.13 we have plotted three 160×240 images. Most of the intensities in the first image are located near the bottom of the $[0, 255]$ intensity scale. In the second image, the intensities are bunched near the top of the intensity scale. The final image has the majority of its intensities located away from the ends of the intensity scale. In the second row of Figure 3.13 we have plotted histograms for each image. These histograms show how often intensities occur in the image.

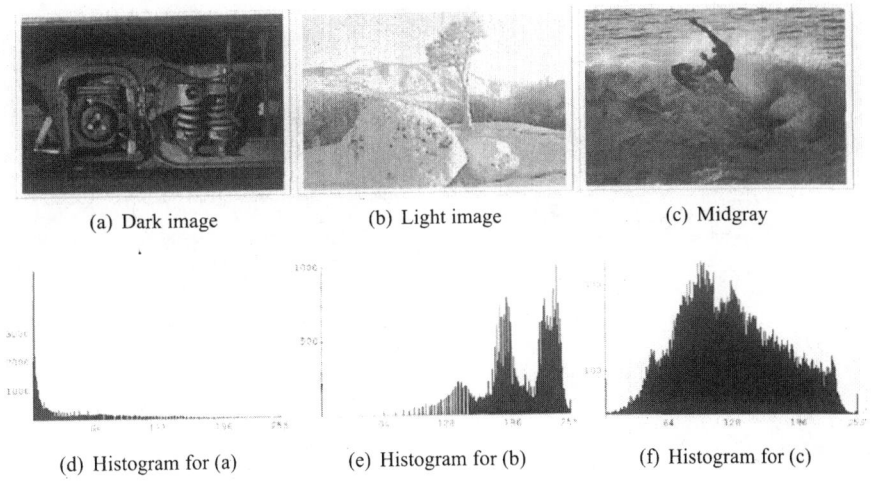

(a) Dark image   (b) Light image   (c) Midgray

(d) Histogram for (a)   (e) Histogram for (b)   (f) Histogram for (c)

Figure 3.13   Images and histograms of their gray-level intensity distributions. The image in Figure 3.13(a) has a large concentration of dark pixels, the image in Figure 3.13(b) has a large concentration of light pixels, and the image in Figure 3.13(c) has a large concentration of midgray pixels.

We will be working with digital images throughout the remainder of the book. You will be encouraged to download or create your own digital images and apply the techniques discussed in this book to them. Some of the images you might want to use may well look like those plotted in Figure 3.13. Thus, we now describe a process called *histogram equalization* that you might apply to your images to enhance them before further processing.

The goal of histogram equalization is to spread the intensities across the entire intensity spectrum $[0, 255]$ so that the resulting histogram for the modified image is more uniform. Figure 3.14 illustrates the desired effect after histogram equalization is applied to the third image in Figure 3.13.

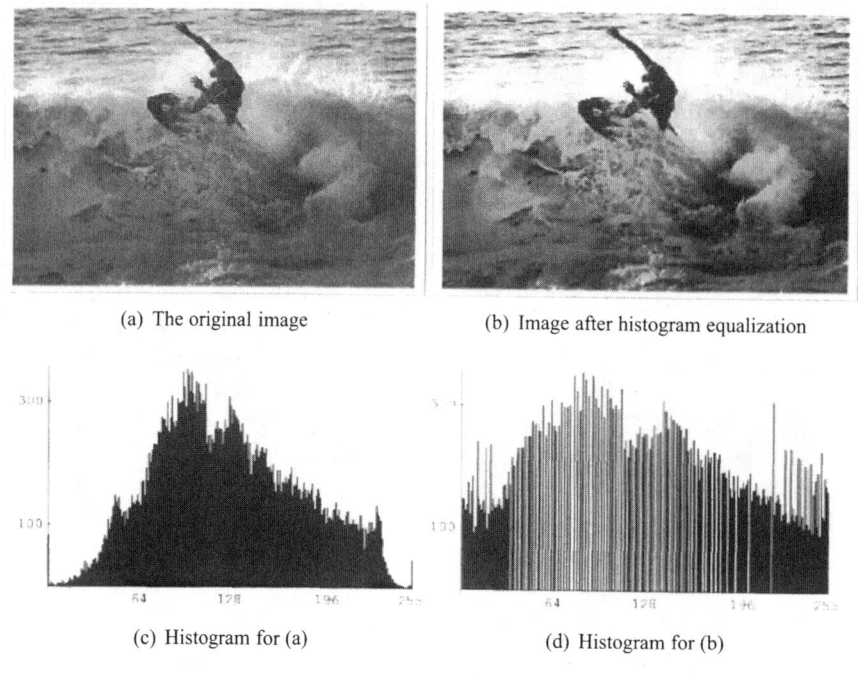

(a) The original image   (b) Image after histogram equalization

(c) Histogram for (a)   (d) Histogram for (b)

Figure 3.14   Histogram equalization applied to a digital image.

How do we get to the image on the upper right from the image on the upper left in Figure 3.14? We define a function $T(k)$ that sends intensity k, $k = 0, \ldots, 255$, to a value $T(k) \in [0, 1]$. The function will depend on the frequency of intensity k.

The value $T(k)$ measures the fraction of intensity k that we wish to retain. The final step in the process is to multiply $T(k)$ by 255. This stretches the values from $[0, 1]$ to $[0, 255]$. Finally, we use the *ceiling function* $\lceil \cdot \rceil$ to convert $255 \cdot T(k)$ to an integer. Mathematically, for $r \in \mathbb{R}$, $\lceil r \rceil = m$, where m is the smallest integer that satisfies $r \leq m$. So once we have established the map $T(k)$, we use the mapping

$$eq(k) = \lceil 255 \cdot T(k) \rceil \tag{3.1}$$

to map intensity k to its new intensity value.

Let's think about the design of $T(k)$. If we want to spread the frequencies across the $[0, 255]$ spectrum, then $T(k)$ should be nondecreasing. We also want $T(k)$ to depend on the image histogram. For an $N \times M$ image, let $h(k)$ be the frequency of intensity k and for $k = 0, 1, \ldots, 255$, and define the function

$$T(k) = \frac{1}{NM} \cdot \sum_{j=0}^{k} h(j) \tag{3.2}$$

We can easily write down the first few values of T.[4] We have $T(0) = h(0)/NM$, $T(1) = \frac{1}{NM}\big(h(0) + h(1)\big)$, $T(2) = \frac{1}{NM}\big(h(0) + h(1) + h(2)\big)$. Thus we see that the summation in (3.2) creates a *cumulative total* of the frequencies. Since the sum must add to the total number of pixels, we divide by NM so that $T(k)$ maps $[0, 255]$ to $[0, 1]$. In Problem 3.9 you will show that $T(k) \leq T(k + 1)$. In Figure 3.15 we have plotted T for the image in Figure 3.13(c).

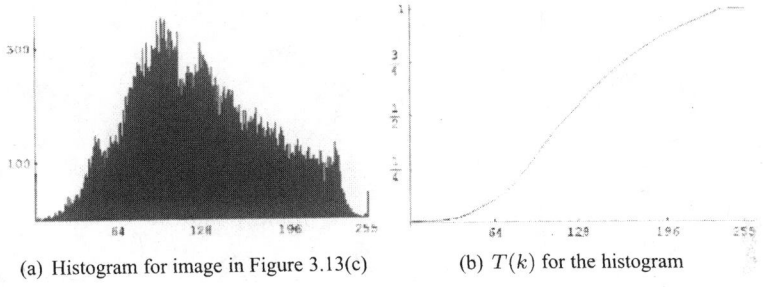

(a) Histogram for image in Figure 3.13(c)   (b) $T(k)$ for the histogram

Figure 3.15   Histogram and the cumulative total function.

A schematic of the histogram equalization process appears in Figure 3.16. The figure on the bottom left is a plot of a histogram of the frequency of the intensities of a 256×256 image. The graph above it shows the function T that represents the cumulative total of the frequencies. The graph on the upper right is a histogram of the frequency of the intensities for the equalized image. The black lines indicate how the function T "stretches" an interval of intensities from the original image. The histogram-equalized versions of the images in Figure 3.13(a) and (b) appear in Figure 3.17.

---

[4]The reader with some background in statistics will recognize T as a cumulative distribution function (CDF) for the probability density function that describes the intensity levels of the pixels in an image. Gonzalez and Woods [39] argue that output intensity levels using T follow a uniform distribution. This is a more rigorous argument for why histogram equalization works than the one we present here.

Figure 3.16   A schematic view that shows how the histogram equalization process works.

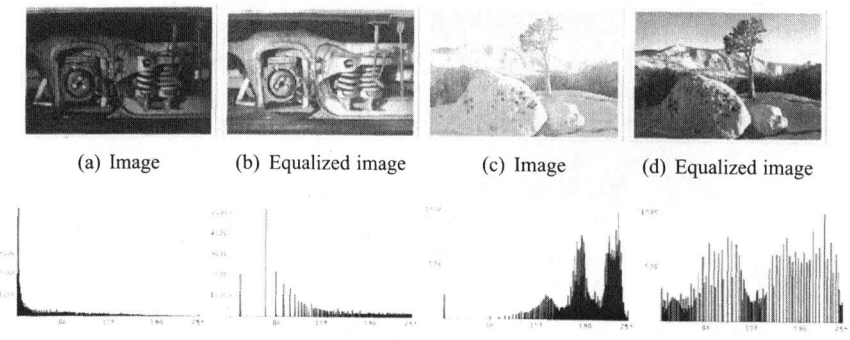

(a) Image   (b) Equalized image   (c) Image   (d) Equalized image

Figure 3.17   Histogram-equalized versions of the images in Figure 3.13(a) and (b). The histograms for each image are shown below the image.

PROBLEMS

3.1   Using the ideas from Example 3.2, write each of the following intensities as two ASCII characters.

(a) 24,301.

(b) 9352.

(c) 1888.

3.2   Find the ASCII characters for each of the following base 2 numbers.

(a) 10010101_2.

(b) 01100001_2.

(c) 11100100_2.

3.3   In Example 3.3 we learned how to produce the negative of an image. For each pixel intensity p, we computed $255 - p$.

(a) The negative of $p = 79$ is 176. Write the base 2 representations for each of these numbers.

(b) Repeat part (a) with $p = 112$.

(c) What is the relationship between the bits of a pixel and its negative?

(d) Prove your assertion from part (c) by writing 255 in base 2 and then subtracting an intensity of the form $(b_7b_6b_5b_4b_3b_2b_1b_0)_2$, where $b_k \in \{0,1\}$ for $k = 0,\ldots,7$.

3.4   In some cases, 24-bit images are desired.

(a) What is the intensity range in this case?

(b) Write 2,535,208 as three ASCII characters.

3.5   Suppose that matrix A represents an 256×256 image. What matrix operation(s) would you use to produce a plot of:

(a) A rotated $90°$ counterclockwise?

(b) A rotated $180°$ counterclockwise?

(c) A rotated $90°$ clockwise?

(d) A with the last 128 rows converted to black?

(e) A with the first 128 columns colored gray (ASCII code 128)?

(f) A with its elements reflected across the main diagonal?

You may need to create special matrices to use in conjunction with A.

3.6   Let $\gamma > 0$ and define $f(t) = t^\gamma$ for $t \in (0, 1]$. Show that $f(t)$ is concave down on $(0, 1]$ whenever $0 < \gamma < 1$ and $f(t)$ is concave up on $(0, 1]$ for $\gamma \geq 1$.

3.7   This problem refers to Example 3.5. Computer Lab 3.2 allows you to further investigate this example.

(a) Can you explain why the bottom row in Figure 3.7(a) looks different from the rest of the image?

(b) Can you explain why the rightmost column in Figure 3.7(b) looks different from the rest of the image?

(c) We decided in Example 3.5 that HM gives a naive method for finding borders along the rows in M and that MH^T is a naive method for finding borders along the columns of M. What would happen if we were to combine the two? That is, can you describe what would result if we computed HMH^T? To get an idea, construct an 8×8 version of H and a generic 8×8 matrix M and write out the product. This should help you see the general pattern.

3.8   Suppose that our grayscale intensity level ranges from 0 (black) to 5 (white) and consider the 6×6 image plotted below (the intensity values are given in each square in lieu of the gray shade).

0	2	2	2	2	4
0	2	1	1	2	3
0	2	1	1	2	3
1	1	1	1	3	3
0	1	1	2	3	4
1	1	2	2	4	5

(a) Fill in the following chart and sketch the associated histogram.

Intensity	Frequency
0	
1	
2	
3	
4	
5	

(b) Using (3.2), fill in the following chart and then plot $T(k)$, $k = 0, \ldots, 5$.

k	T(k)
0	
1	
2	
3	
4	
5	

(c) Using (b) and (3.1), perform histogram equalization and write the new intensity values in the following chart.

You may want to use a **CAS** to plot the "images" from this problem.

(d) Using the new "image" you constructed in part (c), fill in the following chart and sketch the associated histogram.

Intensity	Frequency
0	
1	
2	
3	
4	
5	

3.9 Show that $T(k)$ defined in (3.2) satisfies $T(k) \leq T(k+1)$ for $k = 0, 1, \ldots, 254$.

3.10 What must be true about the histogram of an image to ensure that $T(k) < T(k+1)$?

3.11 *True or False:* If image matrix B is the histogram-equalized version of image A, then at least one element of B has the value 255. If the result is true, explain why. If the result is false, provide a counterexample to support your claim.

3.12   If $eq(\ell) = 0$ for some ℓ with $0 \leq \ell \leq 255$, then what can you claim about the intensities $0, \ldots, \ell$ in the original image?

Computer Labs

3.1   Elementary Image Processing via Matrix Operations. From the text Web site, access the lab basicimaging. In this lab you will investigate basic image processing with matrix operations.

3.2   Edges in Images. From the text Web site, access the lab hmatrixmult. This lab allows you to further investigate Example 3.5 and Problem 3.7.

3.3   Image Processing and Block Matrices. From the text Web site, access the lab blockmatrix. This lab allows you to investigate how block matrices are used in image processing.

♦ **3.4   Software Development: Intensity Transformations.** From the text Web site, access the development package intensitytransforms. In this development problem you will develop functions for gamma correction, logarithmic transformations, and histogram equalization. Instructions are provided that allow you to add all of these functions to the software package DiscreteWavelets.

3.5   Exploring Intensity Transformations. From the text Web site, access the lab intensitytransformstest. In this lab you will test the software modules for various intensity transforms on various grayscale images.

3.2   COLOR IMAGES AND COLOR SPACES

In this section we discuss color images. We begin with a brief description of how they might be stored on a computer and continue by illustrating how to write color images in terms of matrices. In subsequent chapters we apply wavelet transformations to color images. In such applications it is desirable to first transform the image to a different *color space*. Such color spaces allow the wavelet transform to process the data in a setting that better describes how humans might process changes in color images. We conclude the section with a discussion of the YCbCr and HSI color spaces.

The Basics of a Color Image

Pixels in grayscale images assume an integer value g where $0 \leq g \leq 255$. In Section 3.1 we learned how to convert this value to and from an ASCII code and how to write an array of pixels as a matrix. The difference between a grayscale image and a *color image* is that each pixel of a color image now requires three values to describe it. Typically, these three values represent intensity levels for red, green, and blue. Such images are called *RGB color images*. We use a scale where 0 represents no presence of the color and 255 represents the entire presence of the color. So the *RGB triples* $(255, 0, 0)$, $(0, 255, 0)$, and $(0, 0, 255)$ represent bright red, green, and blue, respectively. Black is represented by $(0, 0, 0)$ and white is represented by $(255, 255, 255)$. Red, green, and blue are called the *primary colors* of the *RGB color space*. The *secondary colors* are created by combining exactly two primary colors. Magenta, yellow, and cyan are the secondary colors and their RGB triples are $(255, 0, 255)$, $(255, 255, 0)$ and $(0, 255, 255)$, respectively. As the primary color values range from 0 to 255, different colors in the spectrum are created. For example, $(0, 50, 50)$ represents a dark cyan, brown is $(128, 42, 42)$, and maroon is $(176, 48, 96)$. If the three primary color intensity values are all the same, we obtain a grayscale value. Figure 3.18(a) illustrates the RGB color space, and Figure 3.18(b) gives a schematic of the cube from Figure 3.18(a).

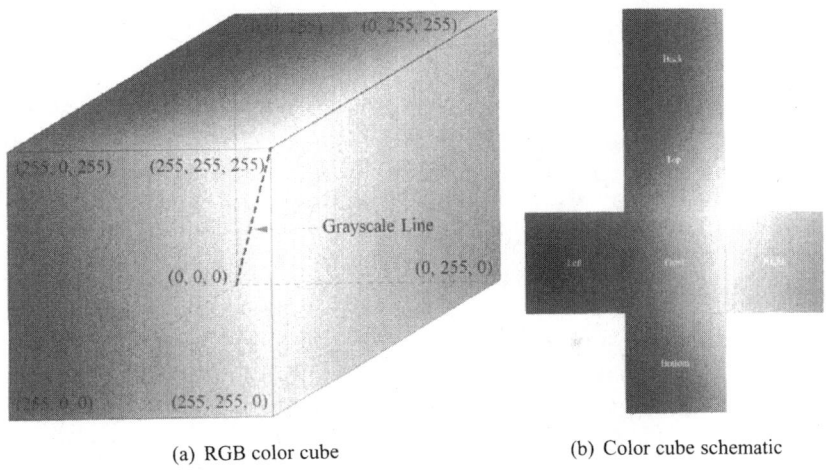

(a) RGB color cube   (b) Color cube schematic

Figure 3.18   The image at left is the RGB color cube. The image on the right is a schematic of the color cube. See the color plates for color versions of these images.

We need three times as much information to store a color image as we do to store a grayscale image. We can still use ASCII codes to represent a color pixel. We just need three codes instead of one. For example, the color *olive* can be represented with the RGB triple $(59, 94, 43)$. If we consult Table 3.1, we see that this triple can be represented by the ASCII code triple ; ^ +. Raw (uncompressed) data file formats such as the *portable pixel map* (PPM) format use these ASCII code triples to write a file to disk. After some header information that lists the bit depth and image dimensions, the pixels are stored as a long string of ASCII characters. Three characters are used for each pixel and they are written to file one row after the other.

We can easily represent color images using matrices, but now we need three matrices. The intensity levels for each primary color are stored in a matrix. Figure 3.19 shows a color image and the three matrices used to represent the image. We have plotted the three matrices using shades of the primary colors, but you should understand that each matrix is simply an integer-valued matrix where the values range from 0 to 255.

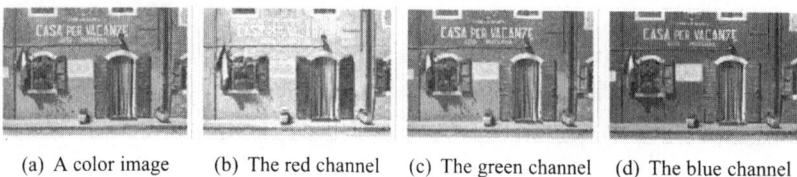

(a) A color image   (b) The red channel   (c) The green channel   (d) The blue channel

Figure 3.19   A color image and the three primary color channels. See the color plates for color versions of these images.

We can perform some basic image processing on color images by simply applying transformations to each color channel. For example, suppose that R, G, and B represent the red, green, and blue intensity matrices, respectively, for the image plotted in Figure 3.19. We can perform image negation by forming the matrix W whose entries are each 255, and then computing $W - R$, $W - G$, and $W - B$. The results are plotted in Figure 3.20. In Problem 3.17 we discuss this process in a different context.

(a) A digital color image.   (b) The negative image.

Figure 3.20   Color image negation. See the color plates for a color versions of these images.

The YCbCr Color Space

Although Figure 3.20 illustrates that image negation operates on the individual red, green, and blue color channels, it is more natural for humans to view images in terms of *luminance* and *chrominance*. In fact, some color image-processing tools do not work well when they process the image in *RGB space* (see Problem 3.20). Thus, in many applications, it is desirable to first convert an image from RGB space to a space that stores information about pixels in terms of luminance and chrominance.

You can think of a luminance channel as one that carries information regarding the brightness and contrast of the pixel. *Chrominance* is the difference between a color and a reference color at the same brightness (see Poynton [61]). Grayscale images contain no chrominance channels since we only need intensity (brightness) values to identify them. The conversions we consider in this book usually quantize chrominance either as color differences or in terms of hue and saturation.

Hue is a value that designates the pure color. Red, purple, and yellow are examples of hues. The *saturation* of a hue measures the relative purity of the hue. For example, pink has the same hue as red but is considered less saturated than red. Figure 3.21 further illustrates the ideas of hue, saturation, and brightness.

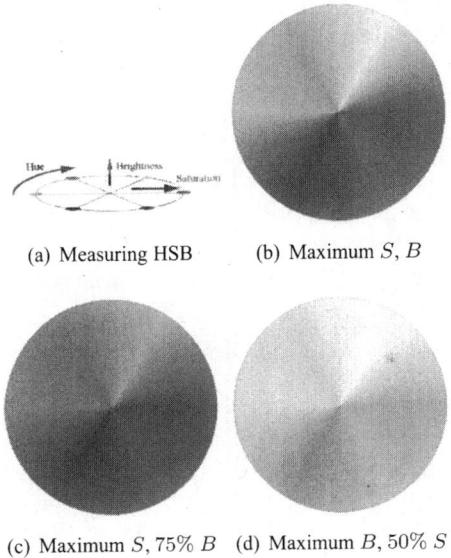

(a) Measuring HSB   (b) Maximum S, B

(c) Maximum $S, 75\% B$ (d) Maximum $B, 50\% S$

Figure 3.21   Hue, saturation (S), and brightness (B). See the color plates for a color versions of these images.

The conversion between RGB space and *HSI space* (hue, saturation, intensity) is complicated and we will only use it when denoising color images (see Chapter 9). We will develop the conversion functions for RGB space and HSI space in Computer Lab 3.6.[5]

---

[5] Make sure to work through this lab if you intend to complete Complete Lab 9.9 on denoising color images in Chapter 9.

Color spaces that describe chrominance via *color difference* channels are typically linear transformations and thus easy to use. One such space is the *YCbCr color space*. There are three channels in the YCbCr color space. Y is the luminance or intensity channel, and the Cb and Cr channels hold the chrominance information.

As pointed out by Varma and Bell [80], images in RGB space typically have a more even distribution of energy than those stored in YCbCr space. Moreover, most of the energy resides in the Y channel. Since the human eye is less sensitive to changes in chrominance, the chrominance channels are often *subsampled* so that less information can be used to adequately display the image. For this reason, the YCbCr color space is typically used in image compression techniques (see Chapter 12).

A straightforward way to describe the luminance would be to average the red, green, and blue channels. However, as you will investigate in Problem 3.13, it is better to use a weighted average when computing the luminance. Suppose that (r, g, b) are the intensity values of red, green, and blue, respectively, and further assume that these values have been scaled by $\frac{1}{255}$ so that they reside in the interval $[0, 1]$. We then define the intensity value y as

$$y = .299r + .587g + .114b \tag{3.3}$$

Note that the minimum value of y is 0 and it occurs when $r = g = b = 0$ and the maximum value of y is 1 and it occurs when $r = g = b = 1$.

Cb and Cr are the color difference channels. Cb is a multiple of the difference between blue and the intensity y given in (3.3), and Cr is a multiple of the difference between red and the intensity y. The exact values can be computed using

$$Cb = \frac{(b - y)}{1.772} \quad \text{and} \quad Cr = \frac{(r - y)}{1.402} \tag{3.4}$$

Equation (3.3) is used to convert color to grayscale in television (see Problem 3.13). The interested reader is encouraged to consult Poynton [61] for some historical information on formulas (3.3) and (3.4).

In Problem 3.15 you will show that the range of Cb and Cr is $[-\frac{1}{2}, \frac{1}{2}]$. In this problem you will also find the matrix C that maps the vector $[r, g, b]^T$ to $[y, Cb, Cr]^T$. The fact that this matrix is invertible gives us a way to convert from YCbCr space to RGB space.

For display purposes, values (y, Cb, Cr) are typically scaled and shifted so that $y \in [16, 235]$, $Cb, Cr \in [16, 240]$. The condensed ranges are used to protect against overflow due to roundoff error (see Poynton [61]). Thus, the transformation from (y, Cb, Cr) to (Y, C_b, C_r) we use for display is

$$\begin{aligned} y' &= 219y + 16 \\ C_b &= 224Cb + 128 \\ C_r &= 224Cr + 128 \end{aligned} \tag{3.5}$$

Thus the algorithm for converting from RGB space to YCbCr space is as follows:

1. Divide the red, green, and blue intensity values by 255 to obtain $r, g, b \in [0, 1]$.

2. Compute the intensity y using (3.3).

3. Compute the chrominance values Cb and Cr using (3.4).

4. If we wish to display the results, employ (3.5) and round the results.

Figure 3.22 shows y', C_b, and C_r for the image plotted in Figure 3.20.

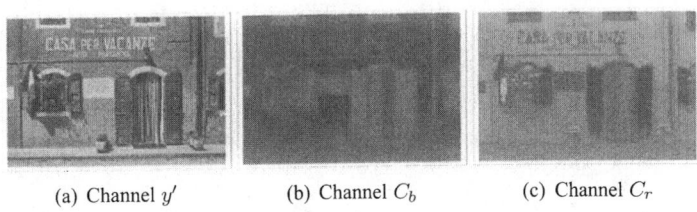

    (a) Channel y' (b) Channel C_b (c) Channel C_r

Figure 3.22 Intensity and chrominance channels for the image in Figure 3.20(a).

Problems 3.17, 3.18, and 3.19 describe other color spaces. The conversions between RGB and YCbCr color spaces as well as RGB and other color spaces are developed in Computer Lab 3.7.

PROBLEMS

3.13 In this problem you will learn how to use some basic matrix arithmetic to convert color images into grayscale images. We can view a color image as three matrices. Suppose that matrices R, G, and B contain the red, green, and blue intensities, respectively, of some color image.

(a) Possibly the most natural way to convert a color image to grayscale is to average corresponding elements of R, G, and B. Let M be the resulting grayscale matrix and write M as a linear combination of R, G, and B. (Recall that a linear combination of matrices R, G, and B is simply any expression of the form $aR + bG + cB$ where a, b, c are real numbers.)

(b) The National Television System Committee (NTSC) uses a different formula to convert color to grayscale. As Poynton points out [61], if red, green, and blue have the same luminescence in the color spectrum, then green will appear brightest, red is less bright, and blue will appear darkest. For this reason, the NTSC uses (3.3) to convert color images to grayscale. If M is the resulting grayscale matrix, write M as a linear combination of R, G, and B.

In Computer Lab 3.8, you can investigate color-to-grayscale image conversions.

3.14   Sometimes it is useful to convert a grayscale image to a color image. In such cases we create a *pseudocolor map* to do the job. Suppose that you are given a grayscale image and wish to convert it to a color image where we use the "front" of the RGB color cube (see Figure 3.18(b)). In particular, we want magenta to represent the grayscale value 0 and yellow to represent the grayscale value 255. Let A be the matrix that represents the grayscale image, and let W be a matrix of the same dimensions as A, whose elements are each 255. Let R, G, and B be the red, green, and blue intensity matrices, respectively, that will house the pseudocolor-mapped result. Write R, G, and B in terms of W and A.

★3.15   In this problem we further investigate the conversion between RGB color space and YCbCr color space. Toward this end, suppose that r, g, and b are the intensities values of red, green, and blue, respectively, for some pixel. Assume further that these intensities have been scaled by $\frac{1}{255}$ so that they reside in the interval $[0, 1]$. Let y, Cb, Cr be the values defined in (3.3) and (3.4).

(a) Use (3.4) to show that if $r, g, b \in [0, 1]$, then $Cb, Cr \in [-\frac{1}{2}, \frac{1}{2}]$.

(b) Using (3.3) and (3.4), find the matrix S so that

$$
\begin{bmatrix} y \\ Cb \\ Cr \end{bmatrix} = S \cdot \begin{bmatrix} r \\ g \\ b \end{bmatrix}
$$

(c) Using a **CAS**, find S^{-1}.

(d) You can also find S^{-1} by first using (3.4) to express b and r in terms of Cb, y and Cr, y, respectively and then using these results in conjunction with (3.3) to solve for g in terms of y, Cb, Cr.

(e) Using (3.5) and (b) to write a matrix equation that expresses y', C_b, C_r in terms of r, g, b.

(f) Use parts (c) and (e) to write a matrix equation that expresses r, g, b in terms of y', C_b, C_r.

3.16   What is the y', C_b, C_r triple that represents black? Which triple represents white?

3.17   Color printers typically use the *CYM color space*. CYM stands for the secondary colors cyan, yellow, and magenta. Suppose that R, G, and B represent the matrices of intensities for red, green, and blue, respectively, and let W be defined as in Problem 3.14. If C, Y, and M represent the matrices of intensities for cyan, yellow, and magenta, respectively, write C, Y, and M in terms of R, G, B, and W. Compare your answer to the process[6] used to create Figure 3.20.

---

[6]In practice, a black channel is added to the model to better display darker colors. This color space is referred to as the *CYMK* color space.

3.18   The *YUV color space* is related to the YCbCr color space. Y is the luminance channel, and the chrominance channels U and V are color difference channels. If (y, u, v) is a triple in YUV space, then y is given by (3.3), $u = .492(b - y)$, and $v = .877(r - y)$, where $r, g, b \in [0,1]$ represent the normalized intensity values of a pixel.

(a) Find matrix T so that

$$\begin{bmatrix} y \\ u \\ v \end{bmatrix} = T \cdot \begin{bmatrix} r \\ g \\ b \end{bmatrix}$$

(b) Use a **CAS** to find T^{-1}.

(c) Use the process described in part (d) of Problem 3.15 with the formulas for u and v to find T^{-1}.

(d) To see how YUV relates to YCbCr, find diagonal matrix D such that

$$\begin{bmatrix} y \\ u \\ v \end{bmatrix} = D \cdot \begin{bmatrix} y \\ Cb \\ Cr \end{bmatrix} = D \cdot S \begin{bmatrix} r \\ g \\ b \end{bmatrix}$$

where S is given in part (b) of Problem 3.15.

3.19   The NTSC also defines the YIQ color space for television sets. This space again uses the *luminance* (Y) channel. The other two channels are *intermodulation* (I) and *quadrature* (Q) as the additional channels.[7] If a pixel in RGB space is given by normalized values $r, g, b \in [0,1]$, then the corresponding triple (y, i, q) has y given by (3.3), $i = .735514(r-y) - .267962(b-y)$, and $q = .477648(r-y) + .412626(b-y)$.

(a) Find matrix V so that

$$\begin{bmatrix} y \\ i \\ q \end{bmatrix} = V \cdot \begin{bmatrix} r \\ g \\ b \end{bmatrix}$$

(b) Use a **CAS** to find V^{-1}.

(c) The I and Q channels of YIQ can be obtained from the U and V channels of YUV by rotating the U and V channels $33°$ counterclockwise and then swapping the results. Use this information in addition to Problem 2.26 from Section 2.2 to find a 2×2 orthogonal matrix R such that

$$\begin{bmatrix} y \\ i \\ q \end{bmatrix} = \begin{bmatrix} 1 & 0 & 0 \\ \hline 0 & & \\ 0 & & R \end{bmatrix} \cdot \begin{bmatrix} y \\ u \\ v \end{bmatrix}$$

[7] A black-and-white television set only uses the luminance channel.

(d) What is the matrix (in block form) that converts y, u, v to y, i, q?

3.20 Color histogram equalization is a bit more complicated than histogram equalization of grayscale images.

(a) Explain what might go wrong if we perform color histogram equalization by simply performing histogram equalization on each of the red, green, and blue intensity matrices.

(b) In some cases, histogram equalization is applied to the matrix $M = (R + G + B)/3$, where R, G, and B are the red, green, and blue intensity matrices, respectively. The elements of R, G, B are modified using h, where h is the histogram returned by performing histogram equalization on M. Explain the advantage that this process has over the process described in part (a).

Computer Labs

♦ **3.6** **Software Development: Color Space Transformations.** From the text Web site, access the development package `colortransformsdev`. In this development lab you will develop functions for converting images between RGB and YUV, and RGB and YCbCr spaces. HSI (hue, saturation, and intensity) space is often used when denoising color images. This development package also contains a section that leads you through a construction of the conversion functions between HSI and RGB. Instructions are provided that allow you to add all of these functions to the software package `DiscreteWavelets`.

3.7 **Exploring Color Space Transformations.** From the text Web site, access the lab `colortransformslab`. This lab utilizes the functions for converting between RGB and other color spaces that are provided in the `DiscreteWavelets` package.

3.8 **Color to Grayscale and Pseudocolor Maps.** From the text Web site, access the lab `colorsandgray`. In this lab you will learn how to convert color maps to grayscale and how to perform simple pseudocolor maps.

3.9 **Performing Color Histogram Equalization.** From the text Web site, access the lab `colorhisteq`. In this lab you will investigate different methods for performing histogram equalization on color images.

3.3 QUALITATIVE AND QUANTITATIVE MEASURES

In many applications of signal and image processing, we want to assign some quantitative measure to our processing tool. For example, if we compress an image for transmission on the Internet, how do we measure how well our compression scheme worked? Suppose that we denoise a noisy audio signal. How can we measure the effectiveness of our technique?

We study three measures in this section. Each measures some quality of an input vector or matrix. Two of the measures return a number and the third returns a vector.

Entropy

We start by defining the *entropy* of a vector. Entropy is used to assess the effectiveness of compression algorithms, and its definition plays an important role in the theory of information theory. Claude Shannon pioneered much of the work in this area, and in his landmark 1948 paper [66] he showed that the best that we can hope for when compressing data via a *lossless*[8] method S is to encode the output of S where the average number of bits per character equals the entropy of S. Let's define entropy and then look at some examples.

Definition 3.1 (Entropy). *Let* $\mathbf{v} = [v_1, v_2, \ldots, v_n]^T \in \mathbb{R}^n$. *Suppose that there are* k *distinct values in* $\mathbf{v}$. *If we denote these distinct values by* $a_1, \ldots, a_k$, *then let* A_i, $i = 1, \ldots, k$, *be the set of elements from* $\mathbf{v}$ *that equal* a_k. *For each* $i = 1, \ldots, k$, *let* $p(a_i)$ *represent the relative frequency of* a_i *in* $\mathbf{v}$. *That is,* $p(a_i)$ *is the number of elements in* A_i *divided by* n. *We define the entropy of* $\mathbf{v}$ *by*

$$Ent(\mathbf{v}) = \sum_{i=1}^{k} p(a_i) \log_2(1/p(a_i)) \qquad (3.6)$$

$\square$

Note that (3.6) is equivalent to

$$Ent(\mathbf{v}) = -\sum_{i=1}^{k} p(a_i) \log_2(p(a_i))$$

Before trying some examples, let's understand what entropy is trying to measure. The factor $\log_2(1/p(a_i))$ is an exponent — it measures the power of 2 we need to represent $1/p(a_i)$. Base 2 is used so that this factor approximates the number of bits we need to represent a_i. We then multiply this factor by $p(a_i)$ — the percentage of the signal comprised of a_i — and then sum up over all distinct a_i to obtain an average of the number of bits needed to encode each of the elements in $\mathbf{v}$.

---

[8]*Lossless compression* is compression whose output can be uncompressed so that the original data are completely recovered.

If all the elements of $\mathbf{v}$ are distinct, then $k = n$, $a_i = v_i$, and $p(a_i) = \frac{1}{n}$, $i = 1, \ldots, n$. In this case,

$$Ent(\mathbf{v}) = -\sum_{i=1}^{n} \frac{1}{n} \log_2(1/n)$$

$$= -n \cdot \frac{1}{n} \cdot \log_2(1/n)$$

$$= \log_2(n) \qquad\qquad (3.7)$$

If all the elements of $\mathbf{v}$ have the same value c, then $k = 1$ with $a_1 = c$, $p(a_1) = 1$, and $\log_2(p(a_1)) = \log_2(1) = 0$ so that

$$Ent(\mathbf{v}) = 0 \qquad\qquad (3.8)$$

In Problem 3.28 you will show that for any vector $\mathbf{v}$, $Ent(\mathbf{v}) \geq 0$ so that constant vectors constitute a set of minimizers for entropy. In Problems 3.29 to 3.32 you will show that (3.7) is the the largest possible value for entropy, so that vectors with distinct components maximize entropy. This should make sense to you — vectors consisting of a single value should be easiest to encode, and vectors with distinct components should require the most information to encode. Let's look at some examples.

Example 3.7 (Entropy). *Find the entropy of each of the following vectors.*

(a) $\mathbf{u} = [1, 1, 1, 1, 2, 2, 3, 4]^T$.

(b) $\mathbf{v} = [1, 0, 0, \ldots, 0]^T \in \mathbb{R}^n$.

(c) $\mathbf{w} = [1, 2, 3, 4, 0, 0, 0, 0]^T$.

Solution. *For* $\mathbf{u}$*, we have* $k = 4$ *with* $p(a_1) = \frac{1}{2}$*,* $p(a_2) = \frac{1}{4}$*, and* $p(a_3) = p(a_4) = \frac{1}{8}$*. We can compute the entropy as*

$$Ent(\mathbf{u}) = \frac{1}{2} \log_2(2) + \cdot\frac{1}{4} \log_2(4) + \frac{1}{8} \log_2(8) + \frac{1}{8} \log_2 8$$

$$= \frac{1}{2} + \frac{1}{2} + \frac{3}{8} + \frac{3}{8}$$

$$= 1.75$$

For $\mathbf{v}$*, we have* $k = 2$ *with* $p(a_1) = \frac{1}{n}$ *and* $p(a_2) = \frac{n-1}{n}$*. The entropy is*

$$Ent(\mathbf{v}) = \frac{1}{n} \log_2(n) + \frac{n-1}{n} \log_2\left(\frac{n}{n-1}\right)$$

$$= \frac{1}{n} \log_2(n) + \left(\frac{n-1}{n}\right)(\log_2(n) - \log_2(n-1))$$

$$= \log_2(n) - \left(\frac{n-1}{n}\right) \log_2(n-1)$$

Finally, for $\mathbf{w}$, *we have* $k = 5$ *with* $p(a_1) = \cdots = p(a_4) = \frac{1}{8}$ *and* $p(a_5) = \frac{1}{2}$. *We compute the entropy as*

$$Ent(\mathbf{w}) = 4 \cdot \frac{1}{8} \log_2(8) + \frac{1}{2} \log_2(2) = \frac{3}{2} + \frac{1}{2} = 2$$

$\square$

Cumulative Energy

We now move to the idea of *cumulative energy*. Roughly speaking, the cumulative energy of vector $\mathbf{v}$ is a new vector, related to $\mathbf{v}$, that indicates how the size of the components of $\mathbf{v}$ are distributed.

Definition 3.2 (Cumulative Energy). *Let* $\mathbf{v} \in \mathbb{R}^n$, $\mathbf{v} \neq \mathbf{0}$. *Form a new vector* $\mathbf{y} \in \mathbb{R}^n$ *by taking the absolute value of each component of* $\mathbf{v}$ *and then sorting them so that* y_1 *is the largest component of* $\mathbf{y}$ *and* y_n *is the smallest component of* $\mathbf{y}$.

Define the cumulative energy *of* $\mathbf{v}$ *as the vector* $\mathbf{C}_e(\mathbf{v}) \in \mathbb{R}^n$ *whose components are given by*

$$\mathbf{C}_e(\mathbf{v})_k = \sum_{i=1}^{k} \frac{|y_i|^2}{\|\mathbf{y}\|^2}, \qquad |y_k| \geq |y_{k+1}|, \; k = 1, \ldots, n \qquad (3.9)$$

$\square$

Some remarks about Definition 3.2 are in order. First, we observe that since the components of $\mathbf{y}$ are formed by taking the absolute value of the components of $\mathbf{v}$ and then ranking them, the norms are the same. That is $\|\mathbf{v}\| = \|\mathbf{y}\|$. Also, if we recall that $\|\mathbf{y}\|^2 = |y_1|^2 + |y_2|^2 + \cdots + |y_n|^2$, we see that

$$\mathbf{C}_e(\mathbf{v})_k = \frac{|y_1|^2 + |y_2|^2 + \cdots + |y_k|^2}{|y_1|^2 + |y_2|^2 + \cdots + |y_k|^2 + \ldots + |y_n|^2}$$

so that $\mathbf{C}_e(\mathbf{v})_k$ is basically the percentage of the largest k components of $\mathbf{v}$ in $\|\mathbf{v}\|$.

We also note that $0 \leq \mathbf{C}_e(\mathbf{v})_k \leq 1$ for every $k = 1, 2, \ldots, n$ with $\mathbf{C}_e(\mathbf{v})_n = 1$. Let's compute the cumulative energy of two vectors.

Example 3.8 (Cumulative Energy). *Find the cumulative energy of*

$$\mathbf{v} = [1, 2, 3, \ldots, 10]^T \qquad and \qquad \mathbf{w} = [1, 3, 5, 7, 9, 1, 1, 1, 1, 1]^T$$

Solution. *Before computing the cumulative energies of these vectors, let's take a closer look at* $\mathbf{w}$. *We've formed* $\mathbf{w}$ *by taking the odd components from* $\mathbf{v}$ *as the first five components of* $\mathbf{w}$. *The reason the last five components of* $\mathbf{w}$ *are set to 1 is that if*

we add 1 to each of the first five components of **w**, *we get the even components of* **v**.
So **w** *can be thought of (albeit crudely!) as a compressed version of* **v**. *We have*

$$\|\mathbf{v}\|^2 = 385 \quad and \quad \|\mathbf{w}\|^2 = 170$$

We compute the cumulative energy of **v** *to be*

$$\mathbf{C}_e(\mathbf{v}) = \frac{1}{385} [100, 181, 245, 294, 330, 355, 371, 380, 384, 385]^T$$

and the cumulative energy of **w** *to be*

$$\mathbf{C}_e(\mathbf{w}) = \frac{1}{170} [81, 130, 155, 164, 165, 166, 167, 168, 169, 170]^T$$

*The components of the energy vectors don't initially provide much information
about* **v** *and* **w**. *But when we plot both vectors in Figure 3.23, we make an interesting
discovery. We see that over 90% (91.1765%) of the energy of* **w** *is contained in the*

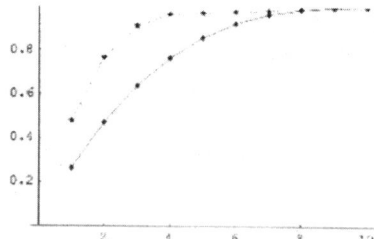

Figure 3.23   A plot of $\mathbf{C}_e(\mathbf{v})$ (solid) and $\mathbf{C}_e(\mathbf{w})$ (dashed).

largest three components of **w**, *whereas it takes the six largest components of* **v** *to
produce over 90% (92.2078%) of the cumulative energy of* **v**.
    We write out the cumulative energy vectors $\mathbf{C}_e(\mathbf{v})$ *and* $\mathbf{C}_e(\mathbf{w})$ *as percents (rounded
to two places) to get a further idea of how the two vectors* **v** *and* **w** *compare.*

$$\mathbf{C}_e(\mathbf{v}) = [25.97\%, 47.01\%, 63.64\%, 76.36\%, 85.71\%,$$
$$92.21\%, 96.36\%, 98.70\%, 99.74\%, 100\%]^T$$
$$\mathbf{C}_e(\mathbf{w}) = [47.65\%, 76.47\%, 91.18\%, 96.47\%, 97.06\%,$$
$$97.65\%, 98.24\%, 98.82\%, 99.41\%, 100\%]^T$$

So we see that the energy is much more spread out in **v** *than is the case in* **w**.   □

Peak Signal-to-Noise Ratio

The last measure we discuss is used primarily in image processing and is called the
peak signal-to-noise ratio (PSNR). In practice we will use the PSNR to compare an
original image with a compressed version of the original.

Suppose that we reconstruct an image by using some compression technique. Then we would want to know how well the reconstructed image approximates the original image. The PSNR is often used for this purpose.

Definition 3.3 (Peak Signal-to-Noise Ratio). *Let A and B be $M \times N$ matrices with entries A_{ij} and B_{ij}, respectively. We first define the error function $Err(A, B)$ as*

$$Err(A, B) = \frac{1}{MN} \sum_{i=1}^{M} \sum_{j=1}^{N} (A_{ij} - B_{ij})^2 \tag{3.10}$$

and next define the peak signal-to-noise ratio *(PSNR) as*

$$PSNR(A, B) = 10 \log_{10} \left(\frac{255^2}{Err(A, B)} \right) \tag{3.11}$$

Here, PSNR is measured in decibels (dB). □

Note that the smaller the error function $Err(A, B)$, the larger the PSNR, so we find larger values of PSNR desirable. Typically, we will use the PSNR to compare different compression methods, and in such a case the measures are relative.

As you might guess, computation of the PSNR is best left to a computer. A 200×300 image consists of 60,000 elements so Example 3.9 below only gives final results. The software used to do the computations is investigated further in Computer Lab 3.11.

Example 3.9 (Peak Signal-To-Noise Ratio). *The leftmost image A in Figure 3.24 is the original 160×240 grayscale image. We have used two different compression methods to produce image B in the middle and image C on the right.*

We utilize (3.10) to compute $Err(A, B) = 21.1124$ and $Err(A, C) = 105.56$. We then use (3.11) and find that $PSNR(A, B) = 34.8854$ and $PSNR(A, C) = 27.8958$. As you can see, the higher PSNR is associated with the middle image B and image B is a better representation of the original image than image C.

(a) The digital image A (b) The approximate B (c) The approximate C

Figure 3.24   Peak signal-to-noise ratio.

□

PROBLEMS

3.21 Find $C_e(\mathbf{v})$ for each of the following. A calculator, **CAS**, or the software package that accompanies this text will be useful.

(a) $\mathbf{v} = [\, 2, -1, 3, 0, 6, -4\,]^T$.

(b) $\mathbf{v} = [\, 1, 1, 1, 1, 1, 1, 1, 1, 1, 1\,]^T$.

(c) $\mathbf{v} = [\, 1, 2, 3, \ldots, 10, \underbrace{0, \ldots, 0}_{10 \text{ zeros}}\,]^T$.

3.22 Find $C_e(\mathbf{v})$ when $\mathbf{v} = [\, v_1, \ldots, v_n\,]^T$, where

(a) $v_k = c, k = 1, \ldots, n$, where $c \neq 0$ is any real number.

(b) $v_k = k, k = 1, \ldots, n$.

(c) $v_k = \sqrt{k}, k = 1, \ldots, n$.

(*Hint:* Problem 2.3 from Section 2.1 will be useful.)

3.23 Let $\mathbf{v} \in \mathbb{R}^n$, $c \neq 0$ any real number, and define $\mathbf{w} = c\mathbf{v}$. Show that $C_e(\mathbf{v}) = C_e(\mathbf{w})$.

3.24 Find the entropy of each of the following vectors. A calculator, **CAS**, or the software package that accompanies this book will be useful.

(a) $\mathbf{a} = [\, 1, 1, 1, 1, 0, 0, 0, 0\,]^T$.

(b) $\mathbf{b} = [\, 1, 1, 0, 0, 1, 0, 1, 0\,]^T$.

(c) $\mathbf{c} = [\, 1, 1, 1, 2, 2, 3\,]^T$.

(d) $\mathbf{d} = [\, 8, 8, 8, 8, 0, 0, 0, 0\,]^T$.

(e) $\mathbf{e} = [\, 1, 2, 3, \ldots, 15, 16\,]^T$.

(f) $\mathbf{g} = [\, 0, 0, 43, 0, 0, 0\,]^T$.

(g) $\mathbf{h} = [\, 0, 0, 0, 51.265, 0, 0\,]^T$.

(h) Can you draw any conclusions about the entropies you have computed?

3.25 Suppose that the $N \times M$ matrix A represents the grayscale intensity values of a digital image. Then the elements a_{jk} of A are integers from the set $\{0, \ldots, 255\}$.

For $k = 0, \ldots, 255$, let n_k denote the number of times intensity level k appears in A, and set $p(k) = n_k/NM$. The entropy is thus defined as

$$Ent(A) = -\sum_{k=0}^{255} p(k) \log_2(p(k)) \tag{3.12}$$

At first glance, there is a problem with this definition. If $p(k) = 0$, then $\log_2(p(k))$ is undefined. However, if we use some elementary calculus, we can see that it is acceptable to omit terms where $p(k) = 0$ and thus show that (3.12) is equivalent to (3.6). Use ideas from calculus to show that

$$\lim_{t \to 0^+} t \log_2(t) = 0$$

3.26 Let $v \in \mathbb{R}^n$ and let c be any real number. Show that $Ent(cv) = Ent(v)$.

3.27 Suppose that v is a vector of length n, where n is an even integer. Moreover, assume that the elements of v are distinct. In this case we know that $Ent(v) = \log_2(n)$ (see (3.7)). Suppose that we form the n-vector w by setting $w_k = (v_{2k-1} + v_{2k})/2$ for $k = 1, 2, \ldots, n/2$ and $w_k = 0$ for $k = n/2 + 1, \ldots, n$. That is, the first $n/2$ elements are averages of consecutive elements in v, and the remaining elements in w are set to 0. Show that $Ent(w) \le Ent(v) - \frac{1}{2}$. We can think of the first half of w as a naive approximation of the elements in v. Thus the average bits per character needed to encode w could possibly be reduced by half a bit. (*Hint:* The largest possible entropy for w will occur when the first $n/2$ elements of w are distinct.)

3.28 In this problem you will show that for all vectors $v \in \mathbb{R}^n$, $Ent(v) \ge 0$.

(a) For $i = 1, \ldots, k$, what can you say about the range of values of $p(a_i)$? The range of $\frac{1}{p(a_i)}$? The range of $\log_2(1/p(a_i))$?

(b) Use part (a) to show that $Ent(v) \ge 0$.

(c) We learned (see the discussion that leads to (3.8)) that vectors $v \in \mathbb{R}^n$ consisting of a single value satisfy $Ent(v) = 0$. Can any nonconstant vector w satisfy $Ent(w) = 0$? Why or why not?

Note: Problems 3.29 to 3.32 were developed from a report by Conrad [21]. In the report, Conrad gives a nice argument that shows the maximum entropy value given in (3.7) is attained if and only if the elements of v are distinct.

3.29 In this problem we show that for $x, y > 0$,

$$y - y \log_2(y) \le x - y \log_2(x) \tag{3.13}$$

with equality if and only if $x = y$ or $x = 2y$. The following steps will help you organize the proof.

(a) First show that (3.13) is equivalent to

$$\log_2(x/y) \leq x/y - 1$$

(b) Substitute $t = x/y$ into the above inequality and write $\log_2(t) \leq t - 1$. Note that $t > 0$ and explain why this inequality is true for $0 < t \leq 1$ with equality at $t = 1$.

3.30   (*Continuation*) Suppose that $p_1, \ldots, p_m$ and $q_1, \ldots, q_m$ are positive numbers with the properties

$$p_k, q_k > 0, \ k = 1, \ldots, m \qquad \text{and} \qquad \sum_{k=1}^{m} p_k = \sum_{k=1}^{m} q_k = 1 \qquad (3.14)$$

Show that

$$-\sum_{k=1}^{m} p_k \log_2(p_k) \leq -\sum_{k=1}^{m} p_k \log_2(q_k) \qquad (3.15)$$

with equality if and only if $p_k = q_k$, $k = 1, \ldots, m$. The following steps will help you organize your work.

(a) Use Problem 3.29 to show that for each $k = 1, \ldots, m$, we have

$$p_k - p_k \log_2(p_k) \leq q_k - p_k \log_2(q_k)$$

(b) Sum the result from part (a) over all $k = 1, \ldots, m$, use (3.14), and simplify to show that the inequality (3.15) holds.

3.31   (*Continuation*) We now show that equality holds in (3.15) if and only if $p_k = q_k$, $k = 1, \ldots, m$. If $p_k = q_k$, for $k = 1, \ldots, m$, then the equality in (3.15) follows immediately. Thus, we need to show that for $k = 1, \ldots, m$,

$$\sum_{k=1}^{m} p_k \log_2(p_k) = \sum_{k=1}^{m} p_k \log_2(q_k) \Rightarrow p_k = q_k \qquad (3.16)$$

We prove this fact by contradiction. The following steps will help you organize your proof.

(a) What do the properties (3.14) imply about the range of p_k, q_k, $k = 1, \ldots, m$?

(b) The ordering of the values does not affect the entropy, so let's assume for the sake of contradiction that there exists a j with $1 \leq j < m$ such that $p_k \neq q_k$ for $k = 1, \ldots, j$. Show that in this case, (3.16) is equivalent to

$$\sum_{k=1}^{j} p_k \log_2(q_k/p_k) = 0$$

(c) Show that the equality in part (b) is equivalent to

$$2^{\left(\sum\limits_{k=1}^{j} p_k \log_2(q_k/p_k)\right)} = 1$$

(d) Show that the equality in part (c) reduces to

$$2^{p_1 \log_2(q_1/p_1)} \cdot 2^{p_2 \log_2(q_2/p_2)} \dots 2^{p_j \log_2(q_j/p_j)} = 1$$

or equivalently,

$$q_1 \cdot q_2 \cdots q_j = 1 \tag{3.17}$$

(e) Use part (a) to explain why (3.17) is a contradiction.

3.32 (*Continuation*) Suppose that p_k, q_k satisfy 3.14 with $q_k = \frac{1}{m}$, for $k = 1, \dots, m$. Use the results of the problems 3.30 and 3.31 to show that

$$-\sum_{k=1}^{m} p_k \log_2(p_k) \le \log_2(m) \tag{3.18}$$

with equality if and only if $p_k = \frac{1}{m}$. We can view the right hand side of (3.18) as the entropy of vector $\mathbf{v} \in \mathbf{R}^n$, where $n \ge m$ with relative frequencies $p_k, k = 1, \dots, m$, and m represents the number of indices in I. Since the log_2 function is increasing, we have

$$Ent(\mathbf{v}) \le \log_2(m) < \log_2(n)$$

If all the values of $\mathbf{v}$ are distinct, then we have $m = n$ and (3.18) becomes an equality.

3.33 Let

$$A = \begin{bmatrix} 2 & -1 & 0 & 4 \\ 1 & 0 & -3 & 2 \\ -3 & 1 & 2 & 1 \\ 0 & 1 & 1 & 0 \end{bmatrix} \quad \text{and} \quad B = \begin{bmatrix} 3 & -1 & 1 & 6 \\ 0 & 0 & -4 & 1 \\ -4 & 3 & 0 & 0 \\ 0 & -1 & -1 & 1 \end{bmatrix}$$

Find $Err(A, B)$ and $PSNR(A, B)$.

3.34 Suppose that A, C_1, and C_2 are $n \times n$ matrices. The elements of A are original grayscale values for a digitial image, while C_1 and C_2 are approximations of A that have been created by two different compression schemes. Suppose that $Err(A, C_1) = \epsilon$ and $Err(A, C_2) = \epsilon/10$ for $\epsilon > 0$. Show that $PSNR(A, C_2) = PSNR(A, C_1) + 10$.

Computer Labs

◆ **3.10   Software Development: Cumulative Energy, Entropy, and PSNR.** From the text Web site, access the development package measurements. In this development lab you will develop functions for cumulative energy, entropy, and PSNR. You will also develop functions that can be used in conjunction with cumulative energy to perform data compression. Instructions are provided that allow you to add all of these functions to the software package DiscreteWavelets.

3.11   Entropy and PSNR. From the text Web site, access the lab entropypsnr. This lab uses the functions for entropy and PSNR included in the DiscreteWavelets package.

3.12   Cumulative Energy and Data Compression. From the text Web site, access the lab cumulativeenergy. This lab will familiarize you with the cumulative energy function included in the DiscreteWavelets package. You will also see how cumulative energy is used to perform naive data compression.

3.4   HUFFMAN ENCODING

In this final section of Chapter 3, we discuss a simple method for reducing the number of bits needed to represent a signal or digital image.

Huffman coding, introduced by David Huffman [48], is an example of lossless compression. The routine can be applied to integer-valued data and the basic idea is quite simple. The method exploits the fact that signals and digital images often contain elements that occur with a much higher frequency than other elements (consider, for example, the digital image and its corresponding histogram in Figure 3.14). Recall that intensity values from grayscale images are integers that range from 0 to 255, and each integer can be represented with an 8-bit ASCII code (see Table 3.1 or Oualline [60]). Thus, if the dimensions of an image are $N \times M$, then we need $8 \cdot NM$ bits to store it. Instead of insisting that each intensity be represented with 8 bits, Huffman suggested that we could use a variable number of bits for each intensity. That is, intensities that occurred the most would be represented with a low number of bits and intensities occurring infrequently would be represented with a larger number of bits.

Generating Huffman Codes

Let's illustrate how to create Huffman codes via an example.

Example 3.10. *Consider the 5×5 digital grayscale image given in Figure 3.25. The bit stream for this image is created by writing each character in binary form and then* *listing them consecutively. Here is the bit stream:*

001000010101000001010000010100000010000101010000001110000111000
01101000001110000101000000111000010100000111111001010000
0011100001010000001110000110100000111000010100000100001
0101000001010000010100000100001

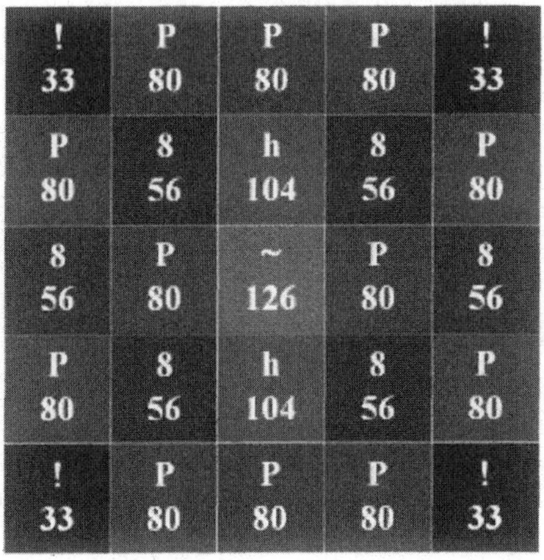

Figure 3.25   A 5 × 5 grayscale image with its intensity levels superimposed.

Table 3.2 lists the binary representations, ASCII codes, frequencies, and relative frequencies for each intensity.

Table 3.2   The ASCII codes, binary representation, frequencies, and relative frequencies for the intensities that appear in the image in Figure 3.25.

Intensity	ASCII Code	Binary Rep.	Frequency	Rel. Frequency
33	!	00100001_2	4	.16
56	8	00111000_2	6	.24
80	P	01010000_2	12	.48
104	h	01101000_2	2	.08
126	~	01111110_2	1	.04

To assign new codes to the intensities, we use a tree diagram. As is evident in Figure 3.26, the first tree is simply one line of nodes that lists the intensities and their relative frequencies in nondecreasing order. Note that for this example, all the relative frequencies were distinct. In the case that some characters have the same relative frequency, the order in which you list these characters does not matter — you need only ensure that all relative frequencies are listed in nondecreasing order.

Figure 3.26 The first tree in the Huffman coding scheme. The characters and their associated relative frequencies are arranged so that the relative frequencies are in nondecreasing order.

The next step in the process is to create a new node from the two leftmost nodes in the tree. The probability of this new node is the sum of the probabilities of the two leftmost nodes. In our case, the new node is assigned the relative frequency .12 = .04 + .08. Thus, the first level of our tree now has four nodes (again arranged in nondecreasing order), and the leftmost node spawns two new nodes on a second level as illustrated in Figure 3.27.

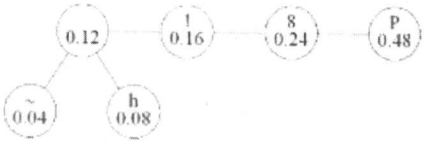

Figure 3.27 The second tree in the Huffman coding scheme.

We repeat the process and again take the two leftmost nodes and replace them with a single node. The relative frequency of this new node is the sum of the relative frequencies of these two leftmost nodes. In our example, the new node is assigned the relative frequency .28 = .12 + .16. We again arrange the first level so that the relative frequencies are nondecreasing. Note that this new node is larger than the node with frequency .24. Our tree now consists of three levels, as illustrated in Figure 3.28(a). We repeat this process for the two remaining characters at the top level. Figure 3.28(b) and (c) illustrate the last two steps in the process and the final Huffman code tree.

Notice that on the final tree in Figure 3.28(c), we label branches at each level using a 0 for left branches and a 1 for right branches. From the top of the tree, we can use these numbers to describe a path to a particular character. This description is nothing more than a binary number, and thus the Huffman code for the particular character. Table 3.3 shows the Huffman codes for each character.

Here is the new bit stream using the Huffman codes:

111000111010110110010011000100101101100111000111

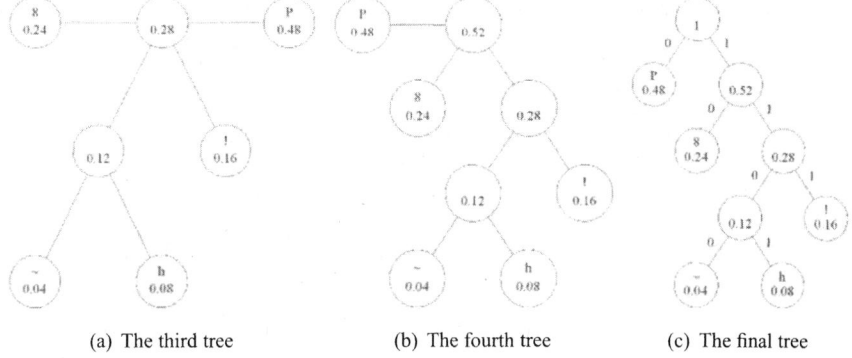

    (a) The third tree   (b) The fourth tree   (c) The final tree

Figure 3.28 The final steps in the Huffman coding algorithm.

Table 3.3 The Huffman code, frequency, and total bits required are given for each character. The total bits needed to encode the image is given as well.

Character	Huffman Code	Frequency	Total Bits
!	111	4	12
8	10	6	12
P	0	12	12
h	1101	2	8
~	1100	1	4
	Bits needed to encode the image:		**48**

Note that the characters that appear the most have the least number of bits assigned to their new code. Moreover, the savings is substantial. Using variable-length codes for the characters, the Huffman scheme needs 48 *bits to represent the image. With no coding, each character requires* 8 *bits. Since our image consists of* 25 *pixels, we would need* 200 *bits to represent it.*   □

Algorithm for Generating Huffman Codes

Next, we review the general algorithm for creating Huffman codes for the elements in a set.

Algorithm 3.1 (Huffman Code Generation). *Given set A, this algorithm generates the Huffman codes for each element in A. Let $|A|$ denote the number of elements in A.*

    Here are the basic steps:

1. *For each distinct element in A, create node n_k with associated relative frequency $p(n_k)$.*

2. *Order the nodes so that the relative frequencies form a nondecreasing sequence.*

3. *While $|A| > 1$:*

   (a) *Find nodes v_j and v_k with the smallest relative frequencies.*

   (b) *Create a new node n_ℓ from nodes n_j and n_k with relative frequency $p(n_\ell) = p(n_j) + p(n_k)$.*

   (c) *Add two descendant branches to node n_ℓ with descendant nodes n_j and n_k. Label the left branch with a 0 and the right branch with a 1.*

   (d) *Drop nodes n_j and n_k from A and add node n_ℓ to A.*

   (e) *Order the nodes so that the relative frequencies form a nondecreasing sequence.*

   EndWhile

4. *For each distinct element in A, create a Huffman code by converting all the 0's and 1's on the descendant branches from the one remaining node in A to the node corresponding to the element into a binary number. Call the set of Huffman codes B.*

5. *Return B.*

□

Implementation of Algorithm 3.1 is straightforward and addressed in Computer Lab 3.13. It should also be pointed out that simply generating the Huffman codes for signal is not the same as compressing it. Although Huffman codes do provide a way to reduce the number of bits needed to store the signal, we still must *encode* these data. The encoded version of the signal not only includes the new bit stream for the signal, but also some header information (essentially the tree generated by Algorithm 3.1) and a pseudo end-of-file character. The encoding process is also addressed in Computer Lab 3.13.

Not only is Huffman coding easy to implement, but as Gonzalez and Woods point out in [39], if we consider all the variable-bit-length coding schemes we could apply to a signal of fixed length, where the scheme is applied to one element of the signal at a time, then the Huffman scheme is optimal. There are several variations of Huffman coding as well as other types of coding. Please see Gonzalez and Wood [39] or Sayood [65] (or references therein) if you are interested in learning more about coding.

Decoding

Decoding Huffman codes is quite simple. You need only the code tree and the bit stream. Consider the following example.

Example 3.11. *Consider the bit stream*

$$100111100110101111001100 \tag{3.19}$$

that is built using the coding tree plotted in Figure 3.29. The stream represents some text. Determine the text.

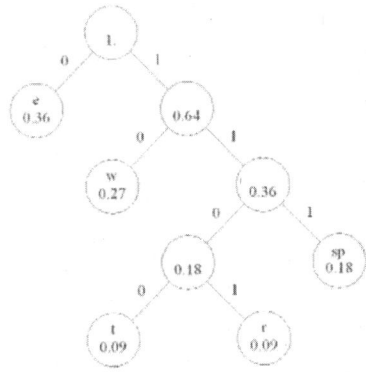

Figure 3.29   The Huffman code tree for Example 3.11. Here *"sp"* denotes a space.

Solution. *From the tree, we note that $e = 0$, $w = 10$, $sp = $ " " $= 111$, $r = 1101$, and $t = 1100$. We begin the decoding process by examining the first few characters until we find a code for one of the letters above. Note that no character has 1 for a code, so we look at the first two characters, 01. This is the code for w, so we now have the first letter. We continue in this manner taking only as many bits as needed to represent a character in the tree. The next bit is 0, which is the code for e. The next code needs three bits — 111 is the space key. The next few characters are w (01), e (0), r (1101), and e (0). The last four characters are the space key (111), then w, e, and t, so that the coded string is* "we were wet." $\quad\square$

In Computer Lab 3.13 you will develop some software to decode Huffman codes, and in Computer Lab 3.14 you will implement these routines to code and decode strings and images.

Bits per Pixel

In this book, we are primarily be interested in using Huffman codes to assess the effectiveness of wavelet transformations in image compression. We usually measure this effectiveness by computing the bits per pixel (bpp) needed to store the image.

Definition 3.4 (Bits per Pixel). *We define the* bits per pixel *of an image, denoted as* bpp, *as the number of bits used to represent the image divided by the number of pixels.*

☐

For example, the 5×5 image in Figure 3.25 is originally coded using 8 bpp. Since Huffman coding resulted in a bit stream of length 48, we say that we can code the image via the Huffman scheme using an average of $48/25 = 1.92$ bpp. As you will soon see, wavelets can drastically reduce this average!

PROBLEMS

3.35   Consider the string *mississippi*.

(a) Generate the Huffman code tree for the string.

(b) Look up the ASCII value for each character in *mississippi*, convert them to base 2, and then write the bit stream for *mississippi*.

(c) Write the bit stream for the string using the Huffman codes.

3.36   Repeat Problem 3.35 for the string *terminal*.

3.37   Consider the 4×4 image whose intensity matrix is

$$
\begin{array}{cccc}
100 & 100 & 120 & 100 \\
100 & 50 & 50 & 40 \\
100 & 40 & 40 & 50 \\
120 & 120 & 100 & 100
\end{array}
$$

(a) Generate the Huffman code tree for the image.

(b) Write the bit stream for the image using Huffman codes.

(c) Compute the bpp for this bit stream.

3.38   Suppose that a and b are two characters in a string. Is it possible for the Huffman code tree to represent a as 00 and b as 000? Why or what not? (*Hint:* If so, how would you interpret a segment of a bit stream that was represented by 00000?)

3.39   As a generalization of Problem 3.38, explain why it is impossible for a Huffman code consisting of n bits to be the first n bits of a longer Huffman code. This is an important feature for decoding — since no code can serve as a *prefix* for another code, we can be assured of no ambiguities when decoding bit streams.

3.40 Suppose that a 40×50 image consists of the four intensities $50, 100, 150$, and 200 with relative frequencies $\frac{1}{8}, \frac{1}{8}, \frac{1}{4}$, and $\frac{1}{2}$, respectively.

(a) Create the Huffman tree for the image.

(b) What is the bpp for the Huffman bit stream representation for the image?

(c) What is the entropy of the image?

(d) Recall in the discussion preceding Definition 3.1 that Shannon [66] showed that the best bpp we could hope for when compressing a signal is the signal's entropy. Do your findings in parts (b) and (c) agree with Shannon's postulate?

3.41 Can you create an image so that the bpp for the Huffman bit stream is equal to the entropy of the image?

3.42 Consider the Huffman codes $e = 0$, $n = 10$, $s = 111$, and $t = 110$.

(a) Draw the Huffman code tree for these codes.

(b) Assume that the frequencies for e, n, s, and t are $4, 2, 2$, and 1, respectively. Find a word that uses all these letters.

(c) Are these the only frequencies that would result in the tree you plotted in part (a)? If not, find some other frequencies for the letters that would result in a tree with the same structure.

3.43 Given the Huffman codes $g = 10$, $n = 01$, $o = 00$, *space key* $= 110$, $e = 1110$, and $i = 1111$, draw the Huffman code tree and use it to decode the bit stream

$$10001111011011010001111011011010000011110$$

3.44 Suppose that you wish to draw the Huffman tree for a set of letters S. Let $s \in S$. Explain why the node for s cannot spawn new branches of the tree. Why is this fact important?

Computer Labs

♦ **3.13 Software Development: Huffman Coding.** From the text Web site, access the development package huffmancodingdev. In this development lab you will develop functions for creating Huffman codes. Instructions are provided that allow you to add all of these functions to the software package DiscreteWavelets.

3.14 Huffman Coding. From the text Web site, access the lab huffmanlab. This lab utilizes the functions for creating Huffman codes and viewing Huffman trees provided in the DiscreteWavelets package.

CHAPTER 4

COMPLEX NUMBERS AND FOURIER SERIES

Complex numbers and Fourier series play vital roles in digital and signal processing. In this chapter we introduce the complex plane and complex numbers and discuss arithmetic with complex numbers. In Section 4.2 we introduce the complex exponential via Euler's famous formula.

As you progress through the book, you will learn why Fourier series are so important. The idea is similar to that of MacLaurin series — we rewrite the given function in terms of basic elements. While the family of functions x^n, $n = 0, 1, \ldots$ serve as building blocks for MacLaurin series, the complex exponentials $e^{ik\omega}$, $k \in \mathbb{Z}$ are used to construct Fourier series. Fourier series coefficients hold much information about the transformations we develop in this book. As you will see in Chapters 8 and 10, once we arrive at a set of desirable conditions for constructing our transformations, it is natural to reformulate the conditions in terms of Fourier series! Thus, a mastery of the material in Section 4.3 is essential for understanding transformation construction that follows later in the text.

Discrete Wavelet Transformations: An Elementary Approach With Applications. By P. J. Van Fleet **97**
Copyright © 2008 John Wiley & Sons, Inc.

4.1 THE COMPLEX PLANE AND ARITHMETIC

In this section we review complex numbers and some of their basic properties. We will also discuss elementary complex arithmetic, modulus, and conjugates. Let's start with the *imaginary number*

$$i = \sqrt{-1}$$

It immediately follows that

$$i^2 = (\sqrt{-1})^2 = -1 \qquad i^3 = i^2 \cdot i = -i \qquad i^4 = i^2 \cdot i^2 = (-1)(-1) = 1$$

In Problem 4.2 you will compute i^n for any nonnegative integer n.

We define a *complex number* to be any number $z = a + bi$, where a and b are any real numbers. We denote the set of real numbers by $\mathbb{R}$ and the set of complex numbers by $\mathbb{C}$. Do you see that $\mathbb{R} \subset \mathbb{C}$? The number a is called the *real part* of z, and b is called the *imaginary part* of z.

It is quite easy to plot complex numbers. Since each complex number has a real and an imaginary part, we simply associate with each complex number $z = a + bi$ the ordered pair (a, b). We use the connection to two-dimensional space and plot complex points in a plane. Our horizontal axis will be used for the real part of z, and the vertical axis will represent the imaginary part of z. Some points are plotted in the complex plane in Figure 4.1.

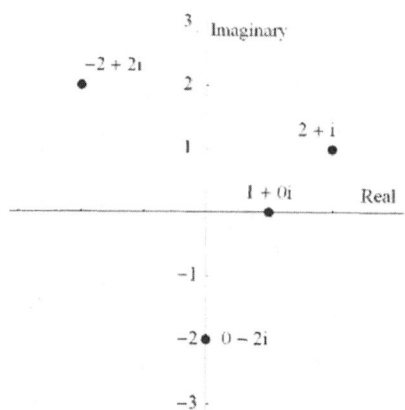

Figure 4.1   Some complex numbers.

It's always good to sprinkle in a little review, especially if it will be of use later. Remember how the trigonometric functions cosine and sine were defined? We draw a circle of radius 1 centered at the origin and then pick any point on the circle. If we draw the right triangle pictured in Figure 4.2, we obtain an angle θ. The length of the horizontal side of this triangle is $\cos \theta$ and the length of the vertical side of this triangle is $\sin \theta$. (Here, we will allow lengths to be negative). So any point on the circle can be determined by an angle θ, and its coordinates are $(\cos \theta, \sin \theta)$.

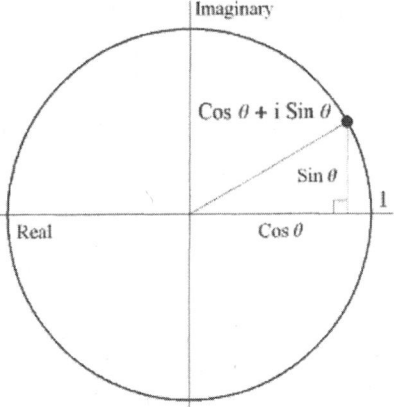

Figure 4.2   Circle and trig functions.

How does this relate to complex numbers? We have already associated complex numbers with two-dimensional space, so it's natural to see that just as any circle of radius 1 centered at the origin has points defined by $(\cos \theta, \sin \theta)$, $0 \le \theta \le 2\pi$, we see that the *function* $f(\theta) = \cos \theta + i \sin \theta$, $0 \le \theta \le 2\pi$ defines a circle in the complex plane.

We have evaluated $f(\theta)$ at three points and plotted the output in Figure 4.3. The points are

$$f(\frac{\pi}{6}) = \cos \frac{\pi}{6} + i \sin \frac{\pi}{6} = \frac{\sqrt{3}}{2} + i\frac{1}{2}$$

$$f(\frac{3\pi}{4}) = \cos \frac{3\pi}{4} + i \sin \frac{3\pi}{4} = -\frac{\sqrt{2}}{2} + i\frac{\sqrt{2}}{2}$$

$$f(\pi) = \cos \pi + i \sin \pi = -1$$

For $0 \le \theta \le 2\pi$, what does $f(2\theta)$ look like? What about $f(4\theta)$ or $f(\frac{\theta}{2})$? See the problems at the end of this section for more functions to plot.

Basic Complex Arithmetic

Now that we have been introduced to complex numbers and the complex plane, let's discuss some basic arithmetic operations.

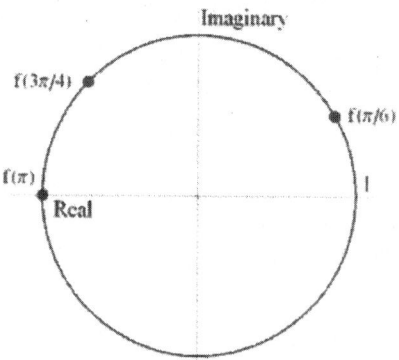

Figure 4.3   The function $f(\theta) = \cos\theta + i\sin\theta$.

Just as with real numbers, we can add, subtract, multiply, and divide complex numbers. Addition and subtraction are very straightforward. To add (subtract) two complex numbers $z = a + bi$ and $w = c + di$, we simply add (subtract) the real parts and add (subtract) the imaginary parts. Thus,

$$z + w = (a + bi) + (c + di) = (a + c) + (b + d)i$$

and

$$z - w = (a + bi) - (c + di) = (a - c) + (b - d)i$$

For example, the sum of $3 + 5i$ and $2 - i$ is $5 + 4i$, and the difference of $3 + 5i$ and $2 - i$ is $1 + 6i$.

Multiplication is also quite straightforward. To multiply $z = a + bi$ and $w = c + di$, we simply multiply the numbers just like we would two algebraic expressions:

$$z \cdot w = (a + bi) \cdot (c + di) = ac + adi + bci + bdi^2 = (ac - bd) + (ad + bc)i$$

For example, the product of $3 + 5i$ and $2 - i$ is $11 + 7i$.

Conjugates

Before we discuss division, we need to introduce the idea of the *conjugate* of a complex number.

Definition 4.1 (Conjugate of a Complex Number). *Let* $z = a + bi \in \mathbb{C}$. *Then the* conjugate *of z, denoted by* $\overline{z}$, *is given by*

$$\overline{z} = a - bi$$

□

As you can see, the conjugate merely serves to negate the imaginary part of a complex number. Let's look at some simple examples and then we'll look at the geometric implications of this operation.

Example 4.1 (Computing Conjugates). *Find the conjugates of* $3 + i$, $-2i$, 5, *and* $\cos \frac{\pi}{4} + i \sin \frac{\pi}{4}$.

Solution.

$\overline{3+i} = 3 - i$ *and the conjugate of* $-2i$ *is* $2i$. *Since* 5 *has no imaginary part, it is its own conjugate and the conjugate of* $\cos \frac{\pi}{4} + i \sin \frac{\pi}{4}$ *is* $\cos \frac{\pi}{4} - i \sin \frac{\pi}{4}$.   □

Some properties of the conjugate are immediate. For example, $\overline{\overline{z}} = z$ and a complex number is real if and only if $\overline{z} = z$. Now let's look at some of the geometry behind the conjugate operator.

To plot a complex number $z = a + bi$ and its conjugate $\overline{z} = a - bi$, we simply reflect the ordered pair (a, b) over the real (horizontal) axis to obtain $(a, -b)$. This is illustrated in Figure 4.4.

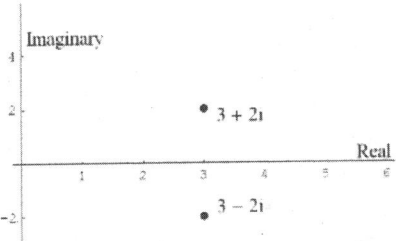

Figure 4.4   A complex number and its conjugate.

Now to get even more geometric insight into the conjugate operator and circles, recall the function $f(\theta) = \cos \theta + i \sin \theta$, $0 \leq \theta \leq 2\pi$. If we start at $\theta = 0$ and plot points in a continuous manner until we reach $\theta = 2\pi$, we'll not only graph a circle but we'll do so in a counterclockwise manner.

What happens if we perform the same exercise with $\overline{f(\theta)}$, $0 \leq \theta \leq 2\pi$? We'll get exactly the same picture, but now the graph will be drawn in a clockwise manner.

One last remark about the conjugate. Consider the product of $z = a + bi$ and its conjugate $\overline{z} = a - bi$:

$$z \cdot \overline{z} = a^2 - bi^2 = a^2 + b^2$$

The Modulus of a Complex Number

Now let's plot $z = a + bi$ and connect it to the origin. We have formed a right triangle whose horizontal side has length a and whose vertical side has length b. By the Pythagorean theorem, the hypotenuse $h = \sqrt{a^2 + b^2}$, so that

$$h^2 = a^2 + b^2 = z \cdot \bar{z}$$

Now the length of h also represents the distance from z to the origin, so we immediately see that this distance is related to the product of z and $\bar{z}$. We are ready for the following definition:

Definition 4.2 (Modulus of a Complex Number). *The* modulus *or the* absolute value *of a complex number $z = a + bi$ is denoted by $|z|$ and computed as*

$$|z| = \sqrt{a^2 + b^2}$$

$\square$

We immediately see that

$$\boxed{|z|^2 = z \cdot \bar{z}} \tag{4.1}$$

Example 4.2 (Computing the Modulus of a Complex Number). *Find the modulus of $3 + 5i$, $-2i$, 5 and $\cos\theta + i\sin\theta$.*
Solution. *$|3 + 5i| = \sqrt{34}$ while $|-2i| = 2$ and $|5| = 5$. To compute $|\cos\theta + i\sin\theta|$, we can think about the point geometrically and since the point is on a circle of radius 1 centered at the origin, arrive at the answer of 1, or we can use a well-known trigonometric identity to compute*

$$|\cos\theta + i\sin\theta| = \sqrt{\cos^2\theta + \sin^2\theta} = 1$$

$\square$

We conclude this section with a discussion of division of complex numbers. The quotient

$$\frac{a + bi}{c + di}$$

can be simplified by multiplying top and bottom by the conjugate of the denominator:

$$\frac{a + bi}{c + di} = \frac{a + bi}{c + di} \cdot \frac{c - di}{c - di} = \frac{(ac + bd) + (bc - ad)i}{c^2 + d^2} = \frac{ac + bd}{c^2 + d^2} + \frac{bc - ad}{c^2 + d^2}i$$

PROBLEMS

4.1   Plot the following points in the complex plane:

(a) $2 + i, 3 + 2i, 5, -4i$.

(b) $\cos \theta + i \sin \theta$ where $\theta = \frac{\pi}{3}, \frac{3\pi}{2}, \frac{7\pi}{6}$.

4.2   Let n be a nonnegative integer. Find a formula that gives i^n in simplest form.

4.3   Consider the function $f(\theta) = \cos \theta + i \sin \theta$. Explain the path traced out by the following functions as θ ranges from 0 to 2π.

(a) $f_1(\theta) = f(-\theta)$.

(b) $f_2(\theta) = f(\frac{\theta}{3})$.

(c) $f_3(\theta) = f(k\theta)$, where k is an integer.

4.4   Plot the following points in the complex plane: $2 + i, 1 - i, 3i, 5, \cos \frac{\pi}{4} + i \sin \frac{\pi}{4},$ $3 - 2i, \frac{1+i}{3+4i}$.

4.5   Find the modulus of each of the points in Problem 4.4.

4.6   Compute the following values:

(a) $(2 + 3i) + (1 - i)$.

(b) $(2 - 5i) - (2 - 2i)$.

(c) $(2 + 3i)\overline{(4 + i)}$.

(d) $\overline{(2 + 3i)(4 + i)}$.

(e) $\overline{(2 + 3i)}(4 + i)$.

(f) $(1 + i) \div (1 - i)$.

★**4.7**   Let $z = a + bi$ and $y = c + di$. Show that $\overline{yz} = \overline{y} \cdot \overline{z}$.

4.8   Let $z = a + bi$. Show that $|z| = |\overline{z}|$.

★**4.9**   Show that z is a real number if and only if $z = \overline{z}$.

★**4.10**   Let $y = a + bi$ and $z = c + di$. Show that $|yz| = |y| \cdot |z|$. Note that in particular, if y is real, say $y = c, c \in \mathbb{R}$, we have $|cz| = |c||z|$.

4.11   Let $y = a + bi$ and $z = c + di$. Show that $\overline{y + z} = \overline{y} + \overline{z} = (a + c) - i(b + d)$.

★**4.12**   This is a generalization Problem 4.11. Suppose that you have n complex numbers $z_k = a_k + ib_k$, $k = 1, 2, \ldots, n$. Show that

$$\overline{\sum_{k=1}^{n} z_k} = \sum_{k=1}^{n} \overline{z_k}$$

$$= \sum_{k=1}^{n} a_k - i \sum_{k=1}^{n} b_k$$

Note: If $\{a_k\}$ and $\{b_k\}$ are convergent sequences then the result holds for infinite sums as well.

4.13   Let $z = a + bi$. Find the real and imaginary parts of $z^{-1} = \frac{1}{z}$.

★**4.14**   Let $f(z)$ be a complex-valued function. Write down all complex numbers w such that $|w| = |f(z)|$.

Computer Lab

4.1   Complex Numbers and Arithmetic. From the text Web site, access the lab `complexarithmetic`. In this lab you will learn how use your **CAS** to perform basic operations involving complex numbers.

4.2   COMPLEX EXPONENTIAL FUNCTIONS

One of the most famous functions in all of mathematics is *Euler's formula*. The result expresses a *complex exponential function* in terms of cosine and sine. The result has numerous applications and we will see that the formula is important as we design tools to process digital signals and images.

MacLaurin Series and Euler's Formula

To develop the basic theory that we need, we must first introduce the concept of a complex exponential function. We need to recall the notion of a Maclaurin series from calculus. If you need to review Maclaurin series, it is strongly recommended that you consult a calculus text (e. g., Stewart [69]) before proceeding with the remainder of the material in this chapter.

You may recall that any function f with continuous derivatives of all order can be expanded as a MacLaurin series:

$$f(t) = f(0) + f'(0)t + \frac{f''(0)}{2!}t^2 + \frac{f'''(0)}{3!}t^3 + \cdots = \sum_{n=0}^{\infty} \frac{f^{(n)}(0)}{n!}t^n$$

Here, $f^{(n)}$ denotes the nth derivative of f for $n = 0, 1, 2, \ldots$ with $f^{(0)} = f$.

We are interested in the MacLaurin series for three well-known functions: $\cos t$, $\sin t$, and e^t. The MacLaurin series for these functions are

$$\cos t = 1 - \frac{t^2}{2!} + \frac{t^4}{4!} - \frac{t^6}{6!} + \frac{t^8}{8!} - + \cdots \tag{4.2}$$

$$\sin t = \frac{t}{1!} - \frac{t^3}{3!} + \frac{t^5}{5!} - \frac{t^7}{7!} + - \cdots \tag{4.3}$$

$$e^t = 1 + \frac{t}{1!} + \frac{t^2}{2!} + \frac{t^3}{3!} + \frac{t^4}{4!} + \frac{t^5}{5!} + \cdots \tag{4.4}$$

It is interesting to note the similarities in these series. The terms in e^t look like a combination of the terms in $\cos t$ and $\sin t$ with some missing minus signs. Undoubtedly, Euler felt that there should be a connection when he discovered the famous result that bears his name.

To derive Euler's formula, we need the MacLaurin series above as well as a bit of knowledge about the powers of i. Let's begin by replacing t with it in (4.4). Using the fact that $i^2 = -1$, $i^3 = -i$, $i^4 = 1$, and so on, we have

$$
\begin{aligned}
e^{it} &= 1 + \frac{it}{1!} + \frac{(it)^2}{2!} + \frac{(it)^3}{3!} + \frac{(it)^4}{4!} + \frac{(it)^5}{5!} + \cdots \\
&= 1 + it - \frac{t^2}{2!} - i\frac{t^3}{3!} + \frac{t^4}{4!} + i\frac{t^5}{5!} - \cdots \\
&= \left(1 - \frac{t^2}{2!} + \frac{t^4}{4!} - \frac{t^6}{6!} + - \cdots\right) \\
&\quad + i\left(\frac{t}{1!} - \frac{t^3}{3!} + \frac{t^5}{5!} - \frac{t^7}{7!} + - \cdots\right) \\
&= \cos t + i\sin t
\end{aligned}
$$

Thus we have *Euler's formula.*

$$\boxed{e^{it} = \cos t + i\sin t} \tag{4.5}$$

There are several problems designed to help you become familiar with Euler's formula. It is perhaps one of the most important results we will use throughout the book. We will spend a little bit of time with Euler's formula before moving on to Fourier series.

At first it seems strange to see a connection between an exponential function and sines and cosines. But look a little closer at the right side of (4.5). Have you seen it before? Recall from Section 4.1, that $f(\theta) = \cos\theta + i\sin\theta$, $0 \le \theta \le 2\pi$ is nothing more than a graph of a circle centered at the origin with radius 1 in the complex plane. In fact, we could let θ be any real number and the effect would simply be a retracing of our circle. So $e^{i\theta}$ is a graph of a circle in the complex plane (Figure 4.5)! What

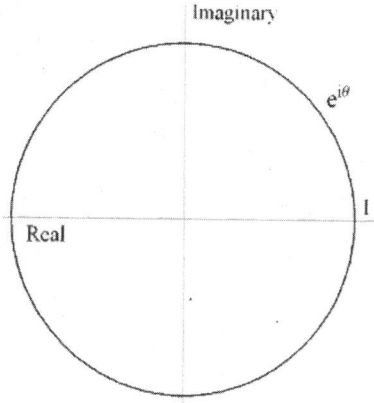

Figure 4.5 A graph of $e^{i\theta} = \cos\theta + i\sin\theta$.

does a picture of $e^{2i\theta}$ from $0 \le \theta \le 2\pi$ look like? How about $e^{i\theta/2}$, $e^{4i\theta}$?

Since cosine and sine are 2π-periodic functions, it follows that $e^{i\theta}$ is also 2π-periodic. What about the periodicity of $e^{i\theta/2}$, $e^{4i\theta}$? What about $\overline{e^{i\theta}}$? We can easily compute the conjugate as follows:

$$\overline{e^{i\theta}} = \overline{\cos\theta + i\sin\theta} = \cos\theta - i\sin\theta$$

Let's take this result a step further. Recall that cosine is an even function so that $\cos\theta = \cos(-\theta)$, and sine is an odd function, so that $\sin(-\theta) = -\sin\theta$. We use these results to compute and simplify $e^{-i\theta}$:

$$e^{-i\theta} = \cos(-\theta) + i\sin(-\theta) = \cos\theta - i\sin\theta = \overline{e^{i\theta}}$$

So we see that

$$\overline{e^{i\theta}} = e^{-i\theta} = \cos\theta - i\sin\theta \tag{4.6}$$

Geometrically, we recall that as θ varies from 0 to 2π, the graph of the circle $\overline{e^{i\theta}}$ is drawn in a clockwise manner. So we see that the same is true for $e^{-i\theta}$.

It is very important that you are familiar with some special values of $e^{i\theta}$.

Example 4.3 (Computing Values of $e^{i\theta}$). *Compute $e^{i\theta}$ for $\theta = 0, \frac{\pi}{2}, \pi$, and 2π.*
Solution. We have

$$e^{i \cdot 0} = 1$$

$$e^{\frac{i\pi}{2}} = \cos\frac{\pi}{2} + i\sin\frac{\pi}{2} = i$$

$$e^{i\pi} = \cos\pi + i\sin\pi = -1$$

$$e^{2\pi i} = \cos 2\pi + i\sin 2\pi = 1$$

For the last two values of θ we first need to note that at any integer multiple of π, the sine function is 0 while the cosine function is ± 1. In particular, if k is even, $\cos k\pi = 1$, and if k is odd, $\cos k\pi = -1$. We can combine these two results to write

$$\cos k\pi = (-1)^k$$

Now we see that

$$\boxed{e^{\pi i k} = \cos k\pi + i \sin k\pi = (-1)^k}$$

and

$$e^{2\pi i k} = \cos 2k\pi + i \sin 2k\pi = 1$$

$\square$

Complex Exponential Functions

Now that we are familiar with Euler's formula for $e^{i\omega}$, it's time to put it to use. To do so, we define a family of *complex exponential functions*:

$$\boxed{e_k(\omega) = e^{ik\omega}, \qquad k \in \mathbb{Z}} \tag{4.7}$$

Note: Hereafter, we use ω as our independent variable when working with complex exponentials and Fourier series — this choice is more in keeping with the majority of the literature on wavelets.

Although it is clear that each function e_k is $\frac{2\pi}{k}$-periodic ($k \neq 0$), it is helpful to view each e_k as 2π-periodic with k copies of cosine and sine in each interval of length 2π. Figure 4.6 illustrates this observation as we graph the real and imaginary parts of $e_k(\omega)$ for different values of k.

Let us now consider the integral

$$\int_{-\pi}^{\pi} e_k(\omega) \overline{e_j(\omega)} \, d\omega$$

for $j, k \in \mathbb{Z}$. We have

$$\int_{-\pi}^{\pi} e_k(\omega) \overline{e_j(\omega)} \, d\omega = \int_{-\pi}^{\pi} e^{ik\omega} e^{(-ij\omega)} \, d\omega = \int_{-\pi}^{\pi} e^{i(k-j)\omega} \, d\omega$$

$$= \int_{-\pi}^{\pi} \cos((k-j)\omega) \, d\omega + i \int_{-\pi}^{\pi} \sin((k-j)\omega) \, d\omega$$

Now if $k = j$, the right-hand side above reduces to

$$\int_{-\pi}^{\pi} e_k(\omega) \overline{e_j(\omega)} \, d\omega = \int_{-\pi}^{\pi} e^{ik\omega} e^{-ik\omega} \, d\omega = \int_{-\pi}^{\pi} 1 \cdot d\omega = 2\pi$$

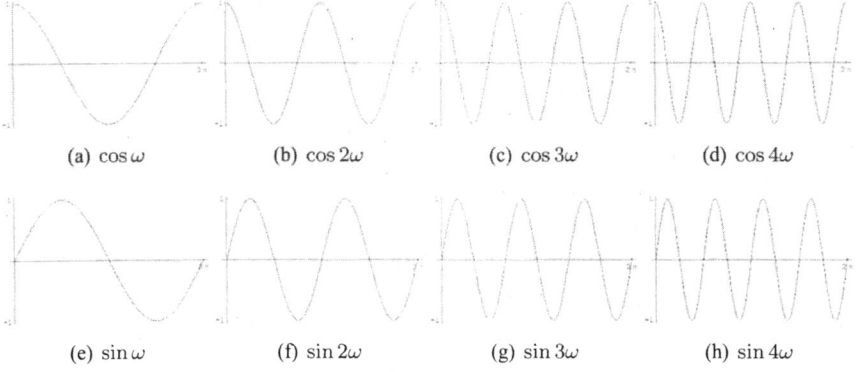

(a) $\cos \omega$ (b) $\cos 2\omega$ (c) $\cos 3\omega$ (d) $\cos 4\omega$

(e) $\sin \omega$ (f) $\sin 2\omega$ (g) $\sin 3\omega$ (h) $\sin 4\omega$

Figure 4.6   A graph of the real (top) and imaginary (bottom) parts of $e_k(\omega)$ for $k = 1,2,3,4$.

If $k \neq j$, we make the substitution $u = (k - j)\omega$ in each integral to obtain

$$\int_{-\pi}^{\pi} e_k(\omega)\overline{e_j(\omega)}\, d\omega = \frac{1}{k-j} \int_{-(k-j)\pi}^{(k-j)\pi} \cos u\, du + \frac{i}{k-j} \int_{-(k-j)\pi}^{(k-j)\pi} \sin u\, du$$

Now sine is an odd function and we are integrating it over an interval $[-(k-j)\pi, (k-j)\pi]$ that is symmetric about the origin, so the value of the second integral on the right-hand side above is zero (see Problem 4.16).

We also utilize the fact that cosine is an even function and since we are also integrating it over an interval $[-(k - j)\pi, (k - j)\pi]$ that is symmetric about the origin, we find that

$$\frac{1}{k-j} \int_{-(k-j)\pi}^{(k-j)\pi} \cos u\, du = \frac{2}{k-j} \int_{0}^{(k-j)\pi} \cos u\, du = \frac{2}{k-j} \sin u \Big|_0^{(k-j)\pi} = 0$$

since sine at any integer multiple of π is zero.

So we have

$$\int_{-\pi}^{\pi} e_k(\omega)\overline{e_j(\omega)}\, d\omega = \begin{cases} 2\pi & k = j \\ 0 & k \neq j \end{cases} \tag{4.8}$$

More generally, we can think of

$$\int_{-\pi}^{\pi} f(\omega)\overline{g(\omega)}\, d\omega \tag{4.9}$$

as the definition of the *inner product* of functions $f, g \colon [-\pi, \pi] \to \mathbb{C}$. It is a quite natural definition — the inner product of two vectors $\mathbf{v}, \mathbf{w} \in \mathbb{R}^n$ is defined to be the sum of componentwise multiplication. That is,

$$\mathbf{v} \cdot \mathbf{w} = v_1 w_1 + v_2 w_2 + \cdots + v_n w_n = \sum_{k=1}^{n} v_k w_k$$

But recall from elementary calculus that the definition of the integral in (4.9) is

$$\int_{-\pi}^{\pi} f(t) \overline{g(t)} \, dt = \lim_{n \to \infty} \sum_{k=1}^{n} f(t_k^*) \overline{g(t_k^*)} \, \Delta t = \lim_{n \to \infty} \Delta t \left(\sum_{k=1}^{n} f(t_k^*) \overline{g(t_k^*)} \right)$$

We can think of the factors $f(t_k^*), \overline{g(t_k^*)}$ in each term of the summand as vector elements — the definition of the integral simply adds a multiplication by the differential Δt and the "length" of the vectors tends to ∞.

Thus, just as $\mathbf{v}, \mathbf{w} \in \mathbb{R}^n$ are called orthogonal vectors if $\mathbf{v} \cdot \mathbf{w} = 0$, we will say that for $j \neq k$, the functions $e_j(\omega)$ and $e_k(\omega)$ are *orthogonal functions* over any interval of length 2π.

PROBLEMS

★4.15   Find $e^{i\theta}$ for $\theta = -\pi, 2\pi, \pm\frac{\pi}{4}, \frac{k\pi}{4}, \frac{\pi}{6}, k\pi, \frac{k\pi}{2}$, and $2k\pi$. Here k is any integer.

4.16   Suppose that $f(\omega)$ is an odd function. That is, $f(-\omega) = -f(\omega)$. Show that for any real number a,

$$\int_{-a}^{a} f(\omega) \, d\omega = 0$$

4.17   Suppose that $f(\omega)$ is a 2π-periodic function and a is any real number. Show that

$$\int_{-\pi-a}^{\pi-a} f(\omega) \, d\omega = \int_{-\pi}^{\pi} f(\omega) \, d\omega$$

(*Hint:* The integral on the left side can be written as

$$\int_{-\pi-a}^{\pi-a} f(\omega) \, d\omega = \int_{-\pi-a}^{-\pi} f(\omega) \, d\omega + \int_{-\pi}^{\pi-a} f(\omega) \, d\omega$$

and the integral on the right side can be written as

$$\int_{-\pi}^{\pi} f(\omega) \, d\omega = \int_{-\pi}^{\pi-a} f(\omega) \, d\omega + \int_{\pi-a}^{\pi} f(\omega) \, d\omega$$

It suffices to show that

$$\int_{-\pi-a}^{-\pi} f(\omega) \, d\omega = \int_{\pi-a}^{\pi} f(\omega) \, d\omega$$

Use the fact that $f(\omega) = f(\omega - 2\pi)$ in the integral on the right-hand side of the above equation and then make an appropriate u-substitution. It is also quite helpful to sketch a graph of the function from $-\pi - a$ to π to see how to solve the problem.)

★**4.18**   Show that

$$
\cos \omega = \frac{e^{i\omega} + e^{-i\omega}}{2} \quad \text{and} \quad \sin \omega = \frac{e^{i\omega} - e^{-i\omega}}{2i}.
$$

Note that these identities are the Fourier series for $\cos \omega$ and $\sin \omega$. (*Hint:* Use Euler's formula (4.5) and (4.6).)

4.19   In a trigonometry class, you learned about half-angle formulas involving sine and cosine. Use the results from Problem 4.18 to prove that

$$
\sin^2 \omega = \frac{1 - \cos 2\omega}{2} \quad \text{and} \quad \cos^2 \omega = \frac{1 + \cos 2\omega}{2}
$$

★**4.20**   Prove *DeMoivre's theorem*. That is, for any nonnegative integer n, show that

$$
(\cos \omega + i \sin \omega)^n = \cos n\omega + i \sin n\omega
$$

4.3 FOURIER SERIES

In this section we introduce one of the most useful tools in applied mathematics — the concept of a Fourier series. Simply put, if f is a 2π-periodic function and *well behaved*, then we can represent it as a linear combination of complex exponential functions. What do we mean by a well behaved function? For our purposes, f will be at the very least a piecewise continuous function with a finite number of jump discontinuities in the interval $[-\pi, \pi]$. Recall that if $f(t)$ has a finite jump discontinuity at $t = a$, then $\lim_{t \to a^+} f(t)$ and $\lim_{t \to a^-} f(t)$ both exist but are not the same. Note that if f is piecewise continuous with a finite number of jump discontinuities on $[-\pi, \pi]$, then we are assured that the integral $\int_{-\pi}^{\pi} f(t)\, dt$ exists.

Fourier Series

As we learned in Section 4.2, the family $\{e_k(\omega)\}_{k \in \mathbb{Z}}$ forms an orthogonal set. More-over, it can be shown that the family forms a *basis* for the space of all 2π-periodic square-integrable[1] functions. That is, any function f in the space can be written as a linear combination of complex exponential functions. Such a linear combination is called a *Fourier series.*[2]

Definition 4.3 (Fourier Series). *Suppose that f is a 2π-periodic, absolutely integrable function. Then the* Fourier series *for f is given by*

$$\boxed{f(\omega) = \sum_{k=-\infty}^{\infty} c_k e^{ik\omega}}$$ (4.10)

□

While we work with Fourier series in a very formal matter in this text, some remarks are in order regarding Definition 4.3.

The right-hand side of (4.10) is an infinite series, and as you learned in calculus, some work is typically required to show convergence of such a series. A simple condition that guarantees convergence of the series is to require that the series of coefficients *converge absolutely* (see Stewart [69]). That is, if

$$\sum_{k=-\infty}^{\infty} |c_k e^{ik\omega}| = \sum_{k=-\infty}^{\infty} |c_k||e^{ik\omega}| = \sum_{k=-\infty}^{\infty} |c_k| < \infty$$

then the series converges absolutely for all $\omega \in \mathbb{R}$. Most of the Fourier series we study in this book satisfy this condition.

Even though the series on the right-hand side of (4.10) might converge for $\omega \in \mathbf{R}$, it is quite possible that the series value is not $f(\omega)$. This statement brings into question the use of the "=" in (4.10)! Indeed, if we were to present a more rigorous treatment of Fourier series, we would learn exactly the context of the "=" in (4.10). This is a fascinating study, but one beyond the scope of this book. The reader new to the topic is referred to Kammler [51] or Boggess and Narcowich [7] and the reader with some background in analysis is referred to Körner [53]. For our purposes, some general statements will suffice.

As long as $f(\omega)$ is a continuous function and the derivative exists at ω, then the series converges to $f(\omega)$. If f is piecewise continuous and ω is a jump discontinuity, then the series converges to one-half the sum of the left and right limits of f at ω. This value may or may not be the value of $f(\omega)$. For example, the Fourier series for the function plotted in Figure 4.7(a) converges to $f(\omega)$ as long as ω is not an odd multiple of π. In the case that ω is an odd multiple of π, the Fourier series converges to 0.

---

[1]Recall that a real-valued function f is square integrable on $[-\pi, \pi]$ if $\int_{-\pi}^{\pi} f^2(t)\, dt < \infty$. It can be shown (see e. g., Bartle and Sherbert [2]) that a piecewise smooth function with a finite number of jump discontinuities in $[-\pi, \pi]$ is square integrable.

[2]Joseph Fourier first studied these series in his attempt to understand how heat disperses through a solid.

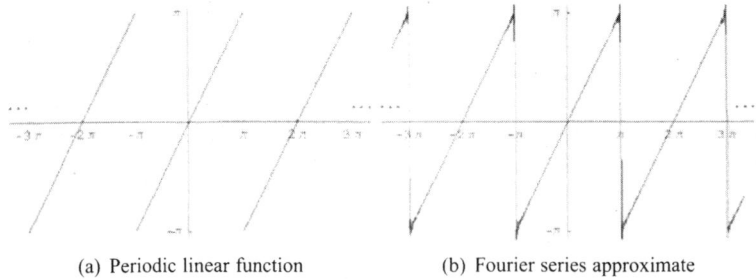

(a) Periodic linear function   (b) Fourier series approximate

Figure 4.7   A graph of a 2π-periodic linear function and its Fourier series approximate.

Fourier Series Coefficients

To understand, use, and analyze Fourier series, we need a way to compute c_k in (4.10). We first rewrite (4.10) using the index variable j and then multiply both sides of the resulting equation by $\overline{e_k(\omega)}$ to obtain

$$f(\omega)\overline{e_k(\omega)} = \sum_{j=-\infty}^{\infty} c_j e^{ij\omega}\overline{e_k(\omega)}$$

Now integrate both sides of the above equation over the interval $[-\pi, \pi]$. Some analysis is required to justify passing the integral through the infinite summation, but we assume that the step is valid and write

$$\int_{-\pi}^{\pi} f(\omega)\overline{e_k(\omega)}\, d\omega = \sum_{j=-\infty}^{\infty} c_j \int_{-\pi}^{\pi} e^{ij\omega}\overline{e_k(\omega)}\, d\omega$$

$$= \sum_{j=-\infty}^{\infty} c_j \int_{-\pi}^{\pi} e_j(\omega)\overline{e_k(\omega)}\, d\omega$$

$$= 2\pi c_k$$

The value $2\pi c_k$ appears by virtue of (4.8). The only integral on the right-hand side above that is not zero is the term where $j = k$. We can easily solve the last equation to obtain a formula for the *Fourier coefficients* c_k. We give this formula in the following definition.

Proposition 4.1 (Fourier Coefficient). *Let f be any integrable, 2π-periodic function. When f is expressed as a Fourier series (4.10), then the Fourier coefficients c_k, $k \in \mathbb{Z}$, satisfy*

$$\boxed{c_k = \frac{1}{2\pi} \int_{-\pi}^{\pi} f(\omega)\overline{e_k(\omega)}\, d\omega = \frac{1}{2\pi} \int_{-\pi}^{\pi} f(\omega) e^{-ik\omega}\, d\omega} \qquad (4.11)$$

□

These c_k's are known as the *Fourier coefficients* for $f(\omega)$. Now let's see how the c_k were computed for the sawtooth function shown in Figure 4.7.

Example 4.4 (Fourier Series for the Sawtooth Function). *Find the Fourier series for the function plotted in Figure 4.7(a).*

Solution. *If we restrict the function shown in Figure 4.7(a) to $(-\pi, \pi)$, we have $f(\omega) = \omega$. So to compute the Fourier coefficients, we must compute the following integral:*

$$c_k = \frac{1}{2\pi} \int_{-\pi}^{\pi} \omega e^{-ik\omega} \, d\omega = \frac{1}{2\pi} \left(\int_{-\pi}^{\pi} \omega \cos(k\omega) \, d\omega - i \int_{-\pi}^{\pi} \omega \sin(k\omega) \, d\omega \right)$$

Now ω is an odd function and $\cos(k\omega)$ is an even function so that the product $\omega \cos(k\omega)$ is odd. Since we are integrating over an interval $(-\pi, \pi)$ that is symmetric about zero, this integral is zero. In a similar way, we see that since $\omega \sin(k\omega)$ is an even function, we have

$$c_k = \frac{-i}{2\pi} \int_{-\pi}^{\pi} \omega \sin(k\omega) \, d\omega = \frac{-i}{\pi} \int_{0}^{\pi} \omega \sin(k\omega) \, d\omega$$

We next integrate by parts. Let $u = \omega$ so that $du = d\omega$, and set $dv = \sin k\omega \, d\omega$ so that $v = -\frac{1}{k} \cos k\omega$.

Note that we can only integrate by parts if $k \neq 0$ (otherwise v is undefined), so let's assume that $k \neq 0$ and return to the $k = 0$ case later. We see that when $k \neq 0$, the integral simplifies to

$$\begin{aligned}
c_k &= \frac{-i}{\pi} \left(-\frac{\omega}{k} \cos(k\omega) \Big|_{0}^{\pi} + \frac{1}{k} \int_{0}^{\pi} \cos(k\omega) \, d\omega \right) \\
&= \frac{-i}{\pi} \left(-\frac{\pi}{k} \cos(k\pi) + \frac{1}{k^2} \sin(k\omega) \Big|_{0}^{\pi} \right)
\end{aligned}$$

Recall that for $k \in \mathbb{Z}$, we have $\sin(k\pi) = 0$ and $\cos(k\pi) = (-1)^k$. Using these facts and simplifying gives

$$c_k = \frac{(-1)^k}{k} i$$

If $k = 0$, we use (4.11) to write

$$c_0 = \frac{1}{2\pi} \int_{-\pi}^{\pi} \omega e^{i \cdot 0 \cdot \omega} \, d\omega = \frac{1}{2\pi} \int_{-\pi}^{\pi} \omega \, d\omega = 0$$

We have

$$c_k = \begin{cases} 0, & k = 0 \\ \frac{(-1)^k}{k} i, & k \neq 0 \end{cases} \tag{4.12}$$

which results in the following Fourier series for f:

$$f(\omega) = \sum_{k \neq 0} \frac{(-1)^k i}{k} e^{ik\omega} \tag{4.13}$$

$\square$

You might find it odd that the Fourier coefficients are purely imaginary for this function.[3] If you work through Problems 4.24 and 4.26, you will see that if a function is odd (even), then its Fourier coefficients are imaginary (real) and the series can be reduced to one of sines (cosines). This should make sense to you — f is an odd function, so we shouldn't need cosine functions to represent it. Indeed, you will show in Problem 4.29 that we can reduce (4.13) to

$$f(\omega) = -2 \sum_{k=1}^{\infty} \frac{(-1)^k}{k} \sin(k\omega)$$

Let's get an idea of what this series looks like. We define the sequence of partial sums by

$$f_n(\omega) = -2 \sum_{k=1}^{n} \frac{(-1)^k}{k} \sin(k\omega)$$

and plot f_n for various values of n. Figure 4.8 shows f, f_1, f_3, f_{10}, f_{20}, and f_{50}.

I still find it amazing that the periodic piecewise linear function can be represented with sine functions! As k gets larger, $\sin(k\omega)$ becomes more and more oscillatory. But the coefficient $\frac{(-1)^k}{k}$ multiplying it becomes smaller and smaller — just the right combination to produce f! You'll notice a little undershoot and overshoot at the points of discontinuity of f. This is the famous Gibbs phenomenon. The topic is fascinating in itself but is beyond the scope of this book. The reader is referred to Kammler [51] for more details.

Let's look at another example.

Example 4.5 (Fourier Series for a Piecewise Constant Function). *For the 2π-periodic piecewise constant function $f(\omega)$ shown in Figure 4.9, find the Fourier series for $f(\omega)$.*

---

[3]If we check the absolute convergence of the Fourier coefficients (see page 111), we find that $\sum_{k \neq 0} \left| \frac{(-1)^k i}{k} \right| = 2 \sum_{k=1}^{\infty} \frac{1}{k}$. This is the harmonic series and we know from calculus that it is a divergent series. Kammler [51, page 43] gives a nice argument that shows the series converges for $\omega \in \mathbb{R}$.

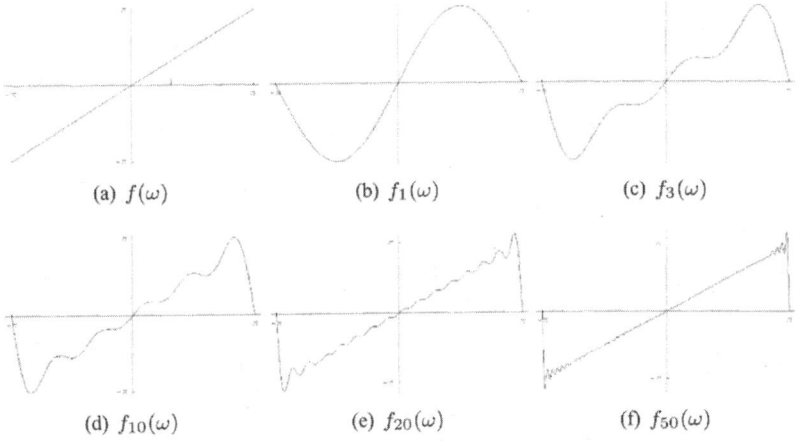

(a) $f(\omega)$ (b) $f_1(\omega)$ (c) $f_3(\omega)$

(d) $f_{10}(\omega)$ (e) $f_{20}(\omega)$ (f) $f_{50}(\omega)$

Figure 4.8   A graph of $f(\omega)$, $f_1(\omega)$, $f_3(\omega)$, $f_{10}(\omega)$, $f_{20}(\omega)$, and $f_{50}(\omega)$ on the interval $[-\pi, \pi]$.

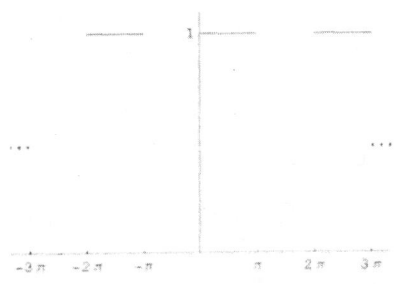

Figure 4.9   Three periods of a 2π-periodic piecewise constant function.

Solution. *When $k \neq 0$, the Fourier coefficients are given by*

$$c_k = \frac{1}{2\pi} \int_{-\pi}^{\pi} f(\omega) e^{-ik\omega} \, d\omega$$

$$= \frac{1}{2\pi} \int_0^{\pi} 1 \cdot e^{-ik\omega} \, d\omega$$

$$= \frac{1}{2\pi} \left(\int_0^{\pi} \cos(k\omega) \, d\omega - i \int_0^{\pi} \sin(k\omega) \, d\omega \right)$$

$$= \frac{1}{2\pi} \left(\frac{1}{k} \sin(k\omega) \Big|_0^{\pi} + \frac{i}{k} \cos(k\omega) \Big|_0^{\pi} \right)$$

$$= \frac{i}{2k\pi} \left((-1)^k - 1 \right)$$

$$= \begin{cases} 0, & k = \pm 2, \pm 4, \ldots \\ \frac{-i}{k\pi}, & k = \pm 1, \pm 3, \pm 5, \ldots \end{cases} \tag{4.14}$$

When $k = 0$, we have $c_0 = \frac{1}{2\pi} \int\limits_{0}^{\pi} 1 \, d\omega = \frac{1}{2}$. So the Fourier series is

$$f(\omega) = \frac{1}{2} - \sum_{k \, odd} \frac{i}{k\pi} e^{ik\omega} \tag{4.15}$$

In Problem 4.29 you will show that this series can be reduced to

$$f(\omega) = \frac{1}{2} + \frac{2}{\pi} \sum_{k=1}^{\infty} \frac{1}{2k-1} \sin((2k-1)\omega)$$

We graph this Fourier series in Figure 4.10. □

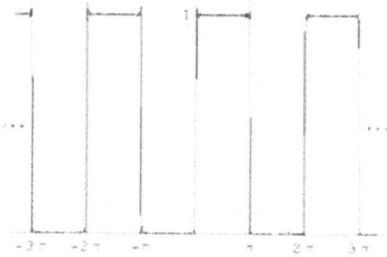

Figure 4.10 Three periods of the Fourier series for the function in Figure 4.9.

As you can see, the computation of Fourier coefficients can be somewhat tedious. Fortunately, there exists an entire calculus for computing Fourier coefficients. The idea is to build Fourier series from a "library" of known Fourier series. Perhaps this concept is best illustrated with an example.

Example 4.6 (Computing One Fourier Series from Another). *Find the Fourier series for the 2π-periodic piecewise constant function $g(\omega)$ shown in Figure 4.11.*
Solution. *Let's denote the Fourier coefficients of $g(\omega)$ with d_k so that*

$$g(\omega) = \sum_{k=-\infty}^{\infty} d_k e^{ik\omega}$$

If we compare $g(\omega)$ to $f(\omega)$ from Example 4.5, we see that $g(\omega) = f(\omega - \frac{\pi}{2})$. Instead of computing the Fourier coefficients for g directly, we consider

$$f\left(\omega - \frac{\pi}{2}\right) = \sum_{k=-\infty}^{\infty} c_k e^{ik(\omega-\pi/2)}$$

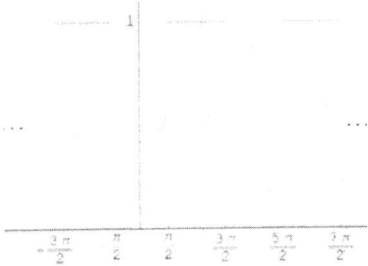

Figure 4.11   Three periods of a 2π-periodic piecewise constant function.

where $c_0 = \frac{1}{2}$ and for $k \neq 0$, c_k is given by (4.14). It is to our advantage not to use the simplified series given by (4.15). Indeed, we have

$$g(\omega) = f\left(\omega - \frac{\pi}{2}\right) = \sum_{k=-\infty}^{\infty} c_k e^{ik(\omega - \pi/2)} = \sum_{k=-\infty}^{\infty} \left(c_k e^{-ik\pi/2}\right) e^{ik\omega}$$

Thus we see that $d_k = e^{-ik\pi/2} c_k$. This is much easier than integrating! We can even simplify. Recall that when $k = 0$, $c_0 = \frac{1}{2}$, so that $d_0 = e^0 c_0 = \frac{1}{2}$. When $k = \pm 2, \pm 4, \pm 6, \ldots$, $c_k = 0$ so that $d_k = 0$ as well. Finally, for k odd, we have $c_k = \frac{i}{k\pi}$ and $e^{ik\pi/2} = i^k$, so that $d_k = \frac{i^{k+1}}{k\pi}(-1)^k$.   □

Rules for Computing Fourier Coefficients

In general we can always build the Fourier series of a function that is expressed as a translation of another function.

Proposition 4.2 (Translation Rule). *Suppose that $f(\omega)$ has the Fourier series representation*

$$f(\omega) = \sum_{k=-\infty}^{\infty} c_k e^{ik\omega}$$

and suppose that $g(\omega) = f(\omega - a)$. If the Fourier series representation for $g(\omega)$ is

$$g(\omega) = \sum_{k=-\infty}^{\infty} d_k e^{ik\omega}$$

then $d_k = e^{-ika} c_k$.   □

Proof of Proposition 4.2. We could prove the result by generalizing the argument given in Example 4.6. As an alternative, we prove the result using the definition of the Fourier coefficient (4.11) for $g(\omega)$. We have

$$d_k = \frac{1}{2\pi} \int_{-\pi}^{\pi} g(\omega)e^{-ik\omega}\, d\omega = \frac{1}{2\pi} \int_{-\pi}^{\pi} f(\omega - a)e^{-ik\omega}\, d\omega$$

Now we do a u-substitution. Let $u = \omega - a$. Then $du = d\omega$ and $\omega = u + a$. Changing endpoints gives the formula

$$\begin{aligned}
d_k &= \frac{1}{2\pi} \int_{-\pi-a}^{\pi-a} f(u)e^{-ik(u+a)}\, du \\
&= \frac{1}{2\pi}e^{-ika} \int_{-\pi-a}^{\pi-a} f(u)e^{-iku}\, du \\
&= \frac{1}{2\pi}e^{-ika} \int_{-\pi}^{\pi} f(u)e^{-iku}\, du \\
&= e^{-ika}c_k
\end{aligned}$$

$\square$

Note that Problem 4.17 from Section 4.2 allows us to change the limits of integration from $-\pi - a, \pi - a$ to $-\pi, \pi$.

Although there are several such rules for obtaining Fourier series coefficients, we only state one more. It will be useful later in the book. In the problems section , other rules are stated and you are asked to provide proofs for them.

Proposition 4.3 (Modulation Rule). *Suppose* $f(\omega)$ *has the Fourier series representation*

$$f(\omega) = \sum_{k=-\infty}^{\infty} c_k e^{ik\omega}$$

and suppose that $g(\omega) = e^{im\omega}f(\omega)$ *for some* $m \in \mathbb{Z}$. *If the Fourier series representation for* $g(\omega)$ *is*

$$g(\omega) = \sum_{k=-\infty}^{\infty} d_k e^{ik\omega}$$

then $d_k = c_{k-m}$. $\square$

Proof of Proposition 4.3. The proof is left as Problem 4.33. $\square$

We could easily spend an entire semester studying the fascinating topic of Fourier series, but we only attempt here to get a basic understanding of how it applies to wavelets.

Finite-Length Fourier Series

You might have decided that it is a little unnatural to construct a series for a function that is already known. There are several reasons for performing such a construction, but admittedly your point would be well taken. In many applications we typically do not know the function that generates a Fourier series — indeed, we usually have only the coefficients c_k or samples of some function.

For our purposes, we are even more constrained. In Chapter 5 we will see how to use a sequence of numbers $\mathbf{h} = (\ldots, h_{-2}, h_{-1}, h_0, h_1, h_2, \ldots)$ to process a signal or an image. We will learn that in order to process the data in certain ways, we need to construct $\mathbf{h}$ in a special way. One method that engineers, scientists, and mathematicians use to analyze such sequences is to construct a Fourier series from them. Invariably, we will insist that all but a finite number of the h_k's are zero. In particular, we will see in the next few chapters that we will assume that $h_k = 0$ for $k < 0$ or $k > L$. So we will form Fourier series of the form

$$H(\omega) = \sum_{k=0}^{L} h_k e^{ik\omega}$$

from these numbers.

What do we stand to gain by forming such a Fourier series? Let's look at an example.

Example 4.7 (Finite-Length Fourier Series). *Let $h_0 = h_2 = \frac{1}{4}$ and $h_1 = \frac{1}{2}$. All other values for h_k are zero. Construct the Fourier series $H(\omega)$ for these h_k, and plot a graph of $|H(\omega)|$, for $-\pi \leq \omega \leq \pi$.*
Solution.

$$H(\omega) = \sum_{k=0}^{2} h_k e^{ik\omega} = \frac{1}{4} + \frac{1}{2}e^{i\omega} + \frac{1}{4}e^{2i\omega}$$

Now let's do a bit of nonconventional factoring. Let's factor out an $e^{i\omega}$ from all three terms and then use Problem 4.18 from Section 4.2. We have

$$H(\omega) = e^{i\omega}\left(\frac{1}{4}e^{-i\omega} + \frac{1}{2} + \frac{1}{4}e^{i\omega}\right)$$

$$= e^{i\omega}\left(\frac{1}{2} + \frac{1}{2}\frac{e^{i\omega} + e^{-i\omega}}{2}\right) = e^{i\omega}\frac{1}{2}(1 + \cos\omega)$$

Now that we have simplified $H(\omega)$, let's compute $|H(\omega)|$. We start by using Problem 4.10 from Section 4.1 to separate the factors in $H(\omega)$

$$|H(\omega)| = |e^{i\omega}| \cdot \left|\frac{1}{2}\right| \cdot |1 + \cos\omega|$$

We know that $e^{i\omega}$ describes a circle of radius 1 centered at the origin, so its absolute value is 1. Also, $-1 \leq \cos\omega \leq 1$ so that $0 \leq 1 + \cos\omega \leq 2$. We can drop the absolute value signs from this factor and write

$$|H(\omega)| = \frac{1}{2}(1 + \cos\omega)$$

The graph of $|H(\omega)|$ for $-\pi \leq \omega \leq \pi$ is shown in Figure 4.12.   □

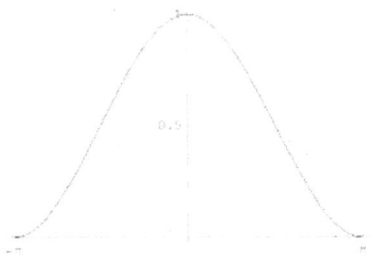

Figure 4.12   A graph of $|H(\omega)|$.

The first thing you might notice about $|H(\omega)|$ is that it is an even function (In Problem 4.32 you will prove that this is always the case). We will learn in Chapter 5 that many scientists and engineers use this plot of $H(\omega)$ to design the process they apply to signals and images. In this setting, ω represents *frequency*, with $\omega = 0$ being the lowest frequency and $\omega = \pi$ being the highest.

We will see that functions $|H(\omega)|$ with "large" values at or near 0 will mean that the process constructed from the h_k's will leave data that are homogeneous (nonoscillatory or similar values) largely unchanged. If values of $|H(\omega)|$ are near zero for values of ω near π, then we will see that the process constructed from the h_k's will take oscillatory data and replace them with very small values.

Alternatively, when values of $|H(\omega)|$ are large at or near π, the process constructed from the h_k's will leave highly oscillatory data unchanged. If values of $|H(\omega)|$ are near 0 for ω near zero, then the process constructed from the h_k's will take homogeneous data and replace them with very small values.

This important example describes in a nutshell exactly how we use Fourier series throughout the text. The numbers $h_0, \ldots, h_L$ are values that we use to process data that comprise signals and images. We have seen that the Fourier series can tell us how these processes will act on homogeneous or highly oscillatory data.

In Chapter 5, we learn the first steps in how the process with the h_k's is constructed.

PROBLEMS

4.21 Use Problem 4.18 from Section 4.2 to write the Fourier coefficients for $\cos \omega$ and the Fourier coefficients for $\sin \omega$.

4.22 Suppose that $\sum\limits_{k=-\infty}^{\infty} |c_k| < \infty$. Show that the Fourier series (4.10) converges absolutely.

4.23 This problem is for students who have taken a real analysis course. Suppose that $\sum\limits_{k=-\infty}^{\infty} |c_k| < \infty$. Show that the Fourier series (4.10) converges uniformly for all $\omega \in \mathbb{R}$.

4.24 In this problem you will show that the Fourier series of an even function can be reduced to a series involving cosines. Assume that f is an even, 2π-periodic function.

(a) Show that the Fourier coefficient $c_k = \frac{1}{\pi} \int\limits_0^\pi f(\omega) \cos(k\omega) \, d\omega$. Thus, c_k is real whenever f is even.

(b) Using part (a), show that $c_k = c_{-k}$.

(c) Using part (b), show that $f(\omega) = c_0 + 2 \sum\limits_{k=1}^{\infty} c_k \cos(k\omega)$. Organize your work using the following steps:

(i) Start with $f(\omega) = \sum\limits_{k \in \mathbb{Z}} c_k e^{ik\omega}$ and write it as $f(\omega) = c_0 + \sum\limits_{k=1}^{\infty} c_k e^{ik\omega} + \sum\limits_{k=-\infty}^{-1} c_k e^{ik\omega}$.

(ii) Change the indices on the last sum to run from $k = 1$ to ∞.

(iii) Combine the two infinite sums into one and use part (b) to replace c_{-k} with c_k.

(iv) Factor out a c_k from each term and then use Problem 4.18 from Section 4.2.

4.25 Assume that $H(\omega) = \sum\limits_{k \in \mathbb{Z}} h_k e^{ik\omega}$ is a Fourier series whose coefficients satisfy $h_k = h_{-k}$ for all $k \in \mathbb{Z}$. Show that $H(\omega) = H(-\omega)$. That is, show that $H(\omega)$ is an even function. (*Hint:* Use Problem 4.24.)

4.26 In this problem you will show that the Fourier series of an odd function can be reduced to a series involving sines. Assume that f is an odd, 2π-periodic function.

(a) Show that the Fourier coefficient $c_k = \frac{-i}{\pi} \int\limits_0^\pi f(\omega) \sin(k\omega) \, d\omega$. Thus, c_k is imaginary whenever f is odd.

(b) Using part (a), show that $c_k = -c_{-k}$. Note that in particular, $c_0 = 0$.

(c) Using part (b), show that $f(\omega) = 2i \sum\limits_{k=1}^{\infty} c_k \sin(k\omega)$. (*Hint:* The ideas from Problem 4.24(c) will be helpful.)

4.27   Assume that $H(\omega) = \sum\limits_{k \in \mathbb{Z}} h_k e^{ik\omega}$ is a Fourier series whose coefficients satisfy $h_k = -h_{-k}$ for all $k \in \mathbb{Z}$. Show that $H(\omega) = -H(-\omega)$. That is, show that $H(\omega)$ is an odd function. (*Hint:* Use Problem 4.26.)

4.28   Use the defining relation (4.11) to find the Fourier series for the function plotted in Figure 4.13. On the interval $[-\pi, \pi]$, the function is $f(\omega) = e^{-|\omega|}$.

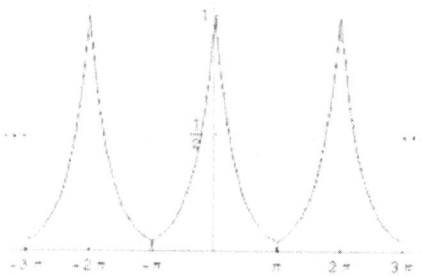

Figure 4.13   The graph of $f(\omega)$ for Problem 4.28.

4.29   In Example 4.4 the Fourier coefficients are given by (4.12) and the resulting Fourier series is given by (4.13). Show that this series can be reduced to the sine series

$$f(\omega) = -2 \sum_{k=1}^{\infty} \frac{(-1)^k}{k} \sin(k\omega)$$

(*Hint:* We know that $c_0 = 0$. Group the $k = \pm 1$ terms together, simplify, and use Problem 4.18 from Section 4.2. Do the same with the $k = \pm 2, \pm 3, \pm 4, \ldots$ terms.)

4.30   In Example 4.5 the Fourier series is given by (4.15). Show that this series can be reduced to the sine series of odd terms

$$f(\omega) = \frac{1}{2} + \frac{2}{\pi} \sum_{k=1}^{\infty} \frac{1}{2k-1} \sin\left((2k-1)\omega\right)$$

(*Hint:* We know that $c_0 = \frac{1}{2}$. Group the $k = \pm 1$ terms together, simplify, and use Problem 4.18 from Section 4.2. Do the same with the $k = \pm 3, \pm 5, \ldots$ terms.)

4.31   Let

$$f_n(\omega) = \frac{1}{2} + \frac{2}{\pi} \sum_{k=1}^{n} \frac{1}{2k-1} \sin((2k-1)\omega)$$

be a sequence of partial sums for the Fourier series given in Problem 4.30. Use a CAS to plot f_n, $n = 1, 2, 5, 10, 50$.

4.32   Suppose that $H(\omega) = \sum_{k=-\infty}^{\infty} h_k e^{ik\omega}$ with h_k a real number for $k \in \mathbb{Z}$.

(a) Show that $H(-\omega) = \overline{H(\omega)}$. (You can "pass" the conjugation operator through the infinite sum without justification — in a complex analysis class, you will learn the conditions necessary to legitimize this operation.)

(b) Show that $|H(\omega)|$ is an even function by verifying that $|H(\omega)| = |H(-\omega)|$. (*Hint:* Use part (a) along with Problem 4.8 from Section 4.1.)

4.33   Prove the modulation rule given by Proposition 4.3. (*Hint:* Start with the definition of d_k.)

4.34   Let $f(\omega)$ have the Fourier series representation $f(\omega) = \sum_{k=-\infty}^{\infty} c_k e^{ik\omega}$ and $g(\omega)$ have the Fourier series representation $g(\omega) = \sum_{k=-\infty}^{\infty} d_k e^{ik\omega}$. Prove the following properties about Fourier series.

(a) If $g(\omega) = f(-\omega)$, show that $d_k = c_{-k}$. (*Hint:* You can prove this either by using the definition of the Fourier coefficients (4.11) or by inserting $-\omega$ into the representation for f above.)

(b) If $g(\omega) = \overline{f(-\omega)}$, show that $d_k = \overline{c_k}$. (*Hint:* Problems 4.7 and 4.12 from Section 4.1 will be helpful here.)

(c) If $g(\omega) = f'(\omega)$, show that $d_k = ikc_k$. (*Hint:* Use (4.11) and integrate by parts. Alternatively, you could differentiate the Fourier series for $f(\omega)$ *if* you can justify differentiating each term of an infinite series. In an analysis class, you will learn how to determine if such a step can be performed.)

4.35   Let $f(\omega)$ be the function given in Example 4.4. The Fourier coefficients for this function are given in (4.12). Find the Fourier coefficients of each of the following functions.

(a) $f_1(\omega) = f(\omega - \pi)$.

(b) $f_2(\omega) = f(-\omega)$.

(c) $f_3(\omega) = e^{2i\omega} f(\omega)$.

4.36   The graph for $f(\omega) = \sum\limits_{k=-\infty}^{\infty} c_k e^{ik\omega}$ is plotted in Figure 4.14. Sketch a graph of g_1, g_2, and g_3. Problem 4.34 and Propositions 4.2 and 4.3 will be helpful.

(a) $g_1(\omega) = \sum\limits_{k=-\infty}^{\infty} (-1)^k c_k e^{ik\omega}$.

(b) $g_2(\omega) = \sum\limits_{k=-\infty}^{\infty} c_{-k} e^{ik\omega}$.

(c) $g_3(\omega) = i \sum\limits_{k=-\infty}^{\infty} k c_k e^{ik\omega}$.

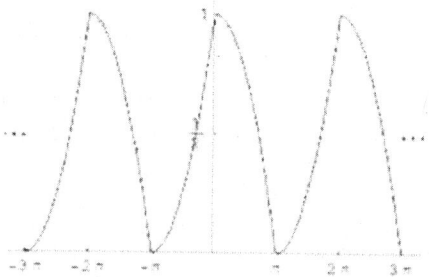

Figure 4.14   A graph of the $f(\omega)$ for Problem 4.36.

4.37   We developed the notion of Fourier series for 2π-periodic functions. It turns out that the Fourier series representation is valid for functions of period $2L$ ($L > 0$). In this problem you will learn about the complex exponential functions and the Fourier coefficients needed for this representation. The following steps will help you organize your work.

(a) Show that the complex exponential function $e_k^L(\omega) = e^{-2\pi i k\omega/2L}$ is a $2L$-periodic function. Here $k \in \mathbb{Z}$ and we add the superscript L to distinguish these complex exponential functions from the ones defined in (4.7) for 2π-periodic functions.

(b) Show that for $j, k \in \mathbb{Z}$, we have

$$\int_0^{2L} e_j^L(\omega)\overline{e_k^L(\omega)}\, d\omega = \int_{-L}^L e^{2\pi i k\omega/2L} e^{-2\pi i k\omega/2L}\, d\omega = \begin{cases} 2L, & j = k \\ 0, & j \neq k \end{cases}$$

(c) It can be shown (see e. g., Boggess and Narcowich [7]) that the set $\{e^{2\pi i k\omega/2L}\}_{k \in \mathbb{Z}}$ forms a basis for $2L$-periodic, integrable (over $[0, 2L]$) functions. Thus, for such a function, we can write

$$f(\omega) = \sum_{k \in \mathbb{Z}} c_k e^{2\pi i k/(2L)}$$

Using an argument similar to that preceding Proposition 4.1 show that

$$c_k = \frac{1}{2L} \int_{-L}^{L} f(\omega) e^{-2\pi i k\omega/(2L)} \, d\omega$$

★**4.38** Let $H(\omega) = \cos^2(\frac{\omega}{2})$. Show that the Fourier series for $H(\omega)$ is $H(\omega) = \frac{1}{4}(e^{i\omega} + 2 + e^{-i\omega}) = \frac{1}{2}(1 + \cos\omega)$.

★**4.39** Let $G(\omega) = \sin^2(\frac{\omega}{2})$. Show that the Fourier series for $G(\omega)$ is $G(\omega) = \frac{1}{4}(-e^{i\omega} + 2 - e^{-i\omega}) = \frac{1}{2}(1 - \cos\omega)$.

★**4.40** Consider the finite-length Fourier series $H(\omega) = \sum_{k=0}^{3} h_k e^{ik\omega}$. Let $G(\omega) = \sum_{k=0}^{3} g_k e^{ik\omega}$ where $g_k = (-1)^k h_{3-k}$ for $k = 0, 1, 2, 3$. Write $G(\omega)$ in terms of $H(\omega)$. The following steps will help you organize your work.

(a) Write down $\overline{G(\omega)}$ and factor $-e^{-3i\omega}$ from the series to obtain

$$\overline{G(\omega)} = -e^{-3i\omega}\left(h_0 - h_1 e^{i\omega} + h_2 e^{2i\omega} - h_3 e^{3i\omega}\right)$$

(b) Chose a value a so that $H(\omega + a)$ gives the four-term series on the right in part (a).

(c) Conjugate the identity in part (b) to obtain the desired result.

★**4.41** This problem requires knowledge of the binomial theorem.

Note: The Binomial Theorem says that for all real numbers a, b and nonnegative integer n, we have

$$(a + b)^n = \sum_{k=0}^{n} \binom{n}{k} a^k b^{n-k}$$

where the *binomial coefficient* is defined as

$$\binom{n}{k} = \frac{n!}{k!\,(n-k)!} \qquad (4.16)$$

with $n! = 1 \cdot 2 \cdots n$ and $0! = 1$.

(a) Let $H(\omega) = (\cos \frac{\omega}{2})^N$ where N is an even, positive integer. Find the Fourier series for $H(\omega)$. (*Hint:* $\cos \frac{\omega}{2} = \frac{1}{2}(e^{i\omega/2} + e^{-i\omega/2}) = \frac{1}{2} e^{-i\omega/2}(e^{i\omega} + 1)$. Use the binomial theorem on $(e^{i\omega} + 1)$.)

(b) Let $G(\omega) = (\sin \frac{\omega}{2})^N$ where N is an even, positive integer. Find the Fourier series for $G(\omega)$. (*Hint:* You can use the hint from part (a) with a half-angle formula for $\sin(\frac{\omega}{2})$.)

★**4.42**   The well-known Pascal's identity is

$$\binom{n+1}{k} = \binom{n}{k-1} + \binom{n}{k} \tag{4.17}$$

where n and k are nonnegative integers with $n \geq k$. Use the definition of the binomial coefficient (4.16) to prove Pascal's identity.

4.43   Use Problem 4.42 and mathematical induction to prove the binomial theorem stated in Problem 4.41.

4.44   In Example 4.7 we created a Fourier series from the three nonzero coefficients $h_0 = h_2 = \frac{1}{4}$ and $h_1 = \frac{1}{2}$. The resulting Fourier series is

$$H(\omega) = e^{i\omega} \frac{1}{2}(1 + \cos \omega)$$

This is counter to having a known function 2π-periodic, integrable function $f(\omega)$ and using it to create Fourier coefficients (see Example 4.4). It is a good exercise to use (4.11) to directly compute the integral and show that

$$h_k = \frac{1}{2\pi} \int_{-\pi}^{\pi} H(\omega)e^{-ik\omega} \, d\omega = \begin{cases} \frac{1}{4}, & k = 0, 2 \\ \frac{1}{2}, & k = 1 \\ 0, & \text{otherwise} \end{cases}$$

Computer Lab

4.2   **Fourier Series.** From the text Web site, access the `fourierseries` lab. In this lab you will further investigate Fourier series. You will construct Fourier coefficients for various functions, create partial sums of Fourier series that allow you to investigate how Fourier series converge, and implement the rules for finding Fourier series. You will also work with finite Fourier series.

CHAPTER 5

CONVOLUTION AND FILTERS

We are interested in processing signals. These signals might come in the form of images or digital sound files. The reason we process signals is so that we can perform tasks such as data compression, image enhancement and audio denoising. A fundamental tool for processing signals is *convolution*.

A convolution product of two sequences of numbers (for the time being we will think of our signals as bi-infinite sequences) h and x results in a new sequence y. Although we can give a formal definition of convolution and we will see that we can convolve any two sequences h and x we desire, it is practical to think of x as input data and h as a processor or *filter* of the input data x.

In this chapter we define the convolution operator $*$ and become familiar with it. We then develop some basic properties of convolution and next, study the interplay between the convolution with filter h and the Fourier series $H(\omega)$ for h. We continue by defining two special types of filters and how they can combine to produce a filter bank. We conclude the chapter by illustrating how to represent convolution as a matrix and discuss the idea of inverting or *deconvolving* the filter process.

Discrete Wavelet Transformations: An Elementary Approach With Applications. By P. J. Van Fleet **127**
Copyright © 2008 John Wiley & Sons, Inc.

Note: All sequences and filters are denoted using bold lowercase letters.

5.1 CONVOLUTION

In Chapter 2 we reviewed some basic ideas involving vectors. Vector addition and subtraction are examples of *binary operations* since they act on two vectors and produce a vector. Convolution is also a binary operator. For the time being we use bi-infinite sequences instead of vectors, but the idea is the same. Convolution takes two bi-infinite sequences h and x and produces a new bi-infinite sequence y.

You should remember sequences from calculus; $a = (a_1, a_2, a_3, \ldots)$ is the standard notation for a sequence. You can think of a as a function on the natural numbers $1, 2, 3, \ldots$ and instead of writing the result using normal function notation $a(1), a(2), \ldots$, we instead write $a_1, a_2, \ldots$. A bi-infinite sequence b is simply a sequence that is a function on the integers with components $\ldots, b_{-2}, b_{-1}, b_0, b_1, b_2, \ldots$, so we'll write $b = (\ldots, b_{-2}, b_{-1}, b_0, b_1, b_2, \ldots)$.

We use bi-infinite sequences to present a formal definition of convolution.

Definition 5.1 (Convolution). *Let* h *and* x *be two bi-infinite sequences. Then the convolution product* y *of* h *and* x*, denoted by* h $*$ x*, is the bi-infinite sequence* y $=$ h $*$ x*, whose nth component is given by*

$$y_n = \sum_{k=-\infty}^{\infty} h_k x_{n-k} \tag{5.1}$$

□

Certainly we can see from (5.1) that unless conditions are placed on the sequences h and x, the series will diverge. In this book, either the components of h will be zero except for a finite number of terms or the terms in h and x will decay rapidly enough to ensure that the convolution product will converge. We assume throughout the book that for all integers n, the right-hand side of (5.1) is a convergent series.

Applications for the convolution operator abound throughout mathematics and engineering. You may not realize it, but you have been convolving sequences ever since you have known how to multiply — the algorithm you learned in elementary school is basically convolution. Using the convolution operator and the fast Fourier transform (see e. g., Kammler [51]), algorithms have been developed to quickly multiply numbers consisting of thousands of digits. Discrete density functions can be constructed using convolution, and many digital filters for processing sounds are built using convolution.

We next present a series of examples of convolution products. We start with some basic computational examples and then proceed to some applications.

Example 5.1 (Convolution Products). *Compute each of the following convolution products.*

(a) $\mathbf{h} * \mathbf{x}$, *where* $\mathbf{x}$ *is any bi-infinite sequence and* $\mathbf{h}$ *is the sequence whose components satisfy*

$$h_k = \begin{cases} 1, & k = 0 \\ -1, & k = 1 \\ 0, & \text{otherwise} \end{cases}$$

(b) $\mathbf{h} * \mathbf{x}$, *where* $\mathbf{x}$ *is any bi-infinite sequence and* $\mathbf{h}$ *is the sequence whose components satisfy*

$$h_k = \begin{cases} \frac{1}{2}, & k = 0, k = 1 \\ 0, & \text{otherwise} \end{cases}$$

(c) Let $\mathbf{h}$ *be the sequence whose components are given by*

$$h_k = \begin{cases} 1, & k = 0, k = 1 \\ 0, & \text{otherwise} \end{cases}$$

Compute $\mathbf{h} * \mathbf{h}$ *and* $\mathbf{h} * \mathbf{h} * \mathbf{h}$.

Solution. *For part (a) we note that*

$$y_n = \sum_{k=-\infty}^{\infty} h_k x_{n-k} = \sum_{k=0}^{1} h_k x_{n-k} = h_0 x_n + h_1 x_{n-1} = x_n - x_{n-1}$$

so that y_n *can be obtained by subtracting* x_{n-1} *from* x_n. *It is worthwhile for you to compare this convolution product to the matrix multiplication HM in Example 3.5.*
For part (b) we have

$$y_n = \sum_{k=-\infty}^{\infty} h_k x_{n-k} = \sum_{k=0}^{1} h_k x_{n-k} = \frac{1}{2} x_n + \frac{1}{2} x_{n-1}$$

so that y_n *can be obtained by averaging* x_{n-1} *with* x_n.
In part (c), we denote $\mathbf{y} = \mathbf{h} * \mathbf{h}$ *and compute*

$$y_n = \sum_{k=-\infty}^{\infty} h_k h_{n-k} = \sum_{k=0}^{1} h_{n-k} = h_n + h_{n-1}$$

Now all the components of $\mathbf{y}$ *will be zero except for* $y_0 = h_0 + h_{-1} = 1$, $y_1 = h_1 + h_0 = 2$, *and* $y_2 = h_2 + h_1 = 1$. *So*

$$\mathbf{y} = (\ldots, 0, 0, 1, 2, 1, 0, 0, \ldots)$$

Now let's compute the components z_n of $\mathbf{z} = \mathbf{h} * \mathbf{y}$. We have

$$z_n = \sum_{k=-\infty}^{\infty} h_k y_{n-k} = \sum_{k=0}^{1} y_{n-k} = y_n + y_{n-1}$$

Now all the components of $\mathbf{z}$ will be zero except for $z_0 = y_0 + y_{-1} = 1$, $z_1 = y_1 + y_0 = 3$, $z_2 = y_2 + y_1 = 3$, and $z_3 = y_3 + y_2 = 1$. So

$$\mathbf{z} = (\ldots, 0, 0, 1, 3, 3, 1, 0, 0, \ldots)$$

$\square$

Convolution as Shifting Inner Products

If the concept of convolution product is still difficult to grasp, perhaps this alternative look at convolution will help.

Recall that if we wish to compute the inner product of two vectors $\mathbf{h} = [h_1, \ldots, h_N]^T$ and $\mathbf{x} = [x_1, \ldots, x_N]^T$, we simply compute $\sum_{k=1}^{N} h_k x_k$.

Now suppose that $\mathbf{h} = (\ldots, h_{-1}, h_0, h_1, \ldots)$ and $\mathbf{x} = (\ldots, x_{-1}, x_0, x_1, \ldots)$ were bi-infinite sequences and we wished to compute their inner product. Assuming the series converges, we have

$$\mathbf{h} \cdot \mathbf{x} = \sum_{k=-\infty}^{\infty} h_k x_k \tag{5.2}$$

Now the inner product in (5.2) looks quite similar to the convolution $\mathbf{h} * \mathbf{x}$ given in Definition 5.1. The only difference is that the subscript of the components of $\mathbf{x}$ in the convolution are $n - k$ rather than k. But we can at least agree that the convolution looks like an inner product.

How do we view the components x_{n-k} in the convolution product? If $n = 0$, then we have components x_{-k}. If we write these components in a bi-infinite sequence, we have

$$(\ldots, x_2, x_1, x_0, x_{-1}, x_{-2}, \ldots) \tag{5.3}$$

This sequence is simply a reflection of $\mathbf{x}$. Thus, to compute $y_0 = \sum_{k=-\infty}^{\infty} h_k x_{0-k}$, we simply reflect $\mathbf{x}$ and dot it with $\mathbf{h}$.

If $n = 1$, then we have components x_{1-k}. It is easy to see what is going on here — we have simply shifted the reflected sequence given in (5.3) 1 unit to the right. So if we want to compute $y_1 = \sum_{k=-\infty}^{\infty} h_k x_{1-k}$, we simply reflect $\mathbf{x}$, shift all the elements 1 unit to the right, and dot the resulting sequence with $\mathbf{h}$.

We can certainly handle negative values of n as well. To compute y_{-13}, we reflect **x**, shift the result 13 units to the *left*, and dot the resulting sequence with **h**.

In general, to compute y_n, we reflect **x**, shift it n units (left if n is negative, right if n is positive), and dot the resulting sequence with **h**. Let's look at an example.

Example 5.2 (Convolution as Shifting Inner Products). *Let* **h** *be the bi-infinite sequence whose only nonzero terms are* $h_0 = 6$, $h_1 = 1$, $h_2 = 2$, *and* $h_3 = 3$. *Suppose that* **x** *is a bi-infinite sequence whose only nonzero components are* $x_{-2} = 3$, $x_{-1} = 2$, $x_0 = 1$, *and* $x_1 = 5$. *Use shifting inner products to compute the convolution product* $\mathbf{y} = \mathbf{h} * \mathbf{x}$.
Solution. *We will compute this product visually. Figure 5.1 shows* **h** *and a reflected version of* **x**. *A sufficient number of* 0*'s have been added to each sequence so that we can shift* **x** *far enough in either direction to see what is happening.*

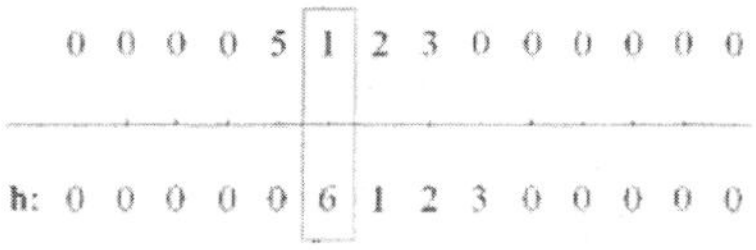

Figure 5.1   The values of **h** on the bottom and reflected values of **x** on top. The box contains h_0 and x_0.

If we dot these two vectors, we obtain $y_0 = 6 \cdot 1 + 1 \cdot 2 + 2 \cdot 3 = 14$. *To compute* y_1, *we simply shift the reflected version of* **x** 1 *unit right and compute the inner product between* **h** *and the resulting sequence. To compute* y_2, *we shift the reflected version of* **x** 2 *units to the right and compute the inner product between* **h** *and the resulting sequence. The sequences needed to compute* y_1, *and* y_2 *are plotted in Figure 5.2. The inner products that result from Figure 5.2 are*

$$y_1 = 6 \cdot 5 + 1 \cdot 1 + 2 \cdot 2 + 3 \cdot 3 = 44$$
$$y_2 = 1 \cdot 5 + 2 \cdot 1 + 3 \cdot 2 = 13$$

If we look at Figure 5.2, we see that we can shift the reflected **x** *sequence two more times to the right — after that, the inner products of subsequent positive shifts and* **h** *will result in zero. From Figure 5.1 we see that we can shift the reflected* **x** *sequence 2 units left before subsequent shifts lead to inner products of zero. The remaining four shifts that result in nonzero inner products are plotted in Figure 5.3.*

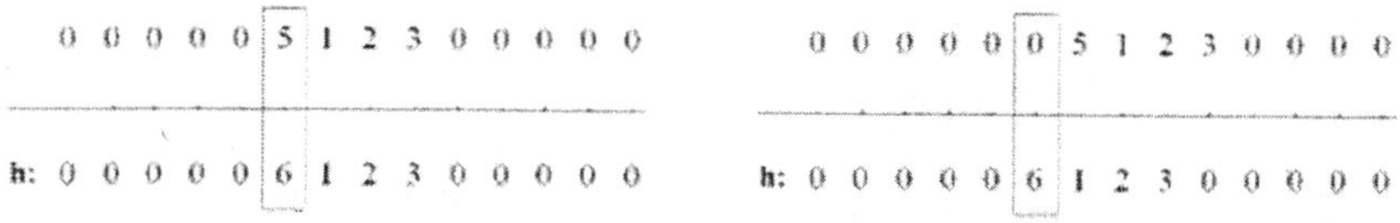

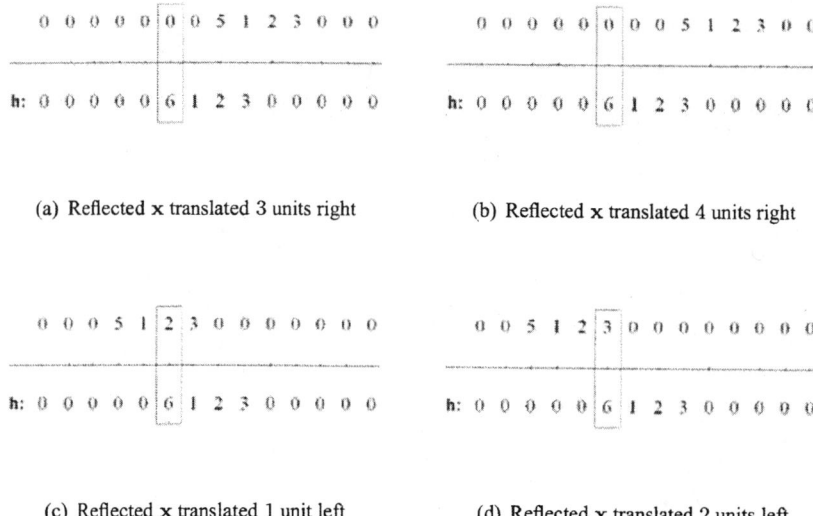

0 0 0 0 0 5 1 2 3 0 0 0 0 0

h: 0 0 0 0 0 6 1 2 3 0 0 0 0 0

(a) Reflected x translated 1 unit right

0 0 0 0 0 0 5 1 2 3 0 0 0 0

h: 0 0 0 0 0 6 1 2 3 0 0 0 0 0

(b) Reflected x translated 2 units right

Figure 5.2   Plots of h (bottom of each graph) and translates of the reflection of x.

0 0 0 0 0 0 0 5 1 2 3 0 0 0

h: 0 0 0 0 0 6 1 2 3 0 0 0 0 0

(a) Reflected x translated 3 units right

0 0 0 0 0 0 0 0 5 1 2 3 0 0

h: 0 0 0 0 0 6 1 2 3 0 0 0 0 0

(b) Reflected x translated 4 units right

0 0 0 5 1 2 3 0 0 0 0 0 0 0

h: 0 0 0 0 0 6 1 2 3 0 0 0 0 0

(c) Reflected x translated 1 unit left

0 0 5 1 2 3 0 0 0 0 0 0 0 0

h: 0 0 0 0 0 6 1 2 3 0 0 0 0 0

(d) Reflected x translated 2 units left

Figure 5.3   Plots of h (bottom of each graph) and translates of the reflection of x.

The inner products resulting from dotting the sequences in Figure 5.3 are

$$y_3 = 5 \cdot 2 + 1 \cdot 3 = 13 \qquad y_4 = 5 \cdot 3 = 15$$
$$y_{-1} = 6 \cdot 2 + 1 \cdot 3 = 15 \qquad y_{-2} = 6 \cdot 3 = 18$$

All other inner products are zero. Thus, the sequence **y** *($y_0 = 14$ is given in bold type) that results from convolving* **h** *and* **x** *is*

$$\mathbf{y} = (\ldots, 0, 0, 0, 18, 15, \mathbf{14}, 44, 13, 13, 15, 0, 0, 0, \ldots)$$

□

Fourier Series from Sequences

It will be important in subsequent chapters to create and analyze Fourier series built from the bi-infinite sequences that are used in problems involving convolution. Suppose that we are given a bi-infinite sequence $\mathbf{h}$. Then we can easily form a Fourier series with the components of $\mathbf{h}$ as the Fourier coefficients

$$H(\omega) = \sum_{k=-\infty}^{\infty} h_k e^{ik\omega}$$

For example, if we consider $\mathbf{h}$ from Example 5.1(a), we have

$$H(\omega) = 1 - e^{i\omega}$$

The Convolution Theorem

We are about to learn that these Fourier series formed from bi-infinite sequences are important tools in the design of filters. There is an important result that characterizes convolution in the Fourier domain. The result is called the convolution theorem and we state and prove it now.

Theorem 5.1 (Convolution Theorem). *Let* $\mathbf{h}$, $\mathbf{x}$, *and* $\mathbf{y}$ *be bi-infinite sequences with* $\mathbf{y} = \mathbf{h} * \mathbf{x}$. *Let* $H(\omega)$, $X(\omega)$, *and* $Y(\omega)$ *denote the Fourier series of* $\mathbf{h}$, $\mathbf{x}$, *and* $\mathbf{y}$, *respectively. Then*

$$Y(\omega) = H(\omega)X(\omega) \tag{5.4}$$

□

Proof of Theorem 5.1. We start with the Fourier series $Y(\omega)$. Our goal will be to manipulate this series so that ultimately we arrive at $H(\omega)X(\omega)$. We have

$$Y(\omega) = \sum_{n=-\infty}^{\infty} y_n e^{in\omega} = \sum_{n=-\infty}^{\infty} \left(\sum_{k=-\infty}^{\infty} h_k x_{n-k} \right) e^{in\omega}$$

What we need to do next is interchange the sums. To interchange infinite sums we need to know that the terms in the series are suitably well behaved (i.e. decay rapidly to zero). We assume in this chapter that all bi-infinite sequence we convolve decay rapidly enough to ensure that the interchanging of sums is valid. The reader interested in the exact details of this part of the argument should review an analysis text such as Rudin [63].

$$Y(\omega) = \sum_{k=-\infty}^{\infty} h_k \left(\sum_{n=-\infty}^{\infty} x_{n-k} e^{in\omega} \right)$$

The inner sum looks a bit like the Fourier series for $\mathbf{x}$ — the indices don't match up. So let's change the index on the inner sum and set $j = n - k$. Then $n = j + k$. We substitute these values and write

$$Y(\omega) = \sum_{k=-\infty}^{\infty} h_k \left(\sum_{n=-\infty}^{\infty} x_{n-k} e^{in\omega} \right)$$

$$= \sum_{k=-\infty}^{\infty} h_k \left(\sum_{j=-\infty}^{\infty} x_j e^{i(j+k)\omega} \right)$$

$$= \sum_{k=-\infty}^{\infty} h_k \left(\sum_{j=-\infty}^{\infty} x_j e^{ij\omega} e^{ik\omega} \right)$$

$$= \sum_{k=-\infty}^{\infty} h_k e^{ik\omega} \left(\sum_{j=-\infty}^{\infty} x_j e^{ij\omega} \right)$$

But the inner sum is simply $X(\omega)$, so that

$$Y(\omega) = \sum_{k=-\infty}^{\infty} h_k e^{ik\omega} X(\omega)$$

and if we factor out the $X(\omega)$, we see that the remaining sum is $H(\omega)$ and the theorem is proved. $\square$

An immediate observation we should make is that the complicated operation of convolution turns into simple multiplication in the Fourier domain. To get a better feel for the convolution theorem, let's look at the following example.

Example 5.3 (Using the Convolution Theorem). *Let* $\mathbf{h}$ *be the bi-infinite sequence whose only nonzero components are* $h_0 = 1$, $h_1 = 2$, *and* $h_2 = 1$. *Let* $\mathbf{g}$ *be the bi-infinite sequence whose terms are all zero except for* $g_0 = g_1 = 1$. *Consider the convolution product* $\mathbf{y} = \mathbf{g} * \mathbf{h}$. *Find the components of the sequence* $\mathbf{y}$ *and then use them to find* $Y(\omega)$. *Then form* $H(\omega)$ *and* $G(\omega)$ *and verify the convolution theorem for these sequences.*
Solution. *We have*

$$y_n = \sum_{k=0}^{2} h_k g_{n-k} = h_0 g_n + h_1 g_{n-1} + h_2 g_{n-2} = g_n + 2g_{n-1} + g_{n-2}$$

Since $g_n = 0$ unless $n = 0, 1$ the only time $y_n \neq 0$ is when $n = 0, 1, 2, 3$. In these cases we have

$$y_0 = g_0 + 2g_{-1} + g_{-2} = 1 \qquad y_1 = g_1 + 2g_0 + g_{-1} = 3$$
$$y_2 = g_2 + 2g_1 + g_0 = 3 \qquad y_3 = g_3 + 2g_2 + g_1 = 1$$

Thus, the Fourier series for **y** *is*

$$Y(\omega) = 1 + 3e^{i\omega} + 3e^{2i\omega} + e^{3i\omega}$$

The Fourier series for **h** *and* **g** *are*

$$H(\omega) = 1 + 2e^{i\omega} + e^{2i\omega} \qquad and \qquad G(\omega) = 1 + e^{i\omega}$$

We compute the product

$$\begin{aligned} H(\omega)G(\omega) &= (1 + 2e^{i\omega} + e^{2i\omega})(1 + e^{i\omega}) \\ &= (1 + 2e^{i\omega} + e^{2i\omega}) + (e^{i\omega} + 2e^{2i\omega} + e^{3i\omega}) \\ &= 1 + 3e^{i\omega} + 3e^{2i\omega} + e^{3i\omega} \end{aligned}$$

to see that, indeed, $Y(\omega) = H(\omega)G(\omega)$. $\qquad\qquad\square$

PROBLEMS

5.1 Let $h^n = \underbrace{h * h * \cdots * h}_{n+1 \text{ times}}$, where $h_0 = h$ and h is the bi-infinite sequence given in Example 5.1(c). In the example we computed h^1 and h^2. Compute h^3 and h^4. Do you see a pattern? Can you write down the components of h^n for general n? Have you seen this pattern in any of your other mathematics courses?

5.2 Let d be the bi-infinite sequence whose terms are given by

$$d_k = \begin{cases} \frac{1}{6}, & 1 \le k \le 6 \\ 0, & \text{otherwise} \end{cases}$$

Compute $y = d * d$. If we think of d as a sequence whose kth component gives the probability of rolling a k with a fair die, how could you interpret y? What about $y * d$?

5.3 This problem is mostly just for fun. Using paper and pencil, verify that $212 \cdot 231 = 48{,}972$. I mostly want you to see that no carrying took place. Now $212 = 2 \cdot 10^0 + 1 \cdot 10^1 + 2 \cdot 10^2$ and we could easily write 231 in an analogous manner. Let's load the coefficient of 10^k from 212 into the kth component of the bi-infinite sequence x ($x_0 = 2$ is given in bold type), so that

$$x = (\ldots, 0, 0, \mathbf{2}, 1, 2, 0, 0, \ldots)$$

Let's do the same for 231 and load these coefficients into y ($y_0 = 1$ is given in bold type), so that

$$y = (\ldots, 0, 0, \mathbf{1}, 3, 2, 0, 0, \ldots)$$

Now convolve $z = x * y$. What can you claim about the components of z?

5.4   Let x and y be bi-infinite sequences with the components of y given by $y_n = x_{n+1}$.

(a) Find $Y(\omega)$ in terms of $X(\omega)$.

(b) Use the result in part (a) to find a sequence h such that $y = h * x$.

5.5   Let e^0 be the bi-infinite sequence defined by

$$e_k^0 = \begin{cases} 1, & k = 0 \\ 0, & \text{otherwise} \end{cases}$$

Show that for any bi-infinite sequence x, we have $e^0 * x = x$.

5.6   Let e^m be the bi-infinite sequence defined by

$$e_k^m = \begin{cases} 1, & k = m \\ 0, & \text{otherwise} \end{cases}$$

and let x be any bi-infinite sequence.

(a) Compute $y = e^m * x$.

(b) Find $E_m(\omega)$ and verify that the convolution theorem holds for these sequences.

5.7   Show that the convolution product is commutative (i.e., $h * x = x * h$).

5.8   Use the method of shifting inner products to compute the following convolution products:

(a) $h * x$, where the only nonzero elements of h are $h_{-1} = -1, h_0 = 2$, and $h_1 = 3$ and the only nonzero elements of x are $x_0 = 5$ and $x_1 = 7$.

(b) $h * x$, where $x_k = 1$ for $k \in \mathbb{Z}$ and the only nonzero elements of h are $h_0 = .299, h_1 = .587$, and $h_2 = .114$.

(c) $h * h$, where the only nonzero elements of h are $h_{-1} = h_1 = \frac{1}{4}$ and $h_0 = \frac{1}{2}$.

(d) In Problem 5.7 you showed that convolution is commutative. Swap the roles of h and x from part (a) in the shifting inner product method and verify the commutative property for this example.

5.9   Suppose that the only nonzero components of h are $h_0, h_1, \ldots, h_n$ and the only nonzero elements of x are $x_0, x_1, \ldots, x_m$, where $m > n > 0$. Suppose further that $h_0, \ldots, h_n, x_0, \ldots, x_m > 0$. Let $y = h * x$. Use the method of shifting inner products to determine which components of y are nonzero.

5.10   Find a bi-infinite sequence h so that when we compute $y = h * x$ (for any bi-infinite sequence x), we have y_n is the average of x_{n-1}, x_{n-2}, and x_{n-3}.

5.11   Suppose that we fix a sequence h and then consider $y = h * x$ as a function of the sequence x. In this problem you will show that in this context, the convolution operator is linear. That is,

(a) for any $a \in \mathbb{R}$, we have

$$y = h * (ax) = (ah) * x$$

(b) for any sequences x^1 and x^2, we have

$$y = h * (x^1 + x^2) = h * x^1 + h * x^2$$

5.12   Let h be the vector defined in Example 5.1(c). Verify the convolution theorem for $y = h * h$.

5.13   Suppose that we create v^m from vector v by translating the components of v right m units if $m > 0$ and left m units if $m < 0$. In the case that $m = 0$, we set $v^0 = v$. Now assume for a given h and x, we compute $y = h * x$. Now form y^m and x^m. Show that $y^m = h * x^m$. That is, show that translating x by m units and convolving it with h results in translating y by m units. Show this result:

(a) by direct computation.

(b) using the convolution theorem. (*Hint:* Problem 5.6 will be helpful.)

5.14   Suppose that $Y(\omega) = \sin(2\omega)$.

(a) Find y.

(b) Write y as a convolution product of two sequences h and g. (*Hint:* Think about a trigonometric identity for $\sin(2\omega)$.)

Computer Lab

5.1   Convolution. From the text Web site, access the convolution lab. In this lab you will learn how to compute convolution products with your **CAS** and see how to use the convolution theorem to compute convolution products. This lab also contains an animation tool that helps you visualize convolution as a process. As a practical application, you will investigate how convolution can be used in audio processing.

5.2 FILTERS

In this section we take the idea of convolution a bit further toward applications in signal and digital image processing. The idea of filtering data is in many ways analogous to what some of us do with our coffeemakers in the morning. We run water through a paper filter containing ground coffee beans with the result that the paper filter allows the molecules of caffeine to pass and holds back the grounds. The result is a pot of coffee. We basically want to do the same with data, but obviously we are not going to make coffee with it! Indeed, we are interested in filtering data so that we can easily identify characteristics such as locally constant segments, noise, large differences, or other artifacts. For example, think about data compression in a naive way. We would like to replace segments of similar data with a much smaller amount of information. Filtering will help us perform this task.

Note: Filters are sequences of numbers. In this book we consider only filters comprised of real numbers.

Let's start with an example of an elementary filter.

Example 5.4 (Averaging Filter). *Given a sequence of numbers*

$$\mathbf{x} = (\ldots, x_{-2}, x_{-1}, x_0, x_1, x_2, \ldots)$$

produce a sequence $\mathbf{y}$ *where each component* y_n *of* $\mathbf{y}$ *is formed by averaging the corresponding element* x_n *of* $\mathbf{x}$ *and the preceding element* x_{n-1} *of* $\mathbf{x}$. *Then*

$$y_n = \frac{x_n + x_{n-1}}{2} \tag{5.5}$$

$\square$

If this example looks familiar to you, it should! If you look back at Example 5.1(b) in Section 5.1, you will see that the same result appeared when we computed $\mathbf{y} = \mathbf{h} * \mathbf{x}$, where $h_0 = h_1 = \frac{1}{2}$ and all other $h_k = 0$.

We will perform filtering by choosing a filter sequence $\mathbf{h}$ and then producing the output $\mathbf{y}$ by convolving it with the input sequence $\mathbf{x}$. So filtering $\mathbf{x}$ by *filter* $\mathbf{h}$ means simply to compute

$$\mathbf{y} = \mathbf{h} * \mathbf{x}$$

Obviously, how we choose $\mathbf{h}$ will entirely influence how our data $\mathbf{x}$ is affected. Think about the filter in Example 5.4. What happens when you average numbers? Generally speaking, you get one number that in some sense represents both original numbers. More specifically, when you average two identical values, the result is the same value. If you average a and $-a$, the result is zero. So we can say that this filter will do a suitably good job at reproducing consecutive values of $\mathbf{x}$ that are similar and returning values close to zero when consecutive values of $\mathbf{x}$ are close to being opposites of each other. This type of filter is very important and we will formally define this class of filters later in the section.

Causal Filter and FIR Filters

Although we have defined the nth component of the output sequence $\mathbf{y}$ by

$$y_n = \sum_{k=-\infty}^{\infty} h_k x_{n-k}$$

we are primarily interested in filters that possess certain properties. We first consider the notion of a causal filter.

Definition 5.2 (Causal Filter). *Let* $\mathbf{h}$ *be a filter. We say that* $\mathbf{h}$ *is a* causal filter *if* $h_k = 0$ *for* $k < 0$. $\qquad\square$

The $\mathbf{h}$ from Example 5.4 is an example of a causal filter. Definition 5.2 tells us how to identify a causal filter, but what does it really mean for a filter to be causal? Let's consider y_n from $\mathbf{y} = \mathbf{h} * \mathbf{x}$, where $\mathbf{h}$ is causal. We have

$$y_n = \sum_{k=-\infty}^{\infty} h_k x_{n-k} = \sum_{k=0}^{\infty} h_k x_{n-k} = h_0 x_n + h_1 x_{n-1} + h_2 x_{n-2} + h_3 x_{n-3} + \cdots$$

So if $\mathbf{h}$ is causal, then component y_n of $\mathbf{y}$ is formed from x_n and components of $\mathbf{x}$ that *precede* x_n. If you think about filtering in the context of sound, the causal requirement on $\mathbf{h}$ is quite natural. If we are receiving a digital sound and filtering it, it would make sense that we have only current and prior information about the signal.

For a large portion of this book, we will consider only causal filters. More specifically, we will primarily be interested in causal filters $\mathbf{h}$, where $h_k = 0$ for some positive integer L and $k > L > 0$. That is, our filters will typically have the form

$$\mathbf{h} = (\dots, 0, 0, h_0, h_1, h_2, \dots, h_L, 0, 0, \dots) \tag{5.6}$$

Thus, the computation of y_n reduces to the finite sum

$$y_n = \sum_{k=0}^{L} h_k x_{n-k} = h_0 x_n + h_1 x_{n-1} + \cdots + h_L x_{n-L} \tag{5.7}$$

Such causal filters with a finite number of nonzero components are often called *finite impulse response* (FIR) filters.

Definition 5.3 (Finite Impulse Response Filter). *Let* h *be a causal filter and assume* $L > 0$, $L \in \mathbb{Z}$. *If* $h_k = 0$ *for* $k > L$, *and* $h_0, h_L \neq 0$, *then we say* h *is a* Finite Impulse Response (FIR) filter *and write*

$$\mathbf{h} = (h_0, h_1, \ldots, h_L)$$

$\square$

Note: Although we have motivated filters as bi-infinite sequences, we often identify only a finite number of filter elements inside ( ). Our convention is that the only possible nonzero elements of a filter will be enclosed in the parentheses.

It is often useful to think of FIR filters as *moving inner products* — we are simply forming y_n by computing an inner product of the FIR filter and the numbers $x_n, x_{n-1}, \ldots, x_{n-L}$.

A Lowpass Filter

In this book we are interested in two types of FIR filters. The first is called a *lowpass filter* and the second is called a *highpass filter*. To properly define these types of filters, it is necessary and important to understand how the Fourier series developed in Section 4.3 relates to the idea of filtering data. We have already seen in Section 5.1 a connection between Fourier series and convolution, so the fact that the Fourier series enters into our analysis of filters should not surprise you.

Consider the filter $\mathbf{h} = (h_0, h_1, h_2) = (\frac{1}{4}, \frac{1}{2}, \frac{1}{4})$. Let's convolve this filter with the special sequences $\mathbf{a} = (\ldots, 1, 1, 1, 1, \ldots)$ and $\mathbf{b} = (\ldots, -1, 1, -1, 1, -1, 1, \ldots)$, or more precisely, $b_n = (-1)^n$, $n \in \mathbb{Z}$. If we call $\mathbf{y} = \mathbf{h} * \mathbf{a}$, we have

$$y_n = \sum_{k=0}^{2} h_k a_{n-k} = \frac{1}{4}a_n + \frac{1}{2}a_{n-1} + \frac{1}{4}a_{n-2} = \frac{1}{4} \cdot 1 + \frac{1}{2} \cdot 1 + \frac{1}{4} \cdot 1 = 1$$

and if we call $\mathbf{z} = \mathbf{h} * \mathbf{b}$, we have

$$z_n = \sum_{k=0}^{2} h_k b_{n-k} = \frac{1}{4}b_n + \frac{1}{2}b_{n-1} + \frac{1}{4}b_{n-2}$$

$$= \frac{1}{4}(-1)^n + \frac{1}{2}(-1)^{n-1} + \frac{1}{4}(-1)^{n-2}$$

$$= \begin{cases} -\frac{1}{4} + \frac{1}{2} - \frac{1}{4}, & n \text{ odd} \\ \frac{1}{4} - \frac{1}{2} + \frac{1}{4}, & n \text{ even} \end{cases} = 0$$

Now $\mathbf{a}$ is a constant sequence. Another way to view the components of $\mathbf{a}$ is to say that they do not oscillate at all. Note that in this case $\mathbf{y} = \mathbf{h} * \mathbf{a} = \mathbf{a}$ — the filter totally preserved this characteristic. On the other hand, the components of **b** oscillate term by term. And in this case, $\mathbf{z} = \mathbf{h} * \mathbf{b} = \mathbf{0}$ — the filter **h** totally annihilated the sequence **b**.

Fourier Series and Lowpass Filters

In Example 4.7 from Section 4.3, we analyzed the Fourier series generated by $\mathbf{h} = (h_0, h_1, h_2) = (\frac{1}{4}, \frac{1}{2}, \frac{1}{4})$. We computed the Fourier series

$$H(\omega) = \frac{1}{4} + \frac{1}{2}e^{i\omega} + \frac{1}{4}e^{2i\omega} = e^{i\omega}\frac{1}{2}(1 + \cos\omega)$$

and plotted the modulus $|H(\omega)|$ from $[-\pi, \pi]$. In addition to being 2π-periodic, we learned in Problem 4.32 in Section 4.3 that $|H(\omega)|$ is an even function. So we have reproduced the plot in Figure 5.4 for the interval $[0, \pi]$.

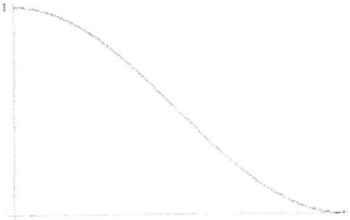

Figure 5.4 $|H(\omega)|$ on $[0, \pi]$.

Now $|H(\omega)|$ is constructed using a linear combination of complex exponential functions $e^{ik\omega}$, which by Euler's formula (4.5) are represented by $\cos(k\omega)$ and $\sin(k\omega)$, $k = 0, 1, 2$. Both k and ω play a role in the oscillatory nature of $\cos(k\omega)$, $\sin(k\omega)$ as ω ranges from 0 to π. The larger the value of k, the more oscillatory the $\cos(k\omega)$ and $\sin(k\omega)$, and at the extreme right of the interval, $\omega = \pi$, we are at a point where $\cos(k\omega)$ and $\sin(k\omega)$ have oscillated the most over the interval $[0, \pi]$. Conversely, at $\omega = 0$, no oscillation has occurred in $\cos(k\omega)$ and $\sin(k\omega)$. Now look at the graph of $|H(\omega)|$. We have

$$|H(0)| = \frac{1}{2}(1 + \cos 0) = 1 \quad \text{and} \quad |H(\pi)| = \frac{1}{2}(1 + \cos \pi) = 0$$

and the way we interpret this is that when **h** is convolved with a sequence **x**, it preserves portions of **x** where the components do not vary much, and it annihilates portions of **x** where the components do vary.

This is essentially the idea of a lowpass filter. A lowpass filter is one that preserves the homogeneity of data and annihilates or dampens portions of highly oscillatory data. You can see by the graph of $|H(\omega)|$ that the term *filter* might better apply in the Fourier domain. Near $\omega = 0$, $|H(\omega)| \approx 1$ and for $\omega \approx \pi$, we have $|H(\omega)| \approx 0$. So you can see that we will completely characterize lowpass filters by looking at a graph of the modulus of their Fourier series on $[0, \pi]$.

Ideal Lowpass Filter

The ideal lowpass filter would be for $|H(\omega)|$ to look like the graph plotted in Figure 5.5. For this filter there is a range $[0, a]$ with $0 < a < \pi$ of low frequencies where the filter h preserves low oscillatory tendencies when convolved with x and a range of high frequencies $[a, \pi]$ where h annihilates high oscillatory tendencies when convolved with x. In Problem 5.16, you are asked to compute the values of a filter h that

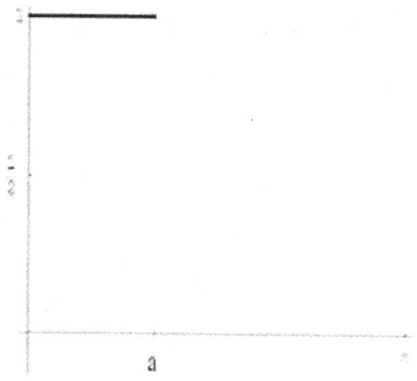

Figure 5.5 $|H(\omega)|$ on $[0, \pi]$.

generates $|H(\omega)|$ and investigate the practical considerations of convolving with h.

Lowpass Filter Defined

Below we define a lowpass filter. You will see that the definition allows for a wide variety of filters to be considered lowpass. What follows might be a definition found in a standard text on signal processing (see, e. g., Smith [67]). To develop some terminology, let's consider the $|H(\omega)|$ plotted in Figure 5.6.

Consider the vertical dashed line segment $\omega = \omega_p$ and note that $1 - \delta \leq |H(\omega)| \leq 1 + \delta$ for $0 \leq \omega \leq \omega_p$. The interval $[0, \omega_p]$ is called the *passband*. Next, note that for $\omega_s \leq \omega \leq \pi$, we have $|H(\omega)| \leq \lambda$. The interval $[\omega_s, \pi]$ is called the *stopband* and λ is referred to as the *stopband attenuation*. Finally, we call the interval $[\omega_p, \omega_s]$ the *transition band*. We now state a formal definition of a lowpass filter.

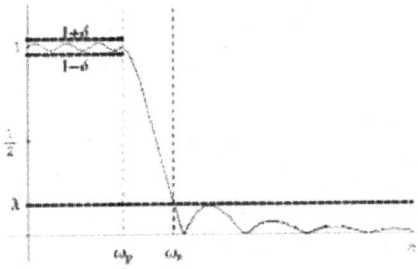

Figure 5.6 $|H(\omega)|$ on $[0, \pi]$.

Definition 5.4 (Lowpass Filter). *Let* h *be some sequence. Let* $0 < \omega_p \leq \omega_s < \pi$ *and suppose that there exists* $0 < \delta < \frac{1}{2}$, *with* $1 - \delta \leq |H(\omega)| \leq 1 + \delta$ *for* $0 \leq \omega \leq \omega_p$ *and an* $0 < \lambda < \frac{1}{2}$, *so that for* $\omega_s \leq \omega \leq \pi$, $|H(\omega)| \leq \lambda$. *Then we call* h *a* lowpass filter. ☐

A good working definition is to say that if **h** is a lowpass filter, then $|H(\omega)| \approx 1$ for $0 \leq \omega \leq \omega_p$ for some $0 < \omega_p < \pi$ and $H(\omega) \approx 0$ for $\omega_s \leq \omega \leq \pi$ where we require that $\omega_p \leq \omega_s < \pi$. For our purposes we make the following requirement of lowpass filters.

Condition 5.1 (Lowpass Filter). *Suppose that* **h** *is a lowpass filter. Then*

$$\boxed{|H(0)| = 1 \qquad and \qquad H(\pi) = 0} \tag{5.8}$$

☐

Before we look at some other examples of lowpass filters, let's take another look at Condition 5.1 with the assumption that **h** is an FIR filter. In this case, for some positive integer L we have

$$H(\omega) = \sum_{k=0}^{L} h_k e^{ik\omega}$$

so that

$$|H(0)| = \left| \sum_{k=0}^{L} h_k e^{ik\cdot 0} \right| = \left| \sum_{k=0}^{L} h_k \right|$$

and we see that

$$\boxed{|H(0)| = 1 \qquad \text{if and only if} \qquad \sum_{k=0}^{L} h_k = \pm 1} \tag{5.9}$$

In Problem 5.15 you will show that

$$
H(\pi) = 0 \quad \text{if and only if} \quad \sum_{k=0}^{L} (-1)^k h_k = 0 \tag{5.10}
$$

Let's look at some examples of lowpass filters that satisfy Condition 5.1.

Example 5.5 (Lowpass Filters). *Show that each of the filters below satisfies Condition 5.1 and plot $|H(\omega)|$ for each filter.*

(a) $\mathbf{h}^1 = (\frac{1}{2}, \frac{1}{2})$.

(b) $\mathbf{h}^2 = (\frac{2}{3}, \frac{1}{2}, -\frac{1}{6})$.

(c) $\mathbf{h}^3 = (\frac{1+\sqrt{3}}{8}, \frac{3+\sqrt{3}}{8}, \frac{3-\sqrt{3}}{8}, \frac{1-\sqrt{3}}{8})$.

Solution. *For part (a), we use Problem 4.18 from Section 4.2 and the ideas from Example 4.7 to write*

$$
H^1(\omega) = \frac{1}{2} + \frac{1}{2}e^{i\omega} = e^{i\omega/2}\cos(\frac{\omega}{2})
$$

Now we can easily verify that

$$
H^1(0) = e^{i\cdot 0/2}\cos(0) = 1 \quad and \quad H^1(\pi) = e^{i\pi/2}\cos(\frac{\pi}{2}) = 0
$$

It also follows that

$$
|H^1(\omega)| = |e^{i\pi/2}| \cdot |\cos(\frac{\omega}{2})| = \cos(\frac{\omega}{2})
$$

and we have plotted it in Figure 5.7(a).
For part (b), we utilize (5.9) and (5.10) to compute

$$
H^2(0) = \frac{2}{3} + \frac{1}{2} + \left(-\frac{1}{6}\right) = 1 \quad and \quad H^2(\pi) = \frac{2}{3} - \frac{1}{2} + \left(-\frac{1}{6}\right) = 0
$$

There is no nice simplification for $|H^2(\omega)|$ in this case — we have plotted it in Figure 5.7(b).
Finally, for part (c) we have, using (5.9) and (5.10),

$$
H^3(0) = \frac{1+\sqrt{3}}{8} + \frac{3+\sqrt{3}}{8} + \frac{3-\sqrt{3}}{8} + \frac{1-\sqrt{3}}{8} = 1
$$

and

$$
H^3(\pi) = \frac{1+\sqrt{3}}{8} - \frac{3+\sqrt{3}}{8} + \frac{3-\sqrt{3}}{8} - \frac{1-\sqrt{3}}{8} = 0
$$

We have plotted $|H^3(\omega)|$ in Figure 5.7(c).   □

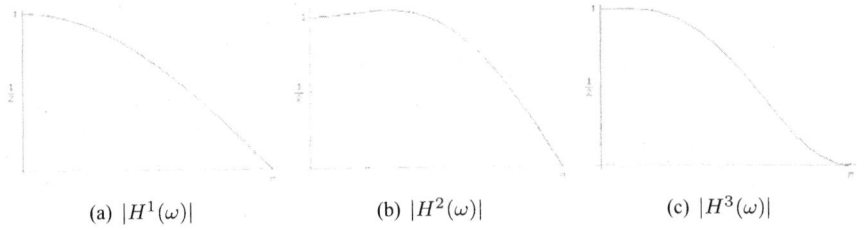

(a) $|H^1(\omega)|$ (b) $|H^2(\omega)|$ (c) $|H^3(\omega)|$

Figure 5.7 Moduli of the Fourier series from Example 5.5 plotted on $[0, \pi]$.

A Highpass Filter

FIR lowpass filters are popular tools in image and signal processing. As we have learned, they preserve locally constant segments of data and annihilate or dampen oscillatory portions of data. Now we study a filter that does just the opposite.

As you might guess, *highpass* filters are designed to do the exact opposite of lowpass filters when convolved with a sequence. We want highpass filters to annihilate locally constant portions of the data and preserve those segments of data that are oscillatory. Here is a simple example of a highpass filter.

Example 5.6 (Highpass Filter). *Let* $\mathbf{g} = (g_0, g_1) = (\frac{1}{2}, -\frac{1}{2})$.

(a) *Compute* $\mathbf{y} = \mathbf{g} * \mathbf{a}$ *and* $\mathbf{z} = \mathbf{g} * \mathbf{b}$, *where* $\mathbf{a} = (\ldots, 1, 1, 1, 1, 1, \ldots)$ *and* $\mathbf{b} = (\ldots, -1, 1, -1, 1, -1, 1, \ldots)$. *More precisely,* $b_n = (-1)^n$.

(b) *Compute* $G(\omega)$ *and plot* $|G(\omega)|$.

Solution. *For part (a) we have*

$$y_n = \sum_{k=0}^{1} g_k a_{n-k} = \frac{1}{2}a_n - \frac{1}{2}a_{n-1} = \frac{1}{2}(1) - \frac{1}{2}(1) = 0$$

and

$$z_n = \sum_{k=0}^{1} g_k b_{n-k} = \frac{1}{2}b_n - \frac{1}{2}b_{n-1} = \frac{1}{2}(-1)^n - \frac{1}{2}(-1)^{n-1}$$

$$= \begin{cases} \frac{1}{2}(-1) - \frac{1}{2}(1), & n \text{ odd} \\ \frac{1}{2}(1) - \frac{1}{2}(-1), & n \text{ even} \end{cases} = \begin{cases} -1, & n \text{ odd} \\ 1, & n \text{ even} \end{cases}$$

$$= (-1)^n = b_n$$

For part (b), we compute $G(\omega)$ by utilizing Problem 4.18 from Section 4.2 and the ideas from Example 4.7 to write

$$G(\omega) = \frac{1}{2} - \frac{1}{2}e^{i\omega} = -ie^{i\omega/2}\sin(\frac{\omega}{2})$$

Since $|-i| = 1$ and $|e^{i\omega/2}| = 1$, we have

$$|G(\omega)| = |\sin(\frac{\omega}{2})| = \begin{cases} \sin(\frac{\omega}{2}), & 0 \le \omega \le \pi \\ -\sin(\frac{\omega}{2}), & -\pi \le \omega < 0 \end{cases}$$

A plot of $|G(\omega)|$ is given in Figure 5.8.   □

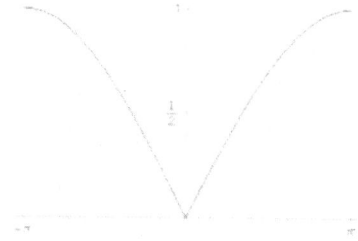

Figure 5.8 $|G(\omega)|$ on $[0, \pi]$.

Highpass Filter Defined

A good working definition is to say that if **g** is a highpass filter, then $|G(\omega)| \approx 0$ for $0 \le \omega \le \omega_p$ for some $0 < \omega_p < \pi$, and $|G(\omega)| \approx 1$ for $\omega_s \le \omega \le \pi$ where we require that $\omega_p \le \omega_s < \pi$. We also list conditions analogous to Condition 5.1 for highpass filters.

Condition 5.2 (Highpass Filter). *Suppose* **g** *is a highpass filter. Then*

$$\boxed{G(0) = 0 \quad and \quad |G(\pi)| = 1}$$ (5.11)

□

For the sake of completeness, we define a highpass filter. A good signal-processing book (see, e. g., Smith [67] for example) will contain more information on highpass filters. The terminology is exactly the same as that developed prior to Definition 5.4.

Definition 5.5 (Highpass Filter). *Let* **g** *be some sequence. Let* $0 < \omega_p \le \omega_s < \pi$ *and suppose that there exists* $0 < \lambda < \frac{1}{2}$ *so that* $|G(\omega)| \le \lambda$ *for* $0 \le \omega \le \omega_p$ *and a* $0 < \delta < \frac{1}{2}$ *with* $1 - \delta \le |G(\omega)| \le 1 + \delta$ *for* $\omega_s \le \omega \le \pi$. *Then we call* **g** *a* highpass filter. $\qquad\square$

As you will see in subsequent chapters, we will use lowpass and highpass filters in tandem. We design lowpass filters to provide an approximation of the locally constant portions of the original signal, and highpass filters to represent the difference between the original signal and the lowpass approximation. Such a tandem is called a *filter bank*. In addition to the filter bank design, we also want our filters to be relatively short. Shorter filters means faster computations. In many applications we need to be able to invert our filter processes — this topic is addressed in Section 5.3. We conclude this section with a brief discussion of an additional property that is important in filter design.

PROBLEMS

★**5.15** Suppose that **h** is an FIR filter that satisfies $H(\pi) = 0$. Show that this condition is equivalent to

$$\sum_{k=0}^{L} h_k(-1)^k = 0$$

5.16 Consider the ideal lowpass filter that generates $|H(\omega)|$ from Figure 5.5. Certainly, $H(\omega)$ could be the following 2π-periodically extended box function:

$$H(\omega) = \sum_{k=-\infty}^{\infty} h_k e^{ik\omega} = \begin{cases} 1, & -a \le 0 \le a \\ 0, & \text{otherwise} \end{cases}$$

where $0 < a < \pi$. We know by (4.11) from Section 4.3 that we can compute the h_k by the formula

$$h_k = \frac{1}{2\pi} \int_{-a}^{a} 1 \cdot e^{-ik\omega} \, d\omega$$

Simplify the integral above to find expressions for h_k. (*Hint:* consider the $k = 0$ case separately). Is **h** an FIR filter? If your answer is no, observe then that such an ideal filter is not computationally tractable.)

5.17 Let $\mathbf{h} = (h_0, h_1, h_2, h_3) = (\frac{1}{4}, \frac{1}{4}, \frac{1}{4}, \frac{1}{4})$.

(a) Given an input vector $\mathbf{x} = (\ldots, x_{-2}, x_{-1}, x_0, x_1, x_2, \ldots)$, let $\mathbf{y} = \mathbf{h} * \mathbf{x}$. Find a simple formula for y_n.

(b) Use your formula from part (a) to find the vector $\mathbf{y}$ when $\mathbf{x}$ is a vector whose components are zero except for $x_1 = 1$, $x_2 = 2$, $x_3 = 3$, $x_4 = 4$.

(c) Write down the Fourier series $H(\omega)$ associated with **h**.

(d) Use part(c) and Conditions 5.1 5.2 to determine whether **h** is a lowpass filter, a highpass filter, or neither.

5.18   Find a causal filter **c** such that $|C(\omega)| = \cos^2(\frac{\omega}{2})$. Is your **c** lowpass, highpass, or neither? Is your choice of **c** unique? (*Hint:* Use Problem 4.18 from Section 4.2 with $\cos(\frac{\omega}{2})$.)

5.19   Suppose that $\mathbf{h} = (h_0, \ldots, h_9)$. Moreover, assume that **h** satisfies Condition 5.1.

(a) Show that

$$h_0 + h_2 + h_4 + h_6 + h_8 = \pm\frac{1}{2} = h_1 + h_3 + h_5 + h_7 + h_9$$

(b) In general, suppose that L is an odd integer with $\mathbf{h} = (h_0, \ldots, h_L)$ and assume that **h** satisfies Condition 5.1. Show that

$$h_0 + h_2 + \cdots + h_{L-1} = \pm\frac{1}{2} = h_1 + h_3 + \cdots + h_L$$

5.20   Suppose that **h** is a causal filter with $h_k = -h_{9-k}$. Can **h** ever be a lowpass filter? Explain.

5.21   Suppose that **h** is a causal filter that satisfies $h_k = h_{7-k}$.

(a) What is the maximum number of nonzero components of **h**?

(b) What is the maximum number of distinct nonzero elements of **h**?

(c) Show that

$$|H(\omega)| = 2|e^{(7i\omega/2)}| \cdot |h_0 \cos(\frac{7\omega}{2}) + h_1 \cos(\frac{5\omega}{2})$$
$$+ h_2 \cos(\frac{3\omega}{2}) + h_3 \cos(\frac{\omega}{2})|$$
$$= 2\left|\sum_{k=0}^{3} h_k \cos(\frac{(7-2k)\omega}{2})\right|$$

(d) Can this filter ever be a highpass filter? Explain.

5.22   Suppose that **h** is a (nonzero) causal filter that satisfies $h_k = h_{L-k}$ for some positive odd integer L. Can **h** ever be highpass? Explain. (*Hint:* Try to generalize the factorization from the preceding problem.)

★**5.23**   Suppose that $\mathbf{h} = (h_0, \ldots, h_L)$ satisfies Condition 5.1, where L is an odd positive integer. Suppose that we construct the filter $\mathbf{g}$ by insisting that $G(\omega) = H(\omega + \pi)$.

(a) Is $\mathbf{g}$ lowpass, highpass, or neither?

(b) What are the coefficents g_k in terms of h_k?

(c) Use your results to find $\mathbf{g}$ when $\mathbf{h} = (\frac{1}{8}, \frac{3}{8}, \frac{3}{8}, \frac{1}{8})$ so that $\mathbf{g}$ is highpass.

(d) Note that $\mathbf{h} \cdot \mathbf{g} = 0$. In general, will we always have $\mathbf{h} \cdot \mathbf{g} = 0$ when we construct $\mathbf{g}$ as suggested by part (b)? Write down a condition for $\mathbf{h} = (h_0, h_1, \ldots, h_L)$ that guarantees $\mathbf{h} \cdot \mathbf{g} = 0$.

5.24   Find a filter $\mathbf{h} = (h_0, h_1, h_2, h_3)$ that satisfies Condition 5.1 and $H'(\pi) = 0$.

★**5.25**   Recall Problem 2.13 from Section 2.1. In terms of filters, we started with $\mathbf{h} = (h_0, h_1, \ldots, h_L)$ where L is odd and constructed $\mathbf{g} = (g_0, g_1, \ldots, g_L)$ by the rule $g_k = (-1)^k h_{L-k}$. We showed in the problem that $\mathbf{h} \cdot \mathbf{g} = 0$.

Now assume that $\mathbf{h}$ satisfies Condition 5.1. Show that $\mathbf{g}$ satisfies Condition 5.2. We use this idea repeatedly in subsequent chapters to construct a highpass filter $\mathbf{g}$ from $\mathbf{h}$ that is orthogonal to $\mathbf{h}$.

5.26   Suppose that $\mathbf{h}^1$ and $\mathbf{h}^2$ are lowpass filters and $\mathbf{g}^1$ and $\mathbf{g}^2$ are highpass filters.

(a) Show that $\mathbf{h} = \mathbf{h}^1 * \mathbf{h}^2$ is a lowpass filter.

(b) Show that $\mathbf{g} = \mathbf{g}^1 * \mathbf{g}^2$ is a highpass filter.

(c) Is $\mathbf{h}^1 * \mathbf{g}^1$ lowpass, highpass, or neither?

5.27   Consider the filter $\mathbf{h}$ where the only nonzero terms are $h_0 = h_2 = \frac{1}{5}$ and $h_1 = \frac{3}{5}$.

(a) Show that $|H(\omega)| = \frac{2}{5}(\frac{3}{2} + \cos\omega)$.

(b) Show that we can write $H(\omega)$ as $H(\omega) = e^{i\omega}|H(\omega)|$.

It turns out that we can write any complex-valued function $f(\omega)$ as

$$f(\omega) = e^{i\rho(\omega)}|f(\omega)|$$

This is analogous to using polar coordinates for representing points in $\mathbb{R}^2$. You can think of $|f(\omega)|$ as the "radius" and $\rho(\omega)$ is called the *phase angle*. The phase angle for this problem is $\rho(\omega) = \omega$, and this is a linear function. In image processing the phase angle plays an important role in filter design. It can be shown that for symmetric filters ($h_k = h_{N-k}$ as in this problem) or antisymmetric filters ($h_k = -h_{N-k}$), the phase angle is a (piecewise) linear function. For more information on phase angles, see Gonzalez and Woods [39] or Embree and Kimble [33].

Computer Lab

5.2   Filters. From the text Web site, access the `filters` lab. In this lab you will further investigate the application of highpass and lowpass filters on data sets.

5.3   CONVOLUTION AS A MATRIX PRODUCT

In the final section of Chapter 5, we look at writing the convolution product $\mathbf{h} * \mathbf{x}$ as a matrix multiplication $H\mathbf{x}$. Note that H must be a matrix of infinite dimensions! The main reason for making this connection is that it will help us with the construction of the wavelet transforms that appear in subsequent chapters. We will learn how the elements in the matrix, the convolution, and the Fourier series coefficients are all connected.

Note: In the development of convolution and filters, we have viewed the filter, input, and output as sequences and used ( ) to enclose entries of these sequences. In what follows, it is sometimes convenient to view these sequences as vectors, and in these cases we employ the vector notation $[\ ]^T$.

In many applications it is necessary not only to transform the data but to apply an inverse transformation as well. In terms of convolution, we are looking for a filter $\tilde{\mathbf{h}}$ so that if $\mathbf{y} = \mathbf{h} * \mathbf{x}$, then $\mathbf{x} = \tilde{\mathbf{h}} * \mathbf{y}$. This is the so-called *deconvolution problem*. We will learn that it is not always possible to find $\tilde{\mathbf{h}}$ so that we can recover $\mathbf{x}$, and in some of the cases where we can find $\tilde{\mathbf{h}}$, the resulting convolution product is not computationally tractable. We will see that the convolution theorem (Theorem 5.1) of Section 5.1 is very useful in helping us understand when we can find $\tilde{\mathbf{h}}$.

Convolution as a Matrix Product

Let's start by writing $\mathbf{y} = \mathbf{h} * \mathbf{x}$ as a matrix H times $\mathbf{x}$. We will consider only FIR filters $\mathbf{h} = (h_0, h_1, \ldots, h_L)$. To see how this is done, let's recall the definition of component y_n of $\mathbf{y}$:

$$y_n = \sum_{k=0}^{L} h_k x_{n-k} = h_0 x_n + h_1 x_{n-1} + h_2 x_{n-2} + \cdots + h_L x_{n-L} \qquad (5.12)$$

Now think about writing $\mathbf{y}$ as a matrix H times $\mathbf{x}$. It is fair to ask if we can even write a convolution product as a matrix. But recall from Problem 5.11 in Section 5.1 that we learned that convolution is a linear operator and a well-known theorem from linear algebra states (see, e. g. Strang [71]) that such a linear operator on a vector space can always be represented by a matrix. Let's consider the components of $\mathbf{y}, \mathbf{x}$ in $\mathbf{y} = H\mathbf{x}$:

$$
\begin{bmatrix}
\vdots \\
y_{-2} \\
y_{-1} \\
y_0 \\
y_1 \\
\vdots \\
y_n \\
\vdots
\end{bmatrix}
= H
\begin{bmatrix}
\vdots \\
x_{-2} \\
x_{-1} \\
x_0 \\
x_1 \\
\vdots \\
x_n \\
\vdots
\end{bmatrix}
$$

We see then that y_n is the result of dotting row n of H with $\mathbf{x}$. But we already know what this inner product is supposed to look like — it is given in (5.12). Consider the first term in this expression: $h_0 x_n$. In terms of an inner product, h_0 must then be the nth element in the row vector, or in terms of a matrix, h_0 is the nth element of row n. This puts h_0 on the diagonal of row n. So every row n has h_0 on the diagonal. Similarly, since x_{n-1} is multiplied by h_1, then h_1 must be the $(n-1)$th element in row n. We say that h_1 is on the first subdiagonal of H. Continuing in this manner, we see that h_{n-L} must be the $(n-L)$th element in row n, so that h_{n-L} is on the Lth subdiagonal of row n. Thus, row n consists entirely of zeros except for column n back to column $n-L$. So to form H, we simply put h_0 on the main diagonal of each row and then left from the diagonal with $h_1, h_2, \ldots, h_L$, respectively. We have

$$
H =
\begin{bmatrix}
\ddots & \ddots & \ddots & \ddots & \ddots & & & & & & & & \\
h_L & \ldots & h_2 & h_1 & \mathbf{h_0} & 0 & 0 & 0 & 0 & 0 & 0 & 0 & 0 \\
0 & h_L & \ldots & h_2 & h_1 & \mathbf{h_0} & 0 & 0 & 0 & 0 & 0 & 0 & 0 \\
\ldots & 0 & 0 & h_L & \ldots & h_2 & h_1 & \mathbf{h_0} & 0 & 0 & 0 & 0 & 0 & 0 & 0 & \ldots \\
0 & 0 & 0 & h_L & \ldots & h_2 & h_1 & \mathbf{h_0} & 0 & 0 & 0 & 0 & 0 \\
0 & 0 & 0 & 0 & h_L & \ldots & h_2 & h_1 & \mathbf{h_0} & 0 & 0 & 0 & 0 \\
& & & & \ddots & \ddots & \ddots & \ddots & \ddots & & & & \ddots
\end{bmatrix}
\tag{5.13}
$$

where we have indicated the main diagonal using bold face symbols.

There are many properties obeyed by H. Since $\mathbf{h}$ is causal, no elements will appear in any row right of the diagonal. We say that H is *lower triangular* since the elements above the diagonal are zeros. Mathematically, we say that H is lower triangular since $H_{ij} = 0$ for $j > i, i, j \in \mathbb{Z}$. H is also called a *banded matrix*. Along with the main diagonal, the first subdiagonal through the Lth subdiagonal have nonzero elements, but those are the only *bands* that are not zero.

Of course, it is computationally untractable to consider utilizing a matrix with infinite dimensions. Moreover, it is realistic to insist that all input sequences $\mathbf{x}$ are of finite length, so at some point we need to think about truncating H. We will discuss this in Chapter 6.

Example 5.7 (Convolution as a Matrix). *Let* $\mathbf{h} = (h_0, h_1, h_2) = (\frac{1}{3}, \frac{1}{2}, \frac{1}{6})$. *Write down the matrix H for this filter.*

Solution. *We have*

$$
H = \begin{bmatrix}
\ddots & & \ddots & & \ddots & & \ddots & & & & & & & \\
& 0 & 0 & \frac{1}{6} & \frac{1}{2} & \frac{1}{3} & 0 & 0 & 0 & 0 & 0 & 0 & 0 & 0 \\
& 0 & 0 & 0 & \frac{1}{6} & \frac{1}{2} & \frac{1}{3} & 0 & 0 & 0 & 0 & 0 & 0 & 0 \\
\cdots & 0 & 0 & 0 & 0 & \frac{1}{6} & \frac{1}{2} & \frac{1}{3} & 0 & 0 & 0 & 0 & 0 & 0 & \cdots \\
& 0 & 0 & 0 & 0 & 0 & \frac{1}{6} & \frac{1}{2} & \frac{1}{3} & 0 & 0 & 0 & 0 & 0 \\
& 0 & 0 & 0 & 0 & 0 & 0 & \frac{1}{6} & \frac{1}{2} & \frac{1}{3} & 0 & 0 & 0 & 0 \\
& & & & & & \ddots & & \ddots & & \ddots & & & \ddots
\end{bmatrix}
$$

We have marked the main diagonal element $h_0 = \frac{1}{3}$ in bold. □

Deconvolution

As we shall see in subsequent chapters, data convolved with suitable filters will realize a compression rate that is better than the rate achieved by compressing the data in its raw form. Suppose that a colleague computed $\mathbf{y} = \mathbf{h} * \mathbf{x}$, used a compression algorithm to compress the elements in sequence $\mathbf{y}$ to obtain $\tilde{\mathbf{y}}$, then sent $\tilde{\mathbf{y}}$ to us via the Internet. We would apply a decompression routine to obtain $\mathbf{y}$. How could we get $\mathbf{x}$? If H were finite-dimensional and nonsingular, we could apply H^{-1} to $\mathbf{y}$ to get $\mathbf{x}$. But the idea of computing H^{-1} in the case when H is infinite-dimensional is a bit daunting. We turn instead to Fourier series for help. Recall the convolution theorem:

$$\text{If } \mathbf{y} = \mathbf{h} * \mathbf{x}, \text{ then } Y(\omega) = H(\omega)X(\omega).$$

If $H(\omega) \neq 0$, we can divide both sides by it and arrive at

$$X(\omega) = \frac{1}{H(\omega)} Y(\omega) \tag{5.14}$$

Since we have $\mathbf{y}$ and know $\mathbf{h}$, we could form and compute $\frac{1}{H(\omega)} Y(\omega)$ to obtain $X(\omega)$. But then we would have to try to write $X(\omega)$ as a Fourier series. This might be problematic, but if we could do it, we would have as coefficients of the Fourier series for $X(\omega)$ the sequence $\mathbf{x}$.

Alternatively, we could try to find a Fourier series representation for the 2π-periodic function $\frac{1}{H(\omega)}$. Suppose that we could write down a Fourier series (with coefficient sequence $\mathbf{m}$) for this function:

$$\frac{1}{H(\omega)} = \sum_{k=-\infty}^{\infty} m_k e^{ik\omega}$$

Then by the convolution theorem we would know that $\mathbf{x}$ could be obtained by convolving $\mathbf{m}$ with $\mathbf{y}$. This process is known as *deconvolution*.

Note that the convolution theorem also tells us *when* it is possible to perform a deconvolution. If $H(\omega) = 0$ for some ω, then we cannot perform the division and arrive at (5.14). In particular, note that if a filter $\mathbf{h}$ satisfies the lowpass filter Condition 5.1, then $H(\pi) = 0$, and if $\mathbf{h}$ satisfies the highpass filter Condition 5.2, then $H(0) = 0$, and in either case, we can't solve for $X(\omega)$.

This makes some practical sense if we consider the averaging filter $\mathbf{h} = (h_0, h_1) = (\frac{1}{2}, \frac{1}{2})$. In this case (see Example 5.5), $\mathbf{y} = \mathbf{h} * \mathbf{x}$ has components $y_n = \frac{x_n}{2} + \frac{x_{n-1}}{2}$ so that y_n is the average of x_n and its previous component. Think about it: If you were told the average of two numbers was 12, could you determine what two numbers produced this average? The answer is no — there are an infinite number of possibilities. So it makes sense that we can't deconvolve with this sequence or any lowpass or highpass sequence.

Example 5.8 (Deconvolution). *Consider the FIR filter $\mathbf{h} = (h_0, h_1) = (2, 1)$. If possible, find the Fourier series for $\frac{1}{H(\omega)}$ and write down the deconvolution filter $\mathbf{m}$.*
Solution. *It is quite simple to draw $H(\omega) = 2 + e^{i\omega}$ in the complex plane. We recognize from Chapter 4 that $e^{i\omega}$ is a circle centered at 0 with radius 1. By adding 2 to it, we're simply moving the circle 2 units right on the real axis. So $H(\omega)$ looks like the graph plotted in Figure 5.9.*

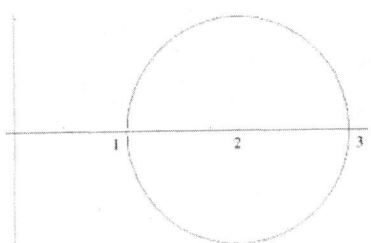

Figure 5.9 $H(\omega) = 2 + e^{i\omega}$.

Since $H(\omega) \neq 0$ for all $\omega \in \mathbb{R}$, we can write

$$X(\omega) = \frac{1}{H(\omega)} Y(\omega) = \frac{1}{2 + e^{i\omega}} Y(\omega)$$

To write $\frac{1}{2+e^{i\omega}}$ as a Fourier series, we recall the MacLaurin's series from calculus for the function $\frac{1}{1+x}$:

$$\frac{1}{1+x} = 1 - x + x^2 - x^3 + x^4 - x^5 \pm \cdots = \sum_{k=0}^{\infty} (-1)^k x^k \qquad (5.15)$$

This series converges as long as $|x| < 1$. We can factor the 2 out of the denominator of $\frac{1}{2+e^{i\omega}}$ to obtain

$$\frac{1}{2 + e^{i\omega}} = \frac{1}{2(1 + \frac{1}{2}e^{i\omega})} = \frac{1}{2} \cdot \frac{1}{1 + \frac{1}{2}e^{i\omega}}$$

Let $t = \frac{1}{2}e^{i\omega}$. Then $|t| = \frac{1}{2} < 1$, so we use (5.15) with this choice of t to write

$$\frac{1}{2 + e^{i\omega}} = \frac{1}{2} \cdot \frac{1}{1 + \frac{1}{2}e^{i\omega}} = \frac{1}{2} \sum_{k=0}^{\infty} (-1)^k \left(\frac{1}{2}e^{i\omega} \right)^k = \sum_{k=0}^{\infty} (-1)^k \frac{1}{2^{k+1}} e^{ik\omega}$$

So we see that

$$m_k = \begin{cases} (-1)^k/2^{k+1}, & k \geq 0 \\ 0, & k < 0 \end{cases}$$

or $\mathbf{m}$ is the infinite-length causal filter $\mathbf{m} = \left(\frac{1}{2}, -\frac{1}{4}, \frac{1}{8}, -\frac{1}{16}, \frac{1}{32}, -\frac{1}{64}, \ldots \right)$.

Although we can, indeed, write down the deconvolution filter $\mathbf{m}$, the fact that $\mathbf{m}$ is of infinite length means that we will have to truncate the convolution $\mathbf{m} * \mathbf{y}$, and thus we will not exactly recover $\mathbf{x}$. $\qquad\square$

Summary

What have we accomplished in this section? We've illustrated how to write a convolution product $\mathbf{h} * \mathbf{x}$ as a matrix H times $\mathbf{x}$. In the case where $\mathbf{h}$ is causal, H is lower triangular.

We also discussed how to deconvolve our convolution product. Unfortunately whenever it is possible to find the deconvolution sequence $\mathbf{m}$, it is not likely that $\mathbf{m}$ will be a FIR filter. To understand this last statement, consider

$$H(\omega) = \sum_{k=0}^{L} h_k e^{ik\omega}$$

and make the substitution $z = e^{i\omega}$. We then have

$$H(\omega) = \sum_{k=0}^{L} h_k z^k$$

so that $H(\omega)$ looks like a polynomial of degree L. Then consider

$$\frac{1}{H(\omega)} = \frac{1}{\displaystyle\sum_{k=0}^{L} h_k z^k}$$

For $\frac{1}{H(\omega)}$ to produce a FIR filter **m**, we need $\frac{1}{\sum\limits_{k=0}^{L} h_k z^k}$ to be a polynomial of some

finite degree. This rarely happens (see Problem 5.32), so in the instances where we can produce **m**, it is usually the case that **m** contains an infinite number of nonzero elements. If we were to implement deconvolution on the computer in such a case, we would need to truncate **m** in some fashion. Thus, we will never exactly reconstruct **x**. What is more frustrating is that lowpass and highpass filters always satisfy $H(\omega) = 0$ (either at $\omega = 0$ or $\omega = \pi$).

The idea of using convolution to process signals is a good one — it just needs some adjustments. The basic idea is to put a lowpass filter and a highpass filter together in what is called a *filter bank*. With some more modifications we will be able to produce highly efficient transformations that are useful in a wide variety of applications.

PROBLEMS

5.28 Consider the matrix

$$
H = \begin{bmatrix}
\ddots & & \ddots & \ddots & \ddots & & & & & \\
 & 0 & 0 & -\frac{1}{4} & \frac{1}{2} & \frac{3}{4} & 0 & 0 & 0 & 0 \\
 & 0 & 0 & 0 & -\frac{1}{4} & \frac{1}{2} & \frac{3}{4} & 0 & 0 & 0 \\
\cdots & 0 & 0 & 0 & 0 & -\frac{1}{4} & \frac{1}{2} & \frac{3}{4} & 0 & 0 & \cdots \\
 & 0 & 0 & 0 & 0 & 0 & -\frac{1}{4} & \frac{1}{2} & \frac{3}{4} & 0 \\
 & 0 & 0 & 0 & 0 & 0 & 0 & -\frac{1}{4} & \frac{1}{2} & \frac{3}{4} & 0 \\
 & & & & & & \ddots & \ddots & \ddots & \ddots
\end{bmatrix}
$$

where the main diagonal is denoted using boldface numbers.

(a) Write down the corresponding filter **h**.

(b) Is **h** lowpass, highpass, or neither?

(c) For infinite-length vector **x**, compute $\mathbf{h} * \mathbf{x}$. Verify that the answer is the same when we compute $H \cdot \mathbf{x}$.

(d) Write down the Fourier series $H(\omega)$.

5.29 Let $\mathbf{h} = (2, 0, 1)$.

(a) Write down the convolution matrix H.

(b) If possible, find the Fourier series for $\frac{1}{H(\omega)}$ and write down the deconvolution filter **m**.

5.30 Repeat Problem 5.29 for $\mathbf{g} = \left(\frac{3}{4}, \frac{1}{4}\right)$.

5.31 Suppose that $\mathbf{h} = (h_0, h_1, \ldots, h_L)$ is an FIR filter with $h_k = h_{L-k}$, $k \in \mathbb{Z}$.

(a) Suppose that L is an odd integer. Show that there is always at least one value for which $H(\omega) = 0$.

(b) Suppose that L is an even integer. Is the claim made in part (a) true in this case as well? Either prove that it is or find a filter so that $H(\omega) \neq 0$ for all $\omega \in \mathbb{R}$.

5.32   In this problem you will learn more about which FIR filters can be used as deconvolution filters.

(a) Let $\mathbf{h} = (h_0, h_1)$, where h_0 and h_1 are nonzero real numbers. Is it possible to construct an FIR filter $\mathbf{m}$ so that the associated Fourier series $M(\omega)$ satisfies $M(\omega) = \frac{1}{H(\omega)}$?

(b) Now suppose that $\mathbf{h} = (h_0, h_1, h_2)$, where at least two of h_0, h_1, and h_2 are nonzero. Explain why the filter $\mathbf{m}$ associated with Fourier series $M(\omega) = \frac{1}{H(\omega)}$ cannot be an FIR filter.

(c) Repeat part (b) using $\mathbf{h} = (h_0, \ldots, h_L)$, where at least two of the elements of $\mathbf{h}$ are nonzero.

(d) Can you now characterize the FIR filters $\mathbf{h}$ for which the filter $\mathbf{m}$ associated with $M(\omega) = \frac{1}{H(\omega)}$ is an FIR filter?

CHAPTER 6

THE HAAR WAVELET TRANSFORMATION

We are now ready to start building wavelet transformations! We have most of the tools we need. We begin with a very simple lowpass filter and an associated highpass filter. Since both can be represented by matrices, we start there. We combine these matrices and after some more straightforward modifications we essentially have our transformation. What's more, the steps we take are utilized in subsequent chapters — we just start with a different lowpass and highpass filter pair(s).

Hopefully, you have an idea of what obstacles we must overcome to build a nice transformation to use in applications. We must deal with the invertibility issue as well as the fact that on a computer, there are no infinite-length sequences — we must think about truncating the matrix we build. Ultimately, it is our goal to produce a transformation that is very useful in applications — toward this end, we insist that our transformation will effectively decompose the input data.

We motivate our basic construction in the next section. In Section 6.2 we fine-tune the transformation so that it is even more useful in applications. Our work in this chapter is to construct a matrix that we can apply to a finite-length sequence or vector. In Section 6.3 we learn that it is very easy to generalize the ideas of the previous two sections to build a two-dimensional transformation. The immediate application here is to apply a two-dimensional transformation to images. We conclude the chapter with two applications: data compression and image edge detection.

6.1 CONSTRUCTING THE HAAR WAVELET TRANSFORMATION

Let's start with the lowpass FIR filter $\mathbf{h} = (h_0, h_1) = (\frac{1}{2}, \frac{1}{2})$. We studied this filter in much detail in Chapter 5. The filter takes an input sequence $\mathbf{x}$ and produces, via $\mathbf{h} * \mathbf{x}$, a sequence $\mathbf{y}$ of averages of values of $\mathbf{x}$. In particular, we have

$$y_n = \frac{1}{2}x_n + \frac{1}{2}x_{n-1}$$

We could write down the convolution matrix

$$H = \begin{bmatrix} \ddots & & \ddots & & \ddots & \ddots & & & & & & & \\ & 0 & 0 & 0 & \frac{1}{2} & \frac{1}{2} & 0 & 0 & 0 & 0 & 0 & 0 & 0 & 0 \\ & 0 & 0 & 0 & 0 & \frac{1}{2} & \frac{1}{2} & 0 & 0 & 0 & 0 & 0 & 0 & 0 \\ \cdots & 0 & 0 & 0 & 0 & 0 & \frac{1}{2} & \frac{1}{2} & 0 & 0 & 0 & 0 & 0 & 0 & \cdots \\ & 0 & 0 & 0 & 0 & 0 & 0 & \frac{1}{2} & \frac{1}{2} & 0 & 0 & 0 & 0 & 0 \\ & 0 & 0 & 0 & 0 & 0 & 0 & 0 & \frac{1}{2} & \frac{1}{2} & 0 & 0 & 0 & 0 \\ & & & & & & & & & \ddots & \ddots & & \ddots \end{bmatrix}$$

but we know that we can't invert the process. As stated in Chapter 5, if $\mathbf{y}$ is a sequence of averages, then there are an infinite number of sequences that could have produced $\mathbf{y}$. Something as simple as $\mathbf{y} = (\ldots, 1, 1, 1, 1, 1, \ldots)$ could have been produced by applying $\mathbf{h}$ to $\mathbf{x} = (\ldots, 1, 1, 1, 1, 1, \ldots)$ or $\mathbf{x} = (\ldots, 2, 0, 2, 0, 2, 0, 2, 0, \ldots)$.

To reiterate, suppose we knew that $y_n = \frac{1}{2}x_n + \frac{1}{2}x_{n-1} = 6$. Then we'd have no way of knowing what x_n and x_{n-1} averaged to 6. We need more information. We know that if we drew x_n and x_{n-1} on a number line, then 6 would be exactly in the middle of the two numbers. So if we also knew the distance from x_n to 6 (or the distance from x_{n-1} to 6), then we could name the values of x_n and x_{n-1}. For example, suppose that in addition to 6, you knew that this distance was 2. Then our two numbers must be $6 - 2 = 4$ and $6 + 2 = 8$.

How do we find this distance? We simply compute

$$z_n = \frac{x_n - x_{n-1}}{2} = \frac{x_n}{2} - \frac{x_{n-1}}{2} = \frac{1}{2}x_n - \frac{1}{2}x_{n-1}$$

Do you recognize z_n? It is nothing more than the components of the sequence $\mathbf{z}$ obtained by convolving $\mathbf{x}$ with the highpass filter $\mathbf{g} = (g_0, g_1) = (\frac{1}{2}, -\frac{1}{2})$ from

Example 5.6 in Section 5.2. The matrix representation for this filter is

$$
G = \begin{bmatrix}
\ddots & & \ddots & \ddots & \ddots & & & & & & & & & \\
& 0 & 0 & 0 & -\frac{1}{2} & \frac{1}{2} & 0 & 0 & 0 & 0 & 0 & 0 & 0 & 0 \\
& 0 & 0 & 0 & 0 & -\frac{1}{2} & \frac{1}{2} & 0 & 0 & 0 & 0 & 0 & 0 \\
\cdots & 0 & 0 & 0 & 0 & 0 & -\frac{1}{2} & \frac{1}{2} & 0 & 0 & 0 & 0 & 0 & \cdots \\
& 0 & 0 & 0 & 0 & 0 & 0 & -\frac{1}{2} & \frac{1}{2} & 0 & 0 & 0 & 0 \\
& 0 & 0 & 0 & 0 & 0 & 0 & 0 & -\frac{1}{2} & \frac{1}{2} & 0 & 0 & 0 & 0 \\
& & & & & & & & \ddots & \ddots & & & \ddots
\end{bmatrix}
$$

So if we compute $\mathbf{y} = \mathbf{h} * \mathbf{x}$, and $\mathbf{z} = \mathbf{g} * \mathbf{x}$ we will at least have enough information to recover $\mathbf{x}$ from $\mathbf{y}$ and $\mathbf{z}$.

Let $\mathbf{x} = (\ldots, x_{-2}, x_{-1}, x_0, x_1, x_2, x_3, x_4, x_5, \ldots)$ be a given sequence. If we compute $\mathbf{y} = \mathbf{h} * \mathbf{x} = H\mathbf{x}$, and $\mathbf{z} = \mathbf{g} * \mathbf{x} = G\mathbf{x}$ we have

$$
\mathbf{y} = \begin{bmatrix} \cdots & y_{-2} & y_{-1} & y_0 & y_1 & y_2 & y_3 & y_4 & y_5 & y_6 & \cdots \end{bmatrix}^T
$$
$$
= H\mathbf{x} = \begin{bmatrix} \cdots & \frac{x_{-2}+x_{-1}}{2}, & \frac{x_{-1}+x_0}{2}, & \frac{x_1+x_0}{2}, & \frac{x_2+x_1}{2}, & \frac{x_3+x_2}{2}, & \frac{x_4+x_3}{2}, & \frac{x_5+x_4}{2}, & \frac{x_6+x_5}{2}, & \cdots \end{bmatrix}^T
$$

and

$$
\mathbf{z} = \begin{bmatrix} \cdots & z_{-2} & z_{-1} & z_0 & z_1 & z_2 & z_3 & z_4 & z_5 & z_6 & \cdots \end{bmatrix}^T
$$
$$
= G\mathbf{x} = \begin{bmatrix} \cdots & \frac{x_{-1}-x_{-2}}{2}, & \frac{x_0-x_{-1}}{2}, & \frac{x_1-x_0}{2}, & \frac{x_2-x_1}{2}, & \frac{x_3-x_2}{2}, & \frac{x_4-x_3}{2}, & \frac{x_5-x_4}{2}, & \frac{x_6-x_5}{2}, & \cdots \end{bmatrix}^T
$$

Using our ideas of matrix block multiplication, we could combine the two output sequences above and write

$$
\begin{bmatrix} H \\ \hline G \end{bmatrix} \mathbf{x} = \begin{bmatrix} H\mathbf{x} \\ \hline G\mathbf{x} \end{bmatrix} = \begin{bmatrix} \mathbf{y} \\ \hline \mathbf{z} \end{bmatrix}
$$

or

$$
\begin{bmatrix}
\ddots \\
& 0 & \frac{1}{2} & \frac{1}{2} & 0 & 0 & 0 & 0 & 0 \\
& 0 & 0 & \frac{1}{2} & \frac{1}{2} & 0 & 0 & 0 & 0 \\
\cdots & 0 & 0 & 0 & \frac{1}{2} & \frac{1}{2} & 0 & 0 & 0 & \cdots \\
& 0 & 0 & 0 & 0 & \frac{1}{2} & \frac{1}{2} & 0 & 0 \\
& 0 & 0 & 0 & 0 & 0 & \frac{1}{2} & \frac{1}{2} & 0 \\
& & & & & & & \ddots & \ddots & \ddots \\
\ddots \\
& 0 & -\frac{1}{2} & \frac{1}{2} & 0 & 0 & 0 & 0 & 0 \\
& 0 & 0 & -\frac{1}{2} & \frac{1}{2} & 0 & 0 & 0 & 0 \\
\cdots & 0 & 0 & 0 & -\frac{1}{2} & \frac{1}{2} & 0 & 0 & 0 & \cdots \\
& 0 & 0 & 0 & 0 & -\frac{1}{2} & \frac{1}{2} & 0 & 0 \\
& 0 & 0 & 0 & 0 & 0 & -\frac{1}{2} & \frac{1}{2} & 0 \\
& & & & & & & \ddots & \ddots & \ddots
\end{bmatrix}
\begin{bmatrix}
\vdots \\
x_{-2} \\
x_{-1} \\
x_0 \\
x_1 \\
x_2 \\
x_3 \\
x_4 \\
x_5 \\
x_6 \\
\vdots
\end{bmatrix}
=
\begin{bmatrix}
\vdots \\
y_{-2} \\
y_{-1} \\
y_0 \\
y_1 \\
y_2 \\
y_3 \\
y_4 \\
y_5 \\
y_6 \\
\vdots \\
z_{-2} \\
z_{-1} \\
z_0 \\
z_1 \\
z_2 \\
z_3 \\
z_4 \\
z_5 \\
z_6 \\
\vdots
\end{bmatrix}
\tag{6.1}
$$

So we have basically combined two related filters. The lowpass filter produces an approximation, if you will, of the original input sequence $\mathbf{x}$, and the highpass filter provides the values that tell us how far the lowpass portion is away from the original. This becomes more apparent when we think about how to recover $\mathbf{x}$ from $\mathbf{y}$ and $\mathbf{z}$.

In ordinary words, if we have an average a of two numbers u and v and the distance d from a to one of the two numbers, then we recover u and v by computing $a + d$ and $a - d$. For example, if $a = 4$ and $d = 2$, then we must have started with $4 + 2 = 6$ and $4 - 2 = 2$.

In terms of $\mathbf{x}, \mathbf{y}$, and $\mathbf{z}$, we have

$$
\vdots
$$
$$
\begin{aligned}
y_1 - z_1 &= \tfrac{x_1 + x_0}{2} - \tfrac{x_1 - x_0}{2} = x_0 \\
y_1 + z_1 &= \tfrac{x_1 + x_0}{2} + \tfrac{x_1 - x_0}{2} = x_1 \\
y_2 - z_2 &= \tfrac{x_2 + x_1}{2} - \tfrac{x_2 - x_1}{2} = x_1 \\
y_2 + z_2 &= \tfrac{x_2 + x_1}{2} + \tfrac{x_2 - x_1}{2} = x_2 \\
y_3 - z_3 &= \tfrac{x_3 + x_2}{2} - \tfrac{x_3 - x_2}{2} = x_2 \\
y_3 + z_3 &= \tfrac{x_3 + x_2}{2} + \tfrac{x_3 - x_2}{2} = x_3 \\
y_4 - z_4 &= \tfrac{x_4 + x_3}{2} - \tfrac{x_4 - x_4}{2} = x_3 \\
y_4 + z_4 &= \tfrac{x_4 + x_3}{2} + \tfrac{x_4 - x_3}{2} = x_4
\end{aligned}
\tag{6.2}
$$
$$
\vdots
$$

So by combining our two filters $\mathbf{h} = (h_0, h_1) = (\frac{1}{2}, \frac{1}{2})$ and $\mathbf{g} = (g_0, g_1) = (\frac{1}{2}, -\frac{1}{2})$, we have the means to recover $\mathbf{x}$ from the *combined* output sequences $\mathbf{y}$ and $\mathbf{z}$.

Downsampling and Matrix Truncation

We are now ready for a crucial observation. Look again at equations in (6.2). Not only can we recover $\mathbf{x}$ but it should be clear that we don't need all the values of $\mathbf{y}$ and $\mathbf{z}$ to do it! For example, if we only used the odd values of $\mathbf{y}$ and $\mathbf{z}$, we could recover $\mathbf{x}$:

$$
\begin{aligned}
y_1 - z_1 &= \frac{x_1 + x_0}{2} - \frac{x_1 - x_0}{2} = x_0 \\
y_1 + z_1 &= \frac{x_1 + x_0}{2} + \frac{x_1 - x_0}{2} = x_1 \\
y_3 - z_3 &= \frac{x_3 + x_2}{2} - \frac{x_3 - x_2}{2} = x_2 \\
y_3 + z_3 &= \frac{x_3 + x_2}{2} + \frac{x_3 - x_2}{2} = x_3
\end{aligned}
$$

In engineering, this step is called *downsampling*. In particular, we have downsampled by a factor of 2 since we discarded every other value of $\mathbf{y}$ and $\mathbf{z}$. Do you see how to adjust the matrix in (6.1) to do the downsampling? We simply discard every even row from the matrix to obtain

$$
\left[
\begin{array}{cccccccc}
\ddots & & & & & & & \\
0 & \frac{1}{2} & \frac{1}{2} & 0 & 0 & 0 & 0 & 0 \\
\dots\ 0 & 0 & 0 & \frac{1}{2} & \frac{1}{2} & 0 & 0 & 0 & \dots \\
0 & 0 & 0 & 0 & 0 & \frac{1}{2} & \frac{1}{2} & 0 \\
& & & & \ddots & \ddots & \ddots & \\
\hline
\ddots & & & & & & & \\
0 & -\frac{1}{2} & \frac{1}{2} & 0 & 0 & 0 & 0 & 0 \\
\dots\ 0 & 0 & 0 & -\frac{1}{2} & \frac{1}{2} & 0 & 0 & 0 & \dots \\
0 & 0 & 0 & 0 & 0 & -\frac{1}{2} & \frac{1}{2} & 0 \\
& & & & \ddots & \ddots & \ddots &
\end{array}
\right]
\left[
\begin{array}{c}
\vdots \\ x_{-2} \\ x_{-1} \\ x_0 \\ x_1 \\ x_2 \\ x_3 \\ x_4 \\ x_5 \\ x_6 \\ \vdots
\end{array}
\right]
=
\left[
\begin{array}{c}
\vdots \\ y_{-2} \\ y_0 \\ y_2 \\ y_4 \\ \vdots \\ \hline z_{-2} \\ z_0 \\ z_2 \\ z_4 \\ \vdots
\end{array}
\right]
\tag{6.3}
$$

We have done some important analysis — we have combined two convolution filters to produce a transformation that provides the means to recover the original input sequence from the output sequence.

In applications, all but a finite number of elements in our input sequence $\mathbf{x}$ would be nonzero. Consider, for instance, that $\mathbf{x}$ represents some digital audio file or a column from an image. We need to think about how to truncate the matrix in (6.3).

Toward this end, let's now consider an N-vector $\mathbf{x}$ and the output that we desire. We'll assume that N is an even number.

Recall that $y_n = \frac{x_n + x_{n-1}}{2}$ and $z_n = \frac{x_n - x_{n-1}}{2}$ so our transformation should take $\mathbf{x} = [x_1, x_2, \ldots, x_N]^T$ and produce the vector

$$
\begin{bmatrix}
\frac{x_1 + x_2}{2} \\
\frac{x_3 + x_4}{2} \\
\vdots \\
\frac{x_{N-1} + x_N}{2} \\
\hline
\frac{x_2 - x_1}{2} \\
\frac{x_4 - x_3}{2} \\
\vdots \\
\frac{x_N - x_{N-1}}{2}
\end{bmatrix}
$$

The matrix we need to do the job is simply an $N \times N$ truncated version of the matrix in (6.3):

$$
\tilde{W}_N =
\left[
\begin{array}{cccccccccc}
\frac{1}{2} & \frac{1}{2} & 0 & 0 & 0 & 0 & & & 0 & 0 \\
0 & 0 & \frac{1}{2} & \frac{1}{2} & 0 & 0 & \cdots & & 0 & 0 \\
\vdots & & & & \ddots & & & & & \vdots \\
0 & 0 & 0 & 0 & & & \frac{1}{2} & \frac{1}{2} & 0 & 0 \\
0 & 0 & 0 & 0 & \cdots & & 0 & 0 & \frac{1}{2} & \frac{1}{2} \\
\hline
-\frac{1}{2} & \frac{1}{2} & 0 & 0 & 0 & 0 & & & 0 & 0 \\
0 & 0 & -\frac{1}{2} & \frac{1}{2} & 0 & 0 & \cdots & & 0 & 0 \\
\vdots & & & & \ddots & & & & & \vdots \\
0 & 0 & 0 & 0 & & & -\frac{1}{2} & \frac{1}{2} & 0 & 0 \\
0 & 0 & 0 & 0 & \cdots & & 0 & 0 & -\frac{1}{2} & \frac{1}{2}
\end{array}
\right]
\tag{6.4}
$$

The matrix $\tilde{W}_N$ is very close to being what we'll call a *wavelet transformation*. The tilde is added to W_N to indicate that there is one more topic to discuss.

Invertibility of the Transform

We are interested in invertibility, so the first question we should answer is whether or not $\tilde{W}_N$ is invertible. If it is, we obviously want an explicit form for $\tilde{W}_N^{-1}$. There are many ways to check for invertibility. You may remember from linear algebra that if the determinant of $\tilde{W}_N$ is nonzero, then $\tilde{W}$ is invertible. We could also use Gaussian elimination and just attempt to find the inverse (if it exists). Given that the matrix varies with N, we'll take a different approach.

Suppose that $\tilde{W}_N$ is invertible and consider the first row of $\tilde{W}_N$: $\tilde{w}^1 = (\frac{1}{2}, \frac{1}{2}, 0, 0, \ldots, 0, 0)$ It must dot with the first column of $\tilde{W}_N^{-1}$ and give one and dot with all other columns of $\tilde{W}_N^{-1}$ and give zero. It turns out that we have some pretty good candidates for columns of $\tilde{W}_N^{-1}$. Think about rows 2 through N of $\tilde{W}_N$. If we view these as columns, each of them dot with $\tilde{w}^1$ to give zero. As a matter of fact, any row $\tilde{w}^j$ of $\tilde{W}_N$ dotted with any other row $\tilde{w}^k$ (viewed as a column), where $k \neq j$, gives zero! So turning the rows of $\tilde{W}_N$ into columns almost works. This is nothing more than the transpose of $\tilde{W}_N$! So what happens if we compute $\tilde{W}_N \tilde{W}_N^T$? We have

$$
\tilde{W}_N \tilde{W}_N^T =
\begin{bmatrix}
\frac{1}{2} & \frac{1}{2} & & 0 & 0 \\
0 & 0 & & 0 & 0 \\
\vdots & & \ddots & & \vdots \\
0 & 0 & \cdots & \frac{1}{2} & \frac{1}{2} \\
\hline
-\frac{1}{2} & \frac{1}{2} & & 0 & 0 \\
0 & 0 & & 0 & 0 \\
\vdots & & \ddots & & \vdots \\
0 & 0 & \cdots & -\frac{1}{2} & \frac{1}{2}
\end{bmatrix}
\cdot
\begin{bmatrix}
\frac{1}{2} & 0 & & 0 & -\frac{1}{2} & 0 & & 0 \\
\frac{1}{2} & 0 & & 0 & \frac{1}{2} & 0 & & 0 \\
0 & \frac{1}{2} & & 0 & 0 & -\frac{1}{2} & & 0 \\
0 & \frac{1}{2} & & 0 & 0 & \frac{1}{2} & & 0 \\
\vdots & & \ddots & \vdots & \vdots & & \ddots & \vdots \\
0 & 0 & & 0 & 0 & 0 & & 0 \\
0 & 0 & & 0 & 0 & 0 & & 0 \\
0 & 0 & & \frac{1}{2} & 0 & 0 & & -\frac{1}{2} \\
0 & 0 & & \frac{1}{2} & 0 & 0 & & \frac{1}{2}
\end{bmatrix}
$$

$$
=
\begin{bmatrix}
\frac{1}{2} & 0 & \cdots & 0 & 0 \\
0 & \frac{1}{2} & & 0 & 0 \\
\vdots & & \ddots & & \vdots \\
0 & 0 & & \frac{1}{2} & 0 \\
0 & 0 & \cdots & 0 & \frac{1}{2}
\end{bmatrix}
\tag{6.5}
$$

So $\tilde{W}_N^T$ is close but not quite right. The problem is clearly with the diagonal elements. But this problem is easily fixed. If we replace $\pm \frac{1}{2}$ by ± 1 in $\tilde{W}_N^T$, we obtain the inverse of $\tilde{W}_N$.

$$
\tilde{W}_N^{-1} =
\begin{bmatrix}
1 & 0 & & 0 & -1 & 0 & & 0 \\
1 & 0 & & 0 & 1 & 0 & & 0 \\
0 & 1 & & 0 & 0 & -1 & & 0 \\
0 & 1 & & 0 & 0 & 1 & & 0 \\
\vdots & & \ddots & \vdots & \vdots & & \ddots & \vdots \\
0 & 0 & & 0 & 0 & 0 & & 0 \\
0 & 0 & & 0 & 0 & 0 & & 0 \\
0 & 0 & & 1 & 0 & 0 & & -1 \\
0 & 0 & & 1 & 0 & 0 & & 1
\end{bmatrix}
\tag{6.6}
$$

Before finalizing our first wavelet transform, we have an excellent opportunity to see how (6.4) and (6.6) work on a vector $\mathbf{v}$. I have no doubt that you can multiply $\mathbf{y} = \tilde{W}_N \mathbf{v}$ as well as $\mathbf{v} = \tilde{W}_N^{-1} \mathbf{y}$, but let's talk about the process.

Suppose that you are given a vector $\mathbf{v}$ of even length N. Then $\tilde{W}_N \mathbf{v}$ produces a vector $\mathbf{y}$ that is comprised of two blocks. The first block is simply the result of taking the components of $\mathbf{v}$ two at a time, and averaging them. The bottom block is the result of taking the components of $\mathbf{v}$ two at a time, and in a directional way, determining the distance from the components to the averages. We'll commonly refer to these blocks as the *averages block* and the *differences block*.

What about $\tilde{W}_N^{-1}$? Do you see how it works? If we are given an average and a difference, how do we get the two original numbers back? We add the average and the difference and we subtract the average and difference. Do you see that this is exactly what $\tilde{W}_N^{-1}$ is doing? Look at the structure of the matrix. We can consider the rows in groups of two and note that the top row of each group picks out the average and associated difference and subtracts them and the bottom row in the group adds the average and difference. Figure 6.1 shows how $\tilde{W}$ and $\tilde{W}^{-1}$ act on consecutive components v_{2k-1} and v_{2k} of $\mathbf{v}$.

$$
\boxed{\begin{array}{c} v_{2k-1} \\ v_{2k} \end{array}} \xrightarrow{\tilde{W}_N} \boxed{\begin{array}{l} (v_{2k-1} + v_{2k})/2 = y_k \\ (-v_{2k-1} + v_{2k})/2 = y_{k+N/2} \end{array}}
$$

$$
\boxed{\begin{array}{c} y_k \\ y_{k+N/2} \end{array}} \xrightarrow{\tilde{W}_N^{-1}} \boxed{\begin{array}{l} y_k - y_{k+N/2} = v_{2k-1} \\ y_k + y_{k+N/2} = v_{2k} \end{array}}
$$

Figure 6.1 $\tilde{W}_N$ takes two consecutive components, v_{2k-1} and v_{2k}, and produces the kth average y_k and the kth difference $y_{k+N/2}$. $\tilde{W}_N^{-1}$ substracts the kth difference $y_{k+N/2}$ from the kth average y_k to recover v_{2k-1} and adds the average and difference to recover v_{2k}.

Why would anyone be interested in applying $\tilde{W}_N$ to a vector $\mathbf{v}$? Here is one quick example. Suppose that you wanted to send me a list of numbers over the Internet and you didn't want to use compression schemes. For the sake of argument, let's say that your list is $\mathbf{v} = [500, 504, 210, 200]^T$. Each number consists of three digits so you are sending me 12 digits in total. But if you were to send me $\tilde{W}_4 \mathbf{v} = [502, 205, 2, -5]^T$ instead, you would be sending me eight digits and a minus sign – a small savings for sure, but consider the savings if $N = 4000$ rather than $N = 4$. One more answer to our question is given in the following example.

Example 6.1 (Naive Data Compression). *Suppose that you want to send a grayscale image over the Internet to a friend and are willing to sacrifice a bit of resolution for speed of transmission. Suppose that*

$$
\mathbf{v} = [200, 200, 200, 210, 40, 80, 100, 102]^T
$$

represents the gray values of the first row of the image. Then

$$
\tilde{W}_8 \mathbf{v} = [200, 205, 60, 101 \mid 0, 5, 20, 1]^T
$$

Now think about the four differences 0, 5, 20, and 1. Can you really see a difference of one or even five levels of gray on a scale that consists of 256 levels? Most people can't, and if you are willing to sacrifice a bit of resolution, you would replace 5 and

1 *by zeros and use the vector* $\mathbf{y} = [200, 205, 60, 101, 0, 0, 20, 0]^T$. *Compression routines work very well when given data with large sets of homogeneous values. So* $\mathbf{y}$ *would compress better than an unaltered version of* $\tilde{W}_8\mathbf{v}$. *What is the price you pay? Your friend would receive a compressed version of* $\mathbf{y}$, *uncompress it, and compute*

$$\mathbf{w} = \tilde{W}_8^{-1}\mathbf{y}$$

$$= \begin{bmatrix} 1 & 0 & 0 & 0 & -1 & 0 & 0 & 0 \\ 1 & 0 & 0 & 0 & 1 & 0 & 0 & 0 \\ 0 & 1 & 0 & 0 & 0 & -1 & 0 & 0 \\ 0 & 1 & 0 & 0 & 0 & 1 & 0 & 0 \\ 0 & 0 & 1 & 0 & 0 & 0 & -1 & 0 \\ 0 & 0 & 1 & 0 & 0 & 0 & 1 & 0 \\ 0 & 0 & 0 & 1 & 0 & 0 & 0 & -1 \\ 0 & 0 & 0 & 1 & 0 & 0 & 0 & 1 \end{bmatrix} \begin{bmatrix} 200 \\ 205 \\ 60 \\ 101 \\ 0 \\ 0 \\ 20 \\ 0 \end{bmatrix}$$

$$= [200, 200, 205, 205, 40, 80, 101, 101]^T$$

Figure 6.2 is an actual pictorial representation of these two rows. The top row is the original $\mathbf{v}$. *The bottom row is the uncompressed version given by* $\mathbf{w}$. *As you can see, there's not much difference in the two rows. We will have a better look at image compression in Section 6.4.*   □

Figure 6.2   The original and compressed versions of the top row of the image.

Orthogonalizing the Transform

We have one task left in this section. Since $\tilde{W}_N^T$ is very close to being the same as $\tilde{W}_N^{-1}$, we want to *orthogonalize* $\tilde{W}_N$ to finalize our wavelet transform.

From (6.5) we know that $\tilde{W}_N\tilde{W}_N^T = \frac{1}{2}I_N$. Recall that the elements on the diagonal of $\frac{1}{2}I_N$ result when we dot row k in $\tilde{W}_N$ with column k of $\tilde{W}_N^T$. But the kth column of $\tilde{W}_N^T$, is just the transpose of row k of $\tilde{W}_N$ so this computation is nothing more than computing the norm of row k in $\tilde{W}_N$.

It is easy to see from (6.4) that if $\mathbf{w}$ is any row in $\tilde{W}_N$, then $\|\mathbf{w}\| = \frac{1}{\sqrt{2}}$. If we were to multiply $\mathbf{w}$ by $\sqrt{2}$, then $\|\sqrt{2}\mathbf{w}\| = 1$. So to orthogonalize $\tilde{W}_N$, we simply multiply it by $\sqrt{2}$! We are ready to define our first wavelet transform.

Definition 6.1 (Haar Wavelet Transformation). *Let N be an even positive integer. Then we define the* Haar wavelet transformation[1] *(HWT) by the matrix*

$$
W_N = \left[
\begin{array}{cccccccc}
\frac{\sqrt{2}}{2} & \frac{\sqrt{2}}{2} & 0 & 0 & & & 0 & 0 \\
0 & 0 & \frac{\sqrt{2}}{2} & \frac{\sqrt{2}}{2} & & & 0 & 0 \\
\vdots & & & & \ddots & & & \vdots \\
0 & 0 & 0 & 0 & \cdots & & \frac{\sqrt{2}}{2} & \frac{\sqrt{2}}{2} \\
\hline
-\frac{\sqrt{2}}{2} & \frac{\sqrt{2}}{2} & 0 & 0 & & & 0 & 0 \\
0 & 0 & -\frac{\sqrt{2}}{2} & \frac{\sqrt{2}}{2} & & & 0 & 0 \\
\vdots & & & & \ddots & & & \vdots \\
0 & 0 & 0 & 0 & \cdots & & -\frac{\sqrt{2}}{2} & \frac{\sqrt{2}}{2}
\end{array}
\right]
\tag{6.7}
$$

$$
= \sqrt{2} \left[
\begin{array}{cccccccc}
\frac{1}{2} & \frac{1}{2} & 0 & 0 & & & 0 & 0 \\
0 & 0 & \frac{1}{2} & \frac{1}{2} & & & 0 & 0 \\
\vdots & & & & \ddots & & & \vdots \\
0 & 0 & 0 & 0 & \cdots & & \frac{1}{2} & \frac{1}{2} \\
\hline
-\frac{1}{2} & \frac{1}{2} & 0 & 0 & & & 0 & 0 \\
0 & 0 & -\frac{1}{2} & \frac{1}{2} & & & 0 & 0 \\
\vdots & & & & \ddots & & & \vdots \\
0 & 0 & 0 & 0 & \cdots & & -\frac{1}{2} & \frac{1}{2}
\end{array}
\right]
$$

The filter

$$
\mathbf{h} = (h_0, h_1) = \left(\frac{\sqrt{2}}{2}, \frac{\sqrt{2}}{2} \right)
\tag{6.8}
$$

is called the Haar filter *and we will call*

$$
\mathbf{g} = (g_0, g_1) = \left(\frac{\sqrt{2}}{2}, -\frac{\sqrt{2}}{2} \right)
\tag{6.9}
$$

the Haar wavelet filter.

□

We have factored out the $\sqrt{2}$ from (6.7) to indicate to you that all we are doing to $\tilde{W}_N$ is simply *scaling* it so that W_N is now an orthogonal matrix. The matrix W_N

---

[1]The Haar wavelet transformation is named in honor of the Hungarian mathematician Alféd Haar (1885–1933), who studied spaces and basis functions that lead to the transform as part of his 1909 doctoral thesis *Zur Theorie der orthogonalen Funktionensysteme* (The Theory of Orthogonal Function Systems).

still essentially computes averages and differences, except that, all these values are scaled by $\sqrt{2}$. The benefit, of course, is that we have an easy formula for the inverse:

$$W_N^{-1} = W_N^T$$

Let's look at an example.

Example 6.2 (Applying the Haar Wavelet Transform). *Consider the vector* **v** *that consists of 64 uniformly spaced samples of the function* $f(t) = \cos(2\pi t)$ *on the interval* $[0,1]$. *That is* $v_1 = f(0)$, $v_2 = f(\frac{1}{64})$, $v_3 = f(\frac{2}{64})$, ..., $v_{64} = f(\frac{63}{64})$. *Compute the Haar wavelet transformation (HWT)* $\mathbf{y} = W_{64}\mathbf{v}$ *and graph the cumulative energy of* **v** *and* **y**.

Solution. *The values of* **v** *are plotted below.*

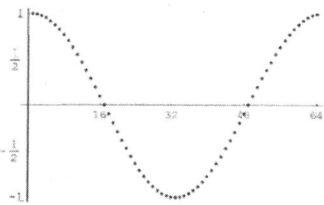

Figure 6.3 The function $f(t) = \cos(2\pi t)$ uniformly sampled 64 times on $[0, 1]$.

We form the matrix W_{64} *and compute* $\mathbf{y} = W_{64}\mathbf{v}$. *Let's think about* **y**. *The top half of* **y** *will be weighted (by* $\sqrt{2}$) *averages of consecutive values of* **v**. *The second half of* **y** *will be weighted (again by* $\sqrt{2}$) *differences between the average value and two values that generated the average value. The result is plotted in Figure 6.4.*

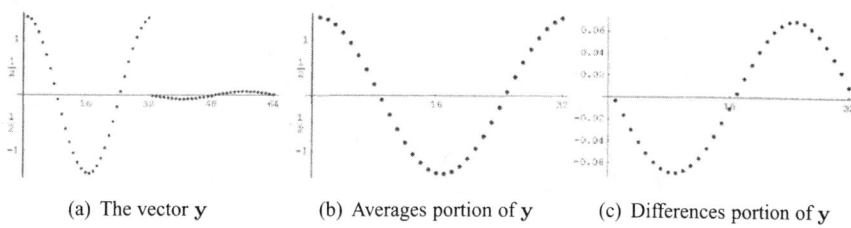

(a) The vector **y**   (b) Averages portion of **y**   (c) Differences portion of **y**

Figure 6.4 Plots of $\mathbf{y} = W_{64}\mathbf{v}$. Note the scale of each graph.

Note that the differences are quite small, and the plot of the averages looks like a scaled version of the original. Figure 6.5 shows the cumulative energy graphs of **v** *and* **y**.

Note that the energy in **y** *approaches* 1 *much faster than the energy in* **v**. *For example, to store* 90% *of the energy, we need* 38 *values from* **v** *but only* 19 *values*

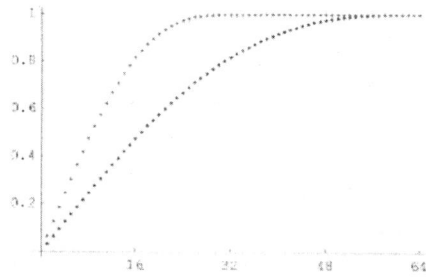

Figure 6.5   Plots of the cumulative energy of **v** (dark dots) and **y** (light dots).

from **y**. *Thus, there are many more values in* **y** *that are zero or approximately zero than in* **v**.   □

Pseudocodes

We conclude this section with the development of pseudocodes for implementing $\dot{W}_N$ and W_N^T. Let's start with the computation $W_N \mathbf{v} = \mathbf{y}$. We are given an N-vector **v** (N even), so we know how to construct W_N. Is this how we want to proceed? Should our pseudocode consist of constructing W_N and then steps to compute $W_N \mathbf{v}$ without regards to the sparse structure of W_N? Of course not — the reason this transform is so fast is that there are only two nonzero elements per row of W_N. If we were just to build W_N with rows $\mathbf{w}^k$, $k = 1, \ldots, N$ and write a routine to compute the N inner products $\mathbf{w}^k \cdot \mathbf{v}$, then we would have to naively compute

$$y_k = w_1^k v_1 + w_2^k v_2 + \cdots + w_N^k v_N \qquad (6.10)$$

for every $k = 1, \ldots, N$. If you look again at (6.10), you will see that we are performing N multiplies. Since there are N rows, we would be doing N^2 multiplies. This is a gross misuse of computational resources given that only two elements of $\mathbf{w}^k$ are nonzero! We know that if we performed all N^2 computations, we would compute the top half of $\mathbf{y}$:

$$y_1 = w_1^1 v_1 + w_2^1 v_2 + w_3^1 v_3 + \cdots + w_N^1 v_N = \frac{\sqrt{2}}{2} v_1 + \frac{\sqrt{2}}{2} v_2$$

$$y_2 = w_1^2 v_1 + w_2^2 v_2 + w_3^2 v_3 + \cdots + w_N^2 v_N = \frac{\sqrt{2}}{2} v_3 + \frac{\sqrt{2}}{2} v_4$$

$$y_3 = w_1^3 v_1 + w_2^3 v_2 + w_3^3 v_3 + \cdots + w_N^3 v_N = \frac{\sqrt{2}}{2} v_5 + \frac{\sqrt{2}}{2} v_6 \qquad (6.11)$$

$$\vdots$$

$$y_k = w_1^k v_1 + w_2^k v_2 + w_3^k v_3 + \cdots + w_N^k v_N = \frac{\sqrt{2}}{2} v_{2k-1} + \frac{\sqrt{2}}{2} v_{2k}$$

for $k = 1, 2, \ldots, \frac{N}{2}$. Also for $k = 1, 2, \ldots, \frac{N}{2}$ we would compute the bottom half of $\mathbf{y}$:

$$y_{k+N/2} = w_1^{k+N/2} v_1 + w_2^{k+N/2} v_2 + w_3^{k+N/2} v_3 + \cdots + w_N^{k+N/2} v_N$$

$$= -\frac{\sqrt{2}}{2} v_{2k-1} + \frac{\sqrt{2}}{2} v_{2k} \qquad (6.12)$$

If we combine (6.11) and (6.12), we can produce a pseudocode for computing $W_N \mathbf{v}$ described below.

Algorithm 6.1 (Haar Wavelet Transform: 1D). *Given vector $\mathbf{v}$ of even length N, this algorithm computes $\mathbf{y} = W_N \mathbf{v}$. The routine simply consists of a loop from $k = 1, \ldots, \frac{N}{2}$. We will do the corresponding lowpass and highpass values of $\mathbf{y}$ in the same trip through the loop. For ease of presentation, we save the multiplication by $\frac{\sqrt{2}}{2}$ until the last step.*

Algorithm: HWT1D1
Input: Vector $\mathbf{v}$ of even length N.
Output: Vector $\mathbf{y}$ where $\mathbf{y} = W_N \mathbf{v}$.

For $[k = 1, \ k \leq \frac{N}{2}, \ k++,$
   $\quad y_k = v_{2k-1} + v_{2k}$
   $\quad y_{k+N/2} = -v_{2k-1} + v_{2k}$
$]$

Return $\left[\frac{\sqrt{2}}{2} \mathbf{y} \right]$

$\square$

Pseudocode for Inverting the HWT

How about the inverse process? We know that W_N is an orthogonal matrix so $W_N^{-1} = W_N^T$. Therefore, for a given N-vector $\mathbf{y}$, we need an algorithm to compute the product $W_N^T \mathbf{y}$. We could unpack the product, but I believe it is easier just to look at the input $\mathbf{v}$ and the resulting output $W_N \mathbf{v} = \mathbf{y}$ and determine the algorithm from that. For example, let's look at the $N = 8$ case:

$$
\begin{bmatrix} v_1 \\ v_2 \\ v_3 \\ v_4 \\ v_5 \\ v_6 \\ v_7 \\ v_8 \end{bmatrix} \longrightarrow \frac{\sqrt{2}}{2} \begin{bmatrix} v_1+v_2 \\ v_3+v_4 \\ v_5+v_6 \\ v_7+v_8 \\ \hline -v_1+v_2 \\ -v_3+v_4 \\ -v_5+v_6 \\ -v_7+v_8 \end{bmatrix} = \begin{bmatrix} y_1 \\ y_2 \\ y_3 \\ y_4 \\ \hline y_5 \\ y_6 \\ y_7 \\ y_8 \end{bmatrix}
$$

How can we build $\mathbf{y}$ from $\mathbf{v}$? First, note that if we add y_1 and $-y_5$ and multiply the result by $\frac{\sqrt{2}}{2}$ we get v_1. Similarly, if we add y_1 and y_5 and then multiply by $\frac{\sqrt{2}}{2}$ we get v_2. This pattern holds throughout — we go through the top half of $\mathbf{y}$ and add or subtract to each component y_k the corresponding component y_{k+4} in the bottom half of $\mathbf{y}$.

For general N we move to the bottom half of $\mathbf{y}$ by shifting $\frac{N}{2}$ components. It is also natural to compute two values of $\mathbf{v}$ at once. We start with v_1 and v_2, then compute $v_3, v_4, \ldots, v_{N-1}, v_N$. In general, we let k run from 1 to $\frac{N}{2}$ and for each k compute v_{2k-1} and v_{2k}. Thus, we compute for $k = 1, \ldots, \frac{N}{2}$,

$$
\frac{\sqrt{2}}{2}(y_k + y_{k+N/2}) = v_{2k} \quad \text{and} \quad \frac{\sqrt{2}}{2}(y_k - y_{k+N/2}) = v_{2k-1} \quad (6.13)
$$

These relations are consistent with those given in (6.11) and (6.12). There we used v_{2k-1}, v_{2k} to build $y_k, y_{k+N/2}$, so it is natural that we would use $y_k, y_{k+N/2}$ to reconstruct v_{2k-1}, v_{2k}. Here is the algorithm for computing the inverse HWT:

Algorithm 6.2 (Inverse Haar Wavelet Transform: 1D). *Given an even-length vector* $\mathbf{y}$ *we compute* $\mathbf{v} = W_N^T \mathbf{y}$. *The routine simply consists of a loop from* $k = 1, \ldots, \frac{N}{2}$, *where we perform the computations given in (6.13).*

Algorithm: IHWT1D1
Input: Vector $\mathbf{y}$ of even length N.
Output: Vector $\mathbf{v}$ where $\mathbf{v} = W_N^T \mathbf{y}$.

For $[k = 1, \, k \leq \frac{N}{2}, \, k{+}{+},$

$$v_{2k} = y_k + y_{k+N/2}$$
$$v_{2k-1} = y_k - y_{k+N/2}$$

]

Return $\left[\frac{\sqrt{2}}{2} \mathbf{v} \right]$

$\square$

PROBLEMS

6.1   Compute the cumulative energy for $\mathbf{v}$ and $\mathbf{w}$ in Example 6.1. You may wish to use a **CAS** or the CE function from the `DiscreteWavelets` package for the computation. Which vector does a better job of conserving energy?

6.2   Consider the product $\tilde{W}_8 \mathbf{v}$ from Example 6.1. In the example we replaced the 1 and 5 by zeros and left the 20 alone. What would we lose if we converted 20 to 0 as well? Set $\mathbf{y} = [200, 205, 60, 101, 0, 0, 0, 0]^T$ and recompute $\mathbf{w} = \tilde{W}_8^{-1} \mathbf{y}$. How does it differ from the original $\mathbf{v}$? Use your **CAS** and draw a grayscale plot of both $\mathbf{v}$ and $\mathbf{w}$. (See the **CAS** code for Example 6.1 on the text Web site for assistance in producing this plot.)

6.3   Let $\mathbf{v} = [2, 6, -4, 2, 400, 402, -8, -6]^T$ and use Algorithm 6.1 to compute the HWT of $\mathbf{v}$. There is a large jump in the values of $\mathbf{v}$ from v_4 to v_5 and from v_6 to v_7. Is this reflected in the difference block of the transformed data?

6.4   Suppose that $\mathbf{v} \in \mathbb{R}^N$ with N even. Let $\mathbf{y}$ denote the HWT of $\mathbf{v}$ (i. e., $\mathbf{y} = W_N \mathbf{v}$, where W_N is given by (6.7)). Show that

(a) if $\mathbf{v}$ is a constant vector (i. e., $v_k = c$ where c is any real number), then the components of the highpass portion of $\mathbf{y}$ are zero.

(b) if $\mathbf{v}$ is a linear vector (i. e., $v_k = mk + b$ for real number m and b), then the components of the highpass portion of $\mathbf{y}$ are constant. Find this constant value.

6.5   Suppose that $\mathbf{v} \in \mathbb{R}^8$ and let W_8 be the 8×8 HWT matrix given by (6.7) with $\mathbf{h} = (h_0, h_1) = (\frac{\sqrt{2}}{2}, \frac{\sqrt{2}}{2})$ and $\mathbf{g} = (g_0, g_1) = (\frac{\sqrt{2}}{2}, -\frac{\sqrt{2}}{2})$.

(a) Find a 4×2 matrix X so that

$$W_8 \mathbf{v} = \left[\frac{X\mathbf{h}}{X\mathbf{g}} \right] \tag{6.14}$$

For a **CAS** such as *Mathematica*, utilizing (6.14) will produce a faster algorithm than the pseudocode provide in Algorithm 6.1. See Computer Lab 6.1 for more details.

(b) Generalize part (a) in the case where $\mathbf{v} \in \mathbb{R}^N$ (N even) and W_N is the $N \times N$ HWT matrix.

(c) Use part (a) to compute the HWT $\mathbf{y}$ of the vector $\mathbf{v}$ given in Problem 6.3.

6.6   Suppose that $\mathbf{y} \in \mathbb{R}^8$ and let W_8^T be the 8×8 inverse HWT matrix given by (6.7) with filters $\mathbf{h} = (h_0, h_1) = (\frac{\sqrt{2}}{2}, \frac{\sqrt{2}}{2})$ and $\mathbf{g} = (g_0, g_1) = (-\frac{\sqrt{2}}{2}, \frac{\sqrt{2}}{2})$. Let $\mathbf{v} = W_8^T \mathbf{y}$.

(a) Find a 4×2 matrix X such that $X\mathbf{g}$ holds the odd entries of $\mathbf{v}$ (i. e., $X\mathbf{g} = [v_1, v_3, v_5, v_7]^T$) and $X\mathbf{h}$ holds the even entries of $\mathbf{v}$ (i. e., $X\mathbf{h} = [v_2, v_4, v_6, v_8]^T$.

(b) Find the 8×8 matrix P such that

$$\mathbf{v} = P \left[\frac{X\mathbf{h}}{X\mathbf{g}} \right]$$

For a **CAS** such as *Mathematica*, utilizing X from part (a) and P from part (b) will produce a faster algorithm than the pseudocode provided in Algorithm 6.2. See Computer Lab 6.1 for more details.

(c) Generalize parts (a) and (b) when $\mathbf{y} \in \mathbb{R}^N$ (N even) and W_N^T is the $N \times N$ inverse HWT matrix.

(d) In part (c) of Problem 6.5, you computed the HWT $\mathbf{y}$ of the vector from Problem 6.3. Use parts (a) and (b) of this problem to find the inverse HWT of $\mathbf{y}$.

6.7   Let $\mathbf{v} = [1, 2, 3, 4, 5, 6, 7, 8]^T$.

(a) Compute $\mathbf{y} = W_8 \mathbf{v}$.

(b) From Problem 3.32 in Section 3.3, we know that $Ent(\mathbf{v}) = 3$. What is $Ent(\mathbf{y})$?

6.8   Suppose that $\mathbf{v} \in \mathbb{R}^N$ where N is even and $v_k = k$, $k = 1, \ldots, N$. Then the elements of $\mathbf{v}$ are distinct so that $Ent(\mathbf{v}) = \log_2(N)$.

(a) Compute $\mathbf{y} = W_N \mathbf{v}$.

(b) Show that $Ent(\mathbf{y}) = \frac{1}{2} \log_2(N) + \frac{1}{2}$.

6.9   Sometimes the result of the HWT can have higher entropy than the original vector. Suppose that $\mathbf{v} \in \mathbb{R}^N$ with N even and $v_k = c$ for $k = 1, \ldots, N$. Then by Problem 3.28, $Ent(\mathbf{v}) = 0$.

(a) Compute $\mathbf{y} = W_N \mathbf{v}$.

(b) What is $Ent(\mathbf{y})$?

Computer Labs

✦ **6.1   Software Development: The 1D Haar Wavelet Transform (one iteration).** From the text Web site, access the development package hwt1d1. In this development lab you will create a function that will compute one iteration of the one-dimensional HWT. You will also construct a function that computes one iteration of the inverse HWT. Instructions are provided that allow you to add all of these functions to the software package DiscreteWavelets.

6.2   Haar Wavelet Transform. From the text Web site, access the lab hwt. This lab utilizes the functions for the HWT and the inverse HWT in the DiscreteWavelets package. You will perform some naive data compression, compute the entropy of transformed vectors, and investigate the difference block of the transform when the input data are obtained by sampling well-known functions from calculus.

6.2   ITERATING THE PROCESS

We now have a transformation, the HWT, that takes an input vector $\mathbf{v} \in \mathbb{R}^N$ and returns $\mathbf{y} = W_N \mathbf{v}$, where the top half of $\mathbf{y}$ consists of weighted averages of the elements in $\mathbf{v}$ and the bottom half is comprised of weighted differences of the elements of $\mathbf{v}$. We have also seen (in Problems 6.7 and 6.8 in Section 6.1) that the application of W_N can result in a vector $\mathbf{y}$ that has lower entropy and more concentrated cumulative energy than the original input $\mathbf{v}$. These measures are both indicators that we might obtain a higher compression ratio using the output vector $\mathbf{y}$ rather than the input $\mathbf{v}$.

It is natural to ask if it is possible to improve these measures even further. The answer is yes, and all that we need to do is make further use of the HWT. Suppose that N is divisible by 4 and let $\mathbf{v} \in \mathbb{R}^N$. We apply W_N to $\mathbf{v}$ and obtain the vector $\mathbf{y}$. Let's name the lowpass and highpass parts of $\mathbf{y}$. We will call the lowpass part $\mathbf{y}^\ell$ and the highpass portion $\mathbf{y}^h$. Then

$$\mathbf{y}^1 = \begin{bmatrix} \mathbf{y}^\ell \\ \mathbf{y}^h \end{bmatrix} \tag{6.15}$$

The vector $\mathbf{y}^\ell \in \mathbb{R}^{\frac{N}{2}}$, and since N is divisible by 4, $\frac{N}{2}$ is divisible by 2. Since $\mathbf{y}^\ell$ is an approximation of the original input $\mathbf{v}$, it makes sense to apply $W_{N/2}$ to the vector $\mathbf{y}^\ell$. The resulting output would be

$$\mathbf{y}^2 = \begin{bmatrix} W_{N/2}\mathbf{y}^\ell \\ \mathbf{y}^h \end{bmatrix} \tag{6.16}$$

We have thus performed a second iteration of the HWT. Let's look at an example to see the effects.

Example 6.3 (Iterating the HWT). *Perform two iterations of the HWT on function samples in Example 6.2 and plot the cumulative energies of the original input vector* **v**, *one iteration of the HWT, and two iterations of the HWT.*
Solution. *Let's introduce some notation:*

$$
\mathbf{y} = \mathbf{y}^1 = \left[\frac{\mathbf{y}^\ell}{\mathbf{y}^h} \right] = \left[\frac{\mathbf{y}^{1\ell}}{\mathbf{y}^{1h}} \right]
$$

So $\mathbf{y}^1$ *denotes one iteration of the HWT and* $\mathbf{y}^{1\ell}$, $\mathbf{y}^{1h}$ *represent the lowpass and highpass portions, respectively, of the first iteration. We then apply* W_{32} *to* $\mathbf{y}^{1\ell}$ *to obtain*

$$
W_{32}\mathbf{y}^{1\ell} = \left[\frac{\mathbf{y}^{2\ell}}{\mathbf{y}^{2h}} \right]
$$

We thus obtain the second iteration of the HWT:

$$
\mathbf{y}^2 = \left[\frac{\mathbf{y}^{2\ell}}{\frac{\mathbf{y}^{2h}}{\mathbf{y}^{1h}}} \right]
$$

Graphs of **v** *and* $\mathbf{y} = \mathbf{y}^1$ *appear in Figures 6.3 6.4, respectively, while* $\mathbf{y}^2$ *and its various components are plotted in Figure 6.6.*

The cumulative energy of **v**, $\mathbf{y}^1$, *and* $\mathbf{y}^2$ *is plotted in Figure 6.7. It is clear that* $\mathbf{y}^2$ *does an even better job than* **y** *of conserving energy.* ☐

Iteration as Products of Matrices

We can certainly iterate again if we desire. Of course, to iterate once (apply the HWT), we need N even. To iterate twice, we need N divisible by 4 and to iterate three times, we need N divisible by 8. In general, to iterate j times, we need N divisible by 2^j.

It is convenient at this time to develop some notation for iterating our transformation. Let $\mathbf{y}^j$ represent the jth iteration of the HWT. We denote by $\mathbf{y}^{j\ell}$ the lowpass portion resulting from the jth iteration of the HWT, and by $\mathbf{y}^{jh}$ the highpass portion resulting from the jth iteration of the HWT. Let's look at an example to help us make sense of this rather complicated notation.

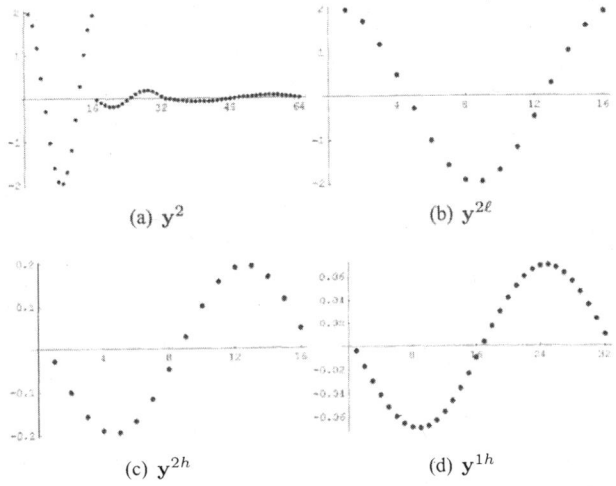

(a) $\mathbf{y}^2$ (b) $\mathbf{y}^{2\ell}$

(c) $\mathbf{y}^{2h}$ (d) $\mathbf{y}^{1h}$

Figure 6.6   Iterating the Haar Wavelet Transformation.

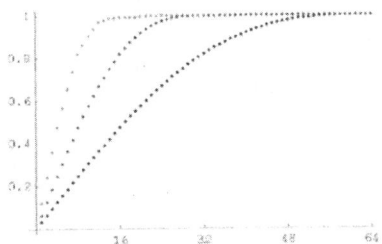

Figure 6.7   Plots of the cumulative energy of $\mathbf{v}$ (dark dots), $\mathbf{y}$ (medium dots), and $\mathbf{y}^2$ (light dots).

Example 6.4 (Iteration as a Product of Matrices). *Suppose that we wish to perform $j = 3$ iterations of the HWT on a vector $\mathbf{v} \in \mathbb{R}^{40}$. According to our notation, $\mathbf{y}^1$ represents the first iteration of the HWT. We have*

$$\mathbf{y}^1 = W_{40}\mathbf{v} = \left[\begin{array}{c} \mathbf{y}^{1\ell} \\ \hline \mathbf{y}^{1h} \end{array} \right]$$

To reiterate, $\mathbf{y}^{1\ell}$ denotes the lowpass portion resulting from the first iteration of the HWT and $\mathbf{y}^{1h}$ represents the highpass portion resulting from the first iteration of the HWT. Note that $\mathbf{y}^{1\ell}, \mathbf{y}^{1h} \in \mathbb{R}^{20}$.

To perform the second iteration $\mathbf{y}^2$, we must apply W_{20} to $\mathbf{y}^{1\ell}$. This gives the vectors $\mathbf{y}^{2\ell}, \mathbf{y}^{2h} \in \mathbb{R}^{10}$. We have

$$\mathbf{y}^2 = \left[\frac{W_{20}\mathbf{y}^{1\ell}}{\mathbf{y}^{1h}} \right] = \begin{bmatrix} \mathbf{y}^{2\ell} \\ \mathbf{y}^{2h} \\ \mathbf{y}^{1h} \end{bmatrix}$$

Our final iteration yields

$$\mathbf{y}^3 = \left[\frac{\frac{W_{10}\mathbf{y}^{2\ell}}{\mathbf{y}^{2h}}}{\mathbf{y}^{1h}} \right] = \begin{bmatrix} \mathbf{y}^{3\ell} \\ \mathbf{y}^{3h} \\ \mathbf{y}^{2h} \\ \mathbf{y}^{1h} \end{bmatrix}$$

where $\mathbf{y}^{3\ell}, \mathbf{y}^{3h} \in \mathbb{R}^5$.   □

For vectors of length N divisible by 2^j, we have

$$\mathbf{y}^j = \begin{bmatrix} \mathbf{y}^{j\ell} \\ \mathbf{y}^{j,\,h} \\ \mathbf{y}^{j-1,\,h} \\ \mathbf{y}^{j-2,\,h} \\ \vdots \\ \mathbf{y}^{2h} \\ \mathbf{y}^{1h} \end{bmatrix}$$

where $\mathbf{y}^{j\ell} \in \mathbb{R}^{N/2^j}$ and $\mathbf{y}^{kh} \in \mathbb{R}^{N/2^k}$, $k = 1, \ldots, j$.

Can we write these output vectors in terms of the matrices $W_N, W_{N/2}, \ldots$? Using our new notation, we have $\mathbf{y}^1 = W_N \mathbf{v}$. We need to apply $W_{N/2}$ to $\mathbf{y}^{1\ell}$ (the top half of $\mathbf{y}^1$). We can write this product using block matrices.

We can start with (6.16). Let $I_{N/2}, 0_{N/2}$ denote the $\frac{N}{2} \times \frac{N}{2}$ identity matrix and zero matrix, respectively. Then we could use our new notation and write (6.16) as

$$\mathbf{y}^2 = \left[\frac{W_{N/2}\mathbf{y}^{1\ell}}{I_{N/2}\mathbf{y}^{1h}} \right] = \left[\frac{W_{N/2}\mathbf{y}^{1\ell} + 0_{N/2}\mathbf{y}^{1h}}{0_{N/2}\mathbf{y}^{1\ell} + I_{N/2}\mathbf{y}^{1h}} \right] \tag{6.17}$$

We can write $W_{N/2}\mathbf{y}^{1\ell} + 0_{N/2}\mathbf{y}^{1h}$ from (6.17) as

$$[W_{N/2} \mid 0_{N/2}] \left[\frac{\mathbf{y}^{1\ell}}{\mathbf{y}^{1h}} \right]$$

In a similar way, we can write $0_{N/2}\mathbf{y}^{1\ell} + I_{N/2}\mathbf{y}^{1h}$ from (6.17) as

$$[0_{N/2} \mid I_{N/2}] \left[\frac{\mathbf{y}^{1\ell}}{\mathbf{y}^{1h}} \right]$$

Finally, we can combine these two products into a single block matrix product:

$$\mathbf{y}^2 = \left[\begin{array}{c|c} W_{N/2} & 0_{N/2} \\ \hline 0_{N/2} & I_{N/2} \end{array} \right] \left[\frac{\mathbf{y}^{1\ell}}{\mathbf{y}^{1h}} \right] = \left[\begin{array}{c|c} W_{N/2} & 0_{N/2} \\ \hline 0_{N/2} & I_{N/2} \end{array} \right] \mathbf{y}^1$$

$$= \left[\begin{array}{c|c} W_{N/2} & 0_{N/2} \\ \hline 0_{N/2} & I_{N/2} \end{array} \right] W_N \mathbf{v} \tag{6.18}$$

We could also write $\mathbf{y}^3$ as a product of three matrices times $\mathbf{v}$:

$$\mathbf{y}^3 = \left[\begin{array}{c|c} \begin{array}{c|c} W_{N/4} & 0_{N/4} \\ \hline 0_{N/4} & I_{N/4} \end{array} & 0_{N/2} \\ \hline & \\ 0_{N/2} & I_{N/2} \end{array} \right] \left[\begin{array}{c|c} W_{N/2} & 0_{N/2} \\ \hline 0_{N/2} & I_{N/2} \end{array} \right] W_N \mathbf{v} \tag{6.19}$$

Notation for Iterated Wavelet Transforms

As we can see, writing down the matrix products gets quite cumbersome. We introduce some new notation so that we can easily write down iterated wavelet transforms in terms of matrices.

We add an index to W_N to identify the iteration. The transform matrix will be indexed both by the length N of the input vector $\mathbf{v}$ and the iteration. Suppose that N is divisible by 2^i for some positive integer i. We set

$$W_{N,1} = W_N \quad \text{and} \quad W_{N,2} = \left[\begin{array}{c|c} W_{N/2} & 0_{N/2} \\ \hline 0_{N/2} & I_{N/2} \end{array} \right]$$

Note that if we apply $W_{N,2}$ to a vector $\mathbf{v} \in \mathbb{R}^N$, then only the first $\frac{N}{2}$ elements of $\mathbf{v}$ are altered – the remaining elements in $\mathbf{v}$ are unchanged (see Problem 2.38 from Section 2.3). In a similar fashion, we have

$$
W_{N,3} = \left[
\begin{array}{cc|c}
\begin{array}{c|c} W_{N/4} & 0_{N/4} \\ \hline 0_{N/4} & I_{N/4} \end{array} & & 0_{N/2} \\
\hline
0_{N/2} & & I_{N/2}
\end{array}
\right]
$$

Using our new notation, (6.18) becomes

$$
\mathbf{y}^2 = W_{N,2} W_N \mathbf{v}
$$

and (6.19) can be expressed as

$$
\mathbf{y}^3 = W_{N,3} W_{N,2} W_N \mathbf{v}
$$

If we were to continue in this fashion, it would be quite difficult to write down $W_{N,k}$, where $1 \le k \le i$. Note that $W_{N,k}$ is a block matrix where the only nonzero blocks occur on the main diagonal. In particular, the first block is $W_{N/2^k}$ and the remaining diagonal blocks are $I_{N/2^k}, I_{N/2^{k-1}}, \ldots, I_{N/4}, I_{N/2}$.

Notation. For the kth iteration, $1 \le k \le i$, iteration of the HWT, we can write

$$
\boxed{W_{N,k} = \mathrm{diag}\left[W_{N/2^k}, I_{N/2^k}, I_{N/2^{k-1}}, \ldots, I_{N/4}, I_{N/2}\right]} \tag{6.20}
$$

where $\mathrm{diag}[C_1, C_2, \ldots, C_n]$ is a matrix that has the blocks $C_1, \ldots, C_n$ on its diagonal and zeros elsewhere.

Thus, if we wished to compute i iterations of the wavelet transform applied to $\mathbf{v} \in \mathbb{R}^N$, where N is divisible by 2^i, we would compute

$$
\mathbf{y}^i = W_{N,i} W_{N,i-1} \cdots W_{N,2} W_N \mathbf{v}
$$

Iterating the Inverse Transform

We can also iterate the inverse process. Let's motivate the iterated inverse by example.

Example 6.5 (Iterating the Inverse HWT). *Suppose that* $\mathbf{v} \in \mathbb{R}^{40}$ *and that we have computed* $\mathbf{y}^3 = W_{40,3} W_{40,2} W_{40} \mathbf{v}$. *We want to recover* $\mathbf{v}$ *from* $\mathbf{y}^3$. *Recall that*

$$\mathbf{y}^3 = \begin{bmatrix} \mathbf{y}^{3\ell} \\ \hline \mathbf{y}^{3h} \\ \hline \mathbf{y}^{2h} \\ \hline \mathbf{y}^{1h} \end{bmatrix}$$

The first thing we need to do is recover $\mathbf{y}^{2\ell} \in \mathbb{R}^{10}$ *from* $\mathbf{y}^{3\ell}, \mathbf{y}^{3h} \in \mathbb{R}^5$. *We must compute*

$$\mathbf{y}^{2\ell} = W_{10}^T \begin{bmatrix} \mathbf{y}^{3\ell} \\ \mathbf{y}^{3h} \end{bmatrix}$$

Thus, we have

$$\mathbf{y}^2 = \begin{bmatrix} \mathbf{y}^{2\ell} \\ \hline \mathbf{y}^{2h} \\ \hline \mathbf{y}^{1h} \end{bmatrix} = \begin{bmatrix} W_{10}^T \begin{bmatrix} \mathbf{y}^{3\ell} \\ \mathbf{y}^{3h} \end{bmatrix} \\ \hline \mathbf{y}^{2h} \\ \hline \mathbf{y}^{1h} \end{bmatrix} = \begin{bmatrix} \begin{bmatrix} W_{10}^T & 0_{10} \\ 0_{10} & I_{10} \end{bmatrix} \cdot \begin{bmatrix} \mathbf{y}^{3\ell} \\ \mathbf{y}^{3h} \\ \hline \mathbf{y}^{2h} \end{bmatrix} \\ \hline \mathbf{y}^{1h} \end{bmatrix}$$

$$= \begin{bmatrix} \begin{bmatrix} W_{10}^T & 0_{10} \\ \hline 0_{10} & I_{10} \end{bmatrix} & 0_{20} \\ \hline 0_{20} & I_{20} \end{bmatrix} \cdot \begin{bmatrix} \begin{bmatrix} \mathbf{y}^{3\ell} \\ \mathbf{y}^{3h} \end{bmatrix} \\ \hline \mathbf{y}^{2h} \\ \hline \mathbf{y}^{1h} \end{bmatrix} = W_{40,3}^T \mathbf{y}^3$$

In a similar fashion,

$$\mathbf{y}^1 = W_{40,2}^T \mathbf{y}^2 = W_{40,2}^T W_{40,3}^T \mathbf{y}^3$$

so that we can recover $\mathbf{v}$ *by computing*

$$\mathbf{v} = W_{40}^T \mathbf{y}^1 = W_{40}^T W_{40,2}^T \mathbf{y}^2 = W_{40}^T W_{40,2}^T W_{40,3}^T \mathbf{y}^3$$

$\square$

To summarize, suppose that N is divisible by 2^i and $\mathbf{v} \in \mathbb{R}^N$. We compute i iterations of the HWT using

$$\mathbf{y}^i = W_{N, 2^i} W_{N, 2^i-1} \cdots W_{N, 2} W_N \, \mathbf{v} \tag{6.21}$$

and we compute i iterations of the inverse HWT using

$$\mathbf{v} = W_N^T W_{N, 2}^T W_{N, 3}^T \cdots W_{N, 2^i-1}^T W_{N, 2^i}^T \, \mathbf{y}^i \tag{6.22}$$

Pseudocodes for the Iterated Transform and Inverse

We conclude this section with algorithms for both the iterated HWT and its inverse. We start with the iterated HWT. In addition to an input vector $\mathbf{v}$, we must also provide the algorithm with the number of iterations i that are to be performed. We assume that N is divisible by 2^i and construct the vector

$$\mathbf{y}^i = \begin{bmatrix} \mathbf{y}^{i\ell} \\ \hline \mathbf{y}^{ih} \\ \hline \mathbf{y}^{i-1,\,h} \\ \hline \mathbf{y}^{i-2,\,h} \\ \hline \vdots \\ \hline \mathbf{y}^{2h} \\ \hline \mathbf{y}^{1h} \end{bmatrix} \tag{6.23}$$

Our first step is to use Algorithm 6.1 with input $\mathbf{v}$ to construct $\begin{bmatrix} \mathbf{y}^{1\ell} \\ \hline \mathbf{y}^{1h} \end{bmatrix}$. We will store $\mathbf{y}^{1h}$ in our final output vector and again use Algorithm 6.1 with input $\mathbf{y}^{1\ell}$. The result of this computation is the vector $\begin{bmatrix} \mathbf{y}^{2\ell} \\ \hline \mathbf{y}^{2h} \end{bmatrix}$. Our final output vector is now $\begin{bmatrix} \mathbf{y}^{2\ell} \\ \hline \mathbf{y}^{2h} \\ \hline \mathbf{y}^{1h} \end{bmatrix}$.

We continue this process a total of i times until we have created the vector $\mathbf{y}^i$ given in (6.23). The last step of the algorithm is to prepend $\mathbf{y}^{j\ell}$ to the final output vector.

A loop will be useful. Inside the loop, we call Algorithm 6.1 and prepend the resulting highpass portion to our final output vector. The pseudocode is given below.

Algorithm 6.3 (Haar Wavelet Transform - 1D). *We are given a vector $\mathbf{v} \in \mathbb{R}^N$ and a number i and we return i iterations of the HWT. The algorithm begins by initializing $\mathbf{y}$ as an empty vector. We add highpass portions to $\mathbf{y}$ at each step of the loop using a function called **Join**. This function takes two vectors (with lengths n, m, respectively) as arguments and joins them to form a vector of length $n + m$. Inside the loop, we first use Algorithm 6.1 to compute the HWT of $\mathbf{v}$ and store the result in a dummy variable $\mathbf{b}$. We then overwrite $\mathbf{v}$ with the lowpass portion of $\mathbf{b}$. This is the vector we need to transform on the next pass through the loop. In the last step of the loop, we prepend the highpass portion of $\mathbf{b}$ to $\mathbf{y}$. When we exit the loop, $\mathbf{y}$ consists of all the highpass portions we need. All that is left to do is to prepend the last lowpass portion we computed. This is stored as $\mathbf{v}$, so we return $\begin{bmatrix} \mathbf{v} \\ \mathbf{y} \end{bmatrix}$.*

Algorithm: HWT1D

Input: Vector $\mathbf{v}$ of length N and the number of iterations i.
It is assumed that N is divisible by 2^i.

Output: Vector $\mathbf{y}$ where $\mathbf{y} = W_{N,\,2^i} W_{N,\,2^{i-1}} \cdots W_{N,\,2} W_N \mathbf{v}$ (see (6.21)).

```
y = [ ]
For[j = 1, j ≤ i, j++,
        b = HWT1D1[v]
        v = bℓ
        y = Join[bʰ, y]
]
y = Join[v, y]
```

Return [y]

☐

The algorithm for the iterated inverse HWT follows in a manner very similar to that of Algorithm 6.3. We start with $\mathbf{y}$ and we know that $\mathbf{y}$ has been obtained by iterating the HWT i times. We must first extract $\begin{bmatrix} \mathbf{y}^{i\ell} \\ \mathbf{y}^{ih} \end{bmatrix}$ from $\mathbf{y}$ and apply Algorithm 6.2 to it. The result is $\mathbf{y}^{i-1,\ell}$. We replace $\mathbf{y}^{i\ell}$ and $\mathbf{y}^{ih}$ by $\mathbf{y}^{i-1,\ell}$ in $\mathbf{y}$ and repeat the process $i - 1$ more times. The pseudocode appears below.

Algorithm 6.4 (Inverse Haar Wavelet Transform: 1D). *We are given a vector $\mathbf{y} \in \mathbb{R}^N$ and a number i and we return i iterations of the inverse HWT. We call the* *final output vector* **v**. *The algorithm begins by first setting* **v** *to* $\mathbf{y}^{i\ell}$. *Then we enter a loop. Note that the loop runs from* $j = 0$ *to* $j = i - 1$. *We still perform* i *total steps, but we start at* $j = 0$ *to simplify notation. Once inside the loop, the first thing we must do is concatenate (using a dummy variable* **b***) the lowpass portion* **v** *with the corresponding highpass portion* $\mathbf{y}^{ih}$. *We then compute the inverse Haar wavelet transformation using Algorithm 6.2 and store the result in* **v**. *This is the next lowpass portion we need. We move back to the top of the loop, create* **b** *from* **v** *and the next highpass portion, and proceed. We repeat this process* i *times and the last inverse transform we compute is our final output* **v**.

Algorithm: IHWT1D

Input: Vector y of length N and the number of iterations i.
It is assumed that N is divisible by 2^i.

Output: Vector v where $\mathbf{v} = W_N^T W_{N,2}^T W_{N,3}^T \cdots W_{N,2^{i-1}}^T W_{N,2^i}^T \mathbf{y}$ (see (6.22)).

$\mathbf{v} = \mathbf{y}^{i\ell}$
For $[j = 0, j \le i, j{+}{+}$,
        $\mathbf{b} = \mathbf{Join}[\mathbf{v}, \mathbf{y}^{i-j, h}]$
        $\mathbf{v} = \mathbf{IHWT1D1}[\mathbf{b}]$
]

Return $[\mathbf{v}]$

□

PROBLEMS

6.10   Let $\mathbf{v} = [1, 2, 3, 4, 5, 6, 7, 8]^T$. Use Algorithm 6.3 with $i = 3$ to form the HWT vector **y**. You should work this problem by hand.

6.11   Use Algorithm 6.4 to compute three iterations of the inverse HWT on the vector **y** from Problem 6.10. You should work this problem by hand.

6.12   Suppose that $\mathbf{v} \in \mathbf{R}^N$ where $N = 2^p$ for some $p = 1, 2, \ldots$ Further suppose that $v_k = c$ for $k = 1, \ldots, N$ where c is any real number. Suppose that we apply p iterations of the HWT to **v**. What is the resulting vector?

6.13   Let $\mathbf{v} \in \mathbb{R}^N$ where N is divisible by 4 and suppose that $v_k = k$ for $k = 1, \ldots, N$. Then by Problem 3.32 in Section 3.3, we know that $Ent(\mathbf{v}) = \log_2(N)$.

(a) Compute two iterations of the HWT.

(b) If we call the result in part (a) **y**, what is $Ent(\mathbf{y})$? Compare this result with Problem 6.8 of Section 6.1.

6.14 Let $\mathbf{v} \in \mathbb{R}^N$ where $N = 2^p$ for some $p = 1, 2, \ldots$, and suppose that $v_k = k$, $k = 1, \ldots, N$. Let $\mathbf{y}$ be the vector that results from computing p iterations of the HWT. What is $Ent(\mathbf{y})$?

6.15 Suppose that $\mathbf{v} \in \mathbb{R}^N$, where N is divisible by 4 and the elements in $\mathbf{v}$ are quadratic (i. e., $v_k = ak^2$ for some real number a). Let $\mathbf{y}$ be the result of applying one iteration of the HWT to $\mathbf{v}$. Show that the elements y_k^{1h} of the first highpass iteration are linear (i. e., $y_k^{1h} = mk + b$ for some real numbers m and b). Find the values of m and b.

Computer Labs

♦ **6.3** **Software Development: 1D Haar Wavelet Transformations and Inverse Wavelet Transformations.** From the text Web site, access the development package hwt1d. In this development lab you will create functions for computing the one-dimensional Haar wavelet transformation and the inverse Haar wavelet transformation. Instructions are provided that allow you to add all of these functions to the software package DiscreteWavelets.

6.4 **Iterated HWT.** From the text Web site, access the lab haariterated1d. This lab utilizes the functions for the iterated HWT and the iterated inverse HWT in the DiscreteWavelets package. In this lab you will compare cumulative energies for various iterated transforms of a given vector and also perform some elementary singularity detection. The lab concludes with an interesting application of the inverse HWT involving digital audio files.

6.3 THE TWO-DIMENSIONAL HAAR WAVELET TRANSFORMATION

In Section 6.2, we learned how to apply the HWT iteratively to a vector. We can use this transformation on a variety of applications involving one-dimensional data. In this section we learn how to extend our transformation so that we can apply it to two-dimensional data. Such a transformation can be used in applications involving images. In Chapter 3 we learned that a grayscale digital image can easily be expressed as a matrix. We first learn how to perform one iteration of the two-dimensional HWT, and the iterative process will follow in a very natural way.

Processing the Columns of a Matrix with the HWT

Let's suppose that we have an $N \times N$ matrix A (later, we consider rectangular matrices), where N is even. We denote the columns of A by $\mathbf{a}^1, \mathbf{a}^2, \ldots, \mathbf{a}^N$. If we compute $W_N \mathbf{a}^1$, we are simply transforming the first column of A into weighted averages and differences. We continue by applying W_N to the other columns of A to obtain $W_N \mathbf{a}^1, W_N \mathbf{a}^2, \ldots, W_N \mathbf{a}^N$. But these computations are precisely what we would get if we computed $W_N A$. So applying W_N to A processes all the columns of A.

Example 6.6 (Processing the Columns of a Matrix with the HWT). *Apply W_4 to the matrix*

$$A = \begin{bmatrix} 16 & 20 & 14 & 8 \\ 8 & 10 & 0 & 12 \\ 11 & 12 & 9 & 20 \\ 9 & 10 & 15 & 10 \end{bmatrix}$$

Solution. *We have*

$$W_4 A = \sqrt{2} \begin{bmatrix} \frac{1}{2} & \frac{1}{2} & 0 & 0 \\ 0 & 0 & \frac{1}{2} & \frac{1}{2} \\ -\frac{1}{2} & \frac{1}{2} & 0 & 0 \\ 0 & 0 & -\frac{1}{2} & \frac{1}{2} \end{bmatrix} \begin{bmatrix} 16 & 20 & 14 & 8 \\ 8 & 10 & 0 & 12 \\ 11 & 12 & 9 & 20 \\ 9 & 10 & 15 & 10 \end{bmatrix}$$

$$= \sqrt{2} \begin{bmatrix} 12 & 15 & 7 & 10 \\ 10 & 11 & 12 & 15 \\ -4 & -5 & -7 & 2 \\ -1 & -1 & 3 & -5 \end{bmatrix}$$

Note that each column of A has indeed been transformed to a corresponding column of weighted averages and differences. □

Now that we see how W_N works on a small example, let's try it on a digital image.

Example 6.7 (Processing the Columns of an Image with the HWT). *A 440×440 digital grayscale image A is plotted on the left in Figure 6.8. The plot to its right is the product $W_{440} A$. Notice that the bottom half of $W_{440} A$ represents weighted differences of each column of A. If there is not much difference in consecutive pixel values, this difference is close to zero and plotted as black. The bigger the difference, the lighter the pixel intensity.* □

Processing the Rows of a Matrix with the HWT

We now know how to process the columns of a digital image. Since our images are two-dimensional, it is natural to expect our transformation to process the rows as well. We learned in Chapter 2 that we can process the rows of $N \times N$ matrix A if we

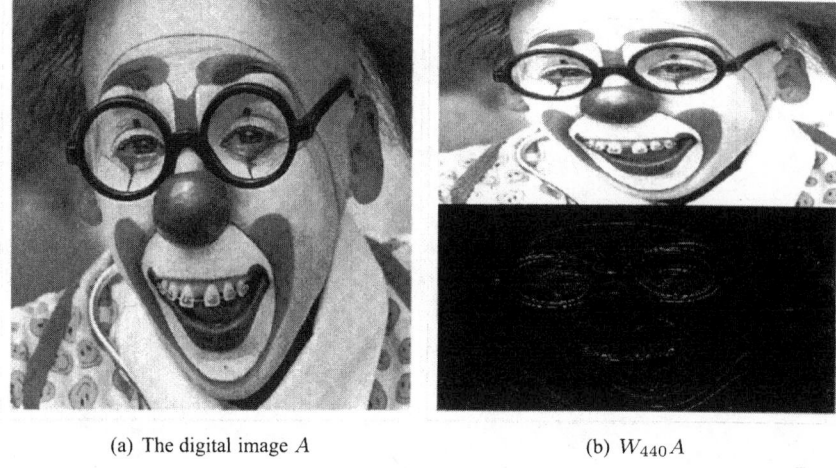

(a) The digital image A (b) $W_{440}A$

Figure 6.8 The HWT applied to the columns of an image.

right-multiply A by some matrix. Since we want averages and differences, a natural choice for this product is W_N.

Let's consider a simple case. Suppose A is a 4×4 matrix. If we compute AW_4, we have

$$
AW_4 = \begin{bmatrix} a_{11} & a_{12} & a_{13} & a_{14} \\ a_{21} & a_{22} & a_{23} & a_{24} \\ a_{31} & a_{32} & a_{33} & a_{34} \\ a_{41} & a_{42} & a_{43} & a_{44} \end{bmatrix} \sqrt{2} \begin{bmatrix} \dfrac{1}{2} & \dfrac{1}{2} & 0 & 0 \\ 0 & 0 & \dfrac{1}{2} & \dfrac{1}{2} \\ -\dfrac{1}{2} & \dfrac{1}{2} & 0 & 0 \\ 0 & 0 & -\dfrac{1}{2} & \dfrac{1}{2} \end{bmatrix}
$$

$$
= \frac{\sqrt{2}}{2} \begin{bmatrix} a_{11} - a_{13} & a_{11} + a_{13} & a_{12} - a_{14} & a_{12} - a_{14} \\ a_{21} - a_{23} & a_{21} + a_{23} & a_{22} - a_{24} & a_{22} - a_{24} \\ a_{31} - a_{33} & a_{31} + a_{33} & a_{32} - a_{34} & a_{32} - a_{34} \\ a_{41} - a_{43} & a_{41} + a_{43} & a_{42} - a_{44} & a_{42} - a_{44} \end{bmatrix}
$$

and this result is not what we want. If we are really going to process rows, our output should be

$$
\frac{\sqrt{2}}{2} \left[\begin{array}{cc|cc} a_{11} + a_{12} & a_{13} + a_{14} & -a_{11} + a_{12} & -a_{13} + a_{14} \\ a_{21} + a_{22} & a_{23} + a_{24} & -a_{21} + a_{22} & -a_{23} + a_{24} \\ a_{31} + a_{32} & a_{33} + a_{34} & -a_{31} + a_{32} & -a_{33} + a_{34} \\ a_{41} + a_{42} & a_{43} + a_{44} & -a_{41} + a_{42} & -a_{43} + a_{44} \end{array} \right]
$$

To obtain this result, we must multiply A on the right by W_4^T instead of W_4. In the case that A is an $N \times N$ matrix, the product AW_N^T processes the rows of A into weighted averages and differences. Let's look at an example.

Example 6.8 (Processing the Rows of an Image with the HWT). *A* 440×440 *digital gray scale image is plotted to the left in Figure 6.9. The plot to its right is the product* AW_{440}^T. *Notice that the right half of* $W_{440}^T A$ *represents weighted differences of each row of A. If there is not much difference in consecutive pixel values, this difference is close to zero and plotted as black. The bigger the difference, the lighter the pixel intensity.* □

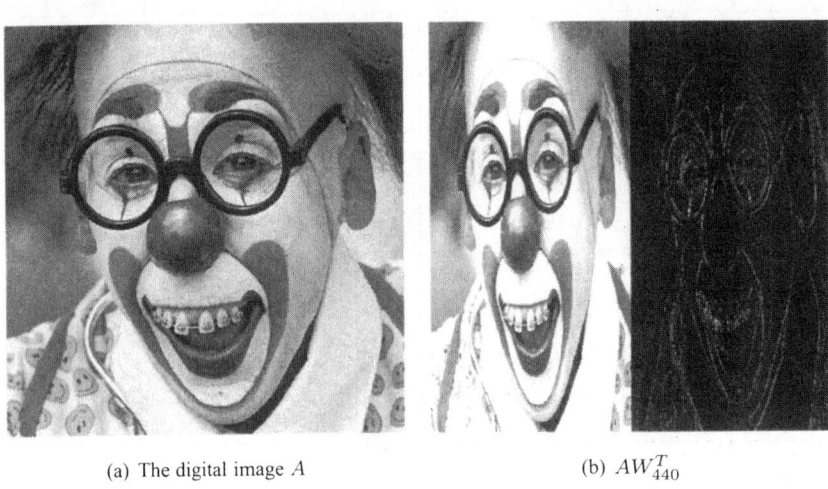

(a) The digital image A (b) AW_{440}^T

Figure 6.9   The HWT applied to the rows of an image.

The Two-Dimensional Transform as a Product of Matrices

How do we put all this together? We want to process the rows and the columns of A so we need to multiply on the left by W_N and on the right by W_N^T. Thus our two-dimensional transformation seems to take the form $W_N A W_N^T$. Does this make sense? Let's try it out on an arbitrary 4×4 matrix and analyze the result.

$$W_4 A W_4^T = \sqrt{2} \begin{bmatrix} \frac{1}{2} & \frac{1}{2} & 0 & 0 \\ 0 & 0 & \frac{1}{2} & \frac{1}{2} \\ -\frac{1}{2} & \frac{1}{2} & 0 & 0 \\ 0 & 0 & -\frac{1}{2} & \frac{1}{2} \end{bmatrix} \begin{bmatrix} a_{11} & a_{12} & a_{13} & a_{14} \\ a_{21} & a_{22} & a_{23} & a_{24} \\ a_{31} & a_{32} & a_{33} & a_{34} \\ a_{41} & a_{42} & a_{43} & a_{44} \end{bmatrix} \sqrt{2} \begin{bmatrix} \frac{1}{2} & 0 & -\frac{1}{2} & 0 \\ \frac{1}{2} & 0 & \frac{1}{2} & 0 \\ 0 & \frac{1}{2} & 0 & -\frac{1}{2} \\ 0 & \frac{1}{2} & 0 & \frac{1}{2} \end{bmatrix}$$

We combine the $\sqrt{2}$'s and multiply $W_4 A$ first to obtain

$$
W_4 A W_4^T = 2
\begin{bmatrix}
\frac{a_{11}+a_{21}}{2} & \frac{a_{12}+a_{22}}{2} & \frac{a_{13}+a_{23}}{2} & \frac{a_{14}+a_{24}}{2} \\
\frac{a_{31}+a_{41}}{2} & \frac{a_{32}+a_{42}}{2} & \frac{a_{33}+a_{43}}{2} & \frac{a_{34}+a_{44}}{2} \\
\frac{-a_{11}+a_{21}}{2} & \frac{-a_{12}+a_{22}}{2} & \frac{-a_{13}+a_{23}}{2} & \frac{-a_{14}+a_{24}}{2} \\
\frac{-a_{31}+a_{41}}{2} & \frac{-a_{32}+a_{42}}{2} & \frac{-a_{33}+a_{43}}{2} & \frac{-a_{34}+a_{44}}{2}
\end{bmatrix}
\begin{bmatrix}
\frac{1}{2} & 0 & -\frac{1}{2} & 0 \\
\frac{1}{2} & 0 & \frac{1}{2} & 0 \\
0 & \frac{1}{2} & 0 & -\frac{1}{2} \\
0 & \frac{1}{2} & 0 & \frac{1}{2}
\end{bmatrix}
$$

Computing the final product is better understood if we group the elements into four 2×2 blocks. Since these blocks are fundamental in understanding the geometric nature of the wavelet transform both here and in subsequent chapters, we use special notation for them. We have

$$
W_4 A W_4^T = 2 \left[\begin{array}{c|c} \mathcal{B} & \mathcal{V} \\ \hline \mathcal{H} & \mathcal{D} \end{array} \right]
$$

where

$$
\mathcal{B} = \frac{1}{4}
\begin{bmatrix}
a_{11} + a_{12} + a_{21} + a_{22} & a_{13} + a_{14} + a_{23} + a_{24} \\
a_{31} + a_{32} + a_{41} + a_{42} & a_{33} + a_{34} + a_{43} + a_{44}
\end{bmatrix}
$$

$$
\mathcal{V} = \frac{1}{4}
\begin{bmatrix}
(a_{12} + a_{22}) - (a_{11} + a_{21}) & (a_{14} + a_{24}) - (a_{23} + a_{13}) \\
(a_{32} + a_{42}) - (a_{31} + a_{41}) & (a_{34} + a_{44}) - (a_{33} + a_{43})
\end{bmatrix}
$$

$$
\mathcal{H} = \frac{1}{4}
\begin{bmatrix}
(a_{21} + a_{22}) - (a_{11} + a_{12}) & (a_{23} + a_{24}) - (a_{13} + a_{14}) \\
(a_{41} + a_{42}) - (a_{31} + a_{32}) & (a_{43} + a_{44}) - (a_{33} + a_{34})
\end{bmatrix}
$$

$$
\mathcal{D} = \frac{1}{4}
\begin{bmatrix}
(a_{11} + a_{22}) - (a_{12} + a_{21}) & (a_{13} + a_{24}) - (a_{14} + a_{23}) \\
(a_{31} + a_{42}) - (a_{32} + a_{41}) & (a_{33} + a_{44}) - (a_{34} + a_{43})
\end{bmatrix}
$$

To analyze these blocks, we organize A into blocks and name each block:

$$
A = \left[\begin{array}{cc|cc}
a_{11} & a_{12} & a_{13} & a_{14} \\
a_{21} & a_{22} & a_{23} & a_{24} \\
\hline
a_{31} & a_{32} & a_{33} & a_{34} \\
a_{41} & a_{42} & a_{43} & a_{44}
\end{array} \right]
= \left[\begin{array}{c|c}
A_{11} & A_{21} \\
\hline
A_{21} & A_{22}
\end{array} \right]
$$

Look at the $(1, 1)$ element in each of $\mathcal{B}, \mathcal{V}, \mathcal{H},$ and $\mathcal{D}$. They are all constructed using the elements in A_{11}. In general, the i, j elements of $\mathcal{B}, \mathcal{V}, \mathcal{H},$ and $\mathcal{D}$ are constructed from elements of $A_{ij}, i, j = 1, 2$.

Let's start with block $\mathcal{B}$. Note that each element b_{ij} of $\mathcal{B}$ is constructed by averaging the values of A_{ij}, $i, j = 1, 2$. For example,

$$A_{21} = \begin{bmatrix} a_{31} & a_{32} \\ a_{41} & a_{42} \end{bmatrix} \longrightarrow \frac{a_{31} + a_{32} + a_{41} + a_{42}}{4} = b_{21}$$

We can thus think of $\mathcal{B}$ as an approximation or *blur* of A.

What about the other blocks? Let's look at the upper right block $\mathcal{V}$. Not only is each element v_{ij} constructed from the elements of A_{ij}, $i, j = 1, 2$, but each v_{ij} is constructed in exactly the same way. We first compute the sum of each of the two columns of A_{ij}. We subtract these sums and divide by 4. For example,

$$A_{12} = \begin{bmatrix} \begin{bmatrix} a_{13} \\ a_{23} \end{bmatrix} & \begin{bmatrix} a_{14} \\ a_{24} \end{bmatrix} \end{bmatrix} \longrightarrow \frac{(a_{13} + a_{23}) - (a_{14} + a_{24})}{4} = v_{21}$$

We can interpret this computation geometrically as a weighted difference of the column sums in the block. With regard to a digital image, we can think of $\mathcal{V}$ as describing the *vertical changes* between the original image A and the blur $\mathcal{B}$.

In a similar way, we can view $\mathcal{H}$ as describing the *horizontal changes* between the original image A and the blur $\mathcal{B}$. Finally, the lower left block $\mathcal{D}$ measures *diagonal differences* between the original image A and the blur $\mathcal{B}$.

To provide a more formal explanation for the geometric significance of these blocks, suppose that A is an $N \times N$ matrix (N even). We write W_N in block format

$$W_N = \begin{bmatrix} H \\ G \end{bmatrix}$$

where H and G are $\frac{N}{2} \times N$ matrices. Matrix H is the lowpass or averaging portion of W_N given in (6.7), and G is the highpass or differencing portion of W_N given in (6.7).

We have

$$W_N A W_N^T = \begin{bmatrix} H \\ G \end{bmatrix} \cdot A \cdot \begin{bmatrix} H \\ G \end{bmatrix}^T = \begin{bmatrix} HA \\ GA \end{bmatrix} \begin{bmatrix} H^T \mid G^T \end{bmatrix}$$

$$= \begin{bmatrix} HAH^T & HAG^T \\ GAH^T & GAG^T \end{bmatrix} = \begin{bmatrix} \mathcal{B} & \mathcal{V} \\ \mathcal{H} & \mathcal{D} \end{bmatrix} \qquad (6.24)$$

Recall that H averages (with weight $\sqrt{2}$) along the columns of A and H^T averages along the rows of A. So, if we multiply left to right, HAH^T first averages the columns of A and then computes the averages along the rows of HA. This creates the approximation (or blur) $\mathcal{B}$ of A.

The matrix G^T differences (with weight $\sqrt{2}$) along the rows of A. So if we again multiply left to right, we see that the product HAG^T first computes the averages along columns of A and then differences along rows of HA. This is exactly the vertical differences described above.

The matrix GAH^T first differences along columns of A and then averages along rows of GA. This product produces the horizontal differences stored in $\mathcal{H}$. Finally the matrix GAG^T first differences along the columns of A and next differences along the rows of GA, forming the diagonal differences found in $\mathcal{D}$. The schematic in Figure 6.10 summarizes our discussion to this point.

$$A \longmapsto W_N A W_N^T = \left[\begin{array}{c|c} \mathcal{B} & \mathcal{V} \\ \hline \mathcal{H} & \mathcal{D} \end{array} \right] = \left[\begin{array}{c|c} \text{blur} & \begin{array}{c}\text{vertical} \\ \text{differences}\end{array} \\ \hline \begin{array}{c}\text{horizontal} \\ \text{differences}\end{array} & \begin{array}{c}\text{diagonal} \\ \text{differences}\end{array} \end{array} \right]$$

Figure 6.10 The two-dimensional HWT takes an image matrix A and maps it to a new matrix consisting of four blocks. The upper left block $\mathcal{B}$ is a blur or approximation of A, the upper right block $\mathcal{V}$ represents the vertical differences between the original and the blur, and the lower blocks $\mathcal{H}$ and $\mathcal{D}$ represent the horizontal and diagonal differences between A and the blur, respectively.

The following example further illustrates the block structure of the two-dimensional HWT.

Example 6.9 (Applying the HWT to a Digital Image). *We return to the image used in Examples 6.7 and 6.8. We plot it and its two-dimensional HWT in Figure 6.11. The upper left block is a blur of the original. The upper right block represents vertical differences between A and the blur, the lower left block gives horizontal differences between A and the blur, and the bottom right block show diagonal differences between A and the blur.* ☐

Rectangular Matrices and the HWT

Rectangular matrices present no problems for the two-dimensional HWT as long as the dimensions are even. Suppose that A is an $N \times M$ matrix with N and M even. Since each column of A has N elements, we multiply on the left by W_N. The result, $W_N A$, is an $N \times M$ matrix. Each row in $W_N A$ has M elements, so we need to process the rows by multiplying $W_N A$ on the right by W_M^T. The result will be an $N \times M$ matrix.

Here is an example.

Example 6.10 (Applying the HWT to a Rectangular Digital Image). *The image A shown on the left in Figure 6.12 is a 200×300 matrix. It is plotted along with its two-dimensional HWT $W_{200} A W_{300}^T$.* ☐

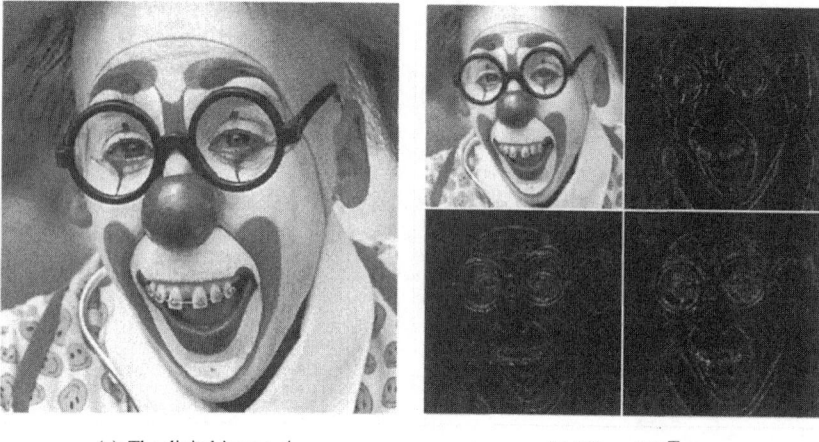

(a) The digital image A.   (b) $W_{440} A W_{440}^T$.

Figure 6.11   The HWT applied to an image.

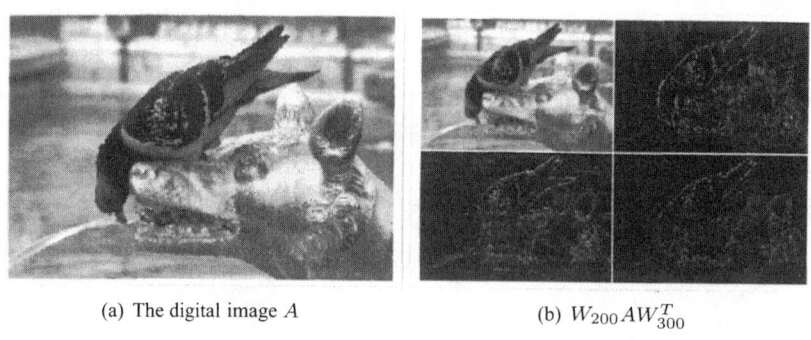

(a) The digital image A (b) $W_{200} A W_{300}^T$

Figure 6.12   The HWT applied to a rectangular image.

Inverting the Two-Dimensional Transform

Inverting the two-dimensional HWT is a simple process. Suppose that A is an $N \times M$ matrix and we have computed the HWT

$$C = W_N A W_M^T. \tag{6.25}$$

We can exploit the fact that W_N and W_M are orthogonal matrices to recover A. We multiply both sides of (6.25) on the left by W_N^T and W_M on the right to obtain

$$W_N^T C W_M = W_N^T W_N A W_M^T W_M = (W_N^T W_N) A (W_M^T W_M) = I_N A I_M = A$$
$$(6.26)$$

Here I_N and I_M are $N \times N$ and $M \times M$ identity matrices, respectively.

Now that we know how to perform a two-dimensional HWT and its inverse, let's write algorithms for both operations. We start with the two-dimensional HWT.

Pseudocode for the Two-Dimensional Transform and Its Inverse

Suppose that we are given the $N \times M$ matrix A and wish to compute its HWT $C = W_N A W_M^T$. We start with the intermediate step $R = W_N A$. To obtain R, we simply apply Algorithm 6.1 to each column of A. We need an algorithm to perform this task.

Algorithm 6.5 (LeftHWT). *Given an $N \times M$ matrix, we compute $R = W_N A$. The routine consists simply of a loop over the M columns of A, where we apply Algorithm 6.1 to each column of A. In the pseudocode that follows, $\mathbf{a}^k$ is the kth column of A and $\mathbf{r}^k$ is the kth column of R.*

Algorithm: LeftHWT
Input: $N \times M$ matrix A.
Output: $N \times M$ matrix $R = W_N A$.

For[$k = 1$, $k \leq M$, $k{+}{+}$,
    $\mathbf{r}^k = $ **HWT1D1**[$\mathbf{a}^k$]
]

Return[R]

□

The remaining task is then to compute $C = RW_M^T$. Rather than writing a new pseudocode to process the rows of R, let's use some elementary linear algebra. Using Problem 2.22 in Section 2.2, we have

$$C^T = (RW_M^T)^T = (W_M^T)^T R^T = W_M R^T$$

So rather than writing a new routine, we simply transpose R, send it to Algorithm 6.5, and then transpose the result! We are now ready to write an algorithm to compute the two-dimensional HWT of $N \times M$ matrix A.

Algorithm 6.6 (Haar Wavelet Transform: 2D). *Given an $N \times M$ matrix A, we compute the two-dimensional HWT $C = W_N A W_M^T$. We first call Algorithm 6.5 to compute $R = W_N A$. We then call Algorithm 6.5 again with R^T and then transpose the result to obtain our final result.*

Algorithm: HWT2D1
Input: $N \times M$ matrix A.
Output: $N \times M$ matrix $C = W_N A W_M^T$.

$R =$ **LeftHWT** $[A]$
$S =$ **LeftHWT** $[R^T]$
$C = S^T$

Return $[C]$

☐

An algorithm for the inverse two-dimensional HWT is easy to write. We wish to recover $N \times M$ matrix A, where $C = W_N A W_M^T$. From (6.26), we have

$$A = W_N^T C W_M \qquad (6.27)$$

As with Algorithm 6.6, we first write an intermediate algorithm to compute $R = W_N^T C$. This algorithm simply applies Algorithm 6.2 to each column of C.

Algorithm 6.7 (LeftIHWT). *Given an $N \times M$ matrix C, we compute $R = W_N^T C$. The routine consists simply of a loop over the M columns of C where we apply Algorithm 6.2 to each column of C.* · *In the pseudocode that follows, $\mathbf{c}^k$ is the kth column of C and $\mathbf{r}^k$ is the kth column of R.*

Algorithm: LeftIHWT
Input: $N \times M$ matrix C.
Output: $N \times M$ matrix $R = W_N^T C$.

For $[k = 1, k \le M, k++,$
    $\mathbf{r}^k = $ **IHWT1D1** $[\mathbf{c}^k]$
$]$

Return $[R]$

☐

Now that we have $R = W_N^T C$, we can easily write a routine to compute the inverse HWT. We have $A = R W_M$. We transpose each side of this equation to obtain $A^T = W_M^T R^T$ and see that we can call Algorithm 6.7 with R^T and obtain A^T. We can complete the process by computing the transpose of A^T.

Algorithm 6.8 (Inverse Haar Wavelet Transform: 2D). *Given an $N \times M$ matrix C, we compute the two-dimensional inverse HWT $A = W_N^T C W_M$. We first compute $R = W_N^T C$ using Algorithm 6.7 and then send R^T as input to algorithm Algorithm 6.7. The result is A^T. We return the transpose of this matrix to complete the process.*

Algorithm: IHWT2D1

Input: $N \times M$ matrix C.

Output: $N \times M$ matrix $A = W_N^T C W_M$.

$R = $**LeftIHWT**$[C]$

$S = $**LeftIHWT**$[R^T]$

$A = S^T$

Return$[A]$

$\square$

Iterating the Process

In Section 6.2 we learned how to iterate the one-dimensional HWT. We applied the HWT successively to the lowpass portion of the output of the preceding step. We can easily extend this iterative process to two-dimensional data.

Consider the $N \times M$ (N, M even) matrix A and its HWT $C_1 = W_N A W_M^T$. In block form,

$$C_1 = W_N A W_M^T = \left[\begin{array}{c|c} \mathcal{B} & \mathcal{V} \\ \hline \mathcal{H} & \mathcal{D} \end{array}\right]$$

where $\mathcal{B}$, $\mathcal{V}$, $\mathcal{H}$, and $\mathcal{D}$ are summarized in Figure 6.10. In particular, $\mathcal{B}$ is an approximation (or blur) of A. We also know that $\mathcal{B}$ is obtained by applying the lowpass portions of W_N and W_M^T to A. We thus perform an iteration by applying the HWT to $\mathcal{B}$. That is, we compute $C_2 = W_{N/2} \mathcal{B} W_{M/2}^T$ and overwrite $\mathcal{B}$ with C_2. This process is entirely invertible — to obtain $\mathcal{B}$, we compute $W_{N/2}^T C_2 W_{M/2}$.

Let's look at an example.

Example 6.11 (Iterating the 2D Haar Wavelet Transformation). *We return to the 256×256 image introduced in Example 6.7. We first plot the original image (denoted by A) and its HWT C_1 in Figure 6.13.*

We now extract the upper left corner of C_1. This is the matrix $\mathcal{B}$, and it represents an approximation of A. We next compute the HWT of $\mathcal{B}$ and call it $C_2 = W_{128} \mathcal{B} W_{128}^T$. The blur $\mathcal{B}$ and the resulting transform are plotted in Figure 6.14.

Finally, we replace the upper left corner of C_1 with C_2 to produce a composite image of two iterations of the HWT. The result is plotted in Figure 6.15(a). The cumulative energies for A, C_1, and C_2 plotted in Figure 6.15(b).

Clearly, C_2 needs fewer elements to store more of the energy contained in the original data. We would probably do better if we were to iterate a third time. $\square$

Let A be an $N \times M$ matrix with N, M divisible by 2^i. We can compute i iterations of the HWT by utilizing the notation given in (6.21). We have

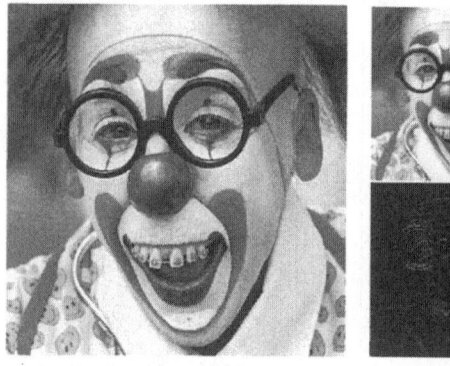
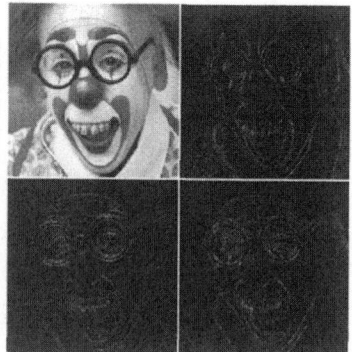

(a) The digital image A (b) One iteration $C_1 = W_{440} A W_{440}^T$

Figure 6.13 One iteration of the HWT applied to an image.

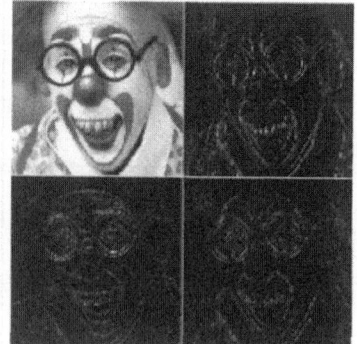

(a) The blur B (b) The product $C_2 = W_{220} B W_{220}^T$

Figure 6.14 A second application of the HWT to B. Note that B has been enlarged for purposes of illustration.

$$C = W_{N,i} W_{N,i-1} \cdots W_{N,2} W_N \, A \, W_M^T W_{M,2}^T \cdots W_{M,i-1}^T W_{M,i}^T \tag{6.28}$$

We can easily recover A. We have

$$A = W_N^T W_{N,2}^T \cdots W_{N,i-1}^T W_{N,i}^T \, C \, W_{M,i} W_{M,i-1} \cdots W_{M,2} W_M \tag{6.29}$$

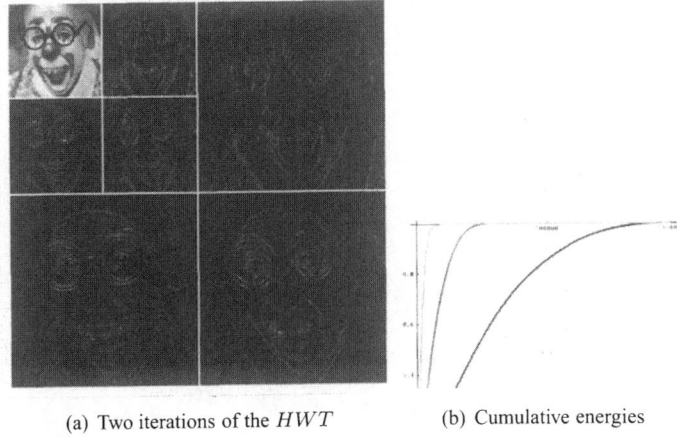

(a) Two iterations of the HWT (b) Cumulative energies

Figure 6.15   Plot of two iterations of the HWT on matrix A and the cumulative energies for A (dark dots), C_1 (medium dots), and C_2 (light dots).

Pseudocodes for the Iterated 2D Transform and Its Inverse

Our next task is to develop algorithms for implementing the iterated two-dimensional HWT and its inverse. We start with an algorithm that will compute i iterations of the two-dimensional HWT.

Algorithm 6.9 (Haar Wavelet Transform: 2D). *Given an $N \times M$ matrix A (with M and N both divisible by 2^i), this algorithm returns C, where C is given by (6.28). This routine uses two algorithms called **GetCorner**$[S, r, c]$ and **PutCorner**$[S,T]$. **Get-Corner** takes as input an $N \times M$ matrix S and values r and c and extracts the upper left corner of S. This new matrix has dimensions $r \times c$. It is assumed that $r \leq N$ and $c \leq M$. The routine **PutCorner** takes as input matrices S and T and replaces the upper left corner of S with T. It is assumed that the dimensions of T are less than or equal to N, M. We need these routines to construct C.*

*We start by assigning $C = A$. We proceed through a loop and ultimately load C with i iterations of the HWT. The loop starts by calling **GetCorner** to extract the upper left $N/2^k \times M/2^k$ matrix from C. Note that on the first pass of this loop ($k = 0$), we are simply extracting the matrix A. Once we have the desired corner, we apply **HWT2D1** to it. We then **PutCorner** this corner back into C and proceed to the next step of the loop.*

Algorithm: HWT2D
Input: Integer i, $N \times M$ matrix A with N, M divisible by 2^i.
Output: Matrix C where C is given by (6.28).

$C = A$
For $[k = 0,\ k < i,\ k++,$
$\qquad R = \textbf{GetCorner}[C, N/2^k, M/2^k]$
$\qquad S = \textbf{HWT2D1}[R]$
$\qquad C = \textbf{PutCorner}[C, S]$
$]$

Return $[C]$

$\square$

The next algorithm performs i iterations of the inverse two-dimensional HWT.

Algorithm 6.10 (Inverse Haar Wavelet Transform: 2D). *Given an $N \times M$ matrix C (with M and N both divisible by 2^i), this algorithm computes A, where A is given by (6.29). This routine is very similar to Algorithm 6.9 and also uses **GetCorner** and **PutCorner**.*

*We start by assigning $A = C$. We proceed through a loop and ultimately load A with i iterations of the inverse HWT. The loop starts by first calling **GetCorner** to extract the upper left $N/2^{i-k} \times M/2^{i-k}$ matrix from A. Note that on the first pass of this loop ($k = 1$), we are simply extracting the last iteration produced by Algorithm 6.9. Once we have the desired corner, we apply **IHWT2D1** to it. We then **PutCorner** this corner back into A and proceed to the next step of the loop.*

Algorithm: IHWT2D
Input: Integer i, $N \times M$ matrix B with N, M divisible by 2^i.
Output: Matrix A where A is given by (6.29).

$A = C$
For $[k = 1,\ k \leq i,\ k++,$
$\qquad R = \textbf{GetCorner}[A, N/2^{(i-k)}, M/2^{(i-k)}]$
$\qquad S = \textbf{IHWT2D1}[R]$
$\qquad A = \textbf{PutCorner}[A, S]$
$]$

Return $[A]$

$\square$

Transforming Color Images

We conclude this section with a discussion of how to apply the HWT to color digital images. Recall from Chapter 3 that a color digital image is constructed using three matrices. The elements of matrix R correspond to the red intensities of the image (with range 0 to 255), elements of matrix G hold the green intensities, and matrix B houses the blue intensities.

There are some applications where i iterations of the HWT are applied to a color digital image simply by applying Algorithm 6.9 to each of R, G, and B. A more common method used in applications such as image compression is to first convert the R, G, and B components to YCbCr space and then apply Algorithm 6.9 to each of Y, Cb, and Cr.

Example 6.12 (Transform of a Color Image). *Consider the* 240×160 *image and red, green, and blue color channels plotted in Figure 3.19. We apply the two-dimensional HWT to each of* R, G, *and* B. *The resulting transformations are plotted in Figure 6.16(a) – 6.16(c).*

For many applications, we first convert the image to YCbCr space. This conversion is plotted in Figure 3.22. In Figure 6.16(d) – 6.16(f) we have plotted the HWT of each of the Y, Cb, *and* Cr *channels.*   □

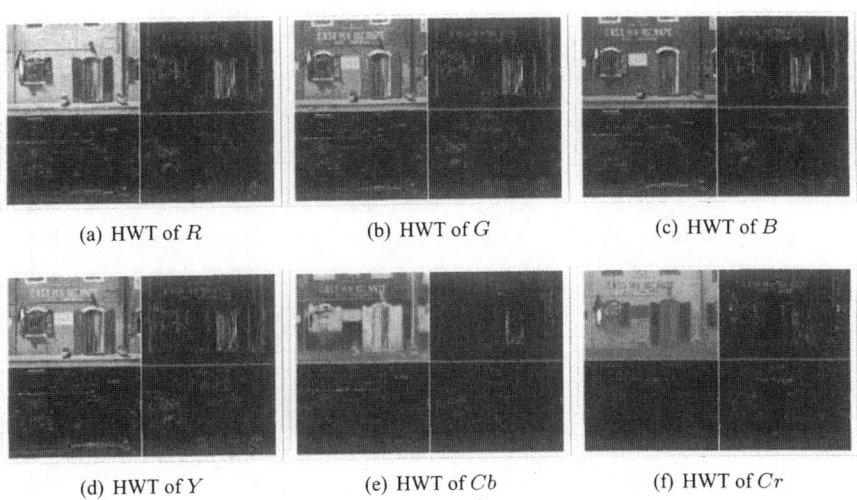

|(a) HWT of R | (b) HWT of G | (c) HWT of B |

|(d) HWT of Y | (e) HWT of Cb | (f) HWT of Cr |

Figure 6.16   The HWT of each of the color channels R, G, and B in Figure 3.19(b) – 3.19(d) and the Y, Cb, and Cr channels. See the color plates for color versions of the top three images.

PROBLEMS

6.16   Consider the matrix

$$A = \begin{bmatrix} 100 & 200 & 100 & 120 \\ 100 & 160 & 120 & 120 \\ 0 & 80 & 100 & 80 \\ 0 & 40 & 80 & 0 \end{bmatrix}$$

Compute two iterations of the two-dimensional HWT. You should do this computation by hand.

6.17   Consider the 6×8 matrix A with entries a_{jk}, $j = 1, \ldots, 6$, $k = 1, \ldots, 8$. Let C denote one iteration of the two-dimensional HWT applied to A. That is, $C = W_6 A W_8^T$. Write down explicit formulas for the elements c_{jk}, $j = 1, \ldots, 6$, $k = 1, \ldots, 8$, of C in terms of the a_{jk}. (*Hint:* You should have four different cases for formulas for c_{jk} depending on how j, k compare to $3, 4$, respectively.)

6.18   Generalize Problem 6.17 to the case where the dimensions of A are $N \times M$, with both N and M even positive integers.

6.19   Suppose that C is a 6×8 matrix with elements c_{jk} where $j = 1, \ldots, 6$ and $k = 1, \ldots, 8$. Let A be the matrix that results when one iteration of the two-dimensional inverse HWT is applied to C. That is $A = W_6^T C W_8$. Write down explicit formulas for the elements a_{jk}, $j = 1, \ldots, 6$, $k = 1, \ldots, 8$, of A in terms of the c_{jk}. (*Hint:* Problem 6.17 should be helpful.)

6.20   Generalize Problem 6.19 to the case where the dimensions of A are $N \times M$, with both N and M even positive integers.

6.21   Suppose that A is an $N \times N$ (N even) matrix that represents the intensity levels of some digital image and we apply one iteration of the two-dimensional HWT to obtain C. In order to crudely compress the image, all elements of vertical, horizontal, and diagonal blocks of C are set to zero. That is, $\tilde{C}$ has the block form

$$\begin{bmatrix} B & 0 \\ \hline 0 & 0 \end{bmatrix}$$

We can approximate A by applying one iteration of the two-dimensional inverse HWT to $\tilde{C}$. If we call the result $\tilde{A}$, write down the elements $\tilde{a}_{jk}$, $j, k = 1, \ldots, N$ of $\tilde{A}$ in terms of the elements $\tilde{c}_{jk}$ of $\tilde{C}$.

6.22   Repeat Problem 6.21, but this time assume that N is divisible by 4 and apply two iterations of the HWT to A. Again, we set the elements of the vertical, horizontal, and diagonal elements (of both iterations) to zero. Thus, the elements of $\tilde{C}$ are 0 save for the $\frac{N}{4} \times \frac{N}{4}$ block in its upper right-hand corner. Write down the elements of $\tilde{A}$ in terms of the elements of $\tilde{C}$. How does this compare to the result from Problem 6.21?

What would happen if N were divisible by 8 and we applied three iterations of the HWT?

Computer Labs

♦ **6.5 Software Development: 2D Haar Wavelet Transformations and Inverse Wavelet Transformations.** From the text Web site, access the development package hwt2d. In this development lab you will create functions for computing the two-dimensional Haar wavelet transformation and the inverse Haar wavelet transformation. Instructions are provided that allow you to add all of these functions to the software package DiscreteWavelets.

6.6 2D Haar Wavelet Transformations and Inverse Wavelet Transformations. From the text Web site, access the lab haariterated2d. This lab utilizes the functions for the two-dimensional iterated HWT and the iterated inverse HWT in the DiscreteWavelets package. In this lab you will compare cumulative energies for various iterated transforms of a given image and apply the HWT to color images. You will also develop code that for a given iterated transform of an image, inverts only a selected part of the image.

6.7 Incremental Image Resolution. From the text Web site, access the lab hwtres. In this lab you will learn how wavelets can be used to display an image at increasingly finer resolution as it is transmitted over the Internet.

6.8 Partial Image Inversion. From the text Web site, access the lab partinv. Images are often stored in terms of i iterations of their wavelet transforms. The transforms are typically compressed to save space. In some applications, we desire only to invert a particular portion of the transformed image. One of the advantages of the Haar wavelet transform is that you need not invert the entire transform to recover a particular portion of the original image. In this lab you create a routine that takes as input i iterations of the Haar wavelet transform of a digital image and the coordinates of a submatrix in the original image and returns that portion of the image.

6.4 APPLICATIONS: IMAGE COMPRESSION AND EDGE DETECTION

We conclude Chapter 6 with two applications of the Haar wavelet transformation. The first application is data compression and the second application is the detection of edges in digital images. It is not our intent to produce state-of-the-art procedures in either case. Indeed, we will see that much better wavelet transforms exist for data compression. The HWT is easy to implement and understand, and thus it is a good tool for providing you with a basic understanding of how the wavelet transform applies to the problems.

Image Compression

We conduct a detailed study of image compression when we learn about the JPEG2000 compression format in Chapter 12. For now, we study the basic algorithm for compressing a signal or image.

The idea is as follows. Suppose that we want to send a digital image file to a friend via the Internet. To expedite the process, we wish to *compress* the contents of the file. There are two forms of compression. The first form is *lossless compression*. Data compressed by lossless schemes can be recovered exactly with no loss of information. *Lossy compression* schemes alter the data in some way. Generally speaking, better results can be obtained using lossy schemes, but the savings come at the expense of information lost from the original data. We demonstrate a naive lossy compression scheme that utilizes the HWT. We will describe the process for two-dimensional data, but the method works just as well for univariate data.

Suppose that the intensities of a grayscale image of dimension $N \times M$ are stored in matrix A. Further assume that for some positive integer p, both N and M are divisible by 2^p. The first step in the process is to compute $1 \leq i \leq p$ iterations of the two-dimensional HWT. The idea is that most of the information in the transform will be stored in the blur, and the horizontal, vertical, and diagonal components will be sparse by comparison. The reason that this process works is that images are typically comprised of large homogeneous regions (think of a background in a still or the sky in an outdoor photograph). If there is little change in some part of an image, we expect the highpass portion for that part to be comprised primarily of small values.

The next step is to *quantize* the components of the HWT. Any values of the transform that are small in absolute value will be converted to zero. The modified transform is then converted to Huffman codes and the resulting bit stream can be encoded. The resulting file should be markedly smaller than the original.

There are a few questionable points with this algorithm. The first is the choice of wavelet transform. For now, we use the HWT since that is what we have developed to this point, but we will learn about some wavelet transforms that do a much better job than the HWT when it comes to compressing data.

The second point is the quantization process. How do we determine if a value is "small enough" to convert it to zero? In Chapters 9 and 12 we learn about some quantization schemes, but for this application we continue to use cumulative energy as our quantization tool. That is, we first choose a percentage of energy that we wish to preserve in the transform and then retain only the m largest elements (in absolute value) that constitute the energy. The other elements will be converted to zero.

The final item we need to address is Huffman coding. Recall that we orthogonalized the HWT W_N by multiplying the matrix $\tilde{W}_N$ in (6.4) by $\sqrt{2}$. Although this action allows us to easily invert the transformation, it results in output that is no longer integer valued. The Huffman coding process is much more effective when we can provide integer values. For this reason we use a modified version of the HWT. Recall the HWT matrix W_N given in (6.7). If we multiply this matrix by $\sqrt{2}$, then the entries are either 0 or ± 1. Thus, if we use the matrix

$$\hat{W}_N = \sqrt{2}\, W_N \tag{6.30}$$

then we can be assured that the elements of

$$B = \hat{W}_N A \hat{W}_M^T \tag{6.31}$$

will be integer valued. Moreover, the inverse process is quite simple — we have

$$\hat{W}_N^{-1} = \frac{1}{\sqrt{2}} W_N^{-1} = \frac{1}{\sqrt{2}} W_N^T$$

For this application we use the modified HWT (6.31). In Problem 6.23 you will be asked to find the integer range of the elements of B. Knowing the range of the transform is useful. If the minimum integer in the range is negative, then we could translate the data by this minimum amount so that all values are now nonnegative. This removes the need to encode negative signs. We would just need to include this translation factor in our compressed file so that the recipient would know to translate the data back to their original state before inverting the compression process.

Image Compression Example

Let's try the procedure on a digital image file.

Example 6.13 (Image Compression with the HWT). *Consider the 256×384 digital image with intensity matrix A plotted in Figure 6.17(a). Let C represent two iterations of the modified HWT.*

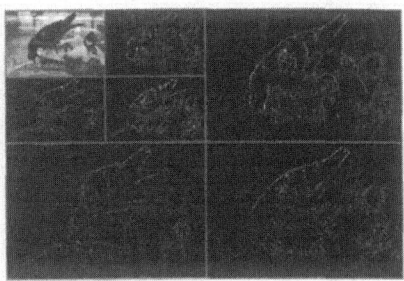

(a) Grayscale image A.   (b) HWT.

Figure 6.17   An image and two iterations of the HWT. A larger version of the image appears in Figure 6.21(a).

*Our goal is to compress A. The number of bits needed to store A is $256 \times 384 \times 8 = 786{,}432$. If we use a **CAS** to apply Huffman coding to the original image, we find that the new bit stream has length $776{,}821$ or a coding rate of 7.90223 bpp. The entropy for the original image is 7.86662. Remember, the entropy serves as a guide for the best bpp we can expect if we encode the given data. We will perform lossy compression in hopes of reducing the bit stream length of A while maintaining an acceptable degree of image resolution.*

In Figure 6.18, the cumulative energy vectors are plotted for both the original image and its wavelet transform. Note that the vectors each have $256 \cdot 384 = 98{,}304$ elements.

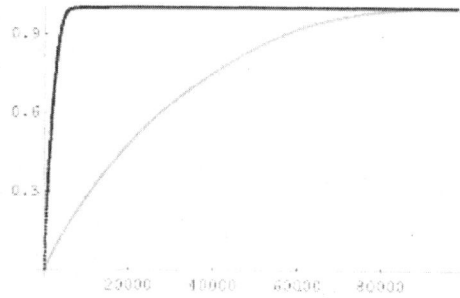

Figure 6.18   The cumulative energy vectors for the original image (gray) and the wavelet transform (black).

*Suppose we choose $r = 99.7\%$ for our energy threshold. We then use a **CAS** to find that the 8842 largest elements (in modulus) in the wavelet transform C produce 99.7% of the energy. We set the remaining 89462 elements of C to zero. Let's call this quantized wavelet transformation C_1. For comparison purposes, we repeat the process with energy levels $r = 99.9\%$ and $r = 99.99\%$ and produce quantized transforms C_2 and C_3. The results are summarized in Table 6.1.*

Table 6.1   Data for the quantized wavelet transformations C_1, C_2, and C_3.[a]

Transform	Energy Level	Nonzero Terms	% Zero
C_1	99.7%	8842	91.01
C_2	99.9%	14410	85.34
C_3	99.99%	34614	64.79

[a]The second column indicates the amount of energy conserved by the transform while the third column lists the number of nonzero terms still present in the quantized transforms, and the last column gives the percentage of elements that are zero in each quantized transform.

The quantized wavelet transformations are plotted in Figure 6.19. Note that the amount of detail increases as we move from C_1 to C_3.

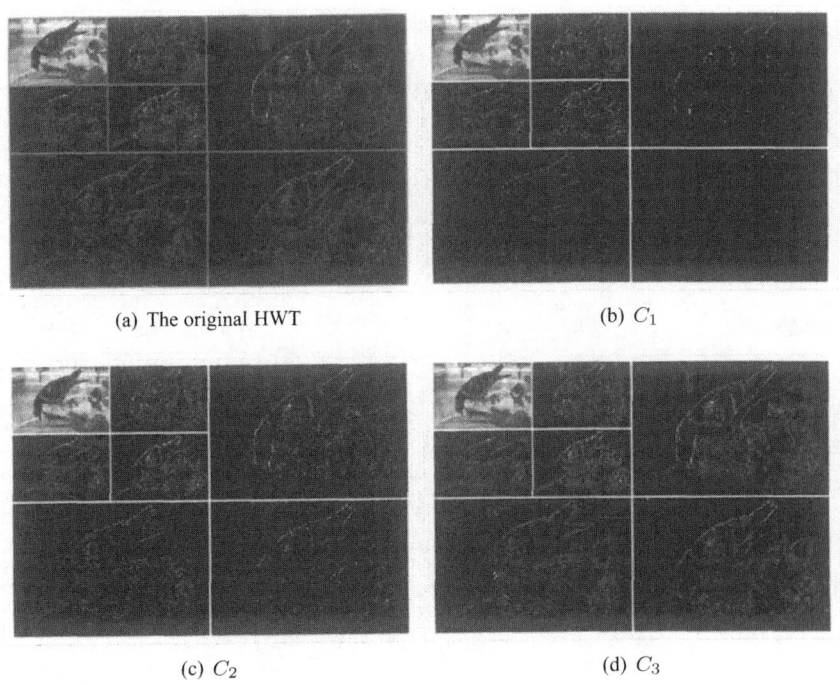

(a) The original HWT

(b) C_1

(c) C_2

(d) C_3

Figure 6.19   The HWT and various quantized versions of the HWT.

In Figure 6.20 we plot the distribution of pixel intensities for the original image and the quantized wavelet transformation C_3. The histograms for C_1 and C_2 are similar to that of C_3.

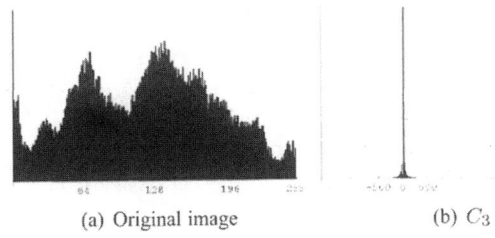

(a) Original image

(b) C_3

Figure 6.20   Histograms of the original image and the quantized HWT C_3.

If we transmit the image over the Internet, we first convert the elements of the quantized wavelet transform to Huffman codes and then create the bit stream for the file. This bit stream is encoded and added to header and footer information to create a compressed file. The length of the bit stream for each transform is given in Table 6.2. With regard to the original bit stream length of $8 \times 98{,}304 = 786{,}432$, the new bit stream lengths suggest that the quantized transforms will create storage sizes of roughly 25%, 32%, and 53%, respectively, of the original. The entropy gives an idea of the best bpp we could hope for when compressing each quantized transform. The bpp is given for each quantized transformation.

Table 6.2 Huffman bit stream length, bpp, and entropy for each quantized transform C_1, C_2, and C_3.

Transform	Bit Stream Length	bpp	Entropy
C_1	198794	2.02224	1.45639
C_2	253706	2.58083	2.17982
C_3	419370	4.26605	4.19548

After inspecting the results in Table 6.2, we might be inclined to use either C_1 or C_2 to compress our image. Of course, it is very important to weigh this selection against the quality of the compressed image.

We now invert each of the quantized transforms to obtain the compressed images A_1, A_2, and A_3. These images are plotted in Figure 6.21.

(a) The original image

(b) A_1 with PSNR 27.1050

(c) A_2 with PSNR 30.8503.

(d) A_3 with PSNR 40.2459.

Figure 6.21   The original images and three compressed images.

As you can see by the plots of the compressed images, C_3 produced a result that most resembles the original image.   One of the most difficult problems in image compression is deciding how and when to sacrifice good compression numbers for resolution. Often, the application drives the decision.   □

Compression of Color Digital Images

In our next example we consider image compression for color images.

Example 6.14 (Color Image Compression). *Consider the* 160×240 *image plotted in Figure 3.19. Rather than perform compression on each of the red, green, and blue channels, we first convert the image to YCbCr space. As discussed in Section 3.2, most of the energy is stored in the* Y *channel and the human eye is less sensitive to changes in the* Cb *and* Cr *channels. The* $Y, Cb,$ *and* Cr *channels for the image are plotted in Figure 3.22.*

What follows is a naive version of color image compression. In Chapter 12 we study the JPEG2000 image compression standard. This standard employs wavelet transforms that are better suited to the task of image compression.

We begin by performing two iterations of the HWT on each of the $Y, Cb,$ *and* Cr *channels. As was the case in Example 6.13, we use the modified HWT given by (6.31). The transforms are plotted in Figure 6.22.*

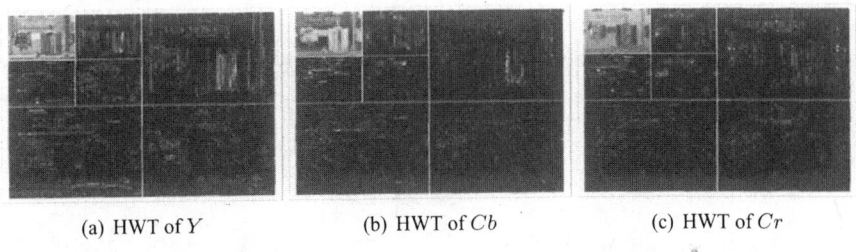

(a) HWT of Y (b) HWT of Cb (c) HWT of Cr

Figure 6.22 Two iterations of the HWT applied to the Y, Cb, and Cr channels plotted in Figure 3.22.

We next compute the cumulative energy vectors for the wavelet transforms of each of the three channels. We keep the largest intensities (in absolute value), that constitute 99.99% of the energy contained in each channel. The remaining intensities will be set to zero. Figure 6.23 shows the cumulative energies for each transformed channel, and Table 6.3 summarizes the number of values we retain for compression. The quantized wavelet transforms are plotted in Figure 6.24.

Each of $Y, Cb,$ *and* Cr *consists of* $38{,}400$ *pixels so without any compression strategy, we must store* $8 \cdot 38{,}400 = 307{,}200$ *bits or* 8 *bpp for each channel. The entropy values of* $Y, Cb,$ *and* Cr *are* 7.4502, 5.6209, *and* 6.1606, *respectively.*

To transmit the image over the Internet, we first construct the quantized wavelet transforms plotted in Figure 6.24, next convert the elements of each quantized trans-form to Huffman codes, and then create a bit stream for each channel. The bit stream length, bpp, and entropy for the quantized wavelet transforms of each channel are given in Table 6.4.

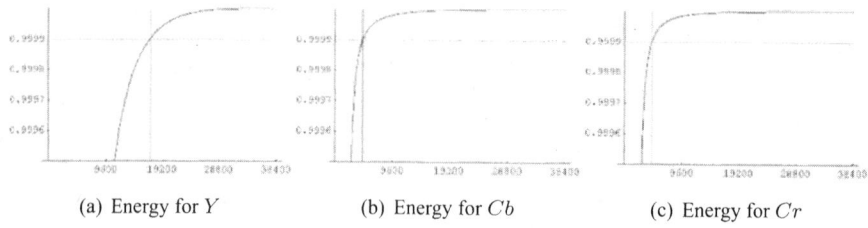

(a) Energy for Y (b) Energy for Cb (c) Energy for Cr

Figure 6.23   The cumulative energy vectors for the modified HWT of each of Y, Cb, and Cr.

Table 6.3   Values retained for compression.[a]

Transform of	Nonzero Terms	Zero Terms	% Zero
Y	17,049	21,351	55.60
Cb	4,629	33,771	87.95
Cr	4,569	33,831	88.10

[a]The second column shows the number of largest elements (in absolute value) that we must retain to preserve 99.99% of the energy of each modified HWT, the third column lists the number of terms we convert to zero in each transform, and the final column gives the percentage of elements in each transform that are zero.

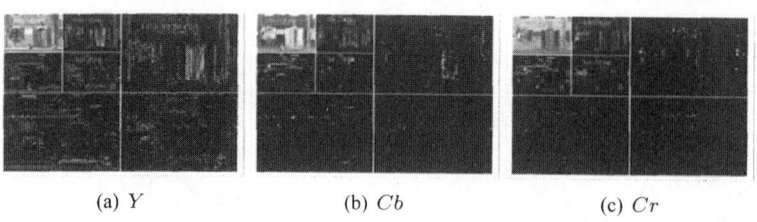

(a) Y (b) Cb (c) Cr

Figure 6.24   The quantized HWT for each of Y, Cb, and Cr.

The total bit stream length for the uncompressed image is $3 \cdot 8 \cdot 160 \cdot 240 = 921,600$. Compression via the HWT results in a bit stream length of $196,908 + 81,542 + 83,280 = 361,730$. This length is just over 39% of the original bit stream length. The uncompressed image is plotted in Figure 6.26. Considering that the bit stream length is about 39% of the original, the reconstructed image does a good job of approximating the original.

If we look at the HWT of the Cb and Cr channels, we can see why it is desirable to convert RGB to YCbCr space. The detail portions of these transforms are quite sparse, so we might be able to obtain even better compression rates if we lower the

Table 6.4 Huffman bit stream length, bpp, and entropy for each quantized transform of Y, Cb, and Cr.

Transform	Bit Stream Length	bpp	Entropy
Y	196,908	5.12781	5.10641
Cb	81,542	2.12349	1.65147
Cr	83,280	2.16875	1.69188

percent of energy we wish to preserve in the transforms. We can make the transforms even more sparse if we perform more iterations.

We set the number of iterations to three and use energy levels 99.97%, 99.7%, and 99.7% for the transforms of Y, Cb, and Cr, respectively. Table 6.5 lists the number of values we will retain for compression, and Figure 6.25 shows the quantized transform for each channel.

Table 6.5 Values retained for compression.[a]

Transform	% Energy	Nonzero Terms	Zero Terms	% Zero
Y	99.97	7,897	30,503	79.43
Cb	99.7	598	37,802	98.44
Cr	99.7	599	37,801	98.44

[a]The third column shows the number of largest elements (in absolute value) that we must retain to preserve the energy (listed in the second column) of each modified HWT, the fourth number is then the the number of terms we convert to zero in each transform, and the last column gives the percentage of elements in each transform that are zero.

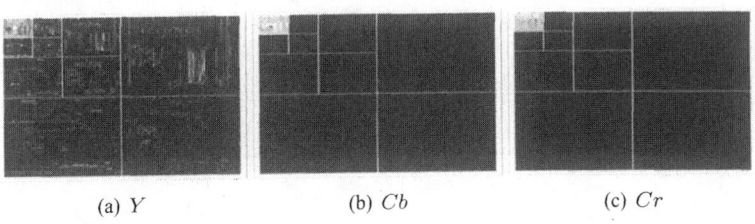

(a) Y (b) Cb (c) Cr

Figure 6.25 The quantized HWT for each of Y, Cb, and Cr for the energy levels given in Table 6.5.

Table 6.6 gives bit stream information for compressing the quantized wavelet transforms into Huffman codes.

The total bit stream length for the uncompressed image is $3 \cdot 8 \cdot 160 \cdot 240 = 921,600$. Compression via the HWT and the varied energy levels results in a bit stream length

Table 6.6   Huffman bit stream length, bpp, and entropy for each quantized transform of Y, Cb, and Cr for the energy levels given in Table 6.5.

Transform	Bit Stream Length	bpp	Entropy
Y	115,983	3.02039	2.7485
Cb	43,818	1.14109	0.2559
Cr	43,842	1.14172	0.2567

of $115{,}983 + 43{,}818 + 43{,}842 = 208{,}541$. *This length is just over* 22% *of the original bit stream length.*

The uncompressed image using the energy levels given in Table 6.5 is plotted in Figure 6.26. While the reconstructed image is not a close likeness to the original, it approximates it remarkably well considering that the bit stream length is about one-fourth of the original bit stream length.   □

(a) Original Image   (b) 99.99% Energy   (c) Variable Energy Rates

Figure 6.26   The original and reconstructed images. The reconstructed image that uses 99.99% energy across each channel has PSNR = 35.235, while the reconstructed image that uses the energy rates given in Table 6.5 has PSNR 23.462. See the color plates for color versions of these images.

In Computer Lab 6.9, you will further explore signal and image compression with the HWT.

Edge Detection

The second application we discuss in this section is the detection of edges in digital images. There are many different ways to perform edge detection. Some of the best known methods are those of Marr and Hildreth [58], Canny [14], and Sobel (see the description in Gonzalez and Woods [39]). Other methods for edge detection are outlined in Gonzalez, Woods, and Eddins [40].

A Brief Review of Popular Edge Detection Methods

Roughly speaking, edges occur when there is an abrupt change in pixel intensity values (see Figure 6.27). As you know from elementary calculus, the derivative is used to measure rate of change. Thus for two-dimensional images, edge detectors attempt to estimate the gradient at each pixel and then identify large gradient values as edges. Recall from multivariate calculus that the gradient of $f(x, y)$ is defined to be

$$\nabla f(x, y) = f_x(x, y) \begin{bmatrix} 1 \\ 0 \end{bmatrix} + f_y(x, y) \begin{bmatrix} 0 \\ 1 \end{bmatrix}$$

where f_x and f_y are the partial derivatives of $f(x, y)$. See Stewart [69] for more details.

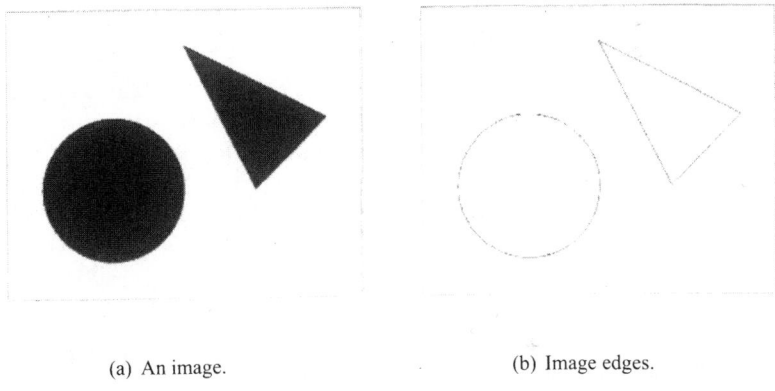

(a) An image.   (b) Image edges.

Figure 6.27   An image and its edges.

The Sobel method is straightforward. The method uses two 3×3 *mask* matrices that are convolved[2] with the image. One mask is designed to approximate the partial derivative in the horizontal (x) direction, while the other approximates the partial derivative in the vertical (y) direction. The results of the convolutions are used to estimate the magnitude of the gradient value at each pixel. Large gradient values correspond to edges.

Many digital images contain noise, and one of the problems with the Sobel detector is that it often identifies noisy pixels as edge pixels. The Canny method [14] addresses this problem by first convolving the image with a *smoothing mask*. The method then estimates the magnitude of the gradient at each pixel. The Canny detector then applies

---

[2]We do not discuss two-dimensional convolution in this book. It is quite straightforward to extend the ideas from Section 5.1 to higher dimensions. See Gonzalez and Woods [39] and Gonzalez, Woods, and Eddins [40] for more details.

an algorithm to refine the edges. Finally, two threshold values are created and used to determine *strong edges* and *weak edges*. Weak edges that are not connected to strong edges are discarded. Since the Canny detector addresses problems such as noise and refinement of edges, it is typically considered one of the most efficient edge detectors.

Rather than search for large magnitudes of the gradient, the Marr-Hildreth detector looks for zero values (called zero crossings) in the Laplacian. Recall the Laplacian (see Stewart [69]) is defined to be

$$\nabla^2 f(x, y) = \frac{\partial^2 f}{\partial x^2} + \frac{\partial^2 f}{\partial y^2}$$

A threshold is applied to the set of zero crossings to determine those pixels that represent edges.

Using Wavelets to Detect Edges

If you worked through Computer Lab 6.2 in Section 6.1, then you learned that the highpass portion of the one-dimensional wavelet transform can be used to approximate the derivative of a sampled function (for further details, see Problem 6.28). The idea extends to two-dimensional wavelet transforms. Information about the edges in an image is contained in $\mathcal{V}$, $\mathcal{H}$, and $\mathcal{D}$. Mallat and Hwang [56] have shown that a maximum of the modulus of the wavelet transform can be used to detect "irregular structures" — in particular, the modulus of the highpass portions of the wavelet transformation can be used to detect edges in images.

In the remainder of this section we give a brief overview of how wavelets can be used in edge detection. In the discussion above, it is noted that digital images are typically denoised or smoothed before application of an edge detector. We do not perform denoising in this application (see Chapter 9 for a detailed discussion of wavelet-based denoising). We also consider only one iteration of the wavelet transform when detecting edges. Wavelet-based edge detection can be improved if more iterations of the transform are performed (see Mallat [56] or Li[55]).

Naive Edge Detection

Possibly the simplest thing we could do to detect edges in a digital image is to compute the HWT, convert the values in the blur block $\mathcal{B}$ to zero, and then apply the inverse HWT. Let's try it on an example.

Example 6.15 (Naive Edge Detection). *Consider the 512×512 image A plotted in Figure 6.28. One iteration of the two-dimensional HWT $C = W_{512} A W_{512}^T$ is also plotted in Figure 6.28.*

We now convert all the values in block $\mathcal{B}$ to zero. Let's call the modified wavelet transform $\tilde{C}$. We then apply the inverse HWT to $\tilde{C}$ to obtain a matrix that is comprised

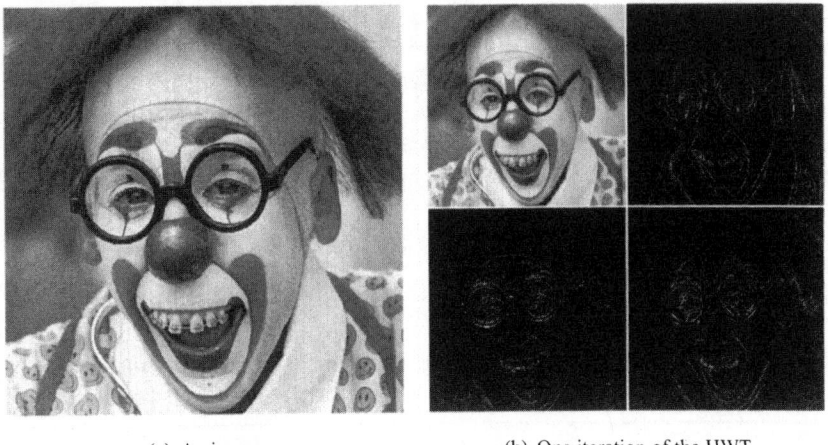

(a) An image   (b) One iteration of the HWT

Figure 6.28   An image and its HWT.

only of the vertical, horizontal, and diagonal differences between A and B. To determine the significance of the edges, we also compute the absolute value of each element of the inverse transform. Let $E = W_{512}^T \tilde{C} W_{512}$ denote the inverse transform of $\tilde{C}$. Both $\tilde{C}$ and E are plotted in Figure 6.29. □

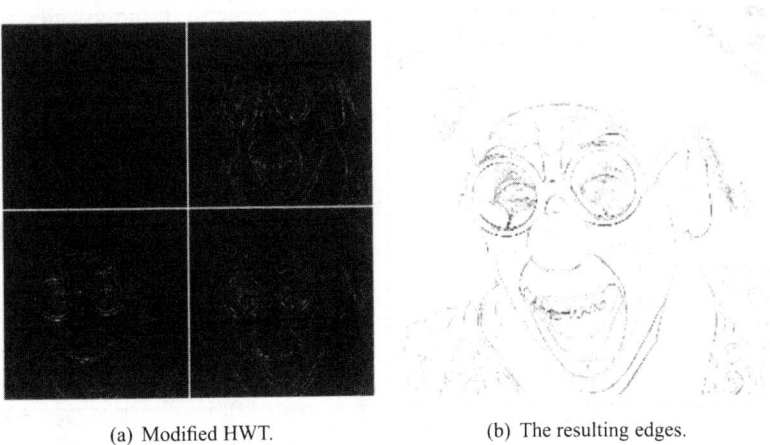

(a) Modified HWT.   (b) The resulting edges.

Figure 6.29   The modified HWT and the resulting edges. Note that we have applied image negation to the edge image so that the boundaries are easier to see.

The naive process does a fair job at detecting edges in the image. Notice that it has problems distinguishing edges in areas where light and dark pixels are interspersed. It is natural to assume that not every element in the detail matrices $V, \mathcal{H}$, and $\mathcal{D}$ constitute an edge in the original image. We can improve our edge detector by keeping only those detail elements that do indeed correspond to edge pixels. The mathematical procedure for selecting only particular elements from a set of numbers is called *thresholding*.

Thresholding

To apply a threshold to the elements of our transform matrix C, we simply choose a value $\tau > 0$ and create a new matrix $\tilde{C}$ whose entries are

$$\tilde{c}_{ij} = \begin{cases} c_{ij}, & |c_{ij}| > \tau \\ 0, & |c_{ij}| \le \tau \end{cases}$$

Unfortunately, there is no proven way to choose τ. We could use trial and error and try different values of τ until we obtain a satisfactory edge matrix. Alternatively, we could employ some elementary statistics to determine τ.

Automating the Threshold Choice

To automate the process, we use an iterative method for determining τ that is described in Gonzalez and Woods [39]. The process creates a series of thresholds $\tau_1, \tau_2, \ldots$ and stops only when the difference $|\tau_{n+1} - \tau_n|$ is smaller than some prescribed $\alpha > 0$.

The method in Gonzalez and Woods [39] for choosing a threshold for a set of numbers S is as follows:

1. Pick a tolerance $\alpha > 0$ that will determine the stopping criteria for the process.

2. Set τ_1 to be the average of the largest and smallest values in S.

3. Divide the values of S into two sets S_1 and S_2. The values in S_1 are all smaller than τ_1 and the values in S_2 are greater than or equal to τ_1.

4. Compute the means $\bar{s}_1$ for S_1 and $\bar{s}_2$ for S_2 and take $\tau_2 = \frac{1}{2}(\bar{s}_1 + \bar{s}_2)$.

5. Repeat steps 3 and 4 replacing τ_1 with the new tolerance obtained from step 4. Continue until $|\tau_{n+1} - \tau_n| < \alpha$.

Let's look at the thresholding process in the following example.

Example 6.16 (Edge Detection with Automated Thresholding). *Consider again the image from Example 6.15. We chose the stopping criteria value to be $\alpha = 1$ pixel* *intensity value. We first compute* $C = W_{384} A W_{512}^T$. *Next, we create a set S that is comprised of the absolute values of all elements of* $\mathcal{V}$, $\mathcal{H}$, *and* $\mathcal{D}$. *We set* τ_1 *to be the average of the largest and smallest values in S. For the image plotted in Figure 6.28, we have* $\max(S) = 255$ *and* $\min(S) = 0$, *so that* $\tau_1 = 127.5$.

We are now ready to employ the method given in [39]. We begin by creating set S_1 *that is comprised of all values from S whose values are smaller than* $\tau_1 = 127.5$. *The set* S_2 *is the complement of* S_1 *in S. The mean of* S_1 *is* $\bar{s}_1 = 5.659841$ *and the mean of* S_2 *is* $\bar{s}_2 = 160.885787$, *so that* $\tau_2 = (\bar{s}_1 + \bar{s}_2)/2 = 83.272814$. *The stopping criterion is* $|\tau_2 - \tau_1| = 44.227186$. *This value is larger than* $\alpha = 1$, *so we repeat the process. As we can see from Table 6.7, we need 10 iterations to reach a threshold of* $\tau = \tau_{10} = 28.520766$.

Table 6.7   Iterations to obtain the threshold value $\tau = 28.520766$.[a]

n	$\bar{s}_1$	$\bar{s}_2$	τ_n	$|\tau_{n+1} - \tau_n|$
1	—	—	127.500000	—
2	5.659841	160.885787	83.272814	44.227186
3	5.359329	114.971271	60.165300	23.107514
4	5.008019	89.846607	47.427312	12.737988
5	4.695660	75.097058	39.896359	7.530953
6	4.448098	66.089429	35.268764	4.627595
7	4.273475	60.694491	32.483983	2.784781
8	4.135717	56.863564	30.499640	1.984343
9	4.034568	54.267145	29.150857	1.348784
10	3.985408	53.056124	28.520766	0.630091

[a] The first column is the iteration number, the next two numbers are the means of S_1 and S_2, respectively, the third value is the average τ_n of $\bar{s}_1$ and $\bar{s}_2$; and the last column gives the stopping criterion $|\tau_{n+1} - \tau_n|$. The method stops when the value in the last column is smaller than $\alpha = 1$.

We now set to zero each element in $\mathcal{V}$, $\mathcal{H}$, *and* $\mathcal{D}$ *whose absolute value is smaller than* $\tau = \tau_{10} = 28.520766$. *We also convert the elements in block B to zero. If we denote the modified HWT by* $\tilde{C}$, *we obtain the edges by using the inverse HWT to compute* $E = W_{512}^T \tilde{C} W_{512}$. *The modified HWT* $\tilde{C}$ *and the edge image E are plotted in Figure 6.30.*

In Figure 6.31 we plot the edge matrix from Figure 6.30 and the edge matrix from Figure 6.29 side by side so that we can compare the results. The automated method does a little better job at clearing up the edges. In Problem 6.30 you are asked to think about the consequences of increasing the size of α.   □

In Lab 6.10, you will investigate edge detection further using the Haar wavelet transformation. In particular, you will look at the effects of using other thresholding methods, thresholding $\mathcal{V}$, $\mathcal{H}$, and $\mathcal{V}$ separately, and performing multiple iterations of the HWT. You will also perform edge detection on color digital images.

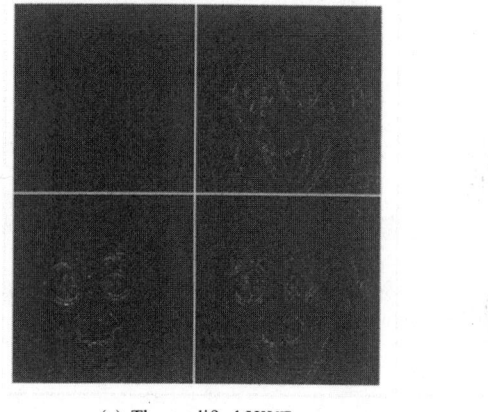 

(a) The modified HWT   (b) The resulting edges

Figure 6.30   The modified HWT and resulting edges. Note that we have applied image negation to the edge image so that the boundaries are easier to see.

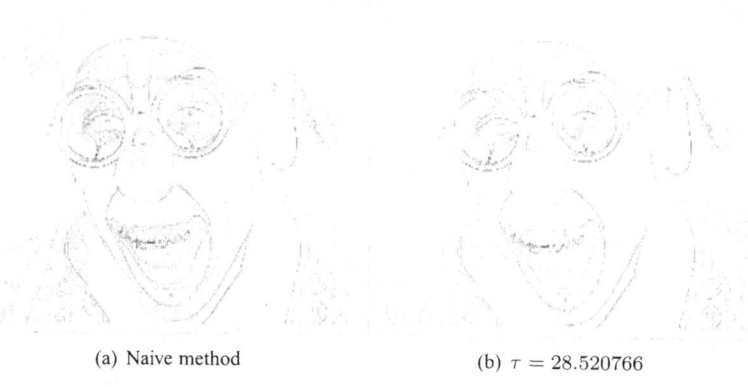

(a) Naive method   (b) $\tau = 28.520766$

Figure 6.31   Different edge matrices. The image on the left is from Example 6.15; the image on the right was constructed in this example.

Although the methods we presented for compression and edge detection are quite naive, they do illustrate what sorts of things we can do when we decompose data into lowpass and highpass portions. In Chapter 7, we learn how to construct wavelet transformations that outperform the HWT.

PROBLEMS

6.23   Determine the range of integers for C where $C = \hat{W}_N A \hat{W}_N^T$ and $\hat{W}$ is given by (6.30).

6.24   Suppose that we apply two iterations of the modified HWT given by (6.31). What is the range of integers in this case?

6.25   In the second half of Example 6.14, a lower energy level was used for the Cb channel. Can you explain why the lower level makes sense for this particular example?

$\star$**6.26**   The one-dimensional Haar transform is constructed by taking two numbers, a and b and mapping them to a lowpass value $\frac{\sqrt{2}}{2}a + \frac{\sqrt{2}}{2}b$ and a highpass value $\frac{\sqrt{2}}{2}b - \frac{\sqrt{2}}{2}a$. For image compression, we multiplied this transform by $\sqrt{2}$ (see 6.30) so that it mapped integers to integers. The modified transform is

$$\begin{bmatrix} a \\ b \end{bmatrix} \rightarrow \begin{bmatrix} a+b \\ b-a \end{bmatrix}$$

The output of this transform is certainly integer valued if a and b are integers. To invert the process, we simply divide the inverse transform by $\sqrt{2}$.

The problem with this transform is that it increases the dynamic range of the lowpass portion of the data. That is, for a, b integers from $\{0, \ldots, 255\}$, the lowpass value could range from 0 to 510. It is desirable to keep the range of lowpass values the same as that of the original input. Thus Sweldens [75] suggests the modified Haar wavelet transform

$$\begin{bmatrix} a \\ b \end{bmatrix} \rightarrow \begin{bmatrix} \frac{a+b}{2} \\ b-a \end{bmatrix} \tag{6.32}$$

Certainly, the average of a and b is contained in $[a, b]$, but the problem is that the lowpass values of this transform are not necessarily integer valued.

In this problem, we see how to compute the transform (6.32) via a process known as *lifting* and then convert the results to integers in such a way that the original input can be recovered. (We will study the concept of lifting in detail in Section 12.3.) Throughout this problem, we assume that $\mathbf{v} = [v_1, \ldots, v_N]^T$ where N is even with $v_k \in \{0, \ldots, 255\}, k = 1, \ldots, N$.

(a)   The transform (6.32) sends

$$\mathbf{v} \rightarrow \left[\frac{v_1 + v_2}{2}, \ldots, \frac{v_{N-1} + v_N}{2} \;\middle|\; v_2 - v_1, \ldots, v_N - v_{N-1} \right]^T$$

If we let $\mathbf{s}$ and $\mathbf{d}$ denote the lowpass and highpass portions of the transform, respectively, then for $k = 1, \ldots, \frac{N}{2}$,

$$s_k = \frac{v_{2k-1} + v_{2k}}{2} \qquad \text{and} \qquad d_k = v_{2k} - v_{2k-1}$$

Let $\mathbf{e} = (e_1, \ldots, e_{N/2})^T = (v_2, v_4, \ldots, v_N)^T$ hold the even-indexed elements of $\mathbf{v}$, and let $\mathbf{o} = (o_1, \ldots, o_{N/2})^T = (v_1, v_3, \ldots, v_{N-1})^T$ hold the odd-indexed elements of $\mathbf{v}$. Show that we can obtain $\mathbf{d}$ and $\mathbf{s}$ by first computing

$$d_k = e_k - o_k \tag{6.33}$$

for $k = 1, \ldots, \frac{N}{2}$, and then computing

$$s_k = o_k + \frac{1}{2} d_k \tag{6.34}$$

for $k = 1, \ldots, \frac{N}{2}$.

(b) What is the process for inverting (6.33) and (6.34)?

(c) The transformation (6.33) and (6.34) in part (a) can be modified so that it maps integers to integers.

Let $\lfloor \cdot \rfloor$ denote the *floor function* or *greatest integer function*. That is, $\lfloor x \rfloor = n$ where n is the integer that satisfies $n \leq x$ and $n + 1 > x$. We compute d_k using (6.33), but we apply the floor function to the division in (6.34) to write

$$s_k^* = o_k + \left\lfloor \frac{1}{2} d_k \right\rfloor \tag{6.35}$$

for $k = 1, \ldots, \frac{N}{2}$. What is the inverse process for this transformation?

(d) Suppose that we apply (6.33) followed by (6.35) to $\mathbf{v}$. Explain why the inverse process you found in part (c) returns the original $\mathbf{e}$ and $\mathbf{o}$.

6.27   Let $\mathbf{v} = [\, 10, 20, 25, 105, 0, 100, 100, 55, 60, 75 \,]^T$.

(a) Use (6.33) and (6.35) to compute an integer-valued Haar wavelet transform.

(b) Use the inverse process you found in part (d) of Problem 6.26 on the transformed data from part (a) to recover $\mathbf{v}$.

6.28   Recall from calculus that one way to define the derivative is

$$f'(t) = \lim_{h \to 0} \frac{f(t) - f(t - h)}{h}$$

So for "small" values of h, the quantity $\frac{f(t) - f(t-h)}{h}$ serves as an approximation of $f'(t)$. Suppose that we form a vector $\mathbf{v}$ of length N (where N is even) by sampling some function $f(t)$. That is,

$$v_k = f(a + (k - 1)\,\Delta x)$$

where $k = 1, \ldots, N$, a is the starting value, and Δx is the sampling width.

(a) Let $\mathbf{y}$ denote one iteration of the HWT. Using the fact that the elements of the highpass portion of $\mathbf{y}$ are given by

$$y_{N/2+k} = \frac{\sqrt{2}}{2}(v_{2k} - v_{2k-1})$$

for $k = 1, \ldots, \frac{N}{2}$, show that

$$y_{N/2+k} \approx \frac{\sqrt{2}}{2}\Delta x\, f'(a + (2k - 1)\,\Delta x)$$

(b) Consider the function

$$f(t) = \begin{cases} 1, & t < 0 \\ t, & t \geq 0 \end{cases}$$

Create the 10-vector $\mathbf{v}$ by sampling $f(t)$ with starting point $a = -1$ and sampling width $\Delta x = 0.1$. Then compute the highpass portion of the HWT of $\mathbf{v}$ and verify the result you obtained in part (a).

(c) How could you use the highpass portion of the HWT to detect jumps in the derivative of $f(t)$?

(d) The function

$$f(t) = \begin{cases} t^2, & t < 0 \\ t^3, & t \geq 0 \end{cases}$$

is continuously differentiable at $a = 0$ but has a jump at 0 in the second derivative. How could you use the HWT to detect this jump?

(d) In practice, we often do not know the sampled function $f(t)$. How could you use the HWT to determine the smoothness of $f(t)$?

Computer Lab 6.11 allows you to implement the results of this problem.

6.29   Use the method outlined on page 214 with $\alpha = 1$ to find the threshold value τ for

$$S = [0, 0, 0, 0, 0, 0, 0, 0, 0, 0, 51, 77, 170, 204, 220, 268, 292, 316, 380, 410]^T$$

6.30   What do you suppose would happen if we increased $\alpha = 1$ to $\alpha = 25$ in Example 6.16? From $\alpha = 1$ to $\alpha = 100$?

6.31   In forming the HWT matrix W_N (6.7), we performed *downsampling*. That is, we removed every other row from the convolution matrices that represented the lowpass and highpass portions of the transform. The reason that we downsampled was to remove redundancy and create an invertible transform. Recall that we set the elements of the approximation portion B of the transform to zero, performed thresholding on the detail matrices, and then inverted the result to obtain an edge matrix.

In this problem we consider a method for edge detection that neither utilizes downsampling nor requires matrix invertibility in the usual sense. A **CAS** will be helpful to complete this problem.

Suppose that A is an $N \times N$ matrix that represents some grayscale image.

(a) Form the *nondecimated* HWT matrix U_N as follows. The odd rows of U_N are simply the rows of W_N. To form an even row of U_N, take the odd row above it and cyclically shift it one unit right (you will have two rows with one non-zero element that "wraps" around to the first column). Write down U_6 and U_{10}.

(b) What are the dimensions of U_N?

(c) The top half of U_N is the "lowpass" portion of the transform and the bottom half of U_N is the "highpass" portion of the transform. Denote these portions by H and G, respectively. What are the dimensions of H and G?

(d) To transform A, we compute

$$B = U_N A U_N^T = \left[\frac{H}{G} \right] A [H^T \mid G^T]$$

Further expansion gives a block form similar to (6.24) with blocks $\mathcal{B}, \mathcal{V}, \mathcal{H}$, and $\mathcal{D}$. What are the dimensions of each of these blocks?

(e) Can you state any advantages that this transformation might have over the conventional HWT in terms of edge detection?

(f) Certainly U_N is not invertible since it is $2N \times N$. Suppose there exists $N \times 2N$ matrix V that satisfies $V U_N = I_N$. Explain how we could use V to recover A from $B = U_N A U_N^T$.

(g) Find V for U_6 so that $V U_6 = I_6$. (*Hint:* You could either solve a large system or start with U_6^T and build V from it.)

(h) Generalize your result in part (g) for arbitrary N.

Note: V is called a *pseudoinverse* of U_N. For a nice introduction to pseudoinverses, please see Strang [72]. Since we know how to "invert" the nondecimated HWT, we can use it in the edge detection methods described in this section. You might also want to try it when you work through Computer Lab 6.10.

Computer Labs

6.9 Image Compression. From the text Web site, access the lab `haarcomp`. In this lab you will use the HWT to perform basic image compression.

6.10 Edge Detection. From the text Web site, access the lab `haaredge`. This lab utilizes the two-dimensional HWT to perform basic edge detection.

6.11 Jump Detection. From the text Web site, access the lab `haar1Ddetect`. In this lab you will learn how to use the HWT to find jumps in digital signals.

6.12 Lifting. From the text Web site, access the lab `haarlifting`. In this lab you will write a module to implement the transform described in Problem 6.26.

6.13 CAPTCHA. From the text Web site, access the lab `haarcaptcha`. This lab is based on the results in [1]. In this paper, the authors show how to use the HWT as part of a method to crack CAPTCHAs (Completely Automated Public Turing tests to tell Computers and Humans Apart). CAPTCHAs are tests that are used by companies that provide e-mail accounts or sell concert tickets to determine if a visitor to their Web site is a real person. In this lab you will learn how the authors used the HWT to crack CAPTCHAs.

CHAPTER 7

DAUBECHIES WAVELET TRANSFORMATIONS

The discrete Haar wavelet transform (HWT) introduced in Chapter 6 serves as an excellent introduction to the theory and construction of wavelet transforms. We were also able to see in Section 6.4 how the transform can be utilized in applications such as image compression and edge detection. There are some limitations to the HWT. The fact that the filters $\mathbf{h} = (h_0, h_1) = (\frac{\sqrt{2}}{2}, \frac{\sqrt{2}}{2})$ and $\mathbf{g} = (g_0, g_1) = (\frac{\sqrt{2}}{2}, -\frac{\sqrt{2}}{2})$ are short leads to a fast algorithm for computation. However, there is a disadvantage with such short filters.

Consider the vector $\mathbf{v} = [\,100, 102, 200, 202\,]^T$. Applying the HWT to $\mathbf{v}$ gives $\mathbf{y} = \sqrt{2}\,[\,101, 201, 1, 1\,]^T$. Now think about the application to edge detection in Section 6.4. We were looking for large values in the highpass (difference) portions of the output and possibly designating them as edges. If we were to employ the same method for designating boundary coefficients as in Example 6.16, then the threshold value would be $\tau = 1$, and since no values in the highpass portion of the transform vector are smaller that τ, the modified wavelet transform vector would be $\tilde{\mathbf{y}} = \sqrt{2}\,[\,0, 0, 1, 1\,]^T$. Application of the inverse HWT gives $\tilde{\mathbf{v}} = [\,-1, 1, -1, 1\,]^T$. The next step is to take the absolute value the elements of $\tilde{\mathbf{v}}$ to determine how far the details are away from zero. The vector we display as edges is $[\,1,1,1,1\,]^T$. Since all values are constant, it would be natural to assume that the original vector had no large jumps between elements even though there is a large jump between the second and third elements of $\mathbf{v}$. The fact that the Haar filters are so short means that the transform did not process enough of the data at a time to catch this jump. The point of this discussion is to show that longer filters might be useful in some applications.

We have to determine what we want from our longer filters. Certainly, we do not want them to be so long as to markedly affect computation speed. We would like to retain desirable properties such as orthogonality of the transform matrix and the lowpass/highpass structure of the transform matrix. Suppose that we were to make $\mathbf{h}$ and $\mathbf{g}$ longer. We are in some sense adding more flexibility to the transform. How can we best take advantage of this flexibility?

In her landmark paper of 1988, Ingrid Daubechies [23] described a family of orthogonal lowpass filters. The first member of this family is the Haar filter $\mathbf{h} = (h_0, h_1) = (\frac{\sqrt{2}}{2}, \frac{\sqrt{2}}{2})$ introduced in Chapter 6. She shows how to construct other family members of arbitrary even length and their accompanying highpass filters. In Section 7.1 we derive Daubechies' orthogonal filters of length 4 and 6. In Section 7.2 we derive Daubechies orthogonal filters where the length can be any even number. We also develop pseudocode for implementing Daubechies filters in Section 7.3.

7.1   DAUBECHIES FILTERS OF LENGTH 4 AND 6

In this section we construct Ingrid Daubechies' family of orthogonal lowpass filters [24] for filter lengths 4 and 6. The filters will be orthogonal in the sense that the wavelet transformation matrix W_N we construct will satisfy $W_N^{-1} = W_N^T$. We can also design the filters to be of any even length we desire. Toward this end, we impose conditions on the Fourier series $H(\omega)$ of filter $\mathbf{h}$.

The construction here is different than the one that Daubechies gives in [23] and [26]. We start with a matrix that has the same structure as that associated with the HWT (6.7), but built with filters $\mathbf{h} = (h_0, h_1, \ldots, h_L)$ and $\mathbf{g} = (g_0, g_1, \ldots, g_L)$, where $L = 3, 5$. We then construct a system of (linear and quadratic) equations that the elements of $\mathbf{h}$ must satisfy. These equations will come from orthogonality conditions and conditions very similar to lowpass filter conditions (recall Condition 5.1 from Section 5.2). Once we have $\mathbf{h}$, we can use it to construct an accompanying highpass filter $\mathbf{g}$.

The Daubechies Four-Term Orthogonal Filter

In the introduction to this chapter, we stated that Daubechies filters are of arbitrary even length. The Haar filter $\mathbf{h} = (h_0, h_1) = (\frac{\sqrt{2}}{2}, \frac{\sqrt{2}}{2})$ is of length 2. The next even number is 4, so let's construct the lowpass filter $\mathbf{h} = (h_0, h_1, h_2, h_3)$ and the highpass (wavelet) filter $\mathbf{g} = (g_0, g_1, g_2, g_3)$. In Problem 7.1 you will investigate why we don't consider odd-length filters for this construction.

To motivate the construction, let's consider a wavelet transform built from $\mathbf{h} = (h_0, h_1, h_2, h_3)$, $\mathbf{g} = (g_0, g_1, g_2, g_3)$ that we can apply to vectors of length $N = 8$. That is, we wish to construct the matrix W_8, where

$$
W_8 = \begin{bmatrix}
h_3 & h_2 & h_1 & h_0 & 0 & 0 & 0 & 0 \\
0 & 0 & h_3 & h_2 & h_1 & h_0 & 0 & 0 \\
0 & 0 & 0 & 0 & h_3 & h_2 & h_1 & h_0 \\
h_1 & h_0 & 0 & 0 & 0 & 0 & h_3 & h_2 \\
g_3 & g_2 & g_1 & g_0 & 0 & 0 & 0 & 0 \\
0 & 0 & g_3 & g_2 & g_1 & g_0 & 0 & 0 \\
0 & 0 & 0 & 0 & g_3 & g_2 & g_1 & g_0 \\
g_1 & g_0 & 0 & 0 & 0 & 0 & g_3 & g_2
\end{bmatrix}
\tag{7.1}
$$

Note that we have built W_8 in much the same way that we constructed the matrix for the HWT. We combined two convolution matrices and downsampled them by a factor of 2. The truncation results in two peculiar rows. The coefficients in the fourth and eighth rows have "wrapped" from the end of the row to the beginning of the row. We did not have to worry about this with the Haar filter — since there were only two coefficients per row, there was no need to wrap. What effect does the wrapping have? Consider $W_8\mathbf{v}$:

$$
W_8\mathbf{v} = \begin{bmatrix}
h_3 & h_2 & h_1 & h_0 & 0 & 0 & 0 & 0 \\
0 & 0 & h_3 & h_2 & h_1 & h_0 & 0 & 0 \\
0 & 0 & 0 & 0 & h_3 & h_2 & h_1 & h_0 \\
h_1 & h_0 & 0 & 0 & 0 & 0 & h_3 & h_2 \\
g_3 & g_2 & g_1 & g_0 & 0 & 0 & 0 & 0 \\
0 & 0 & g_3 & g_2 & g_1 & g_0 & 0 & 0 \\
0 & 0 & 0 & 0 & g_3 & g_2 & g_1 & g_0 \\
g_1 & g_0 & 0 & 0 & 0 & 0 & g_3 & g_2
\end{bmatrix}
\begin{bmatrix} v_1 \\ v_2 \\ v_3 \\ v_4 \\ v_5 \\ v_6 \\ v_7 \\ v_8 \end{bmatrix}
= \begin{bmatrix}
h_3 v_1 + h_2 v_2 + h_1 v_3 + h_0 v_4 \\
h_3 v_3 + h_2 v_4 + h_1 v_5 + h_0 v_6 \\
h_3 v_5 + h_2 v_6 + h_1 v_7 + h_0 v_8 \\
h_3 v_7 + h_2 v_8 + h_1 v_1 + h_0 v_2 \\
g_3 v_1 + g_2 v_2 + g_1 v_3 + g_0 v_4 \\
g_3 v_3 + g_2 v_4 + g_1 v_5 + g_0 v_6 \\
g_3 v_5 + g_2 v_6 + g_1 v_7 + g_0 v_8 \\
g_3 v_7 + g_2 v_8 + g_1 v_1 + g_0 v_2
\end{bmatrix}
\tag{7.2}
$$

The fourth and eighth components of $W_8\mathbf{v}$ are built using v_1, v_2, v_7, v_8. If the elements of $\mathbf{v}$ were sampled from a periodic function, our matrix W_8 would be ideal. Typically, this is not the case, so we might have some problems. Alternatively, we could discard h_0 and h_1 from row 4 and g_0 and g_1 from row 8, but then we would not be using the entire filter to process the data. People build wavelet matrices such as W_8 in different ways (some intertwine the lowpass and highpass rows!), but we will stick with the form in (7.1) for our presentation.

In the lowpass (highpass) portion of W_8, elements of each row overlap with two elements of the rows above and below it (see Figure 7.1). This should help us avoid the problem with the Haar transform that was discussed in the chapter introduction. As we will see, the longer we make $\mathbf{h}$ and $\mathbf{g}$, the more rows in the lowpass (highpass) portion will contain overlapping elements.

$$
\begin{array}{ccccccc}
\cdots & h_3 & h_2 & \boxed{\begin{array}{cc} \mathbf{h_1} & \mathbf{h_0} \\ \mathbf{h_3} & \mathbf{h_2} \end{array}} & \begin{array}{cc} 0 & 0 \\ h_1 & h_0 \end{array} & \cdots \\
 & 0 & 0 & & &
\end{array}
$$

Figure 7.1   Two overlapping rows in the lowpass portion of W_8.

Characterizing the Orthogonality Conditions

We find the elements of W_8 in a couple of stages. Let's first write W_8 in block format and then multiply these blocks to understand what effect the orthogonality of W_8 has on **h**. We write

$$
W_8 = \left[\begin{array}{cccccccc}
h_3 & h_2 & h_1 & h_0 & 0 & 0 & 0 & 0 \\
0 & 0 & h_3 & h_2 & h_1 & h_0 & 0 & 0 \\
0 & 0 & 0 & 0 & h_3 & h_2 & h_1 & h_0 \\
h_1 & h_0 & 0 & 0 & 0 & 0 & h_3 & h_2 \\
g_3 & g_2 & g_1 & g_0 & 0 & 0 & 0 & 0 \\
0 & 0 & g_3 & g_2 & g_1 & g_0 & 0 & 0 \\
0 & 0 & 0 & 0 & g_3 & g_2 & g_1 & g_0 \\
g_1 & g_0 & 0 & 0 & 0 & 0 & g_3 & g_2
\end{array}\right] = \left[\frac{H}{G}\right] \tag{7.3}
$$

Using (7.3) to compute $W_8 W_8^T$ and insisting that W_8 is orthogonal gives

$$
W_8 W_8^T = \left[\frac{H}{G}\right] \left[\, H^T \,\middle|\, G^T \,\right] = \left[\begin{array}{c|c} HH^T & HG^T \\ \hline GH^T & GG^T \end{array}\right] = \left[\begin{array}{c|c} I_4 & 0_4 \\ \hline 0_4 & I_4 \end{array}\right] \tag{7.4}
$$

where I_4 is the 4×4 identity matrix and 0_4 is the 4×4 zero matrix.

Let's analyze HH^T from (7.4). Direct computation gives

$$
I_4 = HH^T = \left[\begin{array}{cccccccc}
h_3 & h_2 & h_1 & h_0 & 0 & 0 & 0 & 0 \\
0 & 0 & h_3 & h_2 & h_1 & h_0 & 0 & 0 \\
0 & 0 & 0 & 0 & h_3 & h_2 & h_1 & h_0 \\
h_1 & h_0 & 3 & 0 & 0 & 0 & h_3 & h_2
\end{array}\right]
\left[\begin{array}{cccc}
h_3 & 0 & 0 & h_1 \\
h_2 & 0 & 0 & h_0 \\
h_1 & h_3 & 0 & 0 \\
h_0 & h_2 & 0 & 0 \\
0 & h_1 & h_3 & 0 \\
0 & h_0 & h_2 & 0 \\
0 & 0 & h_1 & h_3 \\
0 & 0 & h_0 & h_2
\end{array}\right]
= \left[\begin{array}{cccc}
a & b & 0 & b \\
b & a & b & 0 \\
0 & b & a & b \\
b & 0 & b & a
\end{array}\right]
= \left[\begin{array}{cccc}
1 & 0 & 0 & 0 \\
0 & 1 & 0 & 0 \\
0 & 0 & 1 & 0 \\
0 & 0 & 0 & 1
\end{array}\right]
$$

where

$$
a = h_0^2 + h_1^2 + h_2^2 + h_3^2 \qquad b = h_0 h_2 + h_1 h_3 \tag{7.5}
$$

We also have that $a = 1$ and $b = 0$, so we obtain *orthogonality constraints* on **h**:

$$
h_0^2 + h_1^2 + h_2^2 + h_3^2 = 1 \tag{7.6}
$$

$$
h_0 h_2 + h_1 h_3 = 0 \tag{7.7}
$$

We have the three other blocks of (7.4) to consider and we still haven't imposed any lowpass conditions on **h**, but we are in a position to easily find the highpass filter **g**.

Creating the Highpass Filter g.

Assume that **h** is known and that it satisfies (7.6) – (7.7). To obtain **g**, we first recall Problem 2.13 from Section 2.1. In this problem we constructed a vector **v** orthogonal to a given vector **u** by reversing the elements of **u** and alternating the signs. Moreover, in Problem 5.25 in Section 5.2, we showed that if **u** is a lowpass filter, then the **v** we construct is a highpass filter. Thus, we have a good candidate for our highpass filter **g**. We define **g** as

$$\mathbf{g} = (g_0, g_1, g_2, g_3) = (h_3, -h_2, h_1, -h_0)$$

Let's insert these values into G in (7.3) and then compute the product HG^T directly from (7.4). You should verify the easy computation that follows.

$$HG^T = \begin{bmatrix} h_3 & h_2 & h_1 & h_0 & 0 & 0 & 0 & 0 \\ 0 & 0 & h_3 & h_2 & h_1 & h_0 & 0 & 0 \\ 0 & 0 & 0 & 0 & h_3 & h_2 & h_1 & h_0 \\ h_1 & h_0 & 0 & 0 & 0 & 0 & h_3 & h_2 \end{bmatrix} \cdot \begin{bmatrix} -h_0 & 0 & 0 & -h_2 \\ h_1 & 0 & 0 & h_3 \\ -h_2 & -h_0 & 0 & 0 \\ h_3 & h_1 & 0 & 0 \\ 0 & -h_2 & -h_0 & 0 \\ 0 & h_3 & h_1 & 0 \\ 0 & 0 & -h_2 & -h_0 \\ 0 & 0 & h_3 & h_1 \end{bmatrix} = \begin{bmatrix} 0 & 0 & 0 & 0 \\ 0 & 0 & 0 & 0 \\ 0 & 0 & 0 & 0 \\ 0 & 0 & 0 & 0 \end{bmatrix}$$

So $HG^T = 0$ as desired, and we can easily verify $GH^T = 0$ as well:

$$GH^T = (HG^T)^T = 0^T = 0$$

All that remains is to verify $GG^T = I_4$, and we have at least picked **h** and **g** so that W_8 is orthogonal. We compute

$$GG^T = \begin{bmatrix} -h_0 & h_1 & -h_2 & h_3 & 0 & 0 & 0 & 0 \\ 0 & 0 & -h_0 & h_1 & -h_2 & h_3 & 0 & 0 \\ 0 & 0 & 0 & 0 & -h_0 & h_1 & -h_2 & h_3 \\ -h_2 & h_3 & 0 & 0 & 0 & 0 & -h_0 & h_1 \end{bmatrix} \cdot \begin{bmatrix} -h_0 & 0 & 0 & -h_2 \\ h_1 & 0 & 0 & h_3 \\ -h_2 & -h_0 & 0 & 0 \\ h_3 & h_1 & 0 & 0 \\ 0 & -h_2 & -h_0 & 0 \\ 0 & h_3 & h_1 & 0 \\ 0 & 0 & -h_2 & -h_0 \\ 0 & 0 & h_3 & h_1 \end{bmatrix} = \begin{bmatrix} a & b & 0 & b \\ b & a & b & 0 \\ 0 & b & a & b \\ b & 0 & b & a \end{bmatrix} = I_4$$

where a and b are as given in (7.5).

So we can construct an orthogonal matrix by finding $\mathbf{h} = (h_0, h_1, h_2, h_3)$ that satisfy (7.6) – (7.7), and then choosing $\mathbf{g} = (g_0, g_1, g_2, g_3) = (h_3, -h_2, h_1, -h_0)$.

Using the Lowpass Conditions

Now let's impose conditions on **h** so that it is a lowpass filter. We use the Fourier series

$$H(\omega) = h_0 + h_1 e^{i\omega} + h_2 e^{2i\omega} + h_3 e^{3i\omega} \tag{7.8}$$

and Condition 5.1 which states that

$$H(0) = h_0 + h_1 + h_2 + h_3 = 1 \tag{7.9}$$
$$H(\pi) = h_0 - h_1 + h_2 - h_3 = 0 \tag{7.10}$$

Unfortunately, (7.9) has an adverse effect on the orthogonality of W_8. Recall that we had to alter the analogous condition $h_0 + h_1 = 1$ for the HWT. The filter $(\frac{1}{2}, \frac{1}{2})$ did not give an orthogonal transformation — we had to multiply the filter by $\sqrt{2}$, so that for the Haar filter we had $H(0) = h_0 + h_1 = \sqrt{2}$.

The same thing happens here. To see why (7.9) cannot be satisfied if W_8 is an orthogonal matrix, recall Theorem 2.1 from Section 2.2. The contrapositive of Theorem 2.1 says that if $\|W_8 v\| \neq \|v\|$ for some nonzero $v \in \mathbb{R}^8$, then W_8 is not an orthogonal matrix. Choose $v = [\,1, 1, 1, 1, 1, 1, 1, 1\,]^T$ so that $\|v\| = 2\sqrt{2}$. If we compute $W_8 v$, we obtain

$$W_8 v = \begin{bmatrix} h_3 & h_2 & h_1 & h_0 & 0 & 0 & 0 & 0 \\ 0 & 0 & h_3 & h_2 & h_1 & h_0 & 0 & 0 \\ 0 & 0 & 0 & 0 & h_3 & h_2 & h_1 & h_0 \\ h_1 & h_0 & 0 & 0 & 0 & 0 & h_3 & h_2 \\ -h_0 & h_1 & -h_2 & h_3 & 0 & 0 & 0 & 0 \\ 0 & 0 & -h_0 & h_1 & -h_2 & h_3 & 0 & 0 \\ 0 & 0 & 0 & 0 & -h_0 & h_1 & -h_2 & h_3 \\ -h_2 & h_3 & 0 & 0 & 0 & 0 & -h_0 & h_1 \end{bmatrix} v = \begin{bmatrix} h_3 + h_2 + h_1 + h_0 \\ h_3 + h_2 + h_1 + h_0 \\ h_3 + h_2 + h_1 + h_0 \\ h_3 + h_2 + h_1 + h_0 \\ -h_0 + h_1 - h_2 + h_3 \\ -h_0 + h_1 - h_2 + h_3 \\ -h_0 + h_1 - h_2 + h_3 \\ -h_0 + h_1 - h_2 + h_3 \end{bmatrix} = y$$

Using (7.9) – (7.10), we see that $y = [\,1, 1, 1, 1, 0, 0, 0, 0\,]^T$ with $\|y\| = \|W_8 v\| = 2 \neq \|v\|$.

Thus, we will need to change the constant value on the right side of (7.9). In Problem 7.2 you will show that if h_0, h_1, h_2, and h_3 satisfy the orthogonality conditions (7.6) – (7.7), and the second lowpass condition (7.10), then

$$h_0 + h_1 + h_2 + h_3 = \pm\sqrt{2} \tag{7.11}$$

We should note at this time that g is a highpass filter. Indeed,

$$G(\omega) = h_3 - h_2 e^{i\omega} + h_1 e^{2i\omega} - h_0 e^{3i\omega}$$

so that by (7.10),

$$G(0) = h_3 - h_2 + h_1 - h_0 = 0$$

and by (7.11),

$$|G(\pi)| = |h_3 + h_2 + h_1 + h_0| = |\pm\sqrt{2}| = \sqrt{2}$$

Combining the orthogonality conditions (7.6) – (7.7) and the lowpass condition (7.10), we have the following system of equations for determining filter h:

$$h_0^2 + h_1^2 + h_2^2 + h_3^2 = 1 \tag{7.12}$$
$$h_0 h_2 + h_1 h_3 = 0 \tag{7.13}$$
$$h_0 - h_1 + h_2 - h_3 = 0 \tag{7.14}$$

Solving the System Involving Two Orthogonality Conditions and One Lowpass Condition

Let's attempt to solve this system. Equation (7.13) implies that the two-dimensional vectors $[h_0, h_1]^T$ and $[h_2, h_3]^T$ are orthogonal. You will show in Problem 7.4 that in this case we must have

$$[h_2, h_3]^T = c[-h_1, h_0]^T \tag{7.15}$$

for some real number $c \neq 0$. If we insert this identity into (7.12) and simplify, we obtain

$$h_0^2 + h_1^2 = \frac{1}{1 + c^2} \tag{7.16}$$

If we insert (7.15) into (7.14), we can write h_0 in terms of c and h_1:

$$
\begin{aligned}
h_0 - h_1 + h_2 - h_3 &= 0 \\
h_0 - h_1 - ch_1 - ch_0 &= 0 \\
h_0(1 - c) - h_1(1 + c) &= 0
\end{aligned}
\tag{7.17}
$$

$$h_1 = \left(\frac{1 - c}{1 + c}\right) h_0$$

for values $c \neq -1$.

Geometrically, it is easy to see what is happening with this system. Equation (7.16) says that the values h_0 and h_1 must lie on a circle of radius $\frac{1}{\sqrt{1+c^2}}$. Equation (7.17) says that the values h_0 and h_1 must lie on a line through the origin with slope $\frac{1-c}{1+c}$. Certainly, this line intersects the circle at two points (see Figure 7.2). So for each c, there are two possible solutions (in Problem 7.5 you will explain why $c \neq 1$) that satisfy equations (7.12) – (7.14).

To see algebraically that there are infinitely many solutions to (7.12) – (7.14), we insert the value for h_1 given in (7.17) into (7.16). After some simplification we obtain a formula for h_0 in terms of c:

$$h_0^2 + h_1^2 = \frac{1}{1 + c^2}$$

$$h_0^2 + \left(\frac{1 - c}{1 + c}\right)^2 h_0^2 = \frac{1}{1 + c^2}$$

$$h_0^2 \left(1 + \frac{(1 - c)^2}{(1 + c)^2}\right) = \frac{1}{1 + c^2}$$

$$2h_0^2 \left(\frac{1 + c^2}{(1 + c)^2}\right) = \frac{1}{1 + c^2}$$

$$h_0^2 = \frac{(1 + c)^2}{2(1 + c^2)^2}$$

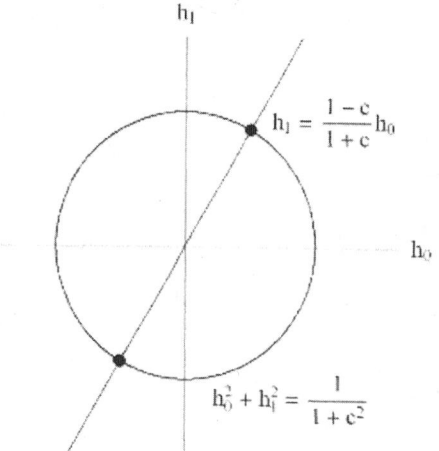

Figure 7.2   Possible values for h_0 and h_1 for a given c.

Taking square roots gives

$$h_0 = \pm \frac{1+c}{\sqrt{2}\,(1+c^2)} \tag{7.18}$$

Let's choose the positive root. Then

$$h_0 = \frac{1+c}{\sqrt{2}\,(1+c^2)} \tag{7.19}$$

and we can find values for h_1, h_2, and h_3. From (7.17) we have

$$h_1 = \frac{1-c}{\sqrt{2}\,(1+c^2)} \tag{7.20}$$

and using (7.15), we find that

$$h_2 = -\frac{c(1-c)}{\sqrt{2}(1+c^2)} \qquad h_3 = \frac{c(1+c)}{\sqrt{2}(1+c^2)} \tag{7.21}$$

Since c can be any real number other than -1, equations (7.19) – (7.21) tells us that there are an infinite number of solutions to the system (7.12)-(7.14). Of course, you can also deduce this fact from Figure 7.2 — for any given $c \neq -1$, the line with slope $(1-c)/(1+c)$ that passes through the origin must intersect the circle twice.

An Additional Lowpass Condition

We use this freedom to further refine our transform. To do so, we impose another condition on the filter **h** or, equivalently, on its Fourier series $H(\omega)$. The condition

$H(\pi) = 0$ will result in h annihilating high oscillations in the data. Also remember that an ideal lowpass filter takes on the value zero for ω in the interval (a, π), $0 < a < \pi$. From Problem 5.16 in Section 5.2, we learned that such an ideal filter is of infinite length. Although we can't produce an entire interval where $H(\omega) = 0$, we can certainly make $H(\omega) \approx 0$ for an interval near $\omega = \pi$. One way to do this is simply to force $|H(\omega)|$ to approach 0 at $\omega = \pi$ tangentially from the left. Our final condition on $H(\omega)$ is from elementary calculus! We insist that

$$H'(\pi) = 0$$

We compute $H'(\omega)$ using (7.8) to obtain

$$H'(\omega) = ih_1 e^{i\omega} + 2ih_2 e^{2i\omega} + 3ih_3 e^{3i\omega}$$

and if we insert $\omega = \pi$ and set the derivative to zero, we have

$$
\begin{aligned}
0 &= ih_1 e^{i\pi} + 2ih_2 e^{2i\pi} + 3ih_3 e^{3i\pi} \\
&= i(h_1(-1) + 2h_2(1) + 3h_3(-1)) \\
&= -i(h_1 - 2h_2 + 3h_3)
\end{aligned}
$$

The imaginary part of the last equation is zero so it must be that the imaginary part of the right side of the last equation is zero as well. We have

$$h_1 - 2h_2 + 3h_3 = 0 \tag{7.22}$$

System for Finding the Daubechies Four-Term Orthogonal Filter

Thus, to find h, we seek a solution to the nonlinear system of equations

$$
\left|
\begin{array}{l}
h_0^2 + h_1^2 + h_2^2 + h_3^2 = 1 \\
h_0 h_2 + h_1 h_3 = 0 \\
h_0 - h_1 + h_2 - h_3 = 0 \\
h_1 - 2h_2 + 3h_3 = 0
\end{array}
\right|
\tag{7.23}
$$

Solving the Daubechies Four-Term System

Let's solve this system. In Problem 7.10 you will employ a solution method that is algebraic in nature. We continue with a geometric solution method. Recall that we used (7.15) in conjunction with the first and third equations of (7.23) to arrive at the two equations

$$h_0^2 + h_1^2 = \frac{1}{1+c^2} \quad \text{and} \quad h_1 = \left(\frac{1-c}{1+c}\right) h_0$$

These equations are plotted in Figure 7.2. Now if we use (7.15) and insert $h_2 = -ch_1$ and $h_3 = ch_0$ into the last equation of (7.23), we obtain

$$0 = h_1 - 2h_2 + 3h_3 = h_1 + 2ch_1 + 3ch_0 = (1 + 2c)h_1 + 3ch_0$$

If we solve this last equation for h_1, we obtain

$$h_1 = -\left(\frac{3c}{1 + 2c}\right) h_0 \tag{7.24}$$

Now (7.24) is a line through the origin. If we are to solve the system, then the slopes of the lines given in (7.17) and (7.24) must be the same. Thus, we seek c, $c \neq -\frac{1}{2}$, and $c \neq -1$, such that

$$\frac{1 - c}{1 + c} = -\frac{3c}{1 + 2c}$$

If we cross-multiply and expand, we obtain

$$1 + c - 2c^2 = -3c - 3c^2$$

This leads to the quadratic equation

$$c^2 + 4c + 1 = 0$$

and using the quadratic formula, we find that

$$c = -2 \pm \sqrt{3} \tag{7.25}$$

If we take $c = -2 + \sqrt{3}$ and insert it into (7.17), we obtain

$$h_1 = \left(\frac{1 - c}{1 + c}\right) h_0 = \left(\frac{3 - \sqrt{3}}{-1 + \sqrt{3}}\right) h_0$$

Multiplying top and bottom of the right side of this last equation by $-1 - \sqrt{3}$ and simplifying gives

$$h_1 = \sqrt{3} h_0 \tag{7.26}$$

We insert this identity into the left side of (7.16) to obtain

$$h_0^2 + h_1^2 = h_0^2 + (\sqrt{3} h_0)^2 = 4h_0^2 \tag{7.27}$$

For $c = -2 + \sqrt{3}$, the right side of (7.16) becomes

$$\frac{1}{1 + c^2} = \frac{1}{1 + (-2 + \sqrt{3})^2} = \frac{1}{4(2 - \sqrt{3})} = \frac{2 + \sqrt{3}}{4} \tag{7.28}$$

Combining (7.27) and (7.28) gives

$$h_0^2 = \frac{2 + \sqrt{3}}{16} \tag{7.29}$$

Now we need only to take square roots of both sides to find values for h_0. If we take square roots of both sides of (7.29), the right side will be quite complicated. Before taking roots, we note that

$$(1 + \sqrt{3})^2 = 4 + 2\sqrt{3} = 2(2 + \sqrt{3})$$

so that

$$2 + \sqrt{3} = \frac{(1 + \sqrt{3})^2}{2}$$

Inserting this identity into (7.29) gives

$$h_0^2 = \frac{(1 + \sqrt{3})^2}{32}$$

Taking square roots of both sides of the previous equation, we obtain

$$h_0 = \pm \frac{1 + \sqrt{3}}{4\sqrt{2}} \tag{7.30}$$

If we take the positive value in (7.30) and use (7.26), we have

$$h_1 = \frac{3 + \sqrt{3}}{4\sqrt{2}} \tag{7.31}$$

Finally, we use (7.30), (7.31), and $c = -2 + \sqrt{3}$ in conjunction with (7.15) to write

$$h_2 = \frac{3 - \sqrt{3}}{4\sqrt{2}} \quad \text{and} \quad h_3 = \frac{1 - \sqrt{3}}{4\sqrt{2}} \tag{7.32}$$

We will take the positive value in (7.30) and the values given by (7.31)-(7.32) as our Daubechies length 4 orthogonal filter.

Definition 7.1 (Daubechies Four-Term Orthogonal Filter). *The Daubechies filter of length 4 (denoted by D4) is the vector* $\mathbf{h} = (h_0, h_1, h_2, h_3)$, *where* h_0, h_1, h_2, *and* h_3 *are given by*

$$
\boxed{
\begin{array}{ll}
h_0 = \frac{1}{4\sqrt{2}}\left(1 + \sqrt{3}\right) & h_1 = \frac{1}{4\sqrt{2}}\left(3 + \sqrt{3}\right) \\[2ex]
h_2 = \frac{1}{4\sqrt{2}}\left(3 - \sqrt{3}\right) & h_3 = \frac{1}{4\sqrt{2}}\left(1 - \sqrt{3}\right)
\end{array}
} \tag{7.33}
$$

$\square$

In Problem 7.7 you will find the other solutions by finding $h_1, h_2,$ and h_3 using the negative value from (7.30) and also the solutions associated with the value $c = -2 - \sqrt{3}$.

The highpass filter **g** for D4 is given by the rule

$$g_k = (-1)^k h_{3-k}, \qquad k = 0, 1, 2, 3.$$

so that

$$g_0 = h_3 = \frac{1}{4\sqrt{2}}\left(1 - \sqrt{3}\right) \qquad g_1 = -h_2 = -\frac{1}{4\sqrt{2}}\left(3 - \sqrt{3}\right)$$

$$g_2 = h_1 = \frac{1}{4\sqrt{2}}\left(3 + \sqrt{3}\right) \qquad g_3 = -h_0 = -\frac{1}{4\sqrt{2}}\left(1 + \sqrt{3}\right)$$

In Figure 7.3, we have plotted $|H(\omega)|$ for the Haar filter $\left(\frac{\sqrt{2}}{2}, \frac{\sqrt{2}}{2}\right)$ and $|H(\omega)|$ for the D4 filter. Note the difference at $\omega = \pi$.

(a) Haar

(b) D4

Figure 7.3   Plots of the modulus of the Fourier series for both the Haar and D4 filter.

The derivation of the lowpass filter D4 was done using an 8×8 matrix. Do you see how to form matrices of larger dimension? Of course, N must be even for the construction of W_N, but the basic form is

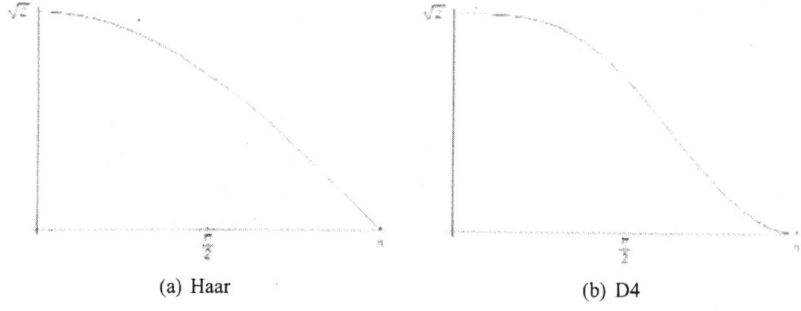

$$
W_N = \begin{bmatrix} H \\ G \end{bmatrix} =
\begin{bmatrix}
h_3 & h_2 & h_1 & h_0 & 0 & 0 & \cdots & 0 & 0 & 0 & 0 \\
0 & 0 & h_3 & h_2 & h_1 & h_0 & & 0 & 0 & 0 & 0 \\
\vdots & & & & & & \ddots & & & & \vdots \\
0 & 0 & 0 & 0 & & & & h_3 & h_2 & h_1 & h_0 \\
h_1 & h_0 & 0 & 0 & & & \cdots & 0 & 0 & h_3 & h_2 \\
-h_0 & h_1 & -h_2 & h_3 & 0 & 0 & \cdots & 0 & 0 & 0 & 0 \\
0 & 0 & -h_0 & h_1 & -h_2 & h_3 & & 0 & 0 & 0 & 0 \\
\vdots & & & & & & \ddots & & & & \vdots \\
0 & 0 & 0 & 0 & & & & -h_0 & h_1 & -h_2 & h_3 \\
-h_2 & h_3 & 0 & 0 & & & \cdots & 0 & 0 & -h_0 & h_1
\end{bmatrix}
\tag{7.34}
$$

Daubechies Orthogonal Filter of Length 6

It is straightforward to construct the Daubechies filter of length 6 now that we have constructed one of length 4. Since the filter we are looking for is longer, let's use a 10×10 matrix this time:

$$W_{10} = \left[\begin{array}{cccccccccc} h_5 & h_4 & h_3 & h_2 & h_1 & h_0 & 0 & 0 & 0 & 0 \\ 0 & 0 & h_5 & h_4 & h_3 & h_2 & h_1 & h_0 & 0 & 0 \\ 0 & 0 & 0 & 0 & h_5 & h_4 & h_3 & h_2 & h_1 & h_0 \\ h_4 & h_5 & 0 & 0 & 0 & 0 & h_5 & h_4 & h_3 & h_2 \\ h_3 & h_2 & h_1 & h_0 & 0 & 0 & 0 & 0 & h_5 & h_4 \\ \hline g_5 & g_4 & g_3 & g_2 & g_1 & g_0 & 0 & 0 & 0 & 0 \\ 0 & 0 & g_5 & g_4 & g_3 & g_2 & g_1 & g_0 & 0 & 0 \\ 0 & 0 & 0 & 0 & g_5 & g_4 & g_3 & g_2 & g_1 & g_0 \\ g_1 & g_0 & 0 & 0 & 0 & 0 & g_5 & g_4 & g_3 & g_2 \\ g_3 & g_2 & g_1 & g_0 & 0 & 0 & 0 & 0 & g_5 & g_4 \end{array}\right]$$

Finding the Orthogonality Conditions

Notice that overlapping rows now come in groups of three. If you think about it, the number of overlapping rows will tell you how many orthogonality conditions you will have. You can verify by direct multiplication of the H and H^T blocks in $W_{10}W_{10}^T$ that the three orthogonality conditions are

$$h_0^2 + h_1^2 + h_2^2 + h_3^2 + h_4^2 + h_5^2 = 1 \tag{7.35}$$
$$h_0 h_2 + h_1 h_3 + h_2 h_4 + h_3 h_5 = 0 \tag{7.36}$$
$$h_0 h_4 + h_1 h_5 = 0 \tag{7.37}$$

We again use the ideas from Problem 2.13 in Section 5.2 to define the highpass filter $\mathbf{g} = (h_5, -h_4, h_3, -h_2, h_1, -h_0)$. We can write W_{10} in block form,

$$W_{10} = \left[\frac{H}{G}\right]$$

and then compute $W_{10}W_{10}^T$. As was the case in (7.4), we see that W_{10} will be orthogonal if

$$HH^T = GG^T = I_5$$
$$HG^T = GH^T = 0_5$$

We verify that if (7.35) – (7.37) are satisfied, then $\mathbf{g} = (h_5, -h_4, h_3, -h_2, h_1, -h_0)$ makes W_{10} orthogonal.

Using the Lowpass Conditions

With regard to the lowpass constraints on **h**, we use

$$H(\pi) = h_0 - h_1 + h_2 - h_3 + h_4 - h_5 = 0 \qquad (7.38)$$

where now

$$H(\omega) = \sum_{k=0}^{5} h_k e^{ik\omega}$$

We will show in Problem 7.12 that just as is the case for the D4 system, the lowpass condition

$$h_0 + h_1 + h_2 + h_3 + h_4 + h_5 = \sqrt{2}$$

can be obtained from (7.38) and the orthogonality conditions (7.35) – (7.37).

If we use a CAS such as *Mathematica* to solve the system of equations (7.35) – (7.38), we again obtain an infinite amount of solutions, but this time there are two degrees of freedom.

Additional Conditions Imposed on $H(\omega)$

We use one of these degrees of freedom to again impose the condition

$$H'(\pi) = 0$$

and further flatten the filter at $\omega = \pi$ by insisting that

$$H''(\pi) = 0$$

as well. We can compute the derivatives of $H(\omega)$ to write these conditions explicitly in terms of the filter elements.

$$H(\omega) = h_0 + h_1 e^{i\omega} + h_2 e^{2i\omega} + h_3 e^{3i\omega} + h_4 e^{4i\omega} + h_5 e^{5i\omega}$$
$$H'(\omega) = ih_1 e^{i\omega} + 2ih_2 e^{2i\omega} + 3ih_3 e^{3i\omega} + 4ih_4 e^{4i\omega} + 5ih_5 e^{5i\omega}$$
$$H''(\omega) = -h_1 e^{i\omega} - 4h_2 e^{2i\omega} - 9h_3 e^{3i\omega} - 16h_4 e^{4i\omega} - 25h_5 e^{5i\omega}$$

Inserting $\omega = \pi$ into $H'(\omega)$, setting the result equal to zero, and simplifying gives

$$H'(\pi) = h_1 - 2h_2 + 3h_3 - 4h_4 + 5h_5 = 0 \qquad (7.39)$$

Inserting $\omega = \pi$ into $H''(\omega)$, setting the result equal to zero, and simplifying gives

$$H''(\pi) = h_1 - 4h_2 + 9h_3 - 16h_4 + 25h_5 = 0 \qquad (7.40)$$

The System for Finding Daubechies Six-Term Orthogonal Filters

We seek a solution to the three orthogonality equations $(7.35) - (7.37)$ and the three lowpass equations $(7.38) - (7.40)$:

$$
\begin{array}{r}
h_0^2 + h_1^2 + h_2^2 + h_3^2 + h_4^2 + h_5^2 = 1 \\
h_0 h_2 + h_1 h_3 + h_2 h_4 + h_3 h_5 = 0 \\
h_0 h_4 + h_1 h_5 = 0 \\
h_0 - h_1 + h_2 - h_3 + h_4 - h_5 = 0 \\
h_1 - 2h_2 + 3h_3 - 4h_4 + 5h_5 = 0 \\
h_1 - 4h_2 + 9h_3 - 16h_4 + 25h_5 = 0
\end{array}
\tag{7.41}
$$

In Problem 7.14 you will produce two real solutions to the system (7.41). We define one of these solutions to be the Daubechies filter of length 6.

Definition 7.2 (Daubechies 6-Term Orthogonal Filter). *The* Daubechies filter of length 6 *(denoted by D6) is the vector* $\mathbf{h} = (h_0, h_1, h_2, h_3, h_4, h_5)$ *where*

$$
\begin{aligned}
h_0 &= \tfrac{\sqrt{2}}{32}\left(1 + \sqrt{10} + \sqrt{5 + 2\sqrt{10}}\right) \approx \;\;\;0.332671 \\[2mm]
h_1 &= \tfrac{\sqrt{2}}{32}\left(5 + \sqrt{10} + 3\sqrt{5 + 2\sqrt{10}}\right) \approx \;\;\;0.806892 \\[2mm]
h_2 &= \tfrac{\sqrt{2}}{32}\left(10 - 2\sqrt{10} + 2\sqrt{5 + 2\sqrt{10}}\right) \approx \;\;\;0.459878 \\[2mm]
h_3 &= \tfrac{\sqrt{2}}{32}\left(10 - 2\sqrt{10} - 2\sqrt{5 + 2\sqrt{10}}\right) \approx -0.135011 \\[2mm]
h_4 &= \tfrac{\sqrt{2}}{32}\left(5 + \sqrt{10} - 3\sqrt{5 + 2\sqrt{10}}\right) \approx -0.085441 \\[2mm]
h_5 &= \tfrac{\sqrt{2}}{32}\left(1 + \sqrt{10} - \sqrt{5 + 2\sqrt{10}}\right) \approx \;\;\;0.035226
\end{aligned}
\tag{7.42}
$$

$\square$

The highpass filter $\mathbf{g}$ for D6 is given by the rule

$$
g_k = (-1)^k h_{5-k}, \qquad k = 0, \ldots, 5
$$

so that

$$g_0 = h_5 = \frac{\sqrt{2}}{32}\left(1 + \sqrt{10} - \sqrt{5 + 2\sqrt{10}}\right) \approx 0.035226$$

$$g_1 = -h_4 = -\frac{\sqrt{2}}{32}\left(5 + \sqrt{10} - 3\sqrt{5 + 2\sqrt{10}}\right) \approx 0.085441$$

$$g_2 = h_3 = \frac{\sqrt{2}}{32}\left(10 - 2\sqrt{10} - 2\sqrt{5 + 2\sqrt{10}}\right) \approx -0.135011$$

$$g_3 = -h_2 = -\frac{\sqrt{2}}{32}\left(10 - 2\sqrt{10} + 2\sqrt{5 + 2\sqrt{10}}\right) \approx -0.459878$$

$$g_4 = h_1 = \frac{\sqrt{2}}{32}\left(5 + \sqrt{10} + 3\sqrt{5 + 2\sqrt{10}}\right) \approx 0.806892$$

$$g_5 = -h_0 = -\frac{\sqrt{2}}{32}\left(1 + \sqrt{10} + \sqrt{5 + 2\sqrt{10}}\right) \approx -0.332671$$

A graph of $|H(\omega)|$ for the D6 filter shows an even flatter curve near $\omega = \pi$ than that of the D4 filter. Both are shown in Figure 7.4.

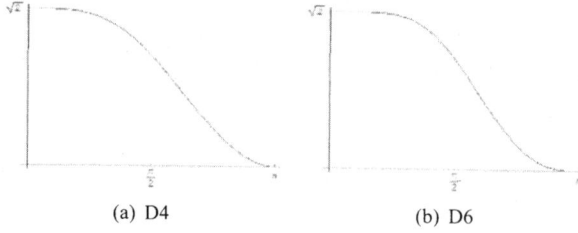

(a) D4   (b) D6

Figure 7.4   Plots of the modulus of the Fourier series for both the D4 and D6 filters.

Example: Data Compression

Let's look at some examples and compare the filters that we have constructed to date. The algorithms used to compute the wavelet transforms are discussed in Section 7.3.

Example 7.1 (Signal Compression Using Haar, D4, and D6). *We consider the digital audio file plotted in Figure 7.5. The file consists of 7,744 samples. The duration of the signal is 0.705273 second.*

In this example we first compute three wavelet transforms (six iterations each) using the Haar filter $\mathbf{h} = (\frac{\sqrt{2}}{2}, \frac{\sqrt{2}}{2})$*, the D4 filter (7.33), and the D6 filter (7.42), respectively.*

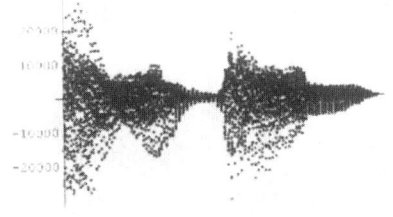

Figure 7.5   Plot of the audio signal for Example 7.1.

We plot the cumulative energy of the original audio file as well as the energy for each of the wavelet transforms in Figure 7.6. As you can see, the D6 slightly outperforms the D4, and both are much more efficient at storing the energy of the signal than the Haar transform.   □

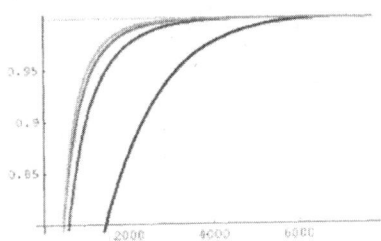

Figure 7.6   Plot of the cumulative energies of the original audio file (black), the Haar transform (dark gray), the D4 transform (gray), and the D6 transform (light gray).

Example: Image Compression

Our final example revisits the image compression application considered in Example 6.13.

Example 7.2 (Image Compression).   *We return to Example 6.13. In that example we used the HWT to perform image compression on the image plotted in Figure 6.17(a). In this example we repeat the process used in Example 6.13, but this time we will use the D4 and D6 filters.*

    Our first step is to compute the wavelet transforms using each of the three filters. We again perform two iterations of each transform. Recall that we modified the HWT so that it mapped integers to integers. The process for modifying D4 and D6 so that they map integers to integers is a difficult problem (see [13] and Problem 7.17), so we have simply rounded the elements in the D4 and D6 transforms in order to compare them with the modified HWT. The results are plotted in Figure 7.7.

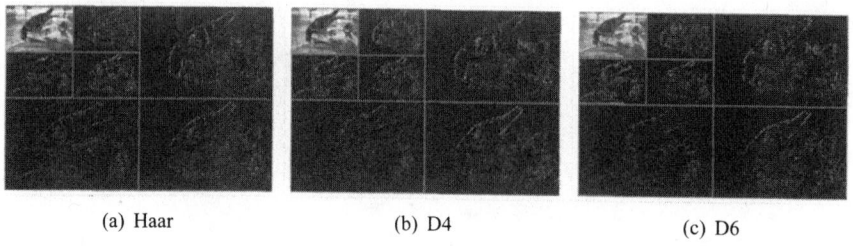

(a) Haar   (b) D4   (c) D6

Figure 7.7   Two iterations of the wavelet transform using the Haar, D4, and D6 filters.

We next compute the cumulative energy for each wavelet transform. These cumulative energies are plotted in Figure 7.8. Note that the Haar transform does a slightly better job of conserving the energy of the transformed image.

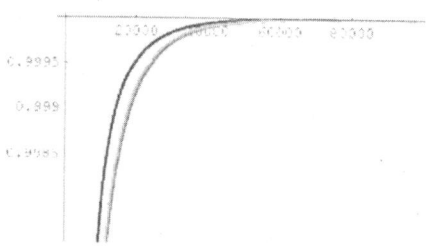

Figure 7.8   The cumulative energies for the Haar transform (black), the D4 filter (middle gray), and the D6 filter (light gray).

*Instead of performing quantization for three different energy levels as we did in Example 6.13, we will use just one energy level. We use a **CAS** to find the largest elements (in modulus) in each wavelet transform that constitutes $r = 99.7\%$ of the energy. All other wavelet coefficients are set to zero. Table 7.1 summarizes the compression information for each wavelet transformation. The quantized transformations are plotted in Figure 7.9.*

Recall from Example 6.13 that the original image consists of 98,304 pixels, so without any compression strategy, we must store pixels at 8 bpp. The entropy for the original image is 7.86662.

Table 7.1   Data for the quantized wavelet transformations using the Haar, D4, and D6 filters.[a]

Filter	Nonzero Terms	% Zero
Haar	8,842	91.01
D4	10,953	88.86
D6	10,504	89.31

[a] The second column indicates the number of nonzero terms still present in the quantized transforms, and the third column lists the percentage of elements that are zero in each quantized transform.

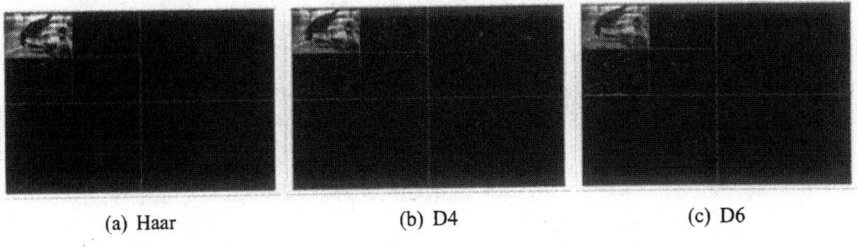

(a) Haar   (b) D4   (c) D6

Figure 7.9   The quantized wavelet transform using the Haar, D4, and D6 filters.

We convert the elements of each quantized transform to Huffman codes and then create a bit stream for each transform. The length of the bit stream for each transform is given in Table 7.2.

Table 7.2   Huffman bit stream length, bpp, and entropy for each quantized transform.

Filter	Bit Stream Length	bpp	Entropy
Haar	198794	2.02224	1.45639
D4	201062	2.04531	1.54619
D6	197448	2.00854	1.49566

Next, we invert each of the quantized transforms to obtain the compressed images. These images are plotted in Figure 7.10.

From Table 7.2 we note that the Haar filter slightly outperforms the other filters in compressing the image. The trade-off, though, is in loss of resolution. The compressed image obtained from the Haar transform is more "blocky" than the other two. The D4 and D6 filters are not without problems. Due to the rows of the D4, D6 transforms that "wrap" (elements in the last and first columns of a row), the pixels near the left side and bottom of the image are different from those in the original. If you work through Computer Lab 7.2, you will learn more about this problem. □

(a) D4 with PSNR 30.7197

(b) D6 with PSNR 30.7185

Figure 7.10 The compressed images using the D4 and D6 filters. Compare these results with the original image and Haar compressed image plotted in Figure 6.21(a) and (b).

PROBLEMS

7.1   In this problem you will show why filters of odd length will not work when constructing Daubechies filters.

   (a) Let's attempt the construction to find filters $\mathbf{h} = (h_0, h_1, h_2)$ and $\mathbf{g} = (g_0, g_1, g_2)$. Write down W_8 in this case. Use (7.1) as a guide.

   (b) We require W_8 to be an orthogonal matrix. If we partition W_8 as in (7.3) and use it to compute $W_8 W_8^T$ (see (7.4)), we see that in particular, $HH^T = I_4$. Write down the equations that h_0, h_1, and h_2 must satisfy so that $HH^T = I_4$.

   (c) What do these equations say about either h_0 or h_2? In either case, what are the values of the other filter coefficients?

   (d) Generalize the argument from parts (a) – (c) to any arbitrary odd length.

7.2   In this problem, you will show that if h_0, h_1, h_2, and h_3 satisfy the orthogonality conditions (7.6) and (7.7) and the lowpass condition (7.10), then

$$h_0 + h_1 + h_2 + h_3 = \pm\sqrt{2}$$

   (a) Use (7.6) and (7.7) to show that

$$(h_0 + h_1 + h_2 + h_3)^2 = 1 + 2(h_0 h_1 + h_0 h_3 + h_1 h_2 + h_2 h_3)$$

   (b) Square both sides of (7.10) and then use (7.6) and (7.7) to show that

$$0 = (h_0 - h_1 + h_2 - h_3)^2 = 1 - 2(h_0 h_1 + h_0 h_3 + h_1 h_2 + h_2 h_3)$$

   (c) Add the equations in parts (a) and (b) to obtain the desired result.

7.3   In this problem you will use Theorem 2.1 from Section 2.2 to show that

$$h_0 + h_1 + h_2 + h_3 = \pm\sqrt{2}$$

Suppose that the matrix W_8 given in (7.1) is orthogonal. Then Theorem 2.1 says that for all nonzero $\mathbf{v} \in \mathbb{R}^8$, we have $\|W_8 \mathbf{v}\| = \|\mathbf{v}\|$. Suppose further that

$$h_0 + h_1 + h_2 + h_3 = c \tag{7.43}$$

for some $c \in \mathbb{R}$.

   (a) For $\mathbf{v} = [\, 1, 1, 1, 1, 1, 1, 1, 1\,]^T$, verify that $\|\mathbf{v}\| = 2\sqrt{2}$.

   (b) For $\mathbf{v}$ in part (a), use (7.43) to compute $\mathbf{y} = W_8 \mathbf{v}$.

(c) Find $\|\mathbf{y}\|$.

(d) Use parts (a) and (c) along with Theorem 2.1 to find all possible values for c.

7.4   Suppose that the two-dimensional vectors $[h_0, h_1]^T$ and $[h_2, h_3]^T$ are orthogonal. Draw a picture in the plane that illustrates this fact. Show that $[h_2, h_3]^T = c[-h_1, h_0]^T$ for any $c \in \mathbb{R}$, $c \neq 0$.

7.5   Suppose that $c = 1$ in (7.17). Use (7.11) – (7.16) to answer the following:

(a) What are the possible values for h_1 and h_2? ·

(b) What are the possible values for h_0 and h_3?

(c) What are the possible highpass filters $\mathbf{g}$?

(d) Write W_8 for each of the filter pairs $\mathbf{h}$ and $\mathbf{g}$. Is W_8 an orthogonal matrix?

7.6   Take the negative root in (7.18) along with $|c| > 1$ and show that we obtain an infinite number of solutions to the system (7.12) – (7.14).

7.7   In this problem, we find the other solutions to the system given in (7.23).

(a) Take the negative root in (7.30) and insert it into (7.16), (7.17), and (7.24) to find another solution to (7.23).

(b) Take $c = -2 - \sqrt{3}$ and use it to find two values for h_0. For each value of h_0, find a solution to the system.

(c) If $\mathbf{h} = (h_0, h_1, h_2, h_3)$ is the filter given in Definition 7.1, how can you characterize the other solutions of (7.23) in terms of $\mathbf{h}$?

7.8   In this problem you will show that the conditions

$$h_0 + h_1 + h_2 + h_3 = \sqrt{2} \tag{7.44}$$
$$h_0 - h_1 + h_2 - h_3 = 0 \tag{7.45}$$
$$h_0 h_2 + h_1 h_3 = 0 \tag{7.46}$$

can be used to obtain

$$h_0^2 + h_1^2 + h_2^2 + h_3^2 = 1$$

To simplify computations, let

$$a = h_0^2 + h_1^2 + h_2^2 + h_3^2$$

(a) Square each of (7.44) and (7.45) and simplify to obtain

$$a + 2\sum_{j \neq k} h_j h_k = 2$$

$$a + 2\sum_{j \neq k} (-1)^{j+k} h_j h_k = 0$$

(b) Use (7.46) to simplify both equations from part (a). That is, show that

$$a + 2(h_0 h_1 + h_0 h_3 + h_1 h_2 + h_2 h_3) = 2$$
$$a - 2(h_0 h_1 + h_0 h_3 + h_1 h_2 + h_2 h_3) = 0$$

(c) Combine the two equations in part (b) to show that $a = 1$.

(d) Replace $\sqrt{2}$ by $-\sqrt{2}$ in (7.44) and repeat the problem.

7.9 In this problem you will show how to use one solution of (7.23) to obtain other solutions.

(a) Suppose that $\mathbf{h} = (a, b, c, d)$ solves (7.23). Show that $-\mathbf{h} = (-a, -b, -c, -d)$ also solves (7.23).

(b) Suppose that $\mathbf{h} = (a, b, c, d)$ solves (7.23). Show that $\tilde{\mathbf{h}} = (d, c, b, a)$ also solves (7.23). (*Hint:* The first three equations in (7.23) are easy to verify. For the fourth equation, write the Fourier series $M(\omega) = d + ce^{i\omega} + be^{2i\omega} + ae^{3i\omega}$ in terms of the Fourier series for $H(\omega) = a + be^{i\omega} + ce^{2i\omega} + de^{3i\omega}$. We know that the fourth condition in (7.23) is $H'(\pi) = 0$ and we can easily show that $H'(-\pi) = 0$ as well. You need to show that $M'(\pi) = 0$. After you write $M(\omega)$ in terms of $H(\omega)$, insert $\omega = \pi$ into the equation and simplify.)

(c) Using parts (a) and (b), how many solutions can we generate from $\mathbf{h}$ if $\mathbf{h}$ solves (7.23)?

7.10 In this problem you will use an algebraic approach to solve for the D4 filter. In Problem 7.8 we showed that the orthogonality condition

$$h_0^2 + h_1^2 + h_2^2 + h_3^2 = 1$$

can be replaced by

$$h_0 + h_1 + h_2 + h_3 = \sqrt{2}$$

in (7.23). Thus, we can find the D4 filter coefficients by solving the system

$$h_0 + h_1 + h_2 + h_3 = \sqrt{2} \qquad (7.47)$$
$$h_0 - h_1 + h_2 - h_3 = 0 \qquad (7.48)$$
$$h_1 - 2h_2 + 3h_3 = 0 \qquad (7.49)$$
$$h_0 h_2 + h_1 h_3 = 0 \qquad (7.50)$$

(a) Show that the first two equations (7.47) and (7.48) can be replaced by the linear equations

$$h_0 + h_2 = \frac{\sqrt{2}}{2} \quad \text{and} \quad h_1 + h_3 = \frac{\sqrt{2}}{2}$$

(b) Use part (a) and show that the linear equations (7.47) – (7.49) can be replaced by the equivalent system

$$h_0 = \frac{\sqrt{2}}{4} - h_3$$

$$h_1 = \frac{\sqrt{2}}{2} - h_3$$

$$h_2 = \frac{\sqrt{2}}{4} + h_3$$

(c) Insert the expressions for h_0, h_1, and h_2 from part (b) into (7.50) to show that

$$h_3^2 - \frac{\sqrt{2}}{4} h_3 - \frac{1}{16} = 0$$

(d) Use the quadratic formula to find two values for h_3. For each value of h_3, find values for h_0, h_1, and h_2.

(e) Replace $\sqrt{2}$ with $-\sqrt{2}$ on the right side of (7.47) and repeat the problem to find two other solutions to the system.

(f) Compare the solutions you obtained in this problem with the results of Problem 7.9.

7.11   In this problem you will show that if $h_0, \ldots, h_5$ satisfy the orthogonality conditions (7.35) – (7.37) and the lowpass condition (7.38), then

$$h_0 + h_1 + h_2 + h_3 + h_4 + h_5 = \pm\sqrt{2}$$

(a) Use (7.35) – (7.37) to show that

$$(h_0 + h_1 + h_2 + h_3 + h_4 + h_5)^2 = 1 + 2b$$

where

$$b = h_0 h_1 + h_0 h_3 + h_0 h_5 + h_1 h_2 + h_1 h_4 + h_2 h_3 + h_2 h_5 + h_3 h_4 + h_4 h_5$$

(b) Square both sides of (7.38) and then use (7.35) – (7.37) to show that

$$0 = (h_0 - h_1 + h_2 - h_3 + h_4 - h_5)^2 = 1 - 2b$$

(c) Add the equations in parts (a) and (b) to obtain the desired result.

7.12   In this problem you will show that we can replace

$$h_0^2 + h_1^2 + h_2^2 + h_3^2 + h_4^2 + h_5^2 = 1$$

in (7.41) with

$$h_0 + h_1 + h_2 + h_3 + h_4 + h_5 = \sqrt{2}$$

provided that the filter coefficients $h_0, \ldots, h_5$ satisfy

$$h_0 + h_1 + h_2 + h_3 + h_4 + h_5 = \sqrt{2} \qquad (7.51)$$
$$h_0 - h_1 + h_2 - h_3 + h_4 - h_5 = 0 \qquad (7.52)$$
$$h_0 h_2 + h_1 h_3 + h_2 h_4 + h_3 h_5 = 0 \qquad (7.53)$$
$$h_0 h_4 + h_1 h_5 = 0 \qquad (7.54)$$

To simplify computations, let

$$a = h_0^2 + h_1^2 + h_2^2 + h_3^2 + h_4^2 + h_5^2$$

(a) Square both sides of (7.51) and (7.52) and simplify to show that

$$a + 2 \sum_{j \neq k} h_j h_k = 2$$

$$a - 2 \sum_{j \neq k} (-1)^{j+k} h_j h_k = 0$$

(b) Use the orthogonality conditions (7.53) and (7.54) to reduce the equations in part (a) to

$$a + 2b = 2$$
$$a - 2b = 0$$

where

$$b = h_0 h_1 + h_0 h_3 + h_0 h_5 + h_1 h_2 + h_1 h_4 + h_2 h_3 + h_2 h_5 + h_3 h_4 + h_4 h_5$$

(c) Combine the equations in part (b) to show that $A = 1$.

(d) Replace $\sqrt{2}$ by $-\sqrt{2}$ on the right side of (7.51) and repeat the problem.

7.13   The fact that **g** is derived from D4 filter **h** $= (h_0, h_1, h_2, h_3)$ via the relation $g_k = (-1)^k h_{3-k}$, $k = 0,1,2,3$, leads us to look for similar relations between $H(\omega)$ and $G(\omega)$.

(a) Using the formula for g_k above, show that $G(\omega) = -e^{3i\omega}\overline{H(\omega + \pi)}$.

(b) Verify that $G(0) = 0$ and $G(\pi) = \sqrt{2}$ so that **g** is indeed a highpass filter.

(c) Let $\mathbf{h} = (h_0, \ldots, h_5)$ be the D6 filter and define **g** by $g_k = (-1)^k h_{5-k}$, $k = 0, \ldots, 5$. Show that $G(\omega) = -e^{5i\omega}\overline{H(\omega + \pi)}$.

(d) Show that the filter $G(\omega)$ from part (c) satisfies $G(0) = 0$ and $G(\pi) = \sqrt{2}$.

7.14   In this problem you will find two real solutions to the system given by (7.41). The process outlined in this problem is algebra intensive. A **CAS** may come in handy with some of the computations. We use Problem 7.12 to replace (7.35) with the condition

$$h_0 + h_1 + h_2 + h_3 + h_4 + h_5 = \sqrt{2}$$

in (7.41).

(a) Show that the system of linear equations in (7.41) is equivalent to

$$h_0 = \frac{\sqrt{2}}{8} - h_4 + 2h_5$$

$$h_1 = \frac{3\sqrt{2}}{8} - 2h_4 + 3h_5$$

$$h_2 = \frac{3\sqrt{2}}{8} - 2h_5$$

$$h_3 = \frac{\sqrt{2}}{8} + 2h_4 - 4h_5$$

(b) Use Problem 7.4 and (7.37) to write

$$h_0 = -ch_5 \qquad h_1 = ch_4 \tag{7.55}$$

where $c \neq 0$.

(c) Plug the results from part (b) in for h_0 and h_1 in part (a) and show that

$$h_4 = \frac{3\sqrt{2}(1 + c)}{8(1 + 4c + c^2)} \qquad h_5 = \frac{\sqrt{2}(1 - c)}{8(1 + 4c + c^2)} \tag{7.56}$$

(d) Plug the results from part (c) into the last two equations from part (a) to show that

$$h_2 = \frac{\sqrt{2}}{8} \cdot \frac{1 + 14c + 3c^2}{1 + 4c + c^2} \qquad h_3 = \frac{\sqrt{2}}{8} \cdot \frac{3 + 14c + c^2}{1 + 4c + c^2} \tag{7.57}$$

(e) We have now written all five variables (7.55) – (7.57) in terms of a parameter c. From part (b) we know that $c \neq 0$. What other nonzero condition must c satisfy? Insert the parametrized versions of $h_0, \ldots, h_5$ from (7.55) – (7.57) into (7.36) to show that

$$3 + 32c + 38c^2 + 32c^3 + 3c^4 = 0$$

(f) Verify that the quartic polynomial

$$p(c) = 3 + 32c + 38c^2 + 32c^3 + 3c^4$$

can be factored as

$$p(c) = 3\left(c^2 + \frac{4}{3}(4 + \sqrt{10})c + 1\right)\left(c^2 + \frac{4}{3}(4 - \sqrt{10})c + 1\right)$$

(g) Use part (f) to find the two real roots of $p(c)$.

(h) For each real value of c from part (g), use (7.55)-(7.57) to compute the values $h_0, \ldots, h_5$.

(i) For each real value of c from part (g), verify that the values for $h_0, \ldots, h_5$ from part (h) satisfy (7.35).

7.15 Suppose that $\mathbf{h} = (a, b, c, d, e, f)$ satisfies (7.41). Use the ideas from Problem 7.9 to show that $-\mathbf{h} = (-a, -b, -c, -d, -e, -f)$ and $\tilde{\mathbf{h}} = (f, e, d, c, b, a)$ both satisfy (7.41).

7.16 Although the lowpass constraint

$$h_0 + h_1 + h_2 + h_3 = \sqrt{2} \tag{7.58}$$

can be derived using other equations from system (7.23), what would be the effect of adding (7.58) to the system (7.23)? (*Hint:* Problem 7.10 will be helpful here.) What happens to the number of solutions to (7.41) if we add

$$h_0 + h_1 + h_2 + h_3 + h_4 + h_5 = \sqrt{2}$$

to (7.41)?

7.17 This problem is adapted from results in [13]. For illustrative purposes, let's consider the 8×8 wavelet transform W_8 given in (7.34) where the values h_0, h_1, h_2, and h_3 are given in Definition 7.1 and $g_k = (-1)^{k+1} h_{3-k}$ for $k = 0, 1, 2, 3$. In many applications, the lowpass and highpass rows of W_8 are often intertwined:

$$
\hat{W}_8 = \begin{bmatrix}
h_3 & h_2 & h_1 & h_0 & 0 & 0 & 0 & 0 \\
g_3 & g_2 & g_1 & g_0 & 0 & 0 & 0 & 0 \\
0 & 0 & h_3 & h_2 & h_1 & h_0 & 0 & 0 \\
0 & 0 & g_3 & g_2 & g_1 & g_0 & 0 & 0 \\
0 & 0 & 0 & 0 & h_3 & h_2 & h_1 & h_0 \\
0 & 0 & 0 & 0 & g_3 & g_2 & g_1 & g_0 \\
h_2 & h_1 & 0 & 0 & 0 & 0 & h_3 & h_2 \\
g_2 & g_1 & 0 & 0 & 0 & 0 & g_3 & g_2
\end{bmatrix}
$$

(a) A permutation matrix is a square matrix with exactly one 1 in each row and column. Find the 8×8 permutation matrix P so that

$$W_8 = P\hat{W}_8$$

(b) Find P^{-1} and describe the actions of P, P^{-1}.

(c) $\hat{W}_8$ has a nice block structure. Let

$$
H_1 = \begin{bmatrix} h_3 & h_2 \\ g_3 & g_2 \end{bmatrix}
\qquad
H_0 = \begin{bmatrix} h_1 & h_0 \\ g_1 & g_0 \end{bmatrix}
\qquad
0_4 = \begin{bmatrix} 0 & 0 \\ 0 & 0 \end{bmatrix}
$$

Using these matrices, write $\hat{W}_8$ in block form.

(d) Verify that

$$
H_1 = U \begin{bmatrix} 1 & 0 \\ 0 & 0 \end{bmatrix} V
\qquad
H_0 = U \begin{bmatrix} 0 & 0 \\ 0 & 1 \end{bmatrix} V
\tag{7.59}
$$

where

$$
U = \frac{1}{2\sqrt{2}} \begin{bmatrix} -1+\sqrt{3} & 1+\sqrt{3} \\ -1-\sqrt{3} & -1+\sqrt{3} \end{bmatrix}
\qquad
V = \frac{1}{2} \begin{bmatrix} -1 & \sqrt{3} \\ \sqrt{3} & 1 \end{bmatrix}
$$

(e) Show that U and V from part (d) are orthogonal matrices. The factorizations in (7.59) are called *singular value decompositions* of H_1 and H_0 (see Strang [71]).

(f) What rotation angle is associated with U? With V? See Problem 2.26 from Section 2.2.

(g) Replace H_0 and H_1 in the block formulation of $\hat{W}_8$ you found in part (c) by the factorizations given in (7.59) and then write $\hat{W}_8$ as $\mathcal{U}P\mathcal{V}$, where $\mathcal{U}$ and $\mathcal{V}$ are block diagonal matrices and P is a permutation matrix.

(h) Show that $\mathcal{U}$ and $\mathcal{V}$ from part (e) are orthogonal and describe in words the action of permutation matrix P.

The factorization in part (e) is used in Calderbank et al. [13] to show how to modify the D4 wavelet transformation so that it maps integers to integers. Recall that an orthogonal matrix rotates the vector to which it is applied and a permutation matrix simply reorders the elements of a vector. Thus, $\mathcal{U}P\mathcal{V}$ applied to vector $\mathbf{v}$ rotates $\mathbf{v}$, permutes $\mathcal{V}\mathbf{v}$, and then rotates the result!

7.18 At the end of Example 7.2, it is observed that the HWT produces some block artifacts in the reconstructed image, while some ringing is evident in images reconstructed using the D4 and D6 filters. Can you explain why this might happen?

7.19 One of the problems with the wavelet transform as we have developed it in this section is the wrapping rows. If our data are periodic, then the wrapping makes sense, but we typically do not think of digital audio or images as periodic data.

In this problem you will learn how to construct a wavelet transform for the D4 filter that has no wrapping rows. The technique is known as *matrix completion*. The resulting transform is orthogonal, and some of the highpass structure is retained. The process is described in Strang and Nguyen [73] and Keinert [52]. [1] A **CAS** will be helpful for calculations in the latter parts of this problem.

Strang and Nguyen [73] suggest adding two rows to each of the lowpass and highpass portions of the wavelet transform matrix. To perform the matrix completion, it is enough if we write down the matrix for processing vectors in $\mathbb{R}^8$. We use $\hat{W}_N$ to denote our modified wavelet transform:

$$\hat{W}_8 = \left[\begin{array}{cccccccc} a_2 & a_1 & a_0 & 0 & 0 & 0 & 0 & 0 \\ 0 & h_3 & h_2 & h_1 & h_0 & 0 & 0 & 0 \\ 0 & 0 & 0 & h_3 & h_2 & h_1 & h_0 & 0 \\ 0 & 0 & 0 & 0 & 0 & b_2 & b_1 & b_0 \\ \hline c_2 & c_1 & c_0 & 0 & 0 & 0 & 0 & 0 \\ 0 & g_3 & g_2 & g_1 & g_0 & 0 & 0 & 0 \\ 0 & 0 & 0 & g_3 & g_2 & g_1 & g_0 & 0 \\ 0 & 0 & 0 & 0 & 0 & d_2 & d_1 & d_0 \end{array}\right] = \left[\begin{array}{ccc|cc|ccc} a_2 & a_1 & a_0 & 0 & 0 & 0 & 0 & 0 \\ 0 & h_3 & h_2 & h_1 & h_0 & 0 & 0 & 0 \\ 0 & 0 & 0 & h_3 & h_2 & h_1 & h_0 & 0 \\ 0 & 0 & 0 & 0 & 0 & b_2 & b_1 & b_0 \\ \hline c_2 & c_1 & c_0 & 0 & 0 & 0 & 0 & 0 \\ 0 & -h_0 & h_1 & -h_2 & h_3 & 0 & 0 & 0 \\ 0 & 0 & 0 & -h_0 & h_1 & -h_2 & h_3 & 0 \\ 0 & 0 & 0 & 0 & 0 & d_2 & d_1 & d_0 \end{array}\right]$$

$$(7.60)$$

(a) Why do you suppose that only three terms are used in each row and that the offset is one position rather than two?

(b) Let $C(\omega) = c_0 + c_1 e^{i\omega} + c_2 e^{2i\omega}$. This is the Fourier series for the first new highpass rows. What conditions must be satisfied by the elements of this row if we insist that $C(0)$?

---

[1] Keinert also gives other methods for creating discrete wavelet transforms with no wrapping rows.

(c) We first complete $\mathbf{a} = (a_0, a_1, a_2)$ and $\mathbf{c} = (c_0, c_1, c_2)$. We know that

$$a_0^2 + a_1^2 + a_2^2 = c_0^2 + c_1^2 + c_2^2 = 1 \qquad (7.61)$$

Using (7.60), write down the other three nontrivial orthogonality conditions that the elements of $\mathbf{c}$ must satisfy.

(d) Two of the equations you found in part (c) involve only c_0 and c_1. Without solving the system, can you think of two values that work? (*Hint:* Think about the orthogonality conditions satisfied by the "conventional rows" in $\hat{W}_8$.)

(e) Use parts (d) and (b) to find c_2.

(f) We know from part (c) that $a_0^2 + a_1^2 + a_2^2 = 1$. Write down the other three orthogonality equations that the elements of $\mathbf{a}$ must satisfy. Two of these equations involve only a_0 and a_1. Use the ideas from part (d) to find a_0 and a_1.

(g) The remaining orthogonality condition is $\mathbf{a} \cdot \mathbf{c} = 0$. You should be able to use this equation and your previous work to show that $a_2 = \frac{1}{\sqrt{2}}$.

(h) The values in $\mathbf{a}$ and $\mathbf{c}$ do not satisfy (7.61). To finish this part of the completion problem, normalize $\mathbf{a}$ and $\mathbf{c}$ (see Problem 2.14 in Section 2.1).

(i) Repeat parts (b) – (h) to find values for $\mathbf{b} = (b_0, b_1, b_2)$ and $\mathbf{d} = (d_0, d_1, d_2)$.

(j) Is this matrix completion unique? That is, can you find other values for $\mathbf{a}, \mathbf{b}, \mathbf{c}$, and $\mathbf{d}$ that preserve orthogonal and the highpass condition $C(0) = D(0) = 0$?

In Computer Lab 7.7 of Section 7.3 you will create modules that implements the transform you have developed in this problem, and in Computer Lab 7.8 of Section 7.3 you will use the transform to perform data compression.

Computer Labs

◆ **7.1   Software Development: D4 and D6 Filters.** From the text Web site, access the lab `daubd4d6filters`. In this developmental lab you will develop code that allows you to solve for the solutions of systems (7.23) and (7.41). You will test the software and plot and compare $|H(\omega)|$ for each filter.

7.2   Image Compression with D4 and D6. From the text Web site, access the lab `daubd4d6compression`. In this lab you will further investigate image compression using the D4 and D6 filters. In particular, you will examine the effects of the transform on the left side and bottom of the uncompressed image.

7.2 DAUBECHIES FILTERS OF EVEN LENGTH

We now derive the system of equations that we must solve in order to find the Daubechies filter $\mathbf{h} = (h_0, h_1, \ldots, h_L)$, where L is an odd positive integer. We denote the filter length by

$$\boxed{M = L + 1}$$

As we have seen with both the D4 and D6 filters, the equations should come from orthogonality and lowpass conditions.

Orthogonality Conditions

The way we found orthogonality conditions for the Haar, D4, and D6 transformations was to write down W_N for sufficiently large N and then analyze $W_N W_N^T = I_N$. We could just as easily look only at the matrix W_N itself. As you can see, we have several wrapped rows in the general case. We study the structure of W_N in more detail in Section 7.3.

$$
W_N = \left[
\begin{array}{cccccccccccc}
h_L & h_{L-1} & h_{L-2} & h_{L-3} & \cdots & h_1 & h_0 & 0 & 0 & \cdots & 0 & 0 \\
0 & 0 & h_L & h_{L-1} & \cdots & h_3 & h_2 & h_1 & h_0 & \cdots & 0 & 0 \\
0 & 0 & 0 & 0 & \cdots & h_5 & h_4 & h_3 & h_2 & \cdots & 0 & 0 \\
 & & & & \ddots & & & & & & & \\
0 & 0 & 0 & 0 & \cdots & h_L & h_{L-1} & h_{L-2} & h_{L-3} & \cdots & 0 & 0 \\
 & & & & \ddots & & & & & & & \\
h_{L-2} & h_{L-3} & h_{L-4} & h_{L-5} & \cdots & & & & & \cdots & h_L & h_{L-1} \\
\hline
g_L & g_{L-1} & g_{L-2} & g_{L-3} & \cdots & g_1 & g_0 & 0 & 0 & \cdots & 0 & 0 \\
0 & 0 & g_L & g_{L-1} & \cdots & g_3 & g_2 & g_1 & g_0 & \cdots & 0 & 0 \\
0 & 0 & 0 & 0 & \cdots & g_5 & g_4 & g_3 & g_2 & \cdots & 0 & 0 \\
 & & & & \ddots & & & & & & & \\
0 & 0 & 0 & 0 & \cdots & g_L & g_{L-1} & g_{L-2} & g_{L-3} & \cdots & 0 & 0 \\
 & & & & \ddots & & & & & & & \\
g_{L-2} & g_{L-3} & g_{L-4} & g_{L-5} & \cdots & & & & & \cdots & g_L & g_{L-1}
\end{array}
\right] \tag{7.62}
$$

We need each row of W_N to dot with itself and produce 1 and dot with any other row and produce 0.

We can further simplify our analysis by recalling that we need only find the filter $\mathbf{h} = (h_0, h_1, \ldots, h_L)$ — the highpass vector $\mathbf{g}$ needed to complete W_N can be found by the rule $g_k = (-1)^k h_{L-k}, \ k = 0, \ldots, L$.

Consider row $\mathbf{w}_i$ from the top half of W_N. The only nonzero elements in $\mathbf{w}_i$ are $h_0, h_1, \ldots, h_L$, so that when we compute $\mathbf{w}_i \cdot \mathbf{w}_i$, we obtain our first orthogonality condition.

$$h_0^2 + h_1^2 + \cdots + h_L^2 = \sum_{k=0}^{L} h_k^2 = 1 \tag{7.63}$$

We now need to establish the *zero orthogonality conditions* that result when we dot row $\mathbf{w}_i$ and $\mathbf{w}_j$, $i \neq j$ from the top half of W_N.

If you recall, there were no zero orthogonality conditions when deriving the HWT (the $L = 1$ case). Multiplying distinct rows in the top half of W_8 (6.7) trivially resulted in inner product equal to zero since no filter coefficients overlapped.

Now think about the zero orthogonality condition for D4. There was only one zero orthogonality condition, equation (7.7), that resulted from one row that overlapped with the first row.

Finally, for D6, there were two zero orthogonality conditions, (7.36) and (7.37), resulting from two overlapping rows that overlapped with the first row.

There seems to be a pattern linking the numbers of zero orthogonality conditions to the filter length. We summarize these observations in Table 7.3.

Table 7.3 The number of zero orthogonality conditions for the Haar, D4, and D6 wavelet transformations.

Transform	Filter Length	Zero Orthogonality Conditions
Haar	2	0
D4	4	1
D6	6	2

Certainly, this pattern is linear and it is easy to guess that if the filter length is M, then the number of zero orthogonality conditions is

$$\frac{M}{2} - 1 = \frac{L+1}{2} - 1 = \frac{L-1}{2}$$

Now we know that we need $\frac{L-1}{2}$ zero orthogonality conditions. We can use the first $\frac{M}{2}$ rows of the top half of W_N

$$
\begin{array}{llllllllll}
\text{Row 1:} & h_L & h_{L-1} & h_{L-2} & h_{L-3} & h_{L-4} & \cdots & h_1 & h_0 & 0 & \cdots \\
\text{Row 2:} & & h_L & h_{L-1} & h_{L-2} & \cdots & h_3 & h_2 & h_1 & & \cdots \\
\text{Row 3:} & & & h_L & \cdots & h_5 & h_4 & h_3 & & \cdots \\
\vdots & & & & & & \vdots & & & \\
\text{Row } M/2: & & & & & \cdots & h_L & h_{L-1} & h_{L-2} & \cdots
\end{array}
$$

(7.64)

to describe them.

We need to write down the dot products of row 1 with rows $2, 3, \ldots, \frac{M}{2}$ from (7.64). As in the cases of the derivation of the D4 and D6 filters, we will obtain all orthogonality conditions for $h_0, \ldots, h_L$ by considering only inner products involving the first row. We begin by dotting row 1 with row 2. This inner product

$$h_0 h_2 + h_1 h_3 + h_2 h_4 + \cdots + h_{L-3} h_{L-1} + h_{L-2} h_L$$

(7.65)

requires $M - 2 = L - 1$ terms, while the inner product of row 3 and row 1

$$h_0 h_4 + h_1 h_5 + h_2 h_6 + \cdots + h_{L-5} h_{L-1} + h_{L-4} h_L \qquad (7.66)$$

needs $M - 4 = L - 3$ terms. Each subsequent inner product requires two fewer terms than the one before it. The final inner product is row 1 with row $\frac{M}{2}$ and this inner product is

$$h_0 h_{L-1} + h_1 h_L \qquad (7.67)$$

and requires only two terms. Any remaining rows in the top half of W_N either don't have filter terms that overlap with the first row or the overlaps are the same as the ones we have recorded.

We have written each inner product (7.65) – (7.67) in "reverse order" such that the first factor in each term is from row 1. Writing the inner products in this way makes it easier to derive a general formula for the $\frac{M}{2} - 1$ zero orthogonality conditions.

For each row $m = 2, \ldots, \frac{M}{2}$, the inner product formula can be written in summation notation. Let's use k as the summing index variable. We need to figure out the starting and stopping points for k as well as the subscripts for the two factors involving elements of $\mathbf{h}$ that appear in each term of the summation. Symbolically, we must find values for each of the boxes in

$$\sum_{k=\square}^{\square} h_\square h_\square$$

The length of each inner product decreases by two terms. We need to incorporate this into our general summation formula. There are a couple of ways to proceed. We can either build our summing variable k to reflect the fact that the first factor in the inner product is always h_0 or we can observe that the second factor in the last term of each inner product is h_L. In the former case we start our sum at $k = 0$ and vary its stopping point, and in the latter, we can vary the starting point and always stop our sum at $k = L$. We adopt the latter scheme and leave the former for Problem 7.20.

Our stopping point will be at $k = L$ and the starting point of the sum depends on the row. Row 2 starts with h_2, row 3 starts with h_4, row 4 starts with h_6 and so on. Clearly, there is a pattern relating the row number to the index of the first element of the row. You should verify that the first factor of row m is $h_{2(m-1)}$ for $m = 2, 3, \ldots, \frac{M}{2}$.

We have determined the sum limits of each inner product as well as the first factor of each inner product. Now we need to determine the second factor. This is easy since we are translating the filter $\mathbf{h}$ in row 2 by 2 units, row 3 by 4 units, and so on. This is exactly the same relation that we had for the index of the first factor! In general, for rows $m = 2, 3, \ldots, \frac{M}{2}$, we are translating $\mathbf{h}$ by $2(m - 1)$ units. Since we always want our second factor to start with h_0 and we know that k starts at $2(m - 1)$, we see that the general form of our second factor is $h_{k-2(m-1)}$.

We thus have the following $\frac{M}{2} - 1$ orthogonality conditions.

$$\sum_{k=2(m-1)}^{L} h_k h_{k-2(m-1)} = 0, \qquad m = 2, 3, \ldots, \frac{M}{2} \tag{7.68}$$

It is a bit less cumbersome if we replace $m - 1$ by m in (7.68). In this case we note that $\frac{M}{2} - 1 = \frac{L-1}{2}$ and write

$$\sum_{k=2m}^{L} h_k h_{k-2m} = 0, \qquad m = 1, 2, \ldots, \frac{L-1}{2} \tag{7.69}$$

Lowpass Conditions

We now need the lowpass conditions. For each filter in Section 7.1, we insisted that $H(\pi) = 0$. We use this constraint again but now

$$H(\omega) = \sum_{k=0}^{L} h_k e^{ik\omega} \tag{7.70}$$

Recall that $e^{ik\pi} = (-1)^k$. If we insert $\omega = \pi$ in (7.70), we obtain

$$0 = H(\pi) = \sum_{k=0}^{L} h_k e^{ik\pi} = \sum_{k=0}^{L} h_k (-1)^k = h_0 - h_1 + h_2 - + \cdots - h_L \tag{7.71}$$

The remaining lowpass conditions are derived by looking at derivatives of $H(\omega)$ at $\omega = \pi$. Recall that for the Haar filter (length 2), we insisted that $H(\pi) = 0$. For the D4 filter (length 4), we asked that

$$H(\pi) = H'(\pi) = 0,$$

and for D6 (length 6) we required that

$$H(\pi) = H'(\pi) = H''(\pi) = 0$$

We can generalize these observations for an arbitrary filter length $M = L + 1$ by stating that for $\mathbf{h} = (h_0, h_1, \ldots, h_L)$, we require that

$$H^{(m)}(\pi) = 0, \qquad m = 1, \ldots, \frac{L-1}{2} \tag{7.72}$$

(Here $H^{(m)}(\omega)$ denotes the mth derivative of $H(\omega)$.)

There are $\frac{L-1}{2} = \frac{M}{2} - 1$ derivative conditions in (7.72). We will now find the general formula for $H^{(m)}(\pi)$, $m = 1, \ldots, \frac{L-1}{2}$. If we differentiate $H(\omega)$, we obtain

$$H'(\omega) = \sum_{k=1}^{L} h_k (ik) e^{ik\omega} = i \sum_{k=1}^{L} h_k k e^{ik\omega}$$

Note that the sum above starts at $k = 1$ since the derivative of the constant term h_0 is zero. Differentiating again gives

$$H''(\omega) = i^2 \sum_{k=1}^{L} h_k k^2 e^{ik\omega}$$

Certainly, we could replace i^2 by -1 above, but to derive a general formula, it is easier to leave it as is. Each time we differentiate, we simply add another power to i outside the summand and another power to k inside the summand. The general formula for $H^{(m)}(\omega)$, $m \geq 1$, is

$$H^{(m)}(\omega) = i^m \sum_{k=1}^{L} h_k k^m e^{ik\omega} \tag{7.73}$$

To obtain our remaining lowpass conditions, we must evaluate (7.73) at $\omega = \pi$ and set the result equal to zero. Again using the fact that $e^{-ik\pi} = (-1)^k$, we have for $m = 1, 2, \ldots, \frac{L-1}{2}$

$$H^{(m)}(\pi) = i^m \sum_{k=1}^{L} h_k k^m e^{ik\pi} = i^m \sum_{k=1}^{L} h_k k^m (-1)^k \tag{7.74}$$

If we set $H^{(m)}(\pi) = 0$ in (7.74), we can divide both sides by i^m to obtain the $\frac{L-1}{2}$ lowpass conditions

$$\sum_{k=1}^{L} (-1)^k k^m h_k = 0, \qquad m = 1, 2, \ldots, \frac{L-1}{2} \tag{7.75}$$

System for Finding Daubechies Filters

Our system thus far consists of $\frac{M}{2}$ orthogonality conditions and $\frac{M}{2}$ lowpass conditions:

$$\sum_{k=0}^{L} h_k^2 = 1 \qquad \text{(orthogonality)} \qquad (7.76)$$

$$\sum_{k=2m}^{L} h_k h_{k-2m} = 0, \quad m = 1, 2, \ldots, \frac{L-1}{2} \quad \text{(orthogonality)} \qquad (7.77)$$

$$\sum_{k=0}^{L} (-1)^k h_k = 0 \qquad (H(\pi) = 0) \qquad (7.78)$$

$$\sum_{k=1}^{L} (-1)^k k^m h_k = 0, \quad m = 1, 2, \ldots, \frac{L-1}{2} \quad (H^{(m)}(\pi) = 0) \qquad (7.79)$$

It should be clear that if $\mathbf{h} = (h_0, \ldots, h_L)$ is a solution to (7.76)– (7.79), then $-\mathbf{h}$ is a solution as well. In Problem 7.21 you will show that if $\mathbf{h}$ satisfies the orthgonality conditions (7.76) and (7.77) and the lowpass condition (7.78), then

$$h_0 + h_1 + \cdots + h_L = \sqrt{2} \qquad (7.80)$$

or $h_0 + h_1 + \cdots + h_L = -\sqrt{2}$.

Now only one of $\mathbf{h}$ or $-\mathbf{h}$ can satisfy (7.80). If we add the extra lowpass condition (7.80) to our system, then we can reduce the number of solutions to our system by a factor of 2.

Thus the $M + 1$ equations we must solve to obtain Daubechies lowpass filter $\mathbf{h} = (h_0, h_1, \ldots, h_L)$, $M = L+1$, are given by (7.76)–(7.80). We state the complete system below.

$$\sum_{k=0}^{L} h_k^2 = 1 \qquad\qquad \text{(orthogonality)}$$

$$\sum_{k=2m}^{L} h_k h_{k-2m} = 0, \qquad m = 1, 2, \ldots, \frac{L-1}{2} \qquad \text{(orthogonality)}$$

$$\sum_{k=0}^{L} h_k = \sqrt{2} \qquad\qquad (H(0) = \sqrt{2})$$

$$\sum_{k=0}^{L} (-1)^k h_k = 0 \qquad\qquad (H(\pi) = 0)$$

$$\sum_{k=1}^{L} (-1)^k k^m h_k = 0, \qquad m = 1, 2, \ldots, \frac{L-1}{2} \qquad (H^{(m)}(\pi) = 0)$$

(7.81)

Characterizing Solutions to the Daubechies System

We conclude this section with a discussion about solving the system (7.81). We have already solved the system for $L = 1, 3, 5$. Can we solve it for larger L? If so, how many solutions exist for a given L?

Ingrid Daubechies [23] gave a complete characterization of the solutions to the system (7.81). Her derivation is a bit different from the one we present here. We state a couple of results and use them to discuss a method that Daubechies developed to choose a particular solution to the system (7.81).

Proofs of Theorems 7.1 and 7.2 are beyond the scope of this book. The reader with some background in advanced analysis is referred to the proofs given by Daubechies in [23, 24].

It turns out that the system (7.81) always has real solutions. Daubechies [24] proved the following result.

Theorem 7.1 (Number of Solutions: Daubechies [24]). *The system (7.81) has* $2^{\lfloor \frac{L+2}{4} \rfloor}$ *real solutions. Here,* $\lfloor \cdot \rfloor$ *is the floor function defined in Problem 6.26(c) in Section 6.4.* □

Table 7.4 shows how the number of real solutions $2^{\lfloor \frac{L+2}{4} \rfloor}$ for the system (7.81) grows as L gets larger.

The solutions to the system (7.81) obey a basic symmetry property. We have the following result.

Table 7.4   The number of real solutions to the system (7.81) for a given L.

L	1	3	5	7	9	11	13
No. of Solutions	1	2	2	4	4	8	8

Proposition 7.1 (Reflections are Solutions). *Suppose that L is an odd positive integer and assume that $\alpha = (\alpha_0, \alpha_1, \ldots, \alpha_L)$ solves the system (7.81). Then $(\alpha_L, \alpha_{L-1}, \ldots, \alpha_1, \alpha_0)$, the reflection of α, solves (7.81) as well.* $\square$

Proof of Proposition 7.1. The proof is left as Problem 7.9. $\square$

Proposition 7.1 tells us that we effectively have $2^{\lfloor \frac{L+2}{4} \rfloor - 1}$ solutions to find. The remaining solutions are simply reflections. Daubechies was interested in choosing a particular solution for the system (7.81). We give an outline of this method now. Before we begin, we need a definition.

Definition 7.3 (Multiplicity of Roots). *Suppose that $p(t)$ is a polynomial of degree n, and let ℓ be a nonnegative integer with $\ell \leq n$. We say that $t = a$ is a root of multiplicity ℓ if $p(t) = (t - a)^\ell q(t)$, where $q(t)$ is a polynomial of degree $n - \ell$ and $q(a) \neq 0$.* $\square$

Example 7.3 (Finding Multiplicity of Roots). *Consider the degree 5 polynomial*

$$p(t) = t^5 - t^4 - 24t^3 + 88t^2 - 112t + 48$$

Since $p(t) = (t - 2)^3(t^2 + 5t - 6)$ and $q(t) = t^2 + 5t - 6 = (t - 1)(t + 6)$ satisfies $q(2) \neq 0$, we see that $a = 2$ is a root of multiplicity 3 for $p(t)$. We can also note that $a = 1$, $a = -6$ are roots of multiplicity 1 for $p(t)$. $\square$

In Problem 7.26 you will show that $t = a$ is a root of multiplicity ℓ if and only if $p^{(m)}(a) = 0$ for $m = 0, 1, \ldots, \ell - 1$.

Returning to Daubechies' selection of a particular solution to (7.81), suppose that $\mathbf{h} = (h_0, h_1, \ldots, h_L)$ is a solution to (7.81). Then we can write down its Fourier series

$$H(\omega) = \sum_{k=0}^{L} h_k e^{ik\omega}$$

and make the change of variable $z = e^{i\omega}$. We then consider the polynomial

$$P(z) = \sum_{k=0}^{L} h_k z^k \tag{7.82}$$

Since $H(\pi) = 0$ and $e^{i\pi} = -1$, we have $P(-1) = 0$. As a matter of fact, using Problem 7.26 we see that $H^{(m)}(\pi) = 0$ for $m = 0, 1, \ldots, \frac{L-1}{2}$ implies $P^{(m)}(-1) =$

0. Thus, $z = -1$ is a root of *at least* multiplicity $\frac{L+1}{2}$ for $P(z)$. This means that

$$P(z) = (z+1)^{\frac{L+1}{2}} Q(z) \tag{7.83}$$

where $Q(z)$ is a polynomial of degree $\frac{L-1}{2}$. Daubechies [23] showed that $Q(-1) \neq 0$ so $z = -1$ is indeed a root of multiplicity $\frac{L+1}{2}$ for $P(z)$. In fact, when she derived the filters, she showed the following result.

Theorem 7.2 (Characterizing Solutions: Daubechies [23]). *For each $L = 1, 3, 5, \ldots$ there is exactly one solution for (7.81) where the roots z_k, $k = 1, 2, \ldots, \frac{L-1}{2}$, of $Q(z)$ satisfy $|z_k| > 1$.* $\square$

Theorem 7.2 gives us a method for choosing a particular solution of (7.81). We know that $P(z)$ is a polynomial of degree L, and we also know that $z = -1$ is a root of multiplicity $\frac{L+1}{2}$, so we find the roots z_k, $k = 1, \ldots, L$ of $P(z)$ and compute $|z_k|$. If $|z_k| \geq 1$ then the solution we used to form $P(z)$ is the Daubechies filter. We demonstrate this result for the D4 filter.

Example 7.4 (P(z) for the D4 Filter). *We first construct $P(z)$ using the coefficients given in (7.33):*

$$P(z) = \frac{1}{4\sqrt{2}} \left(1 + \sqrt{3} + (3 + \sqrt{3})z + (3 - \sqrt{3})z^2 + (1 - \sqrt{3})z^3 \right)$$

Since $H(\pi) = P(-1) = 0$ and $H'(\pi) = P'(-1) = 0$, we know that -1 is a root of multiplicity 2 of $P(z)$, so that

$$P(z) = (z+1)^2 Q(z)$$

so that $Q(z)$ must be a linear polynomial. Basic algebraic long division gives

$$Q(z) = \frac{1 - \sqrt{3}}{4\sqrt{2}} \left(z - (2 + \sqrt{3}) \right)$$

The root of $Q(z)$ is $2 + \sqrt{3}$ and the absolute value of this root is larger than 1.

Using Theorem 7.1, we know that there are two real solutions to the system (7.81) for $L = 3$, and by Proposition 7.1 we know that the other solution is merely a reflection of the D4 filter. If we form $P(z)$ for this solution we have

$$P(z) = \frac{1}{4\sqrt{2}} \left(1 - \sqrt{3} + (3 - \sqrt{3})z + (3 + \sqrt{3})z^2 + (1 + \sqrt{3})z^3 \right)$$

We know that $Q(z)$ is a linear polynomial and if we divide $P(z)$ by $(z+1)^2$, we find that

$$Q(z) = \frac{1 + \sqrt{3}}{4\sqrt{2}} \left(z - (2 - \sqrt{3}) \right)$$

Table 7.5   Solutions to (7.81) for $L = 9$.

	Solution 1	**Solution 2**	**Solution 3**	**Solution 4**
h_0	0.00333573	0.02733307	0.01953888	0.16010240
h_1	-0.01258075	0.02951949	-0.02110183	0.60382927
h_2	-0.00624149	-0.03913425	-0.17532809	0.72430853
h_3	0.07757149	0.19939753	0.01660211	0.13842815
h_4	-0.03224487	0.72340769	0.63397896	-0.24229489
h_5	-0.24229489	0.63397896	0.72340769	-0.03224487
h_6	0.13842815	0.01660211	0.19939753	0.07757149
h_7	0.72430853	-0.17532809	-0.03913425	-0.00624149
h_8	0.60382927	-0.02110183	0.02951949	-0.01258075
h_9	0.16010240	0.01953888	0.02733307	0.00333573

The root in this case is $2 - \sqrt{3}$, and the absolute value of this number is certainly smaller than 1.   □

Let's look at an example using a longer filter.

Example 7.5 (Finding the D10 Filter). *Let's find the Daubechies filter of length 10. By Theorem 7.1 we know that there are four real solutions. We set up the system (7.81) for $L = 9$ and solve it using a **CAS**. The solutions (rounded to eight decimal places) are given in Table 7.5. Now for solution m, $m = 1, \ldots, 4$, we create the degree 9 polynomial $P_m(z) = \sum\limits_{k=0}^{9} h_k z^k$ and then find its roots. We know that -1 is a root of multiplicity $\frac{9+1}{2} = 5$ for each of these polynomials. The roots for each polynomial are plotted in Figure 7.11.*

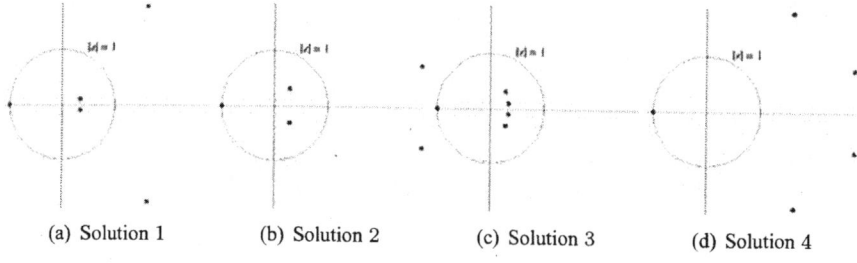

(a) Solution 1   (b) Solution 2   (c) Solution 3   (d) Solution 4

Figure 7.11   The roots of the degree 9 polynomials. Note that -1 is a root of multiplicity 5 for each polynomial.

Note that the roots z_k for solution 4 satisfy $|z_k| \geq 1$, so this is the solution we use. Note that solution 1 is a reflection of solution 4, and solutions 2 and 3 are reflections of each other as well.   □

PROBLEMS

7.20 Show that the orthogonality conditions given by (7.69) can be written as

$$\sum_{k=0}^{L-2m} h_k h_{k+2m}$$

for $m = 1, \ldots, \frac{L-1}{2}$

7.21 In this problem you will show that if $\mathbf{h} = (h_0, h_1, \ldots, h_L)$ satisfies (7.63), (7.69), and the lowpass condition (7.71), then

$$h_0 + h_1 + h_2 + \cdots + h_L = \pm\sqrt{2} \qquad (7.84)$$

(a) Use (7.63) to show that

$$(h_0 + \cdots + h_L)^2 = 1 + 2\sum_{j \neq k} h_j h_k$$

(b) Use (7.63) and (7.71) to show that

$$(h_0 - h_1 + h_2 - h_3 + \cdots - h_L)^2 = 1 + 2\sum_{j \neq k}(-1)^{j+k} h_j h_k$$

(c) Split the summations in parts (a) and (b) into two summations. One summation goes over the $j + k$ odd terms and one over the $j + k$ even terms. What do the orthogonality conditions say about one of these two sums?

(d) Use part (c) to simplify the expressions in parts (a) and (b).

(e) Finally, combine the simplified expressions that you obtained in part (d) to obtain the desired result.

7.22 Suppose that $\mathbf{h} = (h_0, \ldots, h_L)$ satisfies the orthogonality conditions (7.69), the lowpass condition (7.71), and the condition

$$h_0 + h_1 + \cdots + h_L = \sqrt{2} \qquad (7.85)$$

Show that $\mathbf{h}$ satisfies (7.63). The following steps will help you organize your work. To simplify notation, let

$$a = h_0^2 + h_1^2 + \cdots + h_L^2$$

(a) Square both sides of (7.85) and show that

$$2 = a + 2\sum_{j \neq k} h_j h_k$$

(b) Square both sides of (7.71) and show that

$$0 = a + 2\sum_{j \neq k}(-1)^{j+k}h_j h_k$$

(c) Now use the ideas from part (c) of Problem 7.21 to simplify the expressions you verified in parts (a) and (b).

(d) Combine the simplified expressions from part (c) to obtain the desired result.

(e) Replace the $\sqrt{2}$ by $-\sqrt{2}$ on the right side of (7.85) and repeat the problem.

7.23   Prove Proposition 7.1. (Problem 7.9 will be helpful. How would you define $M(\omega)$ in the general case?)

7.24   Let $h = (h_0, \ldots, h_L)$ be the Daubechies filter of length $L + 1$. Then $g = (g_0, \ldots, g_L)$ is defined by $g_k = (-1)^k h_{L-k}$, $k = 0, \ldots, L$. Show that $G(\omega) = -e^{iL\omega}\overline{H(\omega + \pi)}$. (*Hint:* Use the ideas from Problem 7.13 from Section 7.1.)

7.25   In this problem you will find a general formula[2] for $(fg)^{(k)}(t)$, where $k = 0, 1, 2, \ldots$. Use mathematical induction to show that

$$(fg)^{(k)}(t) = \sum_{m=0}^{k}\binom{k}{m}f^{(m)}(t)g^{(k-m)}(t)$$

where the *binomial coefficient* $\binom{k}{m}$ is defined in Problem 4.41 from Section 4.3. (*Hint:* Use Problem 4.42.)

7.26   Suppose that $p(t)$ is a polynomial of degree n and let ℓ be a nonnegative integer satisfying $\ell \leq n$. Show that $t = a$ is a root of multiplicity ℓ if and only if $p^{(m)}(a) = 0$ for $m = 0, 1, \ldots, \ell - 1$. (Hint: For the necessity assertion, you will find it useful to write out a Taylor's series for $p(t)$. For the sufficiency assertion, Problem 7.25 will be helpful).

7.27   Daubechies lowpass filter of length 8.

(a) Set up the system of equations that we must solve in order to find a Daubechies lowpass filter for length 8.

(b) Use a **CAS** to solve your system.

(c) Use a **CAS** to graph $|H(\omega)|$ from $\omega \in [0, \pi]$.

7.28   We saw that for D4 and D6, the lowpass condition $H(0) = 1$ must be changed to $H(0) = \sqrt{2}$ to preserve the orthogonality of W_N. Show that we must require that

---

[2]This rule is often referred to as the *Leibniz generalized product rule*.

$H(0) = \sqrt{2}$ for the general Daubechies lowpass filter $\mathbf{h} = (h_0, h_1, \ldots, h_L)$ for any odd positive integer L.

Computer Labs

♦ **7.3   Software Development: Generating Daubechies Filters.** From the text Web site, access the development package daubfilters. In this development lab you will generate code that will produce solutions to the system (7.81). Your code will incorporate the results of Theorems 7.1 and 7.2. Instructions are provided that allow you to add the code to the software package DiscreteWavelets.

7.4   Daubechies Filters. From the text Web site, access the lab daubfilterstest. In this lab you will use code from the software package DiscreteWavelets to investigate properties of the Daubechies filters of length $L + 1$. You will also plot and compare $|H(\omega)|$ for various filters.

7.3 ALGORITHMS FOR DAUBECHIES WAVELET TRANSFORMATIONS

In this section we develop pseudocode for implementation of the one-dimensional Daubechies wavelet transformation. Code is provided for computing one iteration of the transform (Algorithm 7.1) and its inverse (Algorithm 7.3). Each of our routines will accept an orthogonal lowpass filter $\mathbf{h}$ of even length $L + 1$ in addition to the input vector (or matrix). We can create the highpass filter $\mathbf{g}$ using the relation $g_k = (-1)^k h_{L-k}$, $k = 0, 1, \ldots, L$.

The routines for computing iterated (inverse) transforms and two-dimensional (inverse) transforms are almost identical to those presented in Sections 6.2 and 6.3. We develop these algorithms in Computer Labs 7.5 and 7.6.

Efficient Computation of $\mathbf{W_N v}$

Let's start with the one-dimensional wavelet transform. Suppose that we are given filter $\mathbf{h}$ of even length $L + 1$ and input vector $\mathbf{v}$ with even length N. We must compute the product

$$\mathbf{y} = W_N \mathbf{v} = \left[\frac{H}{G} \right] \mathbf{v} = \left[\frac{H\mathbf{v}}{G\mathbf{v}} \right]$$

where W_N is given by (7.62) and H and G are the lowpass and highpass portions of W_N, respectively. The matrices H and G are identical in structure — they differ only by the filters that we use to build them. Thus, any efficient algorithm we develop that computes $H\mathbf{v}$ can be used to compute $G\mathbf{v}$ as well — we need only replace the filter coefficients h_k with g_k in the latter computation. We now exploit the sparse and cyclic nature of H to design a fast algorithm.

To get an idea of how the algorithm will work, let's look at H for filters of length 4, 6, and 8 and vectors of length 12. Equations (7.86) – (7.88) show these matrices. The subscripts indicate the filter length.

$$H_4 = \left[\begin{array}{cccccccccccc} h_3 & h_2 & h_1 & h_0 & 0 & 0 & 0 & 0 & 0 & 0 & 0 & 0 \\ 0 & 0 & h_3 & h_2 & h_1 & h_0 & 0 & 0 & 0 & 0 & 0 & 0 \\ 0 & 0 & 0 & 0 & h_3 & h_2 & h_1 & h_0 & 0 & 0 & 0 & 0 \\ 0 & 0 & 0 & 0 & 0 & 0 & h_3 & h_2 & h_1 & h_0 & 0 & 0 \\ 0 & 0 & 0 & 0 & 0 & 0 & 0 & 0 & h_3 & h_2 & h_1 & h_0 \\ \hline h_1 & h_0 & 0 & 0 & 0 & 0 & 0 & 0 & 0 & 0 & h_3 & h_2 \end{array}\right] \tag{7.86}$$

$$H_6 = \left[\begin{array}{cccccccccccc} h_5 & h_4 & h_3 & h_2 & h_1 & h_0 & 0 & 0 & 0 & 0 & 0 & 0 \\ 0 & 0 & h_5 & h_4 & h_3 & h_2 & h_1 & h_0 & 0 & 0 & 0 & 0 \\ 0 & 0 & 0 & 0 & h_5 & h_4 & h_3 & h_2 & h_1 & h_0 & 0 & 0 \\ 0 & 0 & 0 & 0 & 0 & 0 & h_5 & h_4 & h_3 & h_2 & h_1 & h_0 \\ \hline h_1 & h_0 & 0 & 0 & 0 & 0 & 0 & 0 & h_5 & h_4 & h_3 & h_2 \\ h_3 & h_2 & h_1 & h_0 & 0 & 0 & 0 & 0 & 0 & 0 & h_5 & h_4 \end{array}\right] \tag{7.87}$$

$$H_8 = \left[\begin{array}{cccccccccccc} h_7 & h_6 & h_5 & h_4 & h_3 & h_2 & h_1 & h_0 & 0 & 0 & 0 & 0 \\ 0 & 0 & h_7 & h_6 & h_5 & h_4 & h_3 & h_2 & h_1 & h_0 & 0 & 0 \\ 0 & 0 & 0 & 0 & h_7 & h_6 & h_5 & h_4 & h_3 & h_2 & h_1 & h_0 \\ \hline h_1 & h_0 & 0 & 0 & 0 & 0 & h_7 & h_6 & h_5 & h_4 & h_3 & h_2 \\ h_3 & h_2 & h_1 & h_0 & 0 & 0 & 0 & 0 & h_7 & h_6 & h_5 & h_4 \\ h_5 & h_4 & h_3 & h_2 & h_1 & h_0 & 0 & 0 & 0 & 0 & h_7 & h_6 \end{array}\right] \tag{7.88}$$

Note the partition lines that appear in each matrix. These lines mark the point where rows whose nonzero elements start wrapping. We can use (7.86) – (7.88) to observe the pattern for the number of wrapping rows when the filter length is $L + 1$. When $L = 3$, there is one wrapping row. For $L = 5, 7$, there are two and three wrapping rows, respectively. In general, for arbitrary odd L, there are $\frac{L+1}{2} - 1 = \frac{L-1}{2}$ wrapping rows. Recall that the Haar transform (see Definition 6.1) had no wrapping rows. In this case, $L = 1$, so that $\frac{L-1}{2} = 0$ and the formula holds here as well.

We divide our algorithm into two parts. In the first part we compute $H\mathbf{v}$ for nonwrapping rows, and in the second part we compute $H\mathbf{v}$ for wrapping rows.

Computing Transform Values For Non-Wrapping Rows

Computation involving the nonwrapping rows is easy. For illustrative purposes, consider the $L = 5$ case (7.87). The computation involving the four nonwrapping rows is given in (7.89).

$$
\begin{bmatrix}
h_5 & h_4 & h_3 & h_2 & h_1 & h_0 & 0 & 0 & 0 & 0 & 0 & 0 \\
0 & 0 & h_5 & h_4 & h_3 & h_2 & h_1 & h_0 & 0 & 0 & 0 & 0 \\
0 & 0 & 0 & 0 & h_5 & h_4 & h_3 & h_2 & h_1 & h_0 & 0 & 0 \\
0 & 0 & 0 & 0 & 0 & 0 & h_5 & h_4 & h_3 & h_2 & h_1 & h_0
\end{bmatrix}
\cdot
\begin{bmatrix}
v_1 \\ v_2 \\ v_3 \\ v_4 \\ v_5 \\ v_6 \\ v_7 \\ v_8 \\ v_9 \\ v_{10} \\ v_{11} \\ v_{12}
\end{bmatrix}
\tag{7.89}
$$

$$
=
\begin{bmatrix}
h_5 v_1 & +h_4 v_2 & +h_3 v_3 & +h_2 v_4 & +h_1 v_5 & +h_0 v_6 \\
h_5 v_3 & +h_4 v_4 & +h_3 v_5 & +h_2 v_6 & +h_1 v_7 & +h_0 v_8 \\
h_5 v_5 & +h_4 v_6 & +h_3 v_7 & +h_2 v_8 & +h_1 v_9 & +h_0 v_{10} \\
h_5 v_7 & +h_4 v_8 & +h_3 v_9 & +h_2 v_{10} & +h_1 v_{11} & +h_0 v_{12}
\end{bmatrix}
$$

We can rewrite the right-hand side of (7.89) as

$$
\begin{bmatrix}
v_1 & v_2 & v_3 & v_4 & v_5 & v_6 \\
v_3 & v_4 & v_5 & v_6 & v_7 & v_8 \\
v_5 & v_6 & v_7 & v_8 & v_9 & v_{10} \\
v_7 & v_8 & v_9 & v_{10} & v_{11} & v_{12}
\end{bmatrix}
\cdot
\begin{bmatrix}
h_5 \\ h_4 \\ h_3 \\ h_2 \\ h_1 \\ h_0
\end{bmatrix}
\tag{7.90}
$$

Equation (7.90) gives us a nice way to view computation of the nonwrapping rows. We simply create four partitions of $\mathbf{v}$. Each partition is of length 6, and the first element in each partition (save the first one) is two elements to the right of the first element in the partition above it. That is, each partition starts with the next odd indexed element in the vector. So nonwrapping row k starts with element v_{2k-1}.

We can easily generalize this to arbitrary odd L. We partition $\mathbf{v}$ into lengths of $L+1$ and start each partition with the next odd index. Each partition is then dotted with a reversed version of $\mathbf{h}$.

Let $\mathbf{y}^\ell$ denote the lowpass portion of the wavelet transform. That is, $H\mathbf{v} = \mathbf{y}^\ell$. Recall that there are $\frac{L-1}{2}$ wrapping rows in $\frac{N}{2} \times N$ matrix H, so if k is our counter for the non-wrapping rows, k runs from 1 to $\frac{N}{2} - \frac{L-1}{2} = \frac{N-L+1}{2}$. We have

$$
\boxed{y_k^\ell = \sum_{j=0}^{L} h_{L-j} v_{2k-1+j}}
\tag{7.91}
$$

where $k = 1, \ldots, \frac{N-L+1}{2}$.

Computing Transform Values For Wrapping Rows

For the wrapping rows, the algorithm is not quite as straightforward. This time, let's use the $L = 7$ case (7.88) and a 12-vector.

$$\begin{bmatrix} h_1 & h_0 & 0 & 0 & 0 & 0 & h_7 & h_6 & h_5 & h_4 & h_3 & h_2 \\ h_3 & h_2 & h_1 & h_0 & 0 & 0 & 0 & 0 & h_7 & h_6 & h_5 & h_4 \\ h_5 & h_4 & h_3 & h_2 & h_1 & h_0 & 0 & 0 & 0 & 0 & h_7 & h_6 \end{bmatrix} \cdot \begin{bmatrix} v_1 \\ v_2 \\ v_3 \\ v_4 \\ v_5 \\ v_6 \\ v_7 \\ v_8 \\ v_9 \\ v_{10} \\ v_{11} \\ v_{12} \end{bmatrix} =$$

$$\begin{bmatrix} (h_1 v_1 & +h_0 v_2 \ ) + (h_7 v_7 & +h_6 v_8 & +h_5 v_9 & +h_4 v_{10} + h_3 v_{11} + h_2 v_{12}) \\ (h_3 v_1 & +h_2 v_2 & +h_1 v_3 & +h_0 v_4 \ ) + (h_7 v_9 & +h_6 v_{10} + h_5 v_{11} + h_4 v_{12}) \\ (h_5 v_1 & +h_4 v_2 & +h_3 v_3 & +h_2 v_4 & +h_1 v_5 & +h_0 v_6 \ ) + (h_7 v_{11} + h_6 v_{12}) \end{bmatrix} \qquad (7.92)$$

There are three wrapping rows in H_8, and each row has some elements at both its beginning and end. In particular, the first wrapping row has two elements at the beginning, the second row has four elements, and the third wrapping row has six elements at the beginning. We break this computation into two sums — the first sum will utilize the h_k that appear at the beginning of the wrapped row, and the second sum will involve the h_k that appear at the end of the wrapped row.

Table 7.6 summarizes the computations (in summation notation) from (7.92) for the wrapped rows in H_8 when they are dotted with a 12-vector.

Table 7.6 The computations in summation notation for the three wrapped rows in $H_8 \mathbf{v}$ for H_8 given in (7.92) and $\mathbf{v}$ a 12-vector.[a]

Wrapped Row	Beginning Sum	End Sum
1	$\sum_{j=1}^{2} h_{2-j} v_j$	$\sum_{j=1}^{6} h_{8-j} v_{6+j}$
2	$\sum_{j=1}^{4} h_{4-j} v_j$	$\sum_{j=1}^{4} h_{8-j} v_{8+j}$
3	$\sum_{j=1}^{6} h_{6-j} v_j$	$\sum_{j=1}^{2} h_{8-j} v_{10+j}$

[a]The second column is the sum of the nonzero elements appearing at the beginning of the wrapped row and the third column is the sum of the nonzero elements appearing at the end of the wrapped row.

Note that some of the values in the sums in Table 7.6 are given in boldface. This is done to help us generalize to the arbitrary odd L case. Consider first the "Beginning Sum" column. The number in bold here is simply the number of filter coefficients at the beginning of each wrapped row. The relationship was stated above — for the first wrapped row, there are two filter coefficients, for the second there are four, and in the general case, there are $2k$ filter coefficients at the beginning of wrapped row k. Thus, for arbitrary odd L, we know that there are $\frac{L-1}{2}$ wrapping rows, and the beginning sum is

$$\sum_{j=1}^{2k} h_{2k-j}v_j, \qquad k = 1, \ldots, \frac{L-1}{2} \tag{7.93}$$

Now we consider generalizing the bold parameters in the "End Sum" column. The 8 that appears in the subscript of h_{8-j} is simply the filter length, so it can be replaced in the general case by $L+1$.

The upper limit of each sum is simply the difference between the filter length and the number of filter coefficients that appear in the beginning sum. In the general case, for wrapping row k, we can replace the upper limit by $L+1-2k$.

The final set of bold parameters appear to subscript elements of $\mathbf{v}$. Recall that these are elements from the end of $\mathbf{v}$, so we need to move back from the end of the vector to find the elements for the computation. In general, how far back we go depends on the number of filter coefficients we need. Recall from the preceding discussion, this number is $L+1-2k$ for wrapping row k. The length of $\mathbf{v}$ is N so the bold subscript can be replaced by the general value $N-(L+1-2k) = N-L-1+2k$. Thus, the end sum is

$$\sum_{j=1}^{L+1-2k} h_{L+1-j}v_{N-L-1+2k+j}, \qquad k = 1, \ldots, \frac{L-1}{2}$$

We can clean up the notation if we replace j by $j+1$ in the sum above. We have

$$\sum_{j=0}^{L-2k} h_{L-j}v_{N-L+2k+j}, \qquad k = 1, \ldots, \frac{L-1}{2} \tag{7.94}$$

We can combine (7.93) and (7.94) to produce a formula for generating the remaining elements of $\mathbf{y}^\ell$. Recall that the values y_k for the nonwrapped rows were given in (7.91) for $k = 1, \ldots, \frac{N-L+1}{2}$. The formula for computing $y_{(N-L+1)/2+1}, \ldots, y_{N/2}$ is given by

$$y_{(N-L+1)/2+k} = \sum_{j=1}^{2k} h_{2k-j}v_j + \sum_{j=0}^{L-2k} h_{L-j}v_{N-L+2k+j} \tag{7.95}$$

where $k = 1, \ldots, \frac{L-1}{2}$.

An Algorithm for One Iteration of the Wavelet Transform

We can now use (7.91) and (7.95) and incorporate similar computations for g to obtain an algorithm for performing one iteration of the wavelet transform using orthogonal lowpass filters.

Algorithm 7.1 (Orthogonal Wavelet Transform: 1D). *We are given a vector* $\mathbf{v} \in \mathbb{R}^N$ *(N even) and an orthogonal lowpass filter* $\mathbf{h} = (h_0, \ldots, h_L)$ *of even length* $L + 1$, *and we return one iteration of the wavelet transform. The algorithm begins by creating* $\mathbf{g}$ *from* $\mathbf{h}$. *We next use (7.91) in a loop to compute* $y_1, \ldots, y_{\frac{N}{2}-p}$ *where* $p = \frac{L-1}{2}$ *is the number of wrapping rows. In the same loop, we use (7.91) with* $\mathbf{h}$ *replaced by* $\mathbf{g}$ *to compute* $y_{\frac{N}{2}}, y_{\frac{N}{2}+1}, \ldots, y_{N-p}$. *In this algorithm, values of* $\mathbf{y}$ *are computed in pairs. The first value is an element of the lowpass portion* $\mathbf{y}^\ell$ *from the discussion above. The second value represents the highpass portion and is simply a shift by* $\frac{N}{2}$ *elements. We next loop over the* p *wrapping rows and compute* $y_{\frac{N}{2}-p+k}$ *and* y_{N-p+k}, $k = 1, \ldots$, *using loops for the beginning sum (7.93) and the end sum (7.94). In each of these two loops we also use* $\mathbf{g}$ *and compute* y_{N-p+k}, $k = 1, \ldots, p$.

Algorithm: WT1D1

Input: Vector $\mathbf{v}$ of even length N and an orthogonal filter $\mathbf{h}$
               of even length $L + 1$. It is assumed that $N \geq L + 1$.

Output: Vector $\mathbf{y}$ where $\mathbf{y} = W_N \mathbf{v}$.

// p denotes the number of wrapping rows.

$p = \frac{L-1}{2}$

// Create the highpass filter.

For $[k = 0, k \leq L, k++,$
$\quad g_k = (-1)^k h_{L-k}$
$]$

// Compute lowpass and highpass values for the non-wrapping rows.
// To index highpass values, add $\frac{N}{2}$ to the subscript.

For $[k = 1, k \leq \frac{N}{2} - p, k++,$
$\quad y_k = 0$
$\quad y_{\frac{N}{2}+k} = 0$
$\quad$ **For** $[j = 0, j \leq L, j++,$
$\quad\quad y_k = y_k + h_{L-j} v_{2k-1+j}$
$\quad\quad y_{\frac{N}{2}+k} = y_{\frac{N}{2}+k} + g_{L-j} v_{2k-1+j}$
$\quad]$
$]$

// Compute lowpass and highpass values for the wrapping rows.

For $[k = 1, \ k \leq p, \ k++,$

$\quad y_{\frac{N}{2}-p+k} = 0$

$\quad y_{N-p+k} = 0$

```
// This loop is the end sum (7.94)
```

For $[j = 0, \ j \leq L - 2k, \ j++,$

$\qquad y_{\frac{N}{2}-p+k} = y_{\frac{N}{2}-p+k} + h_{L-j}v_{N-L+2k+j}$

$\qquad y_{N-p+k} = y_{N-p+k} + g_{L-j}v_{N-L+2k+j}$

$\quad]$

```
// This loop adds the beginning sum (7.93).
```

For $[j = 1, \ j \leq 2k, \ j++,$

$\qquad y_{\frac{N}{2}-p+k} = y_{\frac{N}{2}-p+k} + h_{2k-j}v_j$

$\qquad y_{N-p+k} = y_{N-p+k} + g_{2k-j}v_j$

$\quad]$

$]$

Return $[\mathbf{y}]$

$\qquad\qquad\qquad\qquad\qquad\qquad\qquad\qquad\qquad\qquad\qquad\qquad\qquad\quad \square$

Efficient Computation of $\mathbf{W}_N^T\mathbf{y}$

We now construct pseudocode for computing the inverse wavelet transform $\mathbf{v} = W_N^T\mathbf{y} = \begin{bmatrix} H^T | G^T \end{bmatrix} \mathbf{y}$. Now $\mathbf{y}$ has been processed using Algorithm 7.1, so it is convenient to write $\mathbf{y}$ using block form and the lowpass and highpass portions $\mathbf{y}^{\ell}$ and $\mathbf{y}^h$, respectively. Using block arithmetic, we see that

$$\mathbf{v} = W_N^T\mathbf{y} = \begin{bmatrix} H^T | G^T \end{bmatrix} \cdot \begin{bmatrix} \mathbf{y}^{\ell} \\ \hline \mathbf{y}^h \end{bmatrix} = H^T\mathbf{y}^{\ell} + G^T\mathbf{y}^h$$

An Auxiliary Algorithm for Computing $\mathbf{H}^T\mathbf{y}^{\ell}$ and $\mathbf{G}^T\mathbf{y}^h$

We also know that H^T and G^T have the same structure, so if we can develop an efficient algorithm for $H^T\mathbf{y}^{\ell}$, we can easily modify it to obtain $G^T\mathbf{y}^h$. In fact, we will develop a module that takes as input a filter $\mathbf{f}$, a vector $\mathbf{x}$, and performs the computation $M^T\mathbf{x}$, where M^T is a matrix whose structure is identical to H^T and G^T above and is constructed using filter coefficients from $\mathbf{f}$. We can then call this routine with $\mathbf{y}^\ell$ and $\mathbf{h}$ and then again with $\mathbf{y}^h$ and $\mathbf{g}$ and complete the algorithm by summing the results.

Let's write out $H^T \mathbf{y}^\ell$ for Daubechies filters of length 4, 6, and 8. We choose a suitable size $\frac{N}{2}$ so as to illustrate the structure of $H^T \mathbf{y}^\ell$. For $\mathbf{y}^\ell = (y_1, y_2, \ldots, y_{N/2})$, we have

$$
\begin{bmatrix}
h_3 & & & & h_1 \\
h_2 & & & & h_0 \\
\hline
h_1 & h_3 & & & \\
h_0 & h_2 & & & \\
& h_1 & h_3 & & \\
& h_0 & h_2 & & \\
& & h_1 & h_3 & \\
& & h_0 & h_2 & \\
& & & h_1 & h_3 \\
& & & h_0 & h_2 \\
& & & & h_1 & h_3 \\
& & & & h_0 & h_2
\end{bmatrix}
\cdot
\begin{bmatrix}
y_1 \\ y_2 \\ y_3 \\ y_4 \\ y_5 \\ y_6
\end{bmatrix}
=
\begin{bmatrix}
h_1 y_6 + h_3 y_1 \\
h_0 y_6 + h_2 y_1 \\
\hline
h_1 y_1 + h_3 y_2 \\
h_0 y_1 + h_2 y_2 \\
h_1 y_2 + h_3 y_3 \\
h_0 y_2 + h_2 y_3 \\
h_1 y_3 + h_3 y_4 \\
h_0 y_3 + h_2 y_4 \\
h_1 y_4 + h_3 y_5 \\
h_0 y_4 + h_2 y_5 \\
h_1 y_5 + h_3 y_6 \\
h_0 y_5 + h_2 y_6
\end{bmatrix}
\qquad (7.96)
$$

$$
\begin{bmatrix}
h_5 & & & & h_1 & h_3 \\
h_4 & & & & h_0 & h_2 \\
h_3 & h_5 & & & & h_1 \\
h_2 & h_4 & & & & h_0 \\
\hline
h_1 & h_3 & h_5 & & & \\
h_0 & h_2 & h_4 & & & \\
& h_1 & h_3 & h_5 & & \\
& h_0 & h_2 & h_4 & & \\
& & h_1 & h_3 & h_5 & \\
& & h_0 & h_2 & h_4 & \\
& & & h_1 & h_3 & h_5 \\
& & & h_0 & h_2 & h_4 \\
& & & & h_1 & h_3 & h_5 \\
& & & & h_0 & h_2 & h_4 \\
& & & & & h_1 & h_3 & h_5 \\
& & & & & h_0 & h_2 & h_4
\end{bmatrix}
\cdot
\begin{bmatrix}
y_1 \\ y_2 \\ y_3 \\ y_4 \\ y_5 \\ y_6 \\ y_7 \\ y_8
\end{bmatrix}
=
\begin{bmatrix}
h_1 y_7 + h_3 y_8 + h_5 y_1 \\
h_0 y_7 + h_2 y_8 + h_4 y_1 \\
h_1 y_8 + h_3 y_1 + h_5 y_2 \\
h_0 y_8 + h_2 y_1 + h_4 y_2 \\
\hline
h_1 y_1 + h_3 y_2 + h_5 y_3 \\
h_0 y_1 + h_2 y_2 + h_4 y_3 \\
h_1 y_2 + h_3 y_3 + h_5 y_4 \\
h_0 y_2 + h_2 y_3 + h_4 y_4 \\
h_1 y_3 + h_3 y_4 + h_5 y_5 \\
h_0 y_3 + h_2 y_4 + h_4 y_5 \\
h_1 y_4 + h_3 y_5 + h_5 y_6 \\
h_0 y_4 + h_2 y_5 + h_4 y_6 \\
h_1 y_5 + h_3 y_6 + h_5 y_7 \\
h_0 y_5 + h_2 y_6 + h_4 y_7 \\
h_1 y_6 + h_3 y_7 + h_5 y_8 \\
h_0 y_6 + h_2 y_7 + h_4 y_8
\end{bmatrix}
\qquad (7.97)
$$

$$
\left[
\begin{array}{cccccc}
h_7 & & & & h_1 & h_3 \; h_5 \\
h_6 & & & & h_0 & h_2 \; h_4 \\
h_5 & h_7 & & & & h_1 \; h_3 \\
h_4 & h_6 & & & & h_0 \; h_2 \\
h_3 & h_5 & h_7 & & & h_1 \\
h_2 & h_4 & h_6 & & & h_0 \\
\hline
h_1 & h_3 & h_5 & h_7 & & \\
h_0 & h_2 & h_4 & h_6 & & \\
& h_1 & h_3 & h_5 & h_7 & \\
& h_0 & h_2 & h_4 & h_6 & \\
& & h_1 & h_3 & h_5 & h_7 \\
& & h_0 & h_2 & h_4 & h_6 \\
& & & h_1 & h_3 & h_5 \; h_7 \\
& & & h_0 & h_2 & h_4 \; h_6 \\
& & & & h_1 & h_3 \; h_5 \; h_7 \\
& & & & h_0 & h_2 \; h_4 \; h_6 \\
& & & & & h_1 \; h_3 \; h_5 \; h_7 \\
& & & & & h_0 \; h_2 \; h_4 \; h_6
\end{array}
\right]
\left[
\begin{array}{c}
y_1 \\ y_2 \\ y_3 \\ y_4 \\ y_5 \\ y_6 \\ y_7 \\ y_8 \\ y_9 \\ y_{10}
\end{array}
\right]
=
\left[
\begin{array}{l}
h_1 y_8 + h_3 y_9 + h_5 y_{10} + h_7 y_1 \\
h_0 y_8 + h_2 y_9 + h_4 y_{10} + h_6 y_1 \\
h_1 y_9 + h_3 y_{10} + h_5 y_1 + h_7 y_2 \\
h_0 y_9 + h_2 y_{10} + h_4 y_1 + h_6 y_2 \\
h_1 y_{10} + h_3 y_1 + h_5 y_2 + h_7 y_3 \\
h_0 y_{10} + h_2 y_1 + h_4 y_2 + h_6 y_3 \\
h_1 y_1 + h_3 y_2 + h_5 y_3 + h_7 y_4 \\
h_0 y_1 + h_2 y_2 + h_4 y_3 + h_6 y_4 \\
h_1 y_2 + h_3 y_3 + h_5 y_4 + h_7 y_5 \\
h_0 y_2 + h_2 y_3 + h_4 y_4 + h_6 y_5 \\
h_1 y_3 + h_3 y_4 + h_5 y_5 + h_7 y_6 \\
h_0 y_3 + h_2 y_4 + h_4 y_5 + h_6 y_6 \\
h_1 y_4 + h_3 y_5 + h_5 y_6 + h_7 y_7 \\
h_0 y_4 + h_2 y_5 + h_4 y_6 + h_6 y_7 \\
h_1 y_5 + h_3 y_6 + h_5 y_7 + h_7 y_8 \\
h_0 y_5 + h_2 y_6 + h_4 y_7 + h_6 y_8 \\
h_1 y_6 + h_3 y_7 + h_5 y_8 + h_7 y_9 \\
h_0 y_6 + h_2 y_7 + h_4 y_8 + h_6 y_9 \\
h_1 y_7 + h_3 y_8 + h_5 y_9 + h_7 y_{10} \\
h_0 y_7 + h_2 y_8 + h_4 y_9 + h_6 1 y_{10}
\end{array}
\right]
\qquad (7.98)
$$

Let's look at the right sides of $H^T y^\ell$ in (7.96)–(7.98). There are several observations that we can make. The first is to note that each element of $H^T y^\ell$ is the sum of 2, 3, and 4 elements, respectively. In general, for filter $\mathbf{h} = (h_0, \ldots, h_L)$, the right side of $H^T y^\ell$ will consist of the sum of $\frac{L+1}{2}$ elements.

Now let's look at the terms in the sum. The odd elements of $H^T y^\ell$ can be viewed as a linear combination of the odd elements of filter $\mathbf{h}$, and the even elements of $H^T y^\ell$ can be viewed as a linear combination of the even elements of filter $\mathbf{h}$. Moreover, the same values of y^ℓ are used in consecutive pairs of linear combinations.

Finally, note that the wrapping rows are now located at the top of H^T and come in blocks of two. There is one such 2-block in (7.96), two 2-blocks in (7.97), and three 2-blocks in (7.98). In general, for $\mathbf{h} = (h_0, \ldots, h_L)$ we will have $\frac{L-1}{2}$ such 2-blocks. Thus, there are $L - 1$ total wrapping rows. Since these linear combinations start with elements at the end of y^ℓ, we will use a separate loop to compute them.

Computing Values of $H^T y^\ell$ For Nonwrapping Rows

Let's first develop an efficient way to compute the elements of $H^T y^\ell$ that arise from the nonwrapping rows. Since there are $\frac{L-1}{2}$ 2-blocks of wrapping rows, there are

$$
\frac{N - (L - 1)}{2}
$$

2-blocks of nonwrapping rows.

In order to ease notation, it is best to introduce *even* and *odd* vectors that house the even and odd components of $\mathbf{h}$, respectively. Let

$$
\mathbf{e} = [h_0, h_2, \ldots, h_{L-1}]^T \qquad \text{and} \qquad \mathbf{o} = [h_1, h_3, \ldots, h_L]^T
$$

The computation is straightforward. For each nonwrapping row, we select $\frac{L+1}{2}$ elements from $\mathbf{y}^\ell$ and dot this new vector with each of $\mathbf{o}$ and $\mathbf{e}$. For the kth nonwrapping block, the first element in our vector of length $\frac{L+1}{2}$ is y_k.

Let $\mathbf{v} = H^T \mathbf{y}^\ell$. The nonwrapping row results start at v_L since $v_1, \ldots, v_{L-1}$ hold results for the wrapping rows. For $k = 1, \ldots, \frac{N-L+1}{2}$, we have

$$v_{L-1+2k-1} = v_{L+2k-2} = \sum_{j=k}^{\frac{L+1}{2}+k-1} o_j \, y_j \tag{7.99}$$

and

$$v_{L-1+2k} = v_{L+2k-1} = \sum_{j=k}^{\frac{L+1}{2}+k-1} e_j \, y_j \tag{7.100}$$

Computing Values of $H^T \mathbf{y}^\ell$ For Wrapping Rows

Finally, we discuss the computation of $v_1, \ldots, v_{L-1}$ that results from the wrapping rows. Again, we are dotting a length $\frac{L+1}{2}$ portion of $\mathbf{y}^\ell$ with both $\mathbf{o}$ and $\mathbf{e}$, but now we use some of the elements at the beginning of $\mathbf{y}^\ell$ and some of the elements at the end of $\mathbf{y}^\ell$.

As was the case with the wrapping rows in Algorithm 7.1, we split the computation into two sums. It is best to motivate the computation by looking at a specific case. Table 7.7 expresses the computations given in (7.98) using summation notation and replacing the odd elements of $\mathbf{h}$ with elements from $\mathbf{o}$. We list only computations for the odd rows since the structure of the even rows will be identical save for the use of $\mathbf{e}$ instead of $\mathbf{o}$. In this case, $\mathbf{o} = [o_1, o_2, o_3, o_4]^T = [h_1, h_3, h_5, h_7]^T$.

Table 7.7   The computations, in summation notation, for the three wrapped rows in $H^T \mathbf{y}^\ell$ for H^T given in (7.98) and $\mathbf{y}^\ell$ a 10-vector.[a]

Wrapped Row 2-Block	Beginning Sum	End Sum
1	$\displaystyle\sum_{j=1}^{1} o_{3+j}\, y_j$	$\displaystyle\sum_{j=1}^{3} o_j\, y_{7+j}$
2	$\displaystyle\sum_{j=1}^{2} o_{2+j}\, y_j$	$\displaystyle\sum_{j=1}^{2} o_j\, y_{8+j}$
3	$\displaystyle\sum_{j=1}^{3} o_{1+j}\, y_j$	$\displaystyle\sum_{j=1}^{1} o_j\, y_{9+j}$

[a]The second column is the sum of the nonzero elements appearing at the beginning of the wrapped row, and the third column is the sum of the nonzero elements appearing at the end of the wrapped row.

Note that some of the values in the sums in Table 7.7 are given in boldface. This is done to help us generalize to the arbitrary odd L case. Consider first the "Beginning Sum" column. The bold number that represents each summation's upper limit is simply the current 2-block we are computing. We can replace this upper limit by k in the general case, where $k = 1, \ldots, \frac{L-1}{2}$. The bold number that appears in the subscripts of the odd filter elements represents a starting point for the beginning sum. Each value is the length of o minus the current block. The general form here is $\frac{L+1}{2} - k$. Thus, the general form of the beginning sum is

$$\sum_{j=1}^{k} o_{\frac{L+1}{2} - k + j} \, y_j \tag{7.101}$$

The boldface upper summation limit in the "End Sum" column is simply the length of o minus the current block, so in general, this upper limit can be replaced with $\frac{L+1}{2} - k$. There is a final bold number that appears as a subscript in the elements from $\mathbf{y}^\ell$. Recall that these elements come from the end of $\mathbf{y}^\ell$ and the length of $\mathbf{y}^\ell$ is $\frac{N}{2}$. Since the end sum is over $\frac{L+1}{2} - k$ elements in general, we need to move left $\frac{L+1}{2} - k$ from the end of $\mathbf{y}^\ell$. So the bold number can be replaced in general by $\frac{N}{2} - \frac{L+1}{2} + k$. The general form for the end sum is

$$\sum_{j=1}^{\frac{L+1}{2} - k} o_j \, y_{\frac{N}{2} - \frac{L+1}{2} + k + j} \tag{7.102}$$

where $k = 1, \ldots, \frac{L-1}{2}$.

Combining (7.101), (7.102), and adding the computations that involve e, we have the following formulas for the components of $\mathbf{v}$ that result by applying the wrapping rows of H^T to $\mathbf{y}^\ell$:

$$v_{2k-1} = \sum_{j=1}^{k} o_{r-k+j} \, y_j + \sum_{j=1}^{r-k} o_j \, y_{\frac{N}{2} - r + k + j} \tag{7.103}$$

and

$$v_{2k} = \sum_{j=1}^{k} e_{r-k+j} \, y_j + \sum_{j=1}^{r-k} e_j \, y_{\frac{N}{2} - r + k + j} \tag{7.104}$$

where $k = 1, \ldots, \frac{L-1}{2}$ and $r = \frac{L+1}{2}$.

An Auxiliary Algorithm for Computing $\mathbf{H}^T\mathbf{y}^\ell$ and $\mathbf{G}^T\mathbf{y}^h$

We first present Algorithm 7.2 for computing $H^T\mathbf{y}^\ell$ and $G^T\mathbf{y}^h$.

Algorithm 7.2 (Auxiliary Algorithm for Computing $\mathbf{H}^T\mathbf{y}^\ell$ and $\mathbf{G}^T\mathbf{y}^h$). *We are given a vector* $\mathbf{x} \in \mathbb{R}^{\frac{N}{2}}$ *and an orthogonal filter* $\mathbf{f} = (f_0, \ldots, f_L)$ *of even length* $L + 1$. *We create a matrix* M^T *from the filter* $\mathbf{f}$ *and return the product* $M^T\mathbf{x}$ *(see, e. g., (7.98)).*

The algorithm begins by initializing some constants needed in the loops. The number of wrapping row 2-blocks is given by p, *half the filter length is given by* r, *and to avoid fractions in subscripts, we set* $n = \frac{N}{2}$. *Next, we build the two vectors* $\mathbf{e} = (f_0, f_2, \ldots, f_{L-3}, f_{L-1})$ *and* $\mathbf{o} = (f_1, f_3, \ldots, f_{L-2}, f_L)$ *from* $\mathbf{f}$. *Then we use (7.99) and (7.100) to compute the values that result in applying the nonwrapping rows to* $\mathbf{x}$. *The algorithm is completed when (7.101) and (7.102) are utilized to compute the values* $v_1, \ldots, v_{L-1}$ *that result when* $\mathbf{x}$ *is dotted with the wrapping rows.*

Algorithm: IWTht

Input: Vector x of length $\frac{N}{2}$ and an orthogonal filter f.
            of even length $L + 1$.

Output: Vector v where $\mathbf{v} = M^T\mathbf{x}$.

// For ease of notation, introduce some constants.

$p = \frac{L-1}{2}$
$r = p + 1$ // $\left(= \frac{L+1}{2}\right)$
$n = \frac{N}{2}$

// Initialize the vectors e and o.

For$[k = 0, k \le p, k++,$
        $e_{k+1} = f_{2k}$
        $o_{k+1} = f_{2k+1}$
]

// Compute the nonwrapping row values.

For$[k = 1, k \le n - p, k++,$
        $v_{L+2k-2} = 0$
        $v_{L+2k-1} = 0$
        For$[j = k, j \le r + k - 1, j++,$
                $v_{L+2k-2} = v_{L+2k-2} + o_j x_j$
                $v_{L+2k-1} = v_{L+2k-1} + e_j x_j$

  ]
]

// Compute the wrapping row values.

For$[k = 1,\ k \leq p,\ k++,$
  $v_{2k-1} = 0$
  $v_{2k} = 0$

  // This loop is the beginning sum (7.101).

  For$[j = 1,\ j \leq k,\ j++,$
    $v_{2k-1} = v_{2k-1} + o_{r-k+j} x_j$
    $v_{2k} = v_{2k} + e_{r-k+j} x_j$
  ]

  // This loop adds the end sum (7.102).

  For$[j = 1,\ j \leq r - k,\ j++,$
    $v_{2k-1} = v_{2k-1} + o_j x_{n-r+k+j}$
    $v_{2k} = v_{2k} + e_j x_{n-r+k+j}$
  ]
]

Return[v]

$\square$

An Algorithm for One Iteration of the Inverse Wavelet Transform

Now that we have Algorithm 7.2, we can easily write an algorithm for computing one iteration of the inverse wavelet transform.

Algorithm 7.3 (Orthogonal Inverse Wavelet Transform: 1D). *We are given a vector* $\mathbf{y} \in \mathbb{R}^N$ *(N even) and an orthogonal lowpass filter* $\mathbf{h} = (h_0, \ldots, h_L)$ *of even length* $L + 1$ *and we return one iteration of the inverse wavelet transform.*

  The algorithm begins by creating $\mathbf{y}^\ell$ *and* $\mathbf{y}^h$ *from* $\mathbf{y}$. *We then construct* $\mathbf{g}$ *from the input filter* $\mathbf{h}$. *Finally, we call Algorithm 7.2 to compute* $H^T \mathbf{y}^\ell$ *and* $G^T \mathbf{y}^h$ *and return the sum of these two vectors.*

Algorithm: IWT1D1

Input: Vector **y** of even length N and an orthogonal filter **h**
     of even length $L + 1$. It is assumed that $N \geq L + 1$.

Output: Vector $\mathbf{v}$ where $\mathbf{v} = W_N^T \mathbf{y}$.

For $[k = 0,\ k \leq L,\ k{+}{+},$

$$g_k = (-1)^k h_{L-k}$$

$]$

$$\mathbf{y}^\ell = [y_1, \ldots, y_{\frac{N}{2}}]^T$$
$$\mathbf{y}^h = [y_{\frac{N}{2}+1}, \ldots, y_N]^T$$

$\mathbf{u} = \textbf{IWTht}\,[\mathbf{y}^\ell, \mathbf{h}]$
$\mathbf{w} = \textbf{IWTht}\,[\mathbf{y}^h, \mathbf{g}]$

Return $[\mathbf{u}{+}\mathbf{w}]$

$\square$

PROBLEMS

7.29   Let $\mathbf{v} = \sqrt{2}[\,1, 2, 3, 4, 4, 3, 2, 1\,]^T$. Use Algorithm 7.1 with the D4 filter to compute $W_8 \mathbf{v}$. Do this computation by hand.

7.30   Let $\mathbf{y}^\ell = \sqrt{2}[\,1, 2, 2, 1\,]^T$. Use Algorithm 7.2 with the D4 filter to compute $H^T \mathbf{y}^\ell$, where H is an 8×4 matrix.

7.31   In this problem we develop an alternative to Algorithm 7.1. We assume that we have two routines at our disposal. Suppose that the routine **Select** takes as input a vector, a starting index i, and a stopping index j where $j \geq i$, and returns the new vector $\mathbf{u} = [v_i, \ldots, v_j]^T$. Suppose further that we have access to the routine **Append** that takes as input two vectors and returns a vector that is formed by appending the second vector to the first.

(a) Consider the product (7.89) for applying the D6 filter $\mathbf{h}$ to a 12-vector $\mathbf{v}$. Suppose that we call **Append** with the vectors $\mathbf{v}$ and **Select**$[\mathbf{v}, 1, 2] = [v_1, v_2]^T$ to obtain
$$\mathbf{w} = [v_1, v_2, \ldots, v_{11}, v_{12}, v_1, v_2]^T$$
Write pseudocode that utilizes $\mathbf{w}$ to compute $H_6 \mathbf{v}$.

(b) Modify the pseudocode from part (a) so that it additionally computes the high-pass portion of the transform. Finally, use **Append** one more time to produce pseudocode that performs $W_6 \mathbf{v}$.

(c) Generalize parts (a) to (c) to produce an alternative to Algorithm 7.1. Your algorithm should take as input an N-vector $\mathbf{v}$ and a filter $\mathbf{h}$, use **Select** and **Append**, and return $W_N \mathbf{v}$.

7.32   In this problem you will develop an alternative to Algorithm 7.2. You will use the **Select** and **Append** routines from Problem 7.31.

(a) Consider the product (7.98) for applying the D8 filter $\mathbf{h}$ to a 10-vector $\mathbf{y}^\ell$ to produce $\mathbf{v} = H^T\mathbf{y}^\ell$. Suppose that we call **Append** with the vectors **Select**$[\mathbf{y}^\ell, 8, 10] = [y_8, y_9, y_{10}]^T$ and $\mathbf{y}^\ell$ to obtain

$$\mathbf{w} = [y_8, y_9, y_{10}, y_1, y_2, \ldots, y_9, v_{10}]^T$$

Write pseudocode that utilizes $\mathbf{w}$ to compute $H^T\mathbf{y}^\ell$.

(b) Generalize part (a) to produce an alternative to Algorithm 7.2. Your algorithm should take as input an N-vector $\mathbf{x}$ and a filter $\mathbf{f}$, use **Select** and **Append**, and return $M^T\mathbf{x}$. Here, M has the same structure as H and G from (7.62) but uses filter $\mathbf{f}$.

7.33   In this problem you will write a different algorithm for computing the two-dimensional wavelet transform using the D4 filter. The following steps will help you organize your work. A **CAS** will be very helpful for this problem.

(a) To understand how the algorithm works, let's assume that input matrix A is a square matrix of size 8. In your **CAS**, define a generic 8×8 matrix A. Also create W_8 using the D4 filter. Now have your **CAS** compute $C = W_8 A W_8^T$ and simplify the output.

(b) Partition B into four blocks each of dimension 4×4. The upper left block is the blur B and the upper right, lower left, and lower right are the vertical V, horizontal H, and diagonal D details (see the discussion in Section 6.3). Note that corresponding elements in each of these blocks use the same elements. Also note that each element of a particular block has the same structure — the only thing that changes are the values of A that are used in the computation. For block B, write a formula for the elements b_{ij}, where $i, j = 1, 2, 3, 4$.

(c) Repeat part (b) for each of V, H, and D.

(d) Generalize your formulas from parts (b) and (c) to create formulas in the case where A is a $M \times N$ matrix with both M and N even.

(e) Use part (d) to create an algorithm for computing one iteration of the two-dimensional wavelet transform using the D4 filter.

Computer Labs

♦ **7.5**   **Software Development: 1D Daubechies Wavelet Transforms** From the text Web site, access the development package wt1d. In this development lab you will create a function that will compute one iteration of the one-dimensional wavelet transform using Daubechies orthogonal filters. Additionally, you will construct a function that computes one iteration of the inverse wavelet transform. Construction of iterated transforms and their inverses are also included in this lab. Instructions are provided that allow you to add all of these functions to the software package DiscreteWavelets.

♦ **7.6 Software Development: 2D Daubechies Wavelet Transforms.** From the text Web site, access the development package wt2d. In this development lab you will create the 2D analogs of the routines constructed in Computer Lab 7.5. Instructions are provided that allow you to add all of these functions to the software package DiscreteWavelets.

7.7 Matrix Completion for Daubechies Wavelet Transforms. From the text Web site, access the lab matrixcompletion. In Problem 7.19 you learned how to modify the wavelet transform so that the last rows of the lowpass and highpass portions of the transform no longer "wrapped." In this lab you will create modules for applying this transform and its inverse to one- and two-dimensional data.

7.8 Data Compression. From the text Web site, access the lab daubcomp. In this lab you will use Daubechies filters and the modified wavelet transform from Computer Lab 7.7 to perform basic data compression. You will also compare your results with those you obtained using the Haar filter in Computer Lab 6.9 from Section 6.4.

7.9 Progressive Image Reconstruction. From the text Web site, access the lab imagereconstruction. In many applications, only an approximation of the image is needed (imagine, for example, searching a very large image database). In this lab you will create a routine that takes as input i iterations of the wavelet transform of a digital image and generates a sequence $(\mathcal{B}_i, \mathcal{B}_{i-1}, \ldots, \mathcal{B}_1)$, where $\mathcal{B}_k$, $k = 1, \ldots, i$, is the blur (see the discussion preceding (6.24)) from the kth iteration of the wavelet transform.

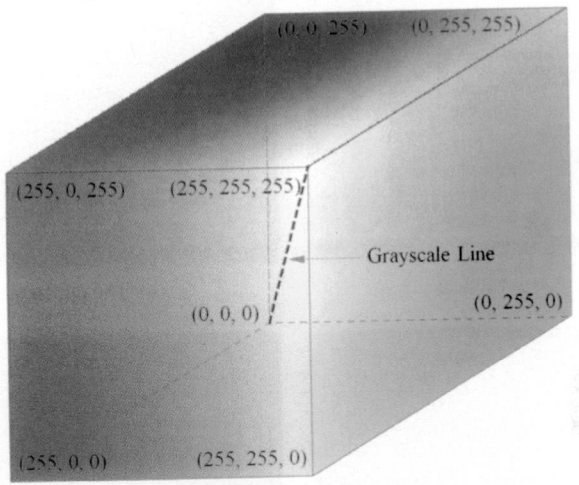

Figure 3.18   The RGB color cube.

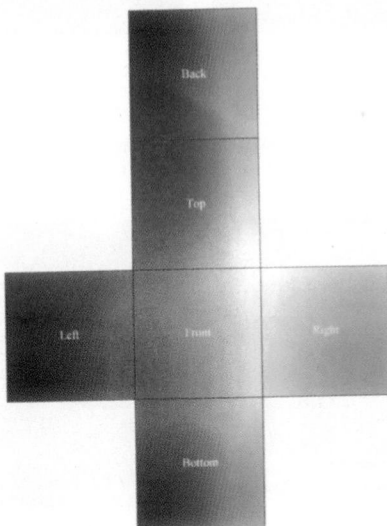

Figure 3.18   The different facets of the RGB color cube.

(a) A color image.

(b) The red channel.

(c) The green channel.

(d) The blue channel.

Figure 3.19 A color image and the three primary color channels.

(a) A digital color image.

(b) The negative image.

Figure 3.20 Color image negation.

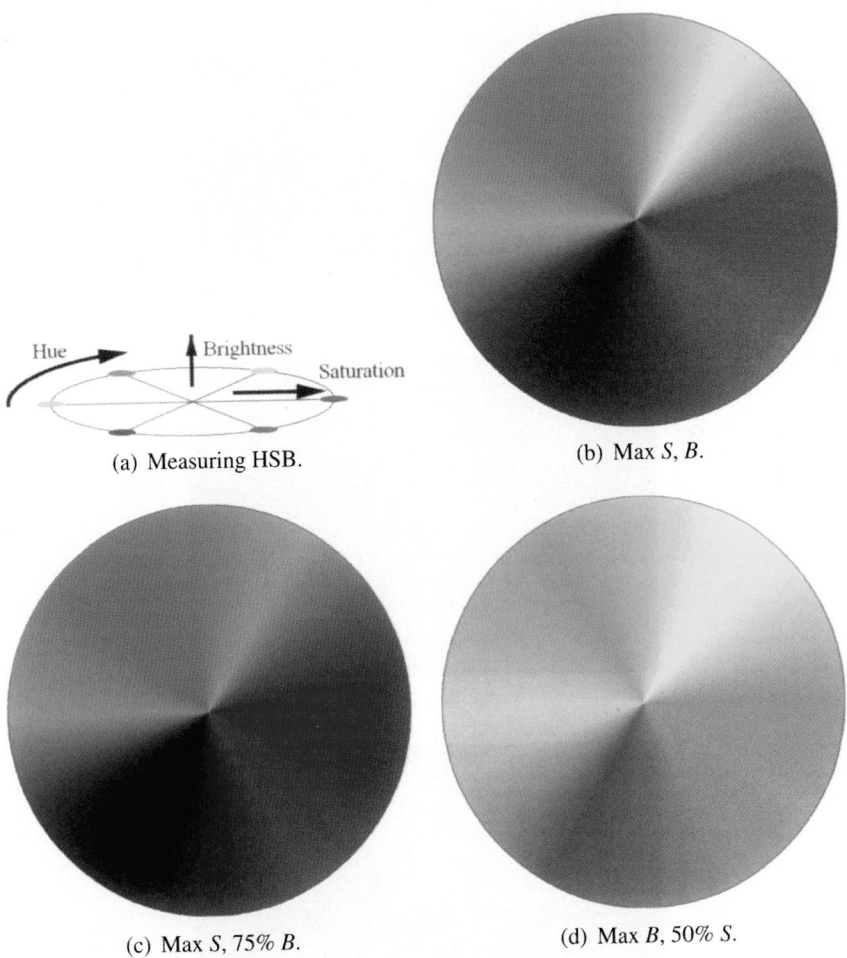

(a) Measuring HSB.

(b) Max S, B.

(c) Max S, 75% B.

(d) Max B, 50% S.

Figure 3.21   Hue, saturation (S), and brightness (B).

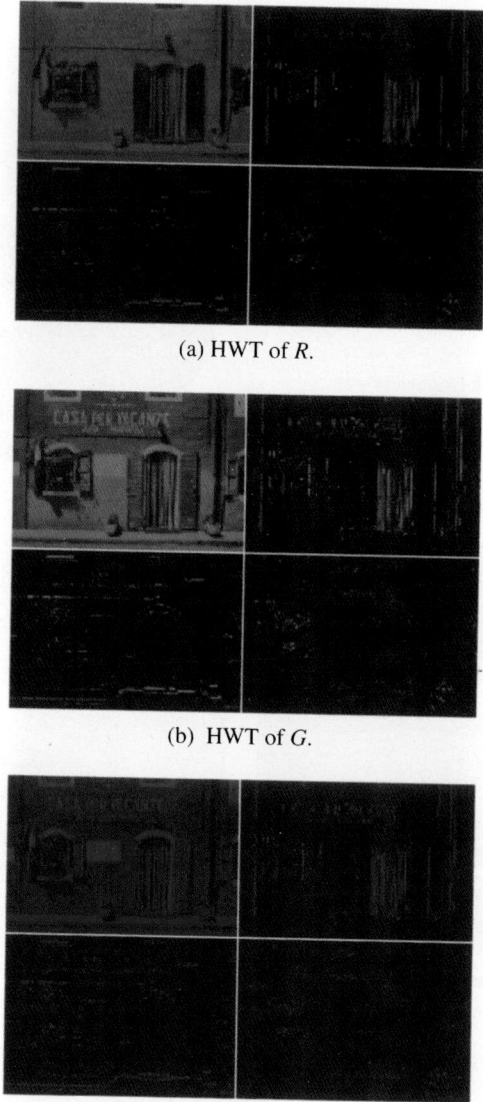

(a) HWT of R.

(b) HWT of G.

(c) HWT of B.

Figure 6.16 The HWT of each of the color channels R, G, and B in Figure 3.19(b)-3.19(d) and the Y, Cb, and Cr channels.

(a) Original Image.

(b) 99.99% Energy.

(c) Variable Energy Rates.

Figure 6.26 The original image and the reconstructed images. The reconstructed image that uses 99.99% energy across each channel has PSNR = 35.235 while the reconstructed image that uses the energy rates given in Table 6.5 has PSNR 23.462.

Figure 12.12   The original color image.

(a) JPEG2000

(b) JPEG

Figure 12.15   The compressed image using JPEG2000 and JPEG.

(a) JPEG2000

(b) JPEG

Figure 12.16   The top left quarters of the images from Figure 12.15.

CHAPTER 8

ORTHOGONALITY AND FOURIER SERIES

In Chapter 7 we constructed wavelet transforms by building orthogonal matrices from a lowpass/highpass filter pair $\mathbf{h}$, $\mathbf{g}$, where the Fourier series $H(\omega)$ satisfied orthogonality conditions and derivative conditions at $\omega = \pi$ (see (7.81)) and the highpass (wavelet) filter $\mathbf{g}$ was constructed from $\mathbf{h}$ using the rule $g_k = (-1)^k h_{L-k}$, $k = 0, \ldots, L$.

While the derivative conditions are in terms of the Fourier series $H(\omega)$, the orthogonality conditions were obtained simply by insisting that the rows of W_N (7.62) are orthonormal. This led to the quadratic part of the system (7.81).

It is possible to develop the quadratic part of system (7.81) using only conditions on the Fourier series $H(\omega)$. In Section 8.1 we learn about these conditions and see how to write the orthogonality conditions entirely in terms of $H(\omega)$. In Chapter 7 we learned how to build the highpass filter $\mathbf{g}$ from the lowpass filter $\mathbf{h}$. As we see in Section 8.2, once we know the Fourier series $H(\omega)$ for the lowpass filter $\mathbf{h}$, we can write down the Fourier series $G(\omega)$. Once we have $G(\omega)$, we can write down the highpass filter $\mathbf{g}$.

Discrete Wavelet Transformations: An Elementary Approach With Applications. By P. J. Van Fleet   **281**
Copyright © 2008 John Wiley & Sons, Inc.

What good is it to write down the orthogonality conditions satisfied by the filters h and g in terms of $H(\omega)$ and $G(\omega)$? One could correctly argue that it offers an alternative construction to that provided in Chapter 7. The strategy is straightforward. The coefficients of a Fourier series are the filter coefficients. So if we wish to constrain our filter coefficients in some way, then we look for the associated constraint on the Fourier series. Once we have this constraint written in terms of Fourier series, we can simply take the series coefficients as our filter coefficients. This is a technique that is used in many areas of applied mathematics.[1] In many cases it is easier to do the work in the Fourier domain.

The real value in describing orthogonality in terms of $H(\omega)$ and $G(\omega)$ is that it allows us to generalize our construction! Once we have the orthogonality in terms of the Fourier series, we can start adding other conditions on $H(\omega)$ (such as the lowpass conditions from (7.81)) to build filters suited for a specific application.

In Section 8.3, we learn how to impose additional conditions on $H(\omega)$ and build orthogonal *Coiflet filters*. These filters, developed by Ingrid Daubechies [24], are well suited for applications such as signal denoising (Chapter 9).

A Look Ahead

Perhaps the biggest reason for connecting the orthogonality of **h** to its Fourier series $H(\omega)$ is that it gives us some insight into how to construct filters that are *not* orthogonal! Thus, the material in this chapter serves as motivation for the ideas presented in Chapter 10 that lead to *biorthogonal filters*. As we learn in Chapter 12, biorthogonal filters are heavily utilized in image compression applications such as the JPEG2000 Image Compression Standard.

8.1 FOURIER SERIES AND LOWPASS FILTERS

Finite-Length Filters

Our development of wavelet transforms in Chapters 6 and 7 utilized finite impulse response (FIR) filters (see Definition 5.3). While the filters we develop in this chapter will be of finite length, it will be convenient to remove the restriction that they also be causal. Toward this end, we have the following definition.

Definition 8.1 (Finite-Length Filters). *Let $\ell, L \in \mathbb{Z}$ with $\ell < L$. We say that $\mathbf{h} = (h_\ell, h_{\ell+1}, \ldots, h_{L-1}, h_L)$ is a finite-length filter provided that $h_\ell \neq 0$, $h_L \neq 0$, and $h_k = 0$ for $k < \ell$ or $k > L$. ℓ is called the starting index and L is called the stopping index.* □

---

[1] We have already employed this technique — we included the derivative conditions $H^{(m)}(\pi) = 0$ in the construction of Daubechies filters in Chapter 7!

Note that Definition 8.1 insists that only $h_\ell, h_L \neq 0$. Elements between h_ℓ and h_L can be zero and the filter is still considered to be a finite-length filter. For example, $\mathbf{h} = (h_{-1}, h_0, h_1, h_2) = (2,0,1,3)$ is a finite-length filter.

The Haar Filter Revisited

Let's begin by recalling the Fourier series $H(\omega)$ for the Haar filter $\mathbf{h} = (h_0, h_1) = (\frac{\sqrt{2}}{2}, \frac{\sqrt{2}}{2})$. From Example 5.5(a) we learned that the modulus of the Fourier series for the filter $(1/2, 1/2)$ is $\cos\left(\frac{\omega}{2}\right)$, so it follows that

$$|H(\omega)| = \sqrt{2}\cos\left(\frac{\omega}{2}\right) \qquad (8.1)$$

on the interval $[-\pi, \pi]$. In the leftmost part of Figure 8.1 we have plotted a few periods of $|H(\omega)|^2$. We translate this graph left by π units to obtain $|H(\omega + \pi)|^2$ and plot several periods of that as well. In the final graph, we plot both functions over one period.

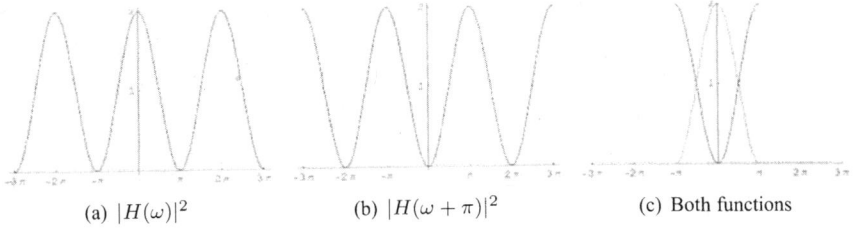

(a) $|H(\omega)|^2$ (b) $|H(\omega + \pi)|^2$ (c) Both functions

Figure 8.1   The functions $|H(\omega)|^2$ and $|H(\omega + \pi)|^2$ over three periods and both functions over the period $[-\pi, \pi]$.

Using Figure 8.1(c), it looks like the sum of the two functions plotted is the constant 2. Indeed, using the trigonometric identity $\cos(\omega + \pi/2) = -\sin(\omega)$ and (8.1), we see that

$$|H(\omega + \pi)|^2 = \left(\sqrt{2}\cos(\frac{\omega + \pi}{2})\right)^2 = 2\cos(\omega/2 + \pi/2)^2 = 2\sin^2\left(\frac{\omega}{2}\right)$$

so that

$$|H(\omega)|^2 + |H(\omega + \pi)|^2 = 2\cos^2\left(\frac{\omega}{2}\right) + 2\sin^2\left(\frac{\omega}{2}\right) = 2$$

In Problem 8.1 you will show that the Fourier series $H(\omega)$ for the D4 filter also satisfies

$$|H(\omega)|^2 + |H(\omega + \pi)|^2 = 2 \qquad (8.2)$$

As you may have guessed, our first result connects the orthogonality conditions of a filter with the general condition (8.2).

Connecting Orthogonal Filters and Their Fourier Series

All the filters we construct in this chapter are finite-length filters. Of course, a finite-length filter can be viewed as a portion of a bi-infinite sequence (see Section 5.1) with finitely many nonzero terms. To give the most general result, we state and prove the main results for bi-infinite sequences and then present the special cases involving finite-length filters as corollaries.

Theorem 8.1 (A Condition That Ensures Orthogonality). *Suppose that* a *is a bi-infinite sequence and denote its Fourier series by*

$$A(\omega) = \sum_{k=-\infty}^{\infty} a_k e^{ik\omega}$$

Then

$$|A(\omega)|^2 + |A(\omega + \pi)|^2 = 2 \qquad (8.3)$$

if and only if

$$\sum_{k=-\infty}^{\infty} a_k^2 = 1 \qquad (8.4)$$

and for $m \in \mathbb{Z}$ *with* $m \neq 0$,

$$\sum_{k=-\infty}^{\infty} a_k a_{k-2m} = 0 \qquad (8.5)$$

$\square$

Proof of Theorem 8.1. We begin by assuming that (8.3) is true and proving (8.4) and (8.5). The proof of the converse of this statement is left as Problem 8.2. We first rewrite (8.3) in terms of conjugates:

$$A(\omega)\overline{A(\omega)} + A(\omega + \pi)\overline{A(\omega + \pi)} = 2$$

Now insert the definition of $A(\omega)$, conjugate, and simplify to obtain

$$2 = A(\omega)\overline{A(\omega)} + A(\omega + \pi)\overline{A(\omega + \pi)}$$
$$= \left(\sum_k a_k e^{ik\omega} \right) \left(\sum_j a_j e^{-ij\omega} \right) + \left(\sum_k a_k e^{ik(\omega+\pi)} \right) \left(\sum_j a_j e^{-ij(\omega+\pi)} \right)$$

Note that we have suppressed notation by using $\sum_k, \sum_j$ for $\sum_{k=-\infty}^{\infty}, \sum_{j=-\infty}^{\infty}$, respectively.

We next use the fact that $e^{i(\omega+\pi)} = (-1)e^{i\omega}$ and simplify the expression above.

$$2 = \sum_k \sum_j a_k a_j \, e^{i\omega(k-j)} + \sum_k \sum_j a_k a_j \, (-1)^{j+k} e^{i\omega(k-j)}$$

$$= \sum_k \sum_j a_k a_j \, e^{i\omega(k-j)} (1 + (-1)^{j+k})$$

We introduce the variable $m = k - j$ and note that $j = k - m$, and $j + k = 2k - m$. For each k, since j ranges from $-\infty$ to ∞, m will range from $-\infty$ to ∞ as well, so we can replace j with m in the inner sum:

$$2 = \sum_k \sum_m a_k a_{k-m} e^{im\omega} (1 + (-1)^{2k-m})$$

$$= \sum_k \sum_m a_k a_{k-m} e^{im\omega} (1 + (-1)^m)$$

Next, we interchange the infinite summations[2] to obtain

$$2 = \sum_m \left(\sum_k a_k a_{k-m} \right) e^{im\omega} (1 + (-1)^m)$$

Now the factor $1 + (-1)^m$ is 2 when m is even and 0 when m is odd so we can replace m by $2m$ in the summand above to obtain

$$2 = 2 \sum_m \left(\sum_k a_k a_{k-2m} \right) e^{2im\omega}$$

$$1 = \sum_m \left(\sum_k a_k a_{k-2m} \right) e^{2im\omega}$$

$$0 = \left(\sum_k a_k^2 - 1 \right) \cdot 1 + \sum_{m \neq 0} \left(\sum_k a_k a_{k-2m} \right) e^{2im\omega}$$

Now the right side of the last equation is a linear combination of the functions $1, e^{\pm 2i\omega}, e^{\pm 4i\omega}, \ldots$, and in Section 4.3 we learned that these functions are linearly independent. So the only way the linear combination can be zero is if each coefficient is zero. We have

$$0 = \sum_{k \in \mathbb{Z}} a_k^2 - 1 \qquad \text{and} \qquad 0 = \sum_{k \in \mathbb{Z}} a_k a_{k-2m}$$

---

[2] Although this step is technically correct, the rigor needed to justify the step is beyond the scope of this course. To learn more about justifying the interchange of such infinite sums, the reader is referred to Rudin [63].

which completes this portion of the proof. □

The following corollary states the results of Theorem 8.1 in the case where a has finitely many nonzero elements.

Corollary 8.1 (Orthogonality for a Finite-Length Fourier Series). *Suppose that $A(\omega)$ is the Fourier series constructed from the finite length filter $\mathbf{a} = (a_\ell, \ldots, a_L)$. Then*

$$|A(\omega)|^2 + |A(\omega + \pi)|^2 = 2$$

if and only if

$$\sum_{k=\ell}^{L} a_k^2 = 1 \tag{8.6}$$

and for $m \in \mathbb{Z}$ with $m \neq 0$,

$$\sum_{k=\ell+2m}^{L} a_k a_{k-2m} = 0 \qquad m = 1, 2, \ldots, \frac{L - \ell - 1}{2} \tag{8.7}$$

□

We leave the proof of Corollary 8.1 as a problem. Note that when $\ell = 0$ and L is an odd integer, the conditions (8.6) and (8.7) reduce to the orthogonality conditions for the Daubechies filters given in (7.81).

Before we proceed with examples, we state a definition that will be useful in the remainder of this chapter and in Chapter 10.

Definition 8.2 (Parity of Two Integers). *Let m and n be two integers. We say that m and n have the same parity if both m and n are odd or both m and n are even. Otherwise, we say that m and n have different parity.* □

Note that (8.7) makes sense only when $L - \ell - 1$ is divisible by 2. Thus, Corollary 8.1 holds some additional information about the first and last nonzero values h_ℓ and h_L of a finite-length filter $\mathbf{h}$. The only way that $L - \ell - 1$ is even is if ℓ and L have different parity. If ℓ and L have the same parity, then their difference less 1 is odd. Let's look at two examples and see why $L - \ell$ must always be odd.

Example 8.1 (Finite-Length Filters That Satisfy (8.3)). *We consider two finite-length filters $\mathbf{h}$ that satisfy (8.3) and the constraints placed on the indices ℓ and L of the first and last nonzero values a_ℓ and a_L, respectively.*

(a) Let's first consider the finite-length filter $\mathbf{h} = (h_{-2}, h_{-1}, h_0, h_1, h_2, h_3)$ so that $\ell = -2$ is even and $L = 3$ is odd. We assume that the Fourier series

$$H(\omega) = h_{-2}e^{-2i\omega} + h_{-1}e^{-i\omega} + h_0 + h_1 e^{i\omega} + h_2 e^{2i\omega} + h_3 e^{3i\omega}$$

satisfies (8.3). We can then use Corollary 8.1 to infer that

$$h_{-2}^2 + h_{-1}^2 + h_0^2 + h_1^2 + h_2^2 + h_3^2 = 1 \qquad (8.8)$$

and since $\frac{L-\ell-1}{2} = 2$,

$$\sum_{k=-2+2m}^{3} h_k h_{k-2m} = 0, \qquad m = 1, 2$$

We can write out both of these equations. When $m = 1$, *we have*

$$\sum_{k=0}^{3} h_k h_{k-2} = 0$$

or

$$h_0 h_{-2} + h_1 h_{-1} + h_2 h_0 + h_3 h_1 = 0 \qquad (8.9)$$

When $m = 2$ *we have*

$$\sum_{k=2}^{3} h_k h_{k-4} = 0 \qquad or \qquad h_2 h_{-2} + h_3 h_{-1} = 0 \qquad (8.10)$$

An alternative way to write all the orthogonality conditions is simply to write down the nonzero elements of the filter as row 1 of a table and then write the shifted rows that overlap with the first row below it.

$\mathbf{h_{-2}}$	$\mathbf{h_{-1}}$	$\mathbf{h_0}$	$\mathbf{h_1}$	$\mathbf{h_2}$	$\mathbf{h_3}$		
		h_{-2}	h_{-1}	h_0	h_1	h_2	h_3
				h_{-2}	h_{-1}	h_0	h_1	h_2	h_3

We can obtain the left side of (8.8) by dotting row 1 with itself, and the left sides of (8.9) and (8.10) are obtained by dotting row 1 with row 2 and row 3, respectively.

(b) *Now let's consider the filter* $\mathbf{h} = (h_{-2}, h_{-1}, h_0, h_1, h_2, h_3, h_4)$ *with* $h_{-2} \neq 0$ *and* $h_4 \neq 0$. *Assume that* $H(\omega)$ *satisfies (8.3). Here* $\ell = -2$ *and* $L = 4$ *are both even so that* $\frac{L-\ell-1}{2} = \frac{5}{2}$. *To see what's going on with the orthogonality conditions, let's write down a table as we did in part (a). We have*

$\mathbf{h_{-2}}$	$\mathbf{h_{-1}}$	$\mathbf{h_0}$	$\mathbf{h_1}$	$\mathbf{h_2}$	$\mathbf{h_3}$	$\mathbf{h_4}$			
		h_{-2}	h_{-1}	h_0	h_1	h_2	h_3	h_4	
				h_{-2}	h_{-1}	h_0	h_1	h_2	h_3	h_4
						$\mathbf{h_{-2}}$	h_{-1}	h_0	h_1	h_2	h_3	h_4

The key observation is what happens when we dot the top row with the last row. We obtain $h_4 h_{-2} = 0$, and this implies that either $h_4 = 0$ or $h_{-2} = 0$. We have contradicted the assumption that both h_{-2} and $h_4 \neq 0$ so either we start with $\ell = -1$ or end with $L = 3$. It is easy to check that when both ℓ and L are odd, we have the same problem.

$\square$

Proposition 8.1 (Constraints for Finite Length Filters That Satisfy (8.3)). *Suppose that $\mathbf{h} = (h_\ell, \ldots, h_L)$ is a finite-length filter. Further assume that the Fourier series*

$$H(\omega) = \sum_{k=\ell}^{L} h_k e^{ik\omega} \text{ satisfies (8.3). Then } \ell \text{ and } L \text{ have different parity.}$$ $\square$

Proof of Proposition 8.1. The proof is left as Problem 8.5. $\square$

PROBLEMS

8.1   Consider the Fourier series for the D4 filter $H(\omega) = h_0 + h_1 e^{i\omega} + h_2 e^{2i\omega} + h_3 e^{3i\omega}$, where the D4 filter is given by Definition 7.1. Show by direct computation that $|H(\omega)|^2 + |H(\omega + \pi)|^2 = 2$.

8.2   Prove the converse of Theorem 8.1 and thereby show that the condition (8.3) is necessary and sufficient for orthogonality conditions (8.4) and (8.5) to hold for filters. To complete the proof, simply work "backward" through the proof of Theorem 8.1.

8.3   Prove Corollary 8.1. (*Hint:* All you have to do for this proof is verify the cutoffs on the limits of summation in (8.6) and (8.7) and the limits on m.)

8.4   Suppose that the Fourier series of each of the following filters satisfy (8.3). Write down the orthogonality conditions (8.6) and (8.7) satisfied by each filter.

(a) $\mathbf{h} = (h_{-3}, h_{-2}, h_{-1}, h_0, h_1, h_2, h_3, h_4)$.

(b) $\mathbf{h} = (h_{-2}, h_{-1}, h_0, h_1)$.

(c) $\mathbf{h} = (h_2, h_3, h_4, h_5, h_6, h_7)$.

8.5   Prove Proposition 8.1.

8.2 BUILDING $G(\omega)$ FROM $H(\omega)$

In Chapters 6 and 7, we learned how to build the highpass filter $\mathbf{g}$ from the lowpass filter $\mathbf{h}$. In this section, we will learn to write the Fourier series $G(\omega)$ associated with highpass filter $\mathbf{g}$ in terms of the Fourier series $H(\omega)$ associated with lowpass filter $\mathbf{h}$.

Orthogonality Conditions and the Wavelet Transform Matrix

Recall the finite-length filter $\mathbf{h} = (h_{-2}, h_{-1}, h_0, h_1, h_2, h_3)$ from Example 8.1. If we assume that the Fourier series $H(\omega)$ satisfies (8.3) and that $\mathbf{h}$ is a lowpass filter (i.e., $H(0) = \sqrt{2}$ and $H(\pi) = 0$), then we might consider using $\mathbf{h}$ in the top half of a wavelet transform matrix. For example, if the vector we wished to process had length 10, the top half of our wavelet transform matrix might be[3]

$$
H = \begin{bmatrix}
h_{-2} & h_{-1} & h_0 & h_1 & h_2 & h_3 & 0 & 0 & 0 & 0 \\
0 & 0 & h_{-2} & h_{-1} & h_0 & h_1 & h_2 & h_3 & 0 & 0 \\
0 & 0 & 0 & 0 & h_{-2} & h_{-1} & h_0 & h_1 & h_2 & h_3 \\
h_2 & h_3 & 0 & 0 & 0 & 0 & h_{-2} & h_{-1} & h_0 & h_1 \\
h_0 & h_1 & h_2 & h_3 & 0 & 0 & 0 & 0 & h_{-2} & h_{-1}
\end{bmatrix}
$$

The fact that $\mathbf{h}$ satisfies the orthogonality conditions (8.8) – (8.10) gives $HH^T = I_6$ where I_6 is the 6×6 identity matrix (you should verify these computations).

Suppose also that the Fourier transform $G(\omega)$ of $\mathbf{g} = (g_{-2}, g_{-1}, g_0, g_1, g_2, g_3)$ satisfies (8.3) and that $\mathbf{g}$ is a highpass filter (i.e., $G(0) = 0$ and $G(\pi) = \sqrt{2}$). Then we could use this filter to create the bottom half of a wavelet transform matrix:

$$
G = \begin{bmatrix}
g_{-2} & g_{-1} & g_0 & g_1 & g_2 & g_3 & 0 & 0 & 0 & 0 \\
0 & 0 & g_{-2} & g_{-1} & g_0 & g_1 & g_2 & g_3 & 0 & 0 \\
0 & 0 & 0 & 0 & g_{-2} & g_{-1} & g_0 & g_1 & g_2 & g_3 \\
g_2 & g_3 & 0 & 0 & 0 & 0 & g_{-2} & g_{-1} & g_0 & g_1 \\
g_0 & g_1 & g_2 & g_3 & 0 & 0 & 0 & 0 & g_{-2} & g_{-1}
\end{bmatrix}
$$

Since $G(\omega)$ satisfies (8.3), then $\mathbf{g}$ satisfies orthogonality conditions (8.8) – (8.10), and we use this fact to conclude that $GG^T = I_6$.

Now if we form the orthogonal wavelet transform

$$
W_{12} = \begin{bmatrix} H \\ G \end{bmatrix}
$$

then $W_{12}W_{12}^T = I_{12}$. If we write out this multiplication in block form, we have

$$
I_{12} = W_{12}W_{12}^T = \begin{bmatrix} H \\ G \end{bmatrix} \cdot \begin{bmatrix} H^T & G^T \end{bmatrix} = \begin{bmatrix} HH^T & HG^T \\ GH^T & GG^T \end{bmatrix} = \begin{bmatrix} I_6 & HG^T \\ GH^T & I_6 \end{bmatrix}
$$

For W_{12} to be orthogonal, we need

$$
HG^T = GH^T = 0_6 \tag{8.11}
$$

---

[3]In Section 8.3 we talk about different ways we might want to write this matrix.

Orthogonality Conditions for $H(\omega)$ and $G(\omega)$

You may recall in Chapter 7 that we constructed **g** from **h** so that (8.11) was satisfied. For our running example, you should verify that $HG^T = 0_6$ is equivalent to **h** and **g** satisfying the system

$$h_{-2}g_{-2} + h_{-1}g_{-1} + h_0 g_0 + h_1 g_1 + h_2 g_2 + h_3 g_3 = 0$$
$$h_0 g_{-2} + h_1 g_{-1} + h_2 g_0 + h_3 g_1 = 0$$
$$h_2 g_{-2} + h_3 g_{-1} = 0$$

Another way to look at this system is to create a table whose first row is **h**, second row is **g**, and remaining rows are 2-translates of **g** and insist that the first row dotted with every other row is zero.

h_{-2}	h_{-1}	h_0	h_1	h_2	h_3		
g_{-2}	g_{-1}	g_0	g_1	g_2	g_3		
		g_{-2}	g_{-1}	g_0	g_1	g_2	g_3
				g_{-2}	g_{-1}	g_0	g_1	g_2	g_3

Mathematically, this system can be described as

$$\sum_{k=-2+2m}^{3} h_k g_{k-2m} = 0, \qquad m = 0, 1, 2 \tag{8.12}$$

Our next result generalizes this running example and gives necessary and sufficient conditions for constructing filters **h** and **g** that can be used in orthogonal wavelet transforms.

Theorem 8.2 (Orthogonality and $\mathbf{H}(\omega)$ and $\mathbf{G}(\omega)$). *Suppose that the bi-infinite sequences* $\mathbf{a} = (\ldots, a_{-2}, a_{-1}, a_0, a_1, a_2, \ldots)$ *and* $\mathbf{b} = (\ldots, b_{-2}, b_{-1}, b_0, b_1, b_2, \ldots)$ *have Fourier series*

$$A(\omega) = \sum_{k=-\infty}^{\infty} a_k e^{ik\omega} \qquad and \qquad B(\omega) = \sum_{k=-\infty}^{\infty} b_k e^{ik\omega}$$

Then

$$A(\omega)\overline{B(\omega)} + A(\omega + \pi)\overline{B(\omega + \pi)} = 0 \tag{8.13}$$

if and only if for all $m \in \mathbb{Z}$

$$\sum_{k=-\infty}^{\infty} a_k b_{k-2m} = 0 \tag{8.14}$$

$\square$

Proof of Theorem 8.2. The proof of this result is quite similar to the proof of Theorem 8.1. We assume that (8.13) and prove (8.14) holds for all $m \in \mathbb{Z}$. The proof of the converse is left as Problem 8.7. We begin by writing

$$0 = A(\omega)\overline{B(\omega)} + A(\omega + \pi)\overline{B(\omega + \pi)}$$
$$= \sum_k a_k\, e^{ik\omega} \cdot \sum_j b_j\, e^{-ij\omega} + \sum_k a_k\, e^{ik(\omega+\pi)} \cdot \sum_j b_j\, e^{-ij(\omega+\pi)}$$
$$= \sum_k a_k\, e^{ik\omega} \cdot \sum_j b_j\, e^{-ij\omega} + \sum_k a_k\,(-1)^k e^{ik\omega} \cdot \sum_j b_j\,(-1)^j e^{-ij\omega}$$

In the last line, we have used the fact that $e^{ik(\omega+\pi)} = e^{ik\omega}e^{ik\pi} = e^{ik\omega}(-1)^k$. In a similar manner, $e^{-ij(\omega+\pi)} = (-1)^j e^{-ij\omega}$. We next multiply the series and simplify:

$$0 = A(\omega)\overline{B(\omega)} + A(\omega + \pi)\overline{B(\omega + \pi)}$$
$$= \sum_k \sum_j a_k\, b_j\, e^{i(k-j)\omega} + \sum_k \sum_j a_k\, b_j\,(-1)^{j+k} e^{i(k-j)\omega}$$
$$= \sum_k \sum_j a_k\, b_j\,(1 + (-1)^{j+k})\, e^{i(k-j)\omega}$$

Now we make the change of variable $m = k - j$. Note that $j = k - m$, so that $j + k = k - m + k = 2k - m$. If we fix the outer sum variable k and let j run over all integers, we see that m runs over all integers as well. Thus, we replace j with m on the inner sum to write

$$0 = A(\omega)\overline{B(\omega)} + A(\omega + \pi)\overline{B(\omega + \pi)}$$
$$= \sum_k \sum_m a_k b_{k-m}(1 + (-1)^{2k-m})e^{im\omega}$$
$$= \sum_k \sum_m a_k b_{k-m}(1 + (-1)^m)e^{im\omega}$$

Now when m is even, $(1 + (-1)^m) = 2$ and when m is odd, $(1 + (-1)^m) = 0$. So we can replace m by $2m$ in the above identity to obtain

$$0 = 2\sum_k \sum_m a_k b_{k-2m} e^{2im\omega}$$

Dividing both sides by 2 and interchanging sums gives

$$0 = \sum_m \left(\sum_k a_k b_{k-2m} \right) e^{2im\omega}$$

Now the functions $e^{2im\omega}$ are linearly independent, so the only way the infinite sum over m could be zero is if the coefficients

$$\sum_k a_k b_{k-2m} = 0$$

for each $m \in \mathbb{Z}$. This completes the proof. $\qquad\square$

We have the following corollary for the case where $\mathbf{a} = (a_\ell, \ldots, a_L)$ and $\mathbf{b} = (b_\ell, \ldots, b_L)$ are finite-length filters. The proof of this corollary is left as Problem 8.8.

Corollary 8.2 (Orthogonality and Finite-Length Fourier Series). *Suppose that* $\mathbf{a} = (a_\ell, \ldots, a_L)$ *and* $\mathbf{b} = (b_\ell, \ldots, b_L)$ *are finite length filters whose Fourier series are* $A(\omega)$ *and* $B(\omega)$, *respectively. Then*

$$A(\omega)\overline{B(\omega)} + A(\omega + \pi)\overline{B(\omega + \pi)} = 0$$

if and only if

$$\sum_{k=\ell+2m}^{L} a_k \, b_{k-2m} = 0, \qquad m = 1, 2, \ldots, \frac{L - \ell - 1}{2} \qquad (8.15)$$

$\qquad\square$

Proof of Corollary 8.2. See Problem 8.8. $\qquad\square$

The result in Corollary 8.2 requires that the finite-length filters $\mathbf{a}$ and $\mathbf{b}$ have the same starting and stopping indices. In practice, we think of $\mathbf{a}$ as our lowpass filter and $\mathbf{b}$ as our highpass filter. All the filters developed in Chapters 6 and 7 satisfied this requirement and recall that the highpass filters were derived in a very specific way from the lowpass filters (i. e. $g_k = (-1)^k h_{L-k}$, $k = 0, \ldots, L$).

As you will see in Problem 8.9, this requirement can be weakened by proving a result similar to Corollary 8.2 for finite-length filters $\mathbf{a} = (a_\ell, \ldots, a_L)$ and $\mathbf{b} = (b_m, \ldots, b_M)$ with $\ell < m$ and $M < L$. The ability to use different-length filters allows a more general construction of the highpass filter and will also be an important idea in our development of biorthogonal filters in Section 10.1.

Constructing $G(\omega)$ from $H(\omega)$

In Chapters 6 and 7 we constructed wavelet transform matrices W_N (7.62) that were built from a lowpass filter $\mathbf{h} = (h_0, \ldots, h_L)$ and a highpass filter $\mathbf{g} = (g_0, \ldots, g_L)$. If

$$H(\omega) = \sum_{k=0}^{L} h_k e^{ik\omega} \qquad G(\omega) = \sum_{k=0}^{L} g_k e^{ik\omega}$$

are the Fourier series for **h** and **g**, respectively, then the lowpass and highpass conditions are

$$H(0) = \sqrt{2} \qquad H(\pi) = 0 \tag{8.16}$$

and

$$G(0) = 0 \qquad G(\pi) = \sqrt{2} \tag{8.17}$$

We also insisted on choosing **h** and **g** in such a way that W_N was an orthogonal matrix.

We ultimately learned that the construction depended entirely on the choice of **h**. That is, if we constructed **h** so that the top half of W_N consisted of orthonormal rows and the lowpass conditions (8.16) were satisfied, then the rule $g_k = (-1)^L h_{L-k}$, $k = 0, \ldots, L$ guaranteed that the bottom half of W_N consisted of orthonormal rows and the highpass conditions (8.17) were satisfied.

Our goal in this subsection is to mimic these steps in the Fourier domain. That is, if we have a lowpass filter **h** (that satisfies $H(0) = \sqrt{2}$, $H(\pi) = 0$) as well as the orthogonality condition

$$|H(\omega)|^2 + |H(\omega + \pi)|^2 = 2 \tag{8.18}$$

can we build a Fourier series $G(\omega)$ from $H(\omega)$ so that $G(0) = 0$, $G(\pi) = \sqrt{2}$,

$$|G(\omega)|^2 + |G(\omega + \pi)|^2 = 2 \tag{8.19}$$

and

$$H(\omega)\overline{G(\omega)} + H(\omega + \pi)\overline{G(\omega + \pi)} = 0 ? \tag{8.20}$$

To answer this question, let's look at what worked for the Daubechies filters from Chapter 7. Once we formed the Daubechies filter $\mathbf{h} = (h_0, \ldots, h_L)$, L odd, we constructed $\mathbf{g} = (g_0, \ldots, g_L)$ by the rule

$$g_k = (-1)^k h_{L-k} \tag{8.21}$$

In Problem 7.24 from Section 7.2, you showed that the Fourier series for **g** defined by (8.21) is

$$G(\omega) = -e^{iL\omega}\overline{H(\omega + \pi)} \tag{8.22}$$

and verified that $G(0) = 0$ and $G(\pi) = \sqrt{2}$. Although we know from our work in Chapter 7 that using **g** to complete the bottom half of W_N (7.62) makes W_N an orthogonal matrix, it is a good exercise to check that $G(\omega)$ given by (8.22) satisfies (8.20) and (8.19). You are asked to do these computations in Problem 8.10.

The formula (8.22) for $G(\omega)$ works well for Daubechies filters, but what about general finite-length filters or even filters of infinite length?

The only dependency on L in (8.22) is the exponential term $e^{iL\omega}$. Let's try something a little more general for $G(\omega)$. Assuming that $H(\omega)$ satisfies the orthogonality condition (8.18), let's take

$$G(\omega) = f(\omega)\overline{H(\omega + \pi)} \tag{8.23}$$

where $f(\omega)$ is some 2π-periodic function. If we insert (8.23) into (8.19), we have

$$
\begin{aligned}
2 &= |G(\omega)|^2 + |G(\omega + \pi)|^2 \\
  &= |f(\omega)\,\overline{H(\omega + \pi)}|^2 + |f(\omega + \pi)\,\overline{H(\omega + 2\pi)}|^2 \\
  &= |f(\omega)|^2\,|H(\omega + \pi)|^2 + |f(\omega + \pi)|^2\,|H(\omega)|^2
\end{aligned} \tag{8.24}
$$

where in the last identity we recall that $H(\omega)$ is a 2π-periodic function and we have used Problem 4.8 from Section 4.1 to remove the conjugation operators.

For (8.24) to hold, we must have $|f(\omega)|^2 = 1$ or $|f(\omega)| = 1$ for all $\omega \in \mathbb{R}$. Thus, it must be that $f(\omega) = Ae^{ip(\omega)}$ for some real-valued function $p(\omega)$ and complex number A with $|A| = 1$.

Since $G(\omega)$ must be a Fourier series, we are limited in our choices for $p(\omega)$. We know that Fourier series are built from complex exponentials $e^{ik\omega}$, where k is an integer, so $e^{ip(\omega)}$ must be of the form $Ae^{in\omega}$. Since $|A| = 1$, it must be of the form $A = e^{ib}$ for some fixed $b \in \mathbb{R}$. Imposing these conditions on $p(\omega)$ and inserting into (8.23) gives

$$G(\omega) = e^{ib}e^{in\omega}\overline{H(\omega + \pi)} = e^{i(n\omega + b)}\overline{H(\omega + \pi)} \tag{8.25}$$

Now we insert (8.25) into (8.20) and use the fact that $e^{-in\pi} = (-1)^n$ to write

$$
\begin{aligned}
0 &= H(\omega)\overline{G(\omega)} + H(\omega + \pi)\overline{G(\omega + \pi)} \\
  &= H(\omega)\overline{e^{i(n\omega + b)}\,\overline{H(\omega + \pi)}} + H(\omega + \pi)\overline{e^{i(n(\omega + \pi) + b)}\,\overline{H(\omega + 2\pi)}} \\
  &= H(\omega)e^{-i(n\omega + b)}\,H(\omega + \pi) + H(\omega + \pi)e^{-i(n(\omega + \pi) + b)}\,H(\omega) \\
  &= H(\omega)H(\omega + \pi)e^{-ib}\left(e^{-in\omega} + e^{-in(\omega + \pi)}\right) \\
  &= H(\omega)H(\omega + \pi)e^{-ib}e^{-in\omega}\left(1 + (-1)^n\right)
\end{aligned} \tag{8.26}
$$

For (8.26) to hold for all $\omega \in \mathbb{R}$, n must be an odd integer. Thus, our choice for $G(\omega)$ is

$$G(\omega) = e^{i(n\omega + b)}\overline{H(\omega + \pi)}$$

where n is an odd integer and b is any real number. This discussion serves as a proof for the following proposition.

Proposition 8.2 (Building G(ω) from H(ω)). *Suppose that* $H(\omega) = \sum\limits_{k \in \mathbb{Z}} h_k e^{ik\omega}$

satisfies

$$|H(\omega)|^2 + |H(\omega + \pi)|^2 = 2$$

Let

$$G(\omega) = e^{i(n\omega + b)}\overline{H(\omega + \pi)} \tag{8.27}$$

where n is any odd integer and b is a real number. Then

$$|G(\omega)|^2 + |G(\omega + \pi)|^2 = 2$$

and

$$H(\omega)\overline{G(\omega)} + H(\omega + \pi)\overline{G(\omega + \pi)} = 0$$

$\square$

The nice feature of Proposition 8.2 is that there are no restrictions on the filter **h** — it can be FIR, finite length, or even a bi-infinite sequence! The next proposition shows us how to write down the elements of **g** when $G(\omega)$ is given by (8.27).

Proposition 8.3 (The Filter Elements for G(ω)). *Suppose that $G(\omega)$ is given by (8.27). Then for all $k \in \mathbb{Z}$,*

$$g_k = -e^{ib}(-1)^k h_{n-k} \tag{8.28}$$

$\square$

Proof of Proposition 8.3. We start by writing down $\overline{H(\omega + \pi)}$ and simplifying

$$\overline{H(\omega + \pi)} = \overline{\sum_{k \in \mathbb{Z}} h_k e^{ik(\omega + \pi)}} = \sum_{k \in \mathbb{Z}} h_k e^{-ik(\omega + \pi)} = \sum_{k \in \mathbb{Z}} (-1)^k h_k e^{-ik\omega}$$

Multiplying by $e^{i(n\omega + b)}$ and simplifying gives

$$G(\omega) = e^{i(n\omega + b)}\overline{H(\omega + \pi)} = e^{ib}e^{in\omega}\sum_{k \in \mathbb{Z}}(-1)^k h_k e^{-ik\omega}$$

$$= e^{ib}\sum_{k \in \mathbb{Z}}(-1)^k h_k e^{i(n-k)\omega}$$

We now make the change of variable $j = n - k$ in the identity above. In this case $k = n - j$ and note that $(-1)^k = (-1)^{n-j} = (-1)^n \cdot (-1)^j = -1 \cdot (-1)^j$. We have

$$G(\omega) = -e^{ib} \sum_{j \in \mathbb{Z}} (-1)^j h_{n-j} e^{ij\omega} = \sum_{j \in \mathbb{Z}} \left(-e^{ib} (-1)^j h_{n-j} \right) e^{ij\omega}$$

Since $G(\omega)$ also can be expressed as $G(\omega) = \sum_{j \in \mathbb{Z}} g_j e^{ij\omega}$, it must be that

$$g_j = -e^{ib} (-1)^j h_{n-j}$$

and the proof is complete. $\qquad\qquad\qquad\qquad\qquad\qquad\qquad\qquad \square$

For the Daubechies filters from Chapter 7 we take $b = \pi$ so that $-e^{ib} = 1$ and $n = L$. For real-valued filters, we must choose b to be some integer multiple of π.

Note: For the remainder of the book, we set $b = \pi$, so that

$$\boxed{g_k = (-1)^k h_{n-k}}$$

In practice, n is often set to 1, so the highpass filter is $g_k = (-1)^k h_{1-k}$ but we are free to pick any odd integer we wish for n. Note that from the modulation rule (Proposition 4.3) in Section 4.3, multiplying a Fourier series by $e^{in\omega}$ simply translates the Fourier coefficients n units to the right if n is positive and n units to the left if n is negative.

Summary

It is important that you be comfortable with the results of Sections 8.1 and 8.2 before moving on. We have completely characterized the construction of wavelet lowpass and highpass filters in terms of their Fourier series. The characterization works for FIR filters, finite-length filters, or bi-infinite filters.

We first saw that if $A(\omega)$ satisfies

$$|A(\omega)|^2 + |A(\omega + \pi)|^2 = 2$$

then we obtain the orthogonality conditions

$$\sum_{k \in \mathbb{Z}} a_k^2 = 1 \qquad \sum_{k \in \mathbb{Z}} a_k a_{k-2m} = 0, \qquad m \neq 0$$

Thus, we ask that the Fourier series $H(\omega)$ and $G(\omega)$ of filters **h** and **g**, respectively, satisfy this condition. When these Fourier series satisfy

$$H(\omega)\overline{G(\omega)} + H(\omega + \pi)\overline{G(\omega + \pi)} = 0$$

the filters meet the orthogonality conditions

$$\sum_{k \in \mathbb{Z}} h_k g_{k-2m} = 0, \qquad m \in \mathbb{Z}.$$

We have learned that if we can construct a Fourier series $H(\omega)$ that satisfies the lowpass conditions

$$H(0) = \sqrt{2} \qquad H(\pi) = 0$$

and the orthogonality condition (8.18), then we can build $G(\omega)$ using the rule

$$G(\omega) = -e^{in\omega} \overline{H(\omega + \pi)}$$

where n is an odd integer. This $G(\omega)$ satisfies the highpass conditions

$$G(0) = 0 \qquad G(\pi) = \sqrt{2}$$

and the orthogonality conditions

$$|G(\omega)|^2 + |G(\omega + \pi)|^2 = 2 \qquad H(\omega)\overline{G(\omega)} + H(\omega + \pi)\overline{G(\omega + \pi)} = 0$$

We can choose n to be any odd integer, but a common choice is $n = 1$.

PROBLEMS

8.6 We can use Example 5.5(a) to see that the Fourier series for the Haar filter $\mathbf{h} = (h_0, h_1) = (\frac{\sqrt{2}}{2}, \frac{\sqrt{2}}{2})$ is $H(\omega) = \sqrt{2}\, e^{i\omega/2} \cos\left(\frac{\omega}{2}\right)$. We can use Example 5.6 to show that the Fourier series for the highpass filter $\mathbf{g} = (g_0, g_1) = (\frac{\sqrt{2}}{2}, -\frac{\sqrt{2}}{2})$ used in the Haar wavelet transform is $G(\omega) = -\sqrt{2}i e^{i\omega/2} \sin\left(\frac{\omega}{2}\right)$. Verify that $H(w)$ and $G(\omega)$ satisfy (8.13).

8.7 Prove the converse of Theorem 8.2. (*Hint:* The ideas from Problem 8.2 will be useful for this problem.)

8.8 Prove Corollary 8.2.

8.9 Suppose that $\mathbf{a} = (a_\ell, \ldots, a_L)$ and $\mathbf{b} = (b_m, \ldots, b_M)$ are finite-length filters with $\ell < m$ and $M < L$. Suppose that $A(\omega)$ and $B(\omega)$ are the Fourier series for filters $\mathbf{a}$ and $\mathbf{b}$, respectively. Show that

$$A(\omega)\overline{B(\omega)} + A(\omega + \pi)\overline{B(\omega + \pi)} = 0$$

if and only if

$$\sum_{k=\ell+2m}^{L} a_k b_{k-2m} = 0, \qquad m = 1, 2, \ldots, \frac{L - \ell - 1}{2} \qquad (8.29)$$

8.10   Let $\mathbf{h} = (h_0, \ldots, h_L)$ be the Daubechies filter of length $L + 1$. Then the resulting Fourier series $H(\omega) = \sum\limits_{k=0}^{L} h_k e^{ik\omega}$ satisfies

$$|H(\omega)|^2 + |H(\omega + \pi)|^2 = 2$$

Show that the Fourier series (8.22) given for **g** satisfies (8.19) and (8.20).

8.11   Suppose that $\mathbf{h} = (h_\ell, \ldots, h_L)$ is a finite-length filter. Use (8.28) with $b = \pi$ and $n = 1$ to write down expressions for all possible nonzero elements of **g**.

8.3   COIFLET FILTERS

In Section 7.2 we learned how to combine the orthogonality conditions imposed by the wavelet matrix W_N and the conditions on the Fourier series $H(\omega)$ of the filter $\mathbf{h}$ to construct Daubechies' family of orthogonal filters.

For filter $\mathbf{h} = (h_0, \ldots, h_L)$, we insisted that the following conditions hold for H and its derivatives:

$$H^{(m)}(\pi) = 0, \qquad m = 0, \ldots, \frac{L-1}{2} \qquad (8.30)$$

The derivative conditions in (8.30) "flatten" $|H(\omega)|$ near $\omega = \pi$. Such a filter will typically annihilate or dampen high oscillations in the data.

We also insisted that $\mathbf{h}$ be an orthogonal filter and constructed it so that the orthogonality conditions

$$\sum_{k=0}^{L} h_k^2 = 1 \quad \text{and} \quad \sum_{k=2m}^{L} h_k h_{k-2m} = 0 \qquad m = 1, 2, \ldots, \frac{L-1}{2} \qquad (8.31)$$

are satisfied.

Flattening the Fourier Series H(ω) at $\omega = 0$

It is worthwhile to think about modifying our filter further to better model the trends in our input vector. For example, is it possible to modify the filter so that the lowpass portion of the output does a better job approximating the original input? Ingrid Daubechies considered this question [24] (the idea was suggested to her by R. Coifman from Yale University) and her solution was to impose derivative conditions on $H(\omega)$ at $\omega = 0$. That is, we wish to construct an orthogonal filter $\mathbf{h}$ that satisfies (8.30) and

$$H^{(m)}(0) = 0, \qquad m = 1, 2, \ldots \qquad (8.32)$$

We will be more specific about the number of derivative conditions in the theorem that follows.

The conditions given in (8.32) will have the same effect on $|H(\omega)|$ at $\omega = 0$ as (8.30) has on $|H(\omega)|$ at $\omega = \pi$. As a result, the lowpass portion of the output should do a good job of modeling smaller oscillation trends in the input vector.

A Theorem for Constructing Coiflet Filters

Before we state the theorem for constructing Coiflet filters, we give a lemma that is used in the proof of both Theorems 8.3 and 10.2.

Lemma 8.1 (A Useful Trigonometric Identity). *For any nonnegative integer* K, *we have*

$$
1 = \cos^{2K}\left(\frac{\omega}{2}\right) \sum_{j=0}^{K-1} \binom{K-1+j}{j} \sin^{2j}\left(\frac{\omega}{2}\right)
$$
$$
+ \sin^{2K}\left(\frac{\omega}{2}\right) \sum_{j=0}^{K-1} \binom{K-1+j}{j} \cos^{2j}\left(\frac{\omega}{2}\right) \tag{8.33}
$$

$\square$

Note: The binomial coefficient $\binom{n}{j}$ is defined in Problem 4.41.

Proof of Lemma 8.1. We give one proof here.[4] Another proof is given in Exercise 8.13.

For convenience, we set $c = \cos\left(\frac{\omega}{2}\right)$ and $s = \sin\left(\frac{\omega}{2}\right)$. We then write the right hand side of (8.33) as

$$
F_K = c^{2K} \sum_{j=0}^{K-1} \binom{K-1+j}{j} s^{2j} + s^{2K} \sum_{j=0}^{K-1} \binom{K-1+j}{j} c^{2j} \tag{8.34}
$$

We will show that $F_K = F_{K-1}$ for any positive integer K. Combining this result with the fact that (8.33) trivially holds for $K = 1$ completes the proof. We compute

$$
F_K - F_{K-1} = c^{2K-2}\left(c^2 \sum_{j=0}^{K-1} \binom{K-1+j}{j} s^{2j} - \sum_{j=0}^{K-2} \binom{K-2+j}{j} s^{2j} \right)
$$
$$
+ s^{2K-2}\left(s^2 \sum_{j=0}^{K-1} \binom{K-1+j}{j} c^{2j} - \sum_{j=0}^{K-2} \binom{K-2+j}{j} c^{2j} \right) \tag{8.35}
$$

---

[4]This proof was shown to me by Yongzhi Yang.

Let's consider the first term on the right hand side of (8.35). Let's call it P_K. We replace c^2 with $1 - s^2$, simplify, and write

$$P_K = c^{2K-2} \left(\sum_{j=0}^{K-1} \binom{K-1+j}{j} s^{2j} - \sum_{j=0}^{K-1} \binom{K-1+j}{j} s^{2j+2} \right.$$
$$\left. - \sum_{j=0}^{K-2} \binom{K-2+j}{j} s^{2j} \right)$$

There are three summations in the identity above. We will group like terms for $j = 1$ to $K - 2$. For the first summation, we have

$$\sum_{j=0}^{K-1} \binom{K-1+j}{j} s^{2j} = 1 + \sum_{j=1}^{K-2} \binom{K-1+j}{j} s^{2j} + \binom{2K-2}{K-1} s^{2K-2} \quad (8.36)$$

For the second summation, we replace j with $j - 1$ and write

$$\sum_{j=0}^{K-1} \binom{K-1+j}{j} s^{2j+2} = \sum_{j=1}^{K} \binom{K-2+j}{j-1} s^{2j}$$
$$= \sum_{j=1}^{K-2} \binom{K-2+j}{j-1} s^{2j} + \binom{2K-3}{K-2} s^{2K-2}$$
$$+ \binom{2K-2}{K-1} s^{2K} \quad (8.37)$$

The third summation is

$$\sum_{j=0}^{K-2} \binom{K-2+j}{j} s^{2j} = 1 + \sum_{j=1}^{K-2} \binom{K-2+j}{j} s^{2j} \quad (8.38)$$

Subtracting (8.37) and (8.38) from (8.36) gives

$$P_K = c^{2K-2} \left(\sum_{j=1}^{K-2} \left(\binom{K-1+j}{j} - \binom{K-2+j}{j-1} - \binom{K-2+j}{j} \right) s^{2j} \right.$$
$$\left. + \left(\binom{2K-2}{K-1} - \binom{2K-3}{K-2} \right) s^{2K-2} - \binom{2K-2}{K-1} s^{2K} \right) \quad (8.39)$$

Now by Pascal's identity (4.17),

$$\binom{K-1+j}{j} - \binom{K-2+j}{j-1} - \binom{K-2+j}{j} = 0$$

so (8.39) reduces to

$$P_K = c^{2K-2} \left(\left[\binom{2K-2}{K-1} - \binom{2K-3}{K-2} \right] s^{2K-2} - \binom{2K-2}{K-1} s^{2K} \right) \quad (8.40)$$

Using Problem 8.12, we see that

$$\binom{2K-2}{K-1} - \binom{2K-3}{K-2} = \frac{1}{2}\binom{2K-2}{K-1}$$

Inserting this identity into (8.40) gives

$$\begin{aligned} P_K &= \frac{1}{2}\binom{2K-2}{K-1} c^{2K-2} s^{2K-2} - \binom{2K-2}{K-1} c^{2K-2} s^{2K} \\ &= \binom{2K-2}{K-1} (cs)^{2K-2} \left(\frac{1}{2} - s^2 \right) \end{aligned} \quad (8.41)$$

In a similar manner, if we let Q_K denote the second term in (8.35), we have

$$Q_K = \binom{2K-2}{K-1} (cs)^{2K-2} \left(\frac{1}{2} - c^2 \right) \quad (8.42)$$

Finally, we add P_K and Q_K to obtain

$$\begin{aligned} F_K - F_{K-1} &= P_K + Q_K \\ &= \binom{2K-1}{K-1} (cs)^{2K-2} \left(\frac{1}{2} - s^2 + \frac{1}{2} - c^2 \right) \\ &= \binom{2K-1}{K-1} (cs)^{2K-2} \left(1 - s^2 - c^2 \right) \\ &= 0 \end{aligned}$$

$\square$

Daubechies gives a result [24] that suggests one way to construct the filter that we present below. We do not give a complete proof of the result, as part of the proof is beyond the scope of this book.

Theorem 8.3 (Coiflet Filters). *Let* $K = 1, 2, \ldots$ *and define the* 2π-*periodic function* $H(\omega)$ *as*

$$\begin{aligned} H(\omega) = \sqrt{2} \cos^{2K}\left(\frac{\omega}{2}\right) \\ \cdot \left(\sum_{j=0}^{K-1} \binom{K-1+j}{j} \sin^{2j}\left(\frac{\omega}{2}\right) + \sin^{2K}\left(\frac{\omega}{2}\right) \sum_{\ell=0}^{2K-1} a_\ell e^{i\ell\omega} \right) \end{aligned} \quad (8.43)$$

Then

$$H(0) = \sqrt{2}$$
$$H^{(m)}(0) = 0, \qquad m = 1, \ldots, 2K - 1 \qquad (8.44)$$
$$H^{(m)}(\pi) = 0, \qquad m = 0, \ldots, 2K - 1$$

Furthermore, real numbers $a_0, \ldots, a_{2K-1}$ exist so that

$$|H(\omega)|^2 + |H(\omega + \pi)|^2 = 2 \qquad (8.45)$$

$\square$

Discussion of the Proof of Theorem 8.3. There are two parts to the proof. The first part is to show that the Fourier series $H(\omega)$ given in (8.43) satisfies the conditions (8.44). We provide this proof below. The second part of the proof shows the existence of real number $a_0, \ldots, a_{2K-1}$ so that (8.43) satisfies (8.45). Recall from Theorem 8.1 in Section 8.1 that if $H(\omega)$ satisfies (8.3), then the corresponding filter h satisfies the orthogonality conditions given by (8.4) and (8.5). The proof that $H(\omega)$ satisfies (8.45) is quite technical and requires some advanced mathematics that we do not cover in this book. The reader with some background in analysis is encouraged to look at Daubechies [24].

To show that $H^{(m)}(\pi) = 0$ for $m = 0, \ldots, 2K - 1$, we note that

$$\cos\left(\frac{\omega}{2}\right) = \frac{e^{i\omega/2} + e^{-i\omega/2}}{2} = \frac{1}{2} e^{-i\omega/2} \left(e^{i\omega} + 1 \right) \qquad (8.46)$$

If we insert (8.46) into (8.43), we have

$$H(\omega) = \frac{\sqrt{2}}{2^{2K} e^{iK\omega}} (1 + e^{i\omega})^{2K} \cdot$$
$$\left(\sum_{j=0}^{K-1} \binom{K-1+j}{j} \sin^{2j}\left(\frac{\omega}{2}\right) + \sin^{2K}\left(\frac{\omega}{2}\right) \sum_{\ell=0}^{2K-1} a_\ell e^{i\ell\omega} \right)$$

Since $H(\omega)$ contains the factor $(1 + e^{i\omega})^{2K}$, we see that it has a root of multiplicity at least $2K$ at $\omega = \pi$. Using the ideas from Problem 7.26 from Section 7.2, we can infer that $H^{(m)}(\pi) = 0$, $m = 0, \ldots, 2K - 1$.

Using Lemma 8.1 and Problem 8.14, we can rewrite (8.43) as

$$H(\omega) = \sqrt{2} + \sqrt{2} \sin^{2K}\left(\frac{\omega}{2}\right)$$
$$\cdot \left(-\sum_{j=0}^{K-1} \binom{K-1+j}{j} \cos^{2j}\left(\frac{\omega}{2}\right) + \cos^{2K}\left(\frac{\omega}{2}\right) \sum_{\ell=0}^{2K-1} a_\ell e^{i\ell\omega} \right) \qquad (8.47)$$

We can rewrite $\sin\left(\frac{\omega}{2}\right)$ as

$$\sin\left(\frac{\omega}{2}\right) = \frac{e^{i\omega/2} - e^{-i\omega/2}}{2i} = \frac{1}{2i}e^{-i\omega/2}\left(e^{i\omega} - 1\right) = \frac{i}{2}e^{-i\omega/2}\left(1 - e^{i\omega}\right) \quad (8.48)$$

If we insert (8.48) into (8.47), we have

$$H(\omega) = \sqrt{2} + \sqrt{2}\left(\frac{i}{2}\right)^{2K} e^{-iK\omega}\left(1 - e^{i\omega}\right)^{2K}$$
$$\cdot \left(-\sum_{j=0}^{K-1}\binom{K-1+j}{j}\cos^{2j}(\frac{\omega}{2}) + \cos^{2K}(\frac{\omega}{2})\sum_{\ell=0}^{2K-1}a_\ell e^{i\ell\omega}\right)$$

Now if we let $Q(\omega) = H(\omega) - \sqrt{2}$, we see that $Q(\omega)$ contains the factor $(1 - e^{i\omega})^{2K}$ and thus has a root of multiplicity at least $2K$ at $\omega = 0$. Thus, $Q^{(m)}(0) = 0$ for $m = 0, \ldots, 2K - 1$.

For $m = 0$, we have $Q(0) = H(0) - \sqrt{2} = 0$ or $H(0) = \sqrt{2}$. It is easy to see that $Q^{(m)}(\omega) = H^{(m)}(\omega)$ for $m = 1, \ldots, 2K - 1$, so we can infer that $H^{(m)}(0) = 0$ for $m = 1, \ldots, 2K - 1$. $\qquad\square$

The Coiflet Filter for $K = 1$

Let's use Theorem 8.3 to find a Coiflet filter for the case $K = 1$. Using (8.43), we see that

$$H(\omega) = \sqrt{2}\cos^2\left(\frac{\omega}{2}\right)\left(1 + \sin^2\left(\frac{\omega}{2}\right)\sum_{\ell=0}^{1}a_\ell e^{i\ell\omega}\right)$$
$$= \sqrt{2}\left(\cos^2\left(\frac{\omega}{2}\right) + \cos^2\left(\frac{\omega}{2}\right)\sin^2\left(\frac{\omega}{2}\right)(a_0 + a_1 e^{i\omega})\right)$$
$$= \sqrt{2}\left(\cos^2\left(\frac{\omega}{2}\right) + \frac{\sin^2(\omega)}{4}(a_0 + a_1 e^{i\omega})\right) \quad (8.49)$$

where we use the double-angle formula $\sin(2\omega) = 2\sin(\omega)\cos(\omega)$ in the last step.

We can use Problem 4.38 to write $\cos^2\left(\frac{\omega}{2}\right) = \frac{1}{4}(e^{i\omega} + 2 + e^{-i\omega})$ and Problem 4.39 to write $\sin^2(\omega) = \frac{1}{4}(-e^{2i\omega} + 2 - e^{-2i\omega})$. We insert these identities into (8.49) and expand to obtain a Fourier series for $H(\omega)$:

$$H(\omega) = \sqrt{2}\left(\frac{1}{4}(e^{i\omega} + 2 + e^{-i\omega}) + \frac{1}{16}(-e^{2i\omega} + 2 - e^{-2i\omega})(a_0 + a_1 e^{i\omega})\right)$$
$$= \frac{\sqrt{2}}{16}\left(-a_0 e^{-2i\omega} + (4 - a_1)e^{-i\omega} + (8 + 2a_0) + (4 + 2a_1)e^{i\omega}\right.$$
$$\left. - a_0 e^{2i\omega} - a_1 e^{3i\omega}\right)$$

If we gather like terms in the identity above, we have the following for $H(\omega)$:

$$H(\omega) = -\frac{\sqrt{2}\,a_0}{16}e^{-2i\omega} + \frac{\sqrt{2}(4-a_1)}{16}e^{-i\omega} + \frac{\sqrt{2}(8+2a_0)}{16}$$
$$+ \frac{\sqrt{2}(4+2a_1)}{16}e^{i\omega} - \frac{\sqrt{2}\,a_0}{16}e^{2i\omega} - \frac{\sqrt{2}\,a_1}{16}e^{3i\omega} \qquad (8.50)$$

We can now read our filter coefficients from (8.50). We have

$$h_{-2} = \frac{-\sqrt{2}\,a_0}{16} \qquad h_{-1} = \frac{\sqrt{2}(4-a_1)}{16} \qquad h_0 = \frac{\sqrt{2}(8+2a_0)}{16}$$
$$h_1 = \frac{\sqrt{2}(4+2a_1)}{16} \qquad h_2 = \frac{-\sqrt{2}\,a_0}{16} \qquad h_3 = \frac{-\sqrt{2}\,a_1}{16} \qquad (8.51)$$

Now we must find a_0 and a_1 so that the (noncausal!) filter

$$\mathbf{h} = (h_{-2}, h_{-1}, h_0, h_1, h_2, h_3)$$

satisfies the orthogonality condition (8.45). At this point, we can use Corollary 8.1 from Section 8.1 (with $\ell = -2$ and $L = 3$) to see that the filter must satisfy

$$h_{-2}^2 + h_{-1}^2 + h_0^2 + h_1^2 + h_2^2 + h_3^2 = 1$$

$$\sum_{k=-2+2m}^{3} h_k h_{k-2m} = 0, \qquad m = 1, 2$$

or

$$h_{-2}^2 + h_{-1}^2 + h_0^2 + h_1^2 + h_2^2 + h_3^2 = 1$$
$$h_{-2}h_0 + h_{-1}h_1 + h_0 h_2 + h_1 h_3 = 0 \qquad (8.52)$$
$$h_{-2}h_2 + h_{-1}h_3 = 0$$

If we insert the values from (8.51) into (8.52) and simplify, we obtain the following system (you are asked to verify the algebra in Problem 8.15) with unknowns a_0 and a_1:

$$16a_0 + 3a_0^2 + 4a_1 + 3a_1^2 = 16$$
$$4 - 4a_0 - a_0^2 - a_1^2 = 0 \qquad (8.53)$$
$$-4a_1 + a_0^2 + a_1^2 = 0$$

Adding the last two equations of (8.53) together and solving for a_0 gives $a_0 = 1 - a_1$. Putting $a_0 = 1 - a_1$ into the first equation of (8.53), we obtain

$$16(1-a_1) + 3(1-a_1)^2 + 4a_1 + 3a_1^2 = 16$$
$$2a_1^2 - 6a_1 + 1 = 0$$

so that $a_1 = \frac{1}{2}(3 \pm \sqrt{7})$. This leads to the following two solutions to (8.53):

$$a_0 = \tfrac{1}{2}(\sqrt{7} - 1) \quad a_1 = \tfrac{1}{2}(3 - \sqrt{7})$$
$$a_0 = -\tfrac{1}{2}(\sqrt{7} + 1) \quad a_1 = \tfrac{1}{2}(3 + \sqrt{7}) \tag{8.54}$$

Coiflet Length 6 Filter

We can plug either of these solutions into (8.51) to find the filter coefficients. For the first pair in (8.54), we have

$$
\boxed{
\begin{array}{ll}
h_{-2} = \frac{1}{16\sqrt{2}} (1 - \sqrt{7}), & h_{-1} = \frac{1}{16\sqrt{2}} (5 + \sqrt{7}) \\[2mm]
h_0 = \frac{1}{8\sqrt{2}} (7 + \sqrt{7}), & h_1 = \frac{1}{8\sqrt{2}} (7 - \sqrt{7}) \\[2mm]
h_2 = \frac{1}{16\sqrt{2}} (1 - \sqrt{7}), & h_3 = \frac{1}{16\sqrt{2}} (-3 + \sqrt{7})
\end{array}
}
\tag{8.55}
$$

In Figure 8.2 we plot $|H(\omega)|$ for the Coiflet filter given in (8.55). You are asked to find the second solution in Problem 8.16. We use (8.55) to formulate our next definition.

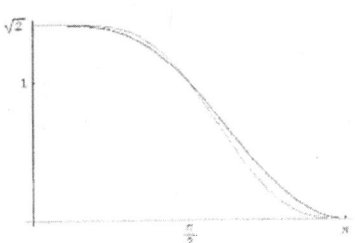

Figure 8.2 $|H(\omega)|$ for the C6 filter (black). A plot of $|H(\omega)|$ for the D6 filter (gray) is provided for comparative purposes.

Definition 8.3 (Coiflet Length 6 Filter). *We define the* Coiflet six-term orthogonal filter *by the values given in (8.55). We denote this filter the* C6 filter. ☐

An Alternative Method for Finding the C6 Filter

You can probably imagine that to find **h** for $K = 2, 3, \ldots$, the steps (analogous to those used to find the Coiflet six-term filter) are quite tedious. The Fourier series (and thus the number of unknowns a_ℓ) grows as K increases, and once we have expressed our filter in terms of the a_ℓ, we must insert these identities into the orthogonality conditions (8.6) and (8.7) and solve the resulting quadratic system.

We present next an alternative form of solution that at least cuts out the need to solve for the a_ℓ. We use the $K = 1$ case to illustrate.

After expanding in the $K = 1$ case, we saw that $H(\omega)$ has the form given by (8.50). Rather than worrying about finding the values of a_0 and a_1, let's instead note that the Fourier series starts at $\ell = -2$ and terminates at $L = 3$ and try to find numbers $h_{-2}, \ldots, h_3$, that satisfy the orthogonality conditions (8.52) and the derivative conditions (8.44). We have written the orthogonality conditions in (8.52), so let's write down the equations for the derivative conditions. We have

$$H(\omega) = h_{-2}\,e^{-2i\omega} + h_{-1}\,e^{-i\omega} + h_0 + h_1\,e^{i\omega} + h_2\,e^{2i\omega} + h_3\,e^{3i\omega} \qquad (8.56)$$

If we plug in $\omega = 0$ into (8.56) and set the result equal to $\sqrt{2}$, we have

$$h_{-2} + h_{-1} + h_0 + h_1 + h_2 + h_3 = \sqrt{2}$$

Inserting $\omega = \pi$ into (8.56) and setting the result equal zero gives

$$h_{-2} - h_{-1} + h_0 - h_1 + h_2 - h_3 = 0$$

Differentiating (8.56) gives

$$H'(\omega) = -2ih_{-2}\,e^{-2i\omega} - ih_{-1}\,e^{-i\omega} + ih_1\,e^{i\omega} + 2ih_2\,e^{2i\omega} + 3ih_3\,e^{3i\omega} \qquad (8.57)$$

The derivative conditions (8.44) are $H'(0) = H'(\pi) = 0$. Plugging 0 and π into (8.57), simplifying, and setting the results to zero gives

$$-2h_{-2} - h_{-1} + h_1 + 2h_2 + 3h_3 = 0$$
$$-2h_{-2} + h_{-1} - h_1 + 2h_2 - 3h_3 = 0$$

Thus, the system we seek to solve is

$$
\begin{aligned}
h_{-2}^2 + h_{-1}^2 + h_0^2 + h_1^2 + h_2^2 + h_3^2 &= 1 \\
h_{-2}h_0 + h_{-1}h_1 + h_0h_2 + h_1h_3 &= 0 \\
h_{-2}h_2 + h_{-1}h_3 &= 0 \\
h_{-2} + h_{-1} + h_0 + h_1 + h_2 + h_3 &= \sqrt{2} \\
h_{-2} - h_{-1} + h_0 - h_1 + h_2 - h_3 &= 0 \\
-2h_{-2} - h_{-1} + h_1 + 2h_2 + 3h_3 &= 0 \\
-2h_{-2} + h_{-1} - h_1 + 2h_2 - 3h_3 &= 0
\end{aligned}
\qquad (8.58)
$$

Table 8.1   Two Coiflet filters for the case $K = 1$.

	Solution 1[a]	Solution 2
h_{-2}	-0.072733	0.161121
h_{-1}	0.337898	0.104044
h_0	0.852572	0.384865
h_1	0.384865	0.852572
h_2	-0.072733	0.161121
h_3	-0.015656	-0.249509

[a] This is the Daubechies C6 filter.

If we use a **CAS** to solve (8.58), we obtain the two solutions (rounded to six digits) listed in Table 8.1.

It is easy to check that the first system is the C6 filter given in (8.55) and that the second solution comes from plugging the second solution pair from (8.54) into (8.51) (see Problem 8.16). In Problem 8.18 you can learn how to solve the system (8.58) by hand.

A General Method for Finding Coiflet Filters

We can use the ideas outlined in the preceding subsection to suggest a general method for finding Coiflet filters. The key is knowing the starting and stopping indices ℓ and L, respectively, of the finite-length filter h associated with the Fourier series $H(\omega)$ given by (8.43). Once we have ℓ and L, we can write down the orthogonality conditions and the derivative conditions and then use a **CAS** to solve the system.

In Problem 8.19(b) you will show that when $K = 2$, the associated finite-length filter is $h = (h_{-4}, h_{-3}, \ldots, h_6, h_7)$, so that $\ell = -4$ and $L = 7$. In Problem 8.19(c), you will show that for $K = 3$, the starting index is $\ell = -6$ and the stopping index is $L = 11$. Finally, in Problem 8.20, you will provide a proof for the following proposition.

Proposition 8.4 (Finite-Length Coiflet Filter). *Suppose that* h *is a finite-length filter associated with the Fourier series* $H(\omega)$ *given by Theorem 8.3. Then the length of* h *is* $6K$ *and*

$$h = (h_\ell, h_{\ell+1}, \ldots, h_{L-1}, h_L)$$

where $\ell = -2K$ *and* $L = 4K - 1$.

Proof of Proposition 8.4. The proof is left as Problem 8.20.   □

Coiflets for $K = 2$

When $K = 2$, we have $\ell = -4$, $L = 7$, and $(L - \ell - 1)/2 = 5$. We use Corollary 8.1 to write down the orthogonality conditions that $\mathbf{h} = (h_{-4}, \ldots, h_7)$ must satisfy:

$$\sum_{k=-4}^{11} h_k^2 = 1 \quad \text{and} \quad \sum_{k=-4+2m}^{11} h_k h_{k-2m} = 0, \qquad m = 1, \ldots, 5 \quad (8.59)$$

Since $2K - 1 = 3$, the derivative conditions (8.44) are

$$\begin{aligned} H(0) &= \sqrt{2} \\ H(\pi) &= 0 \\ H^{(m)}(0) &= 0 \qquad m = 1, \ldots, 3 \\ H^{(m)}(\pi) &= 0 \qquad m = 1, \ldots, 3 \end{aligned} \qquad (8.60)$$

In Problem 8.21 you are asked to explicitly write down the system for finding Coiflet filters when $K = 2$. Solving this system numerically yields four solutions. They are listed (rounded to six digits) in Table 8.2.

Table 8.2   Four Coiflet filters for $K = 2$.

	Solution 1[a]	Solution 2	Solution 3	Solution 4
h_{-4}	0.016387	−0.001359	−0.021684	−0.028811
h_{-3}	−0.041465	−0.014612	−0.047599	0.009542
h_{-2}	−0.067373	−0.007410	0.163254	0.113165
h_{-1}	0.386110	0.280612	0.376511	0.176527
h_0	0.812724	0.750330	0.270927	0.542555
h_1	0.417005	0.570465	0.516748	0.745265
h_2	−0.07649	−0.071638	0.545852	0.102774
h_3	−0.059434	−0.155357	−0.239721	−0.296788
h_4	0.023680	0.050024	−0.327760	−0.020498
h_5	0.005611	0.024804	0.136027	0.078835
h_6	−0.001823	−0.012846	0.076520	−0.002078
h_7	−0.000721	0.001195	−0.034858	−0.006275

[a] This is the Coiflet C12 filter.

We have the following definition.

Definition 8.4 (Coiflet Length 12 Filter). *We define the* Coiflet 12-term orthogonal filter *by the first solution in Table 8.2. We call this filter the* C12 *filter.*   □

We will not cover the method by which Daubechies chooses a particular filter. The interested reader is referred to Daubechies [24].

A Wavelet Filter for the Coiflet Filter

Now that we have the means to construct a Coiflet filter **h**, we need to build the associated wavelet filter **g**. For the causal filters from Chapters 6 and 7, we used the rule $g_k = (-1)^k h_{L-k}$. In this case, the elements g_k are obtained by reversing the elements in **h** and multiplying by $(-1)^k$. You could also think of this operation as reflecting the filter about some center. In particular, the operation $h_k \mapsto h_{L-k}$ has the effect of reflecting the numbers about the line $k = L/2$. As an example, consider the D6 filter $h_0, h_1, h_2, h_3, h_4, h_5$. Here $L = 5$ and $L/2 = 5/2 = 2.5$. If you think about constructing **g** by the rule $g_k = (-1)^k h_{5-k}$, we could instead reflect the elements of **h** about the midpoint of the list $0, \ldots, 5$ in order to reverse the filter and then multiply by ± 1.

Our convention for Coiflet filters will be the same. However, these filters are noncausal, so we will have to think about the line of reflection.

In the case $K = 1$, the reflection line would be $k = \frac{1}{2}$, since three of the filter coefficient indices are to the left of $\frac{1}{2}$ and the other three are to the right. It is easy to see where the $\frac{1}{2}$ comes from — we simply find the midpoint of the list $\ell, \ldots, L$. In order to reflect about the line $k = \frac{1}{2}$, we use the rule

$$g_k = (-1)^k h_{1-k}$$

so that $g_{-2} = h_3, g_{-1} = -h_2, g_0 = h_1, g_1 = -h_0, g_2 = -h_{-1}$, and $g_3 = h_{-2}$.

In the case $K = 2$, the reflection line would be $k = \frac{3}{2}$ since six of the filter coefficient indices are to the left of $\frac{3}{2}$ and six are to the right of $\frac{3}{2}$. The rule we would then use in the $K = 2$ case is

$$g_k = (-1)^k h_{3-k}$$

For general K, Proposition 8.4 tells us that $\ell = -2K$ and $L = 4K - 1$, so that the midpoint of the list $-2K, \ldots, 4K - 1$ is $\frac{-2K+4K-1}{2} = \frac{2K-1}{2}$. Thus we will reflect about the line $k = \frac{2K-1}{2}$. The rule

$$g_k = (-1)^k h_{2K-1-k} \tag{8.61}$$

will incorporate the desired reflection.

From Proposition 8.3 we know that the Fourier series $G(\omega)$ must be

$$G(\omega) = -e^{i(2K-1)\omega} \overline{H(\omega + \pi)}$$

Note that $2K - 1$ is always an odd integer, so it is permissible to use it in this formula.

We also comment that we are free to pick any odd index that we like — reflecting about the midpoint of the indices simply keeps the wavelets associated with Coiflet filters consistent with the highpass filters we constructed for Daubechies filters.

The Coiflet Filter in the Wavelet Transform Matrix

Proposition 8.4 tells us that the Coiflet filters are not causal filters. How do we implement such filters in our wavelet transform algorithms? One answer is simply to treat these filters just like causal filters. That is, if we call Algorithm 7.1 to apply the C6 filter to vector $\mathbf{x}$, then the associated matrix W_N is

$$
W_N = \left[
\begin{array}{cccccccc|cccccc}
h_3 & h_2 & h_1 & h_0 & h_{-1} & h_{-2} & 0 & 0 & \cdots & 0 & 0 & 0 & 0 & 0 & 0 \\
0 & 0 & h_3 & h_2 & h_1 & h_0 & h_{-1} & h_{-2} & & 0 & 0 & 0 & 0 & 0 & 0 \\
\vdots & & & & & & & & \ddots & & & & & & \vdots \\
0 & 0 & 0 & 0 & & & & & & h_3 & h_2 & h_1 & h_0 & h_{-1} & h_{-2} \\
h_{-1} & h_{-2} & 0 & 0 & & & & \cdots & & 0 & 0 & h_3 & h_2 & h_1 & h_0 \\
h_1 & h_0 & h_{-1} & h_{-2} & & & & & & & & 0 & 0 & h_3 & h_2 \\
\hline
-h_{-2} & h_{-1} & -h_0 & h_1 & -h_2 & h_3 & 0 & 0 & \cdots & 0 & 0 & 0 & 0 & 0 & 0 \\
0 & 0 & -h_{-2} & h_{-1} & -h_0 & h_1 & -h_2 & h_3 & & 0 & 0 & 0 & 0 & 0 & 0 \\
\vdots & & & & & & & & \ddots & & & & & & \vdots \\
0 & 0 & 0 & 0 & & & & & & -h_{-2} & h_{-1} & -h_0 & h_1 & -h_2 & h_3 \\
-h_2 & h_3 & 0 & 0 & & & & \cdots & & 0 & 0 & -h_{-2} & h_{-1} & -h_0 & h_1 \\
-h_0 & h_1 & -h_2 & h_3 & & & & & & & & 0 & 0 & -h_{-2} & h_{-1}
\end{array}
\right]
$$
$$\tag{8.62}$$

An Alternative Form of the Wavelet Transform Matrix

If you look at other texts or journal articles about discrete wavelet transforms, you will often see the filters in reverse order as well as shifted so that terms with negative indices appear at the far right of the first row. For example, with the C6 filter, W_N might be constructed as follows:

$$
W_N = \left[
\begin{array}{cccccccc|cccccc}
h_0 & h_1 & h_2 & h_3 & 0 & 0 & 0 & 0 & & 0 & 0 & 0 & 0 & h_{-2} & h_{-1} \\
h_{-2} & h_{-1} & h_0 & h_1 & h_2 & h_3 & 0 & 0 & \cdots & 0 & 0 & 0 & 0 & 0 & 0 \\
0 & 0 & h_{-2} & h_{-1} & h_0 & h_1 & h_2 & h_3 & & & & 0 & 0 & 0 & 0 \\
\vdots & & & & & & & & \ddots & & & & & & \vdots \\
0 & 0 & 0 & 0 & & & & \cdots & & h_{-2} & h_{-1} & h_0 & h_1 & h_2 & h_3 \\
h_2 & h_3 & 0 & 0 & & & & & & 0 & 0 & h_{-2} & h_{-1} & h_0 & h_1 \\
\hline
h_1 & -h_0 & h_{-1} & -h_{-2} & 0 & 0 & 0 & 0 & & 0 & 0 & 0 & 0 & h_3 & -h_2 \\
h_3 & -h_2 & h_1 & -h_0 & h_{-1} & -h_{-2} & 0 & 0 & \cdots & 0 & 0 & 0 & 0 & 0 & 0 \\
0 & 0 & h_3 & -h_2 & h_1 & -h_0 & h_{-1} & -h_{-2} & & & & 0 & 0 & 0 & 0 \\
\vdots & & & & & & & & \ddots & & & & & & \vdots \\
0 & 0 & 0 & 0 & & & & \cdots & & h_3 & -h_2 & h_1 & -h_0 & h_{-1} & -h_{-2} \\
h_{-1} & -h_{-2} & 0 & 0 & & & & & & 0 & 0 & h_3 & -h_2 & h_1 & -h_0
\end{array}
\right]
$$
$$\tag{8.63}$$

This form of the transform is especially useful for application of symmetric biorthogonal wavelet filters in applications such as image compression.

It is easy to see that the rows in (8.63) are still orthogonal to each other (assuming that W_N from (8.62) is an orthogonal matrix) — what we've effectively done is move the last row from the lowpass (highpass) portion of W_N in (8.62) to the top of the lowpass (highpass) portion of W_N in (8.63).

We learned in Problem 7.23 that if $\mathbf{h} = (h_0, \ldots, h_L)$ solves (7.62), then its reflection $(h_L, h_{L-1}, \ldots, h_1, h_0)$ solves (7.62) as well. Note that none of the solutions in Table 8.1 or 8.2 are reflections of other solutions, yet we are proposing the use of reflections in (8.63). In Problem 8.24 you will show that reflections of Coiflet filters also satisfy the orthogonality condition (8.45) and the derivative conditions (8.44) and see why these reflections didn't appear as solutions in the $K = 1, 2$ cases.

In Computer Lab 8.2 you will revise the software you created for computing one- and two-dimensional wavelet transforms by adding an *offset parameter* that allows you to create transform matrices such as (8.63).

Example: Signal Compression

We conclude this section by returning to Example 7.1.

Example 8.2 (Signal Compression Using C6). *We perform six iterations of the wavelet transform on the 7744-sample audio file from Example 7.1. In Figure 8.3 we have reproduced Figure 7.6 and included in it the cumulative energy for the transform using the C6 filter. Note that the C6 transform is comparable to the D4 transform in this example.* □

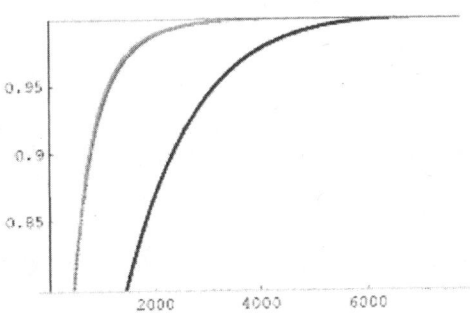

Figure 8.3   Plot of the cumulative energies of the original audio file (black), the D4 transform (gray) and the C6 transform (light gray).

PROBLEMS

8.12   For integers $K \geq 2$, use the definition of the binomial coefficient (4.16) to show that

$$\binom{2K-2}{K-1} = 2\binom{2K-3}{K-2} \tag{8.64}$$

The following steps will help you organize your work.

(a) Show that

$$\binom{2K-2}{K-1} = \frac{K \cdot (K+1) \cdots (2K-2)}{1 \cdot 2 \cdots (K-1)}$$

and

$$\binom{2K-3}{K-2} = \frac{K \cdot (K+1) \cdots (2K-3)}{1 \cdot 2 \cdots (K-2)}$$

(b) Multiply the top and bottom of the second identity in part (a) by $K - 1$ and observe that the denominators for both identities are now the same.

(c) Show that $2(K-1) \cdot K \cdots (2K-3) = K \cdots (2K-2)$ to complete the proof.

★8.13   In this problem you will provide an alternative proof to Lemma 8.1. This identity plays an important role in the proof of Theorems 8.3 and 10.2. The argument is a bit tedious and recommended to those students who are familiar with proof by induction.[5] The following steps will help you organize your work.

(a) Show directly that the result holds when ω is any integer multiple of π. (*Hint:* You might want to consider separately the cases $\omega = 2k\pi$ and $\omega = (2k+1)\pi$, where $k \in \mathbb{Z}$.)

For the remainder of the proof, we assume that ω is not an integer multiple of π.

(b) Show that the result holds for $K = 1$.

The induction part of the proof relies on the identity

$$\sum_{\ell=0}^{m}\binom{n+\ell}{\ell} = \binom{n+m+1}{m} \tag{8.65}$$

You can either prove (8.65) yourself by repeated use Pascal's identity (4.17) or consult an elementary text on combinatorics (e. g., Brualdi [9]).

To better understand the induction step, it is helpful to derive the result for $K = 4$ assuming that it holds for $K = 3$.

(c) Let $c = \cos(\frac{\omega}{2})$ and $s = \sin(\frac{\omega}{2})$. Write down (8.33) for $K = 3$.

[5]This argument was shown to me by Bo Green.

(d) Write down the right-hand side of (8.33) for $K = 4$. Use (8.65) to show that in this case we have

$$c^8 \sum_{j=0}^{3} \sum_{\ell=0}^{j} \binom{2+\ell}{\ell} s^{2j} + s^8 \sum_{j=0}^{3} \sum_{\ell=0}^{j} \binom{2+\ell}{\ell} c^{2j}$$

(e) Consider the first term in part (d). Show that we can rewrite this term as

$$c^8 \sum_{j=0}^{3} \sum_{\ell=0}^{j} \binom{2+\ell}{\ell} s^{2j} = c^8 \sum_{\ell=0}^{3} \binom{2+\ell}{\ell} \sum_{j=\ell}^{3} s^{2j}$$

$$= c^8 \sum_{\ell=0}^{3} \binom{2+\ell}{\ell} s^{2\ell} \sum_{j=0}^{3-\ell} s^{2j}$$

(f) Separate out the $\ell = 3$ term in part (e) to show that

$$c^8 \sum_{j=0}^{3} \sum_{\ell=0}^{j} \binom{2+\ell}{\ell} s^{2j} = \binom{5}{3} c^8 s^6 + c^8 \sum_{\ell=0}^{2} \binom{2+\ell}{\ell} s^{2\ell} \sum_{j=0}^{3-\ell} (s^2)^j$$

(g) Now use the identity

$$1 + t + t^2 + \cdots + t^n = \frac{1 - t^{n+1}}{1 - t}$$

for $t \neq 1$ to show that

$$c^8 \sum_{j=0}^{3} \sum_{\ell=0}^{j} \binom{2+\ell}{\ell} s^{2j} = \binom{5}{3} c^8 s^6 + c^8 \sum_{\ell=0}^{2} \binom{2+\ell}{\ell} s^{2\ell} \frac{1 - s^{8-2\ell}}{1 - s^2}$$

$$= \binom{5}{3} c^8 s^6 + c^6 \sum_{\ell=0}^{2} \binom{2+\ell}{\ell} s^{2\ell} (1 - s^{8-2\ell})$$

$$= \binom{5}{3} c^8 s^6 + c^6 \sum_{\ell=0}^{2} \binom{2+\ell}{\ell} s^{2\ell} - c^6 s^8 \sum_{\ell=0}^{2} \binom{2+\ell}{\ell}$$

Note that since ω is not an odd-integer multiple of π, $1 - s^2 \neq 0$.

(h) Using the ideas from parts (e) – (g) show that the second term in part (d) can be written as

$$s^8 \sum_{j=0}^{3} \sum_{\ell=0}^{j} \binom{2+\ell}{\ell} c^{2j} = \binom{5}{3} s^8 c^6 + s^6 \sum_{\ell=0}^{2} \binom{2+\ell}{\ell} c^{2\ell} - s^6 c^8 \sum_{\ell=0}^{2} \binom{2+\ell}{\ell}$$

(i) Now use (8.65) and part (c) to show that the terms in part (g) and (h) add to 1.

(j) For the general induction step, assume that (8.33) holds and mimic steps (d)-(i) to show that

$$
1 = \cos^{2K+2}(\frac{\omega}{2}) \sum_{j=0}^{K} \binom{K+j}{j} \sin^{2j}(\frac{\omega}{2})
$$
$$
+ \sin^{2K+2}(\frac{\omega}{2}) \sum_{j=0}^{K} \binom{K+j}{j} \cos^{2j}(\frac{\omega}{2})
$$

8.14 Use Lemma 8.1 to verify (8.47).

8.15 Verify the algebra necessary to show that when we insert the filter coefficients given by (8.51) into (8.52), we obtain the system (8.53).

8.16 Find the filter coefficients $\mathbf{h} = (h_{-2}, h_{-1}, h_0, h_1, h_2, h_3)$ for the second pair of a_0, a_1 in (8.54).

8.17 Let $\mathbf{h} = (h_0, h_1, h_2)$ where $h_0 = h_2 = \frac{\sqrt{2}}{4}$ and $h_1 = \frac{\sqrt{2}}{2}$.

(a) Verify that $H(0) = \sqrt{2}$ and that $H(\pi) = H'(\pi) = 0$.

(b) Show that $H(\omega) = \sqrt{2} e^{i\omega} \cos^2(\frac{\omega}{2})$. (*Hint:* Write down $H(\omega)$ and then factor out $e^{i\omega}$.)

(c) Use part (b) to show that $|H(\omega)|^2 = 2\cos^4(\frac{\omega}{2})$.

(d) Use part (c) to show that $|H(\omega + \pi)|^2 = 2\sin^4(\frac{\omega}{2})$. (*Hint:* The identity $\cos(t + \frac{\pi}{2}) = -\sin(t)$ will be useful.)

(e) Use parts (c) and (d) to show that

$$
|H(\omega)|^2 + |H(\omega + \pi)|^2 \neq 2
$$

so that by the converse of Corollary 8.1, $\mathbf{h}$ does not satisfy the orthogonality conditions (8.31).

8.18 In this problem you will solve the system (8.58) algebraically. The following steps will help you organize your work.

(a) The last four equations of (8.58) are linear. Row reduce these linear equations to show that $h_{-2}, \ldots, h_1$ can be expressed in terms of h_2 and h_3. In particular show that

$$
h_{-2} = h_2 \qquad\qquad h_{-1} = \frac{\sqrt{2}}{4} + h_3
$$

$$
h_0 = \frac{\sqrt{2}}{2} - 2h_2 \qquad\qquad h_1 = \frac{\sqrt{2}}{4} - 2h_3
$$

(b) Insert the values for $h_{-2}, \ldots, h_1$ you found in part (a) into the third equation of (8.58) and simplify to show that

$$h_2^2 + \frac{\sqrt{2}}{4}h_3 + h_3^2 = 0 \tag{8.66}$$

(c) Insert the values for $h_{-2}, \ldots, h_1$ you found in part (a) into the second equation of (8.58) and simplify to show that

$$\sqrt{2}h_2^2 - 4h_2^2 + 4h_3^2 = -\frac{1}{8} \tag{8.67}$$

(d) Insert the values for $h_{-2}, \ldots, h_1$ you found in part (a) into the first equation of (8.58) and simplify to show that

$$-2\sqrt{2}h_2 + 6h_2^2 - \frac{\sqrt{2}}{2}h_3 + 6h_3^2 = -\frac{1}{4} \tag{8.68}$$

(e) Combine (8.66) – (8.68) in an appropriate way to show that $h_2 = -(h_3 + \frac{\sqrt{2}}{16})$.

(f) Using parts (a) and (e), write $h_{-2}, \ldots, h_2$ in terms of h_3.

(g) Insert the results from part (f) into the first equation in (8.58) and simplify to obtain

$$12h_3^2 + \frac{9\sqrt{2}}{4}h_3 + \frac{3}{64} = 0 \tag{8.69}$$

(h) Use the quadratic formula to solve (8.69) and show that $h_3 = \frac{\sqrt{2}}{32}(-3 \pm \sqrt{7})$.

(i) Insert the values from part (h) into the identities you obtained in part (f) to find the two solutions listed in Table 8.1. You will need a calculator or a **CAS** to verify your solutions match those given in the table.

8.19   In this problem you will analyze the finite-length filter **h** associated with Fourier series $H(\omega)$ from Theorem 8.3 in the cases $K = 1, 2$.

(a) Let $K = 2$. Use (8.43) and the identities given by (8.46) and (8.48) to write $H(\omega)$ as a linear combination of complex exponentials of the form $e^{ik\omega}$.

(b) Use the result in part (a) to show that the finite filter **h** has starting index $\ell = -4$ and stopping index $L = 7$, so that the length of **h** is 12.

(c) Repeat parts (a) and (b) in the case $K = 3$.

8.20   Prove Proposition 8.4.

8.21 Write down the explicit system (8.59) – (8.60) you must solve to find Coiflet filters in the case $K = 2$.

8.22 Let $K = 1, 2, \ldots$ Write down the explicit system the Coiflet filter $h = (h_{-2K}, h_{-2K+1}, \ldots, h_{4K-2}, h_{4K-1})$ must satisfy. That is, use Corollary 8.1 to write down the general orthogonality conditions and compute the mth derivative of $H(\omega)$, evaluate it at zero and π, and simplify to obtain explicit forms of the derivative conditions (8.44).

8.23 The C6 filter has $H(0) = \sqrt{2}$, $H'(0) = H(\pi) = H'(\pi) = 0$ plus the orthogonality condition (8.3). Suppose that $h = (h_{-2}, h_{-1}, h_0, h_1, h_2, h_3)$ is *any* filter that satisfies the orthogonality condition (8.3).

(a) Can we find h that satisfies

$$H(0) = \sqrt{2} \quad H'(0) = 0 \quad H''(0) = 0 \quad H(\pi) = 0?$$

(b) Why might someone want to use this filter?

8.24 Suppose that $h = (h_{-2K}, \ldots, h_{4K-1})$ is a Coiflet filter whose Fourier transform is given by (8.43). Let $p = (h_{4K-1}, \ldots, h_{-2K})$ be the reflection of h. Show that p satisfies (8.45) and (8.44). (*Hint:* The ideas from Problem 7.9 in Section 7.1 and Problem 7.23 in Section 7.2 will be useful.) After working this problem, you should be able to explain why p for $K = 1, 2$ did not show up in Table 8.1 or 8.2.

Computer Labs

♦ **8.1 Software Development: Generating Coiflet Filters.** From the text Web site, access the development package `coiffilters`. In this development lab you will generate code that will produce Coiflet filters. Your code will incorporate results from Theorem 8.3 as well as the lowpass conditions (8.30) and the conditions (8.32) that flatten $H(\omega)$ at $\omega = 0$. Instructions are provided that allow you to add the code to the software package `DiscreteWavelets`.

♦ **8.2 Generalizing the Wavelet Transform.** From the text Web site, access the lab `wtoffset`. In this lab you will modify the routines you developed in `wt1d` and `wt2d` so that you can also pass an offset parameter. This parameter will allow you to construct wavelet transforms like those described in (8.63).

8.3 Data Compression. From the text Web site, access the lab `coifdatacomp`. In this lab you will use Daubechies filters to perform basic data compression. You will also compare your results with those you obtained using the Haar filter in Computer Lab 6.9 from Section 6.4 and using Daubechies filters in Computer Lab 7.8 from Section 7.3.

CHAPTER 9

WAVELET SHRINKAGE: AN APPLICATION TO DENOISING

We now consider the problem of denoising a digital image or audio sample. We present a method for denoising called *wavelet shrinkage*. This method was developed largely by Stanford statistician David Donoho and his collaborators. In a straightforward argument, Donoho [28] explains why wavelet shrinkage works well for denoising problems, and the advantages and disadvantages of wavelet shrinkage have been discussed by Taswell [76]. Vidakovic [81] has authored a nice book that discusses wavelet shrinkage in detail and also covers several other applications of wavelets in the area of statistics.

Note: The material developed in this section makes heavy use of ideas from statistics. If you are not familiar with concepts such as random variables, distributions, and expected values, you might first want to read Appendix A. You might also wish to work through the problems at the end of each section of the appendix. Two good sources for the material we use in this section are DeGroot and Schervish [27] and Wackerly et al. [82].

For ease of presentation, we develop the wavelet shrinkage method for vectors (signals and audio samples), but it is quite simple to adapt wavelet shrinkage to work on matrices (images) as well. After we have developed the wavelet shrinkage method, we use it to denoise some signals and digital images. Since we can't *hear* the effects of the wavelet shrinkage method on digital audio files, we present these applications in the computer labs, where a **CAS** can be utilized.

In the next section we present a basic overview of wavelet shrinkage and its application to audio and image denoising. In the final two sections of this chapter, we discuss two methods used to denoise images and audio. The VisuShrink method [29] is described in Section 9.2, and the *SureShrink* method [30] is developed in Section 9.3. We discuss some variations of both VisuShrink and SureShrink in the Problems.

9.1 AN OVERVIEW OF WAVELET SHRINKAGE

The simple process of taking a picture can create noise. Scanning a picture can produce a digital image where noise is present. Recording a voice or instrument to an audio file often produces noise. Noise can also be incurred when we transmit a digital file.

Let's suppose that the true signal is stored in an N-vector $\mathbf{v}$. In practice, we never know $\mathbf{v}$. Through acquisition or transmission, we create the noisy N-vector

$$\mathbf{y} = \mathbf{v} + \mathbf{e}$$

Gaussian White Noise

The N-vector $\mathbf{e} = (e_1, e_2, \ldots, e_N)$ is the noise vector. We assume that the entries e_k are *Gaussian white noise*. That is, you can think of the e_k as *independent samples* that are *normally distributed* with mean zero and variance σ^2. See Definition A.10 in Section A.3 for a definition of an independent random variable and Definition A.14 in Section A.5 for a definition of the normal distribution.

The mean is zero since we would expect the noisy components y_k, $k = 1, \ldots, N$, of $\mathbf{y}$ to both underestimate and overestimate the true values. The variance σ^2 estimates how much the data are spread out. For this reason we also refer to σ as the *noise level*. Of course, σ is usually unknown. In some variations of wavelet shrinkage, we will need to estimate σ.

Examples of Noisy Signals and Images

For an example of a noisy signal, consider the N-vector $\mathbf{v}$ formed by uniformly sampling the *heavisine function* (see Donoho and Johnstone [29]) $f(t) = 4\sin(4\pi t) -$

$\text{sgn}(t - .3) - \text{sgn}(.72 - t)$ on the interval $[0, 1]$. The *sign function*, denoted $\text{sgn}(t)$, is defined in (9.1). It is simply a function that indicates the sign of the given input.

$$\text{sgn}(t) = \begin{cases} 1, & t > 0 \\ 0, & t = 0 \\ -1, & t < 0 \end{cases} \tag{9.1}$$

Since the sgn function has a jump discontinuity at $t = 0$, it is often used to introduce jump discontinuities in a function. The function $f(t)$ has jumps at .3 and at .72.

The vector **v** and a noisy version **y** are plotted in Figure 9.1. Our task is to produce an N-vector $\hat{\mathbf{v}}$ that does a good job estimating **v**. A noisy version V of image A is plotted in Figure 9.2.

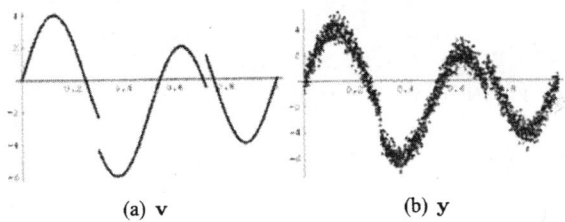

(a) **v** (b) **y**

Figure 9.1 The heavisine function is uniformly sampled and stored in **v**. The vector **y** is formed by adding noise to **v** with $\sigma = .5$.

The Wavelet Shrinkage Algorithm

The wavelet shrinkage method for denoising can be outlined as follows:

Algorithm 9.1 (Wavelet Shrinkage Algorithm). *This algorithm takes a noisy N-vector $\mathbf{y} = \mathbf{v} + \mathbf{e}$, where **v** is the true (unknown) signal and **e** is the noise vector, and returns the (denoised) estimate $\hat{\mathbf{v}}$ of **v**.*

1. *Compute i iterations of the wavelet transformation on **y** to obtain the transformed vector **z**. Call the lowpass (approximation) portion ℓ and the highpass (details) portions **d**.*

2. *Apply a threshold rule (described below) to the highpass portions **d** of **z**. The rule will either "shrink" or set to zero values in **d**. Denote the modified highpass portions by $\hat{\mathbf{d}}$.*

3. *Rejoin these modified highpass portions with the original lowpass portion of **z** to form a modified transform vector $\hat{\mathbf{z}} = [\ell \mid \hat{\mathbf{d}}]^T$.*

4. *Compute i iterations of the inverse wavelet transformation of $\hat{\mathbf{z}}$ to obtain $\hat{\mathbf{v}}$.*

□

(a) A

(b) V

Figure 9.2   An image A and a noisy version V. For this example, we use noise level $\sigma = 18$.

Measuring the Effectiveness of the Denoising Method

How do we decide if the shrinkage method does a good job denoising the signal? The shrinkage method, in particular the threshold rule, is constructed so that the estimator $\hat{\mathbf{v}}$ of $\mathbf{v}$ is optimal in the sense of the *mean squared error*.

Suppose that $\mathbf{v} = (v_1, \ldots, v_N)$ and $\hat{\mathbf{v}} = (\hat{v}_1, \ldots, \hat{v}_N)$. Statisticians define the mean squared error as the *expected value* of the quantity

$$\sum_{k=1}^{N}(v_k - \hat{v}_k)^2 \qquad (9.2)$$

The quantity in (9.2) is nothing more than the square of the norm of $\mathbf{v} - \hat{\mathbf{v}}$, so we can write the mean squared error as

$$\text{mean squared error} = E(\|\mathbf{v} - \hat{\mathbf{v}}\|^2) \qquad (9.3)$$

Here, $E(\cdot)$ is the notation we use for the expected value. See Definition A.11 in Section A.4 of Appendix A for a definition of the expected value of a random variable.

The closer $\mathbf{v}$ is to $\hat{\mathbf{v}}$, the lower the expected value. The expected value can then be interpreted as the mean value of the summands $(v_k - \hat{v}_k)^2$. The threshold rule we describe below attempts to minimize this average value.

Threshold Rules

There are two types of threshold rules. The simplest rule is a *hard threshold rule*. To apply a hard threshold rule to vector $\mathbf{z}$, we first choose a tolerance $\lambda > 0$. We retain z_k if $|z_k| > \lambda$ and we set $z_k = 0$ if $|z_k| \leq \lambda$. Note that we used the hard threshold rule in Example 7.2 in Section 7.1. When we performed image compression, we applied the hard threshold rule to the cumulative energy vector of the image with the threshold set to $\lambda = 0.995$. Although denoising is certainly an application different than compression, we can see in Figure 7.10 that the compressed images either exhibit ringing or are somewhat blocky. To avoid this problem we use a *soft threshold rule*.

The Soft Threshold Function

To apply the soft threshold rule to vector $\mathbf{z}$, we first choose a tolerance $\lambda > 0$. Like the hard threshold rule, we set to zero any element z_k where $|z_k| \leq \lambda$. The soft threshold rule differs from the hard threshold rule in the case where $|z_k| > \lambda$. Here we *shrink* the value z_k so that its value is λ units closer to zero.

Recall that $|z_k| > \lambda$ means that either $z_k > \lambda$ or $z_k < -\lambda$. If $z_k > \lambda$, we replace z_k with $z_k - \lambda$. If $z_k < -\lambda$, then we replace z_k with $z_k + \lambda$ to bring it λ units closer to zero.

We can easily write down the function $s_\lambda(t)$ that describes the soft threshold rule:

$$s_\lambda(t) = \begin{cases} t - \lambda, & t > \lambda \\ 0, & -\lambda \leq t \leq \lambda \\ t + \lambda, & t < -\lambda \end{cases} \qquad (9.4)$$

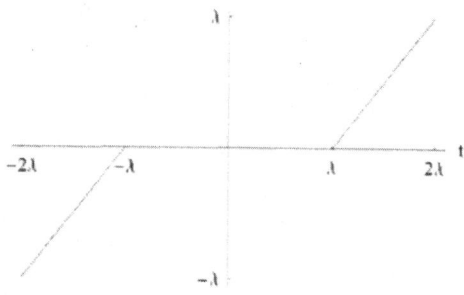

Figure 9.3   The soft threshold rule function $s_\lambda(t)$.

The soft threshold function $s_\lambda(t)$ is plotted in Figure 9.3.

Note that $s_\lambda(t)$ is a piecewise linear function. The piece between $-\lambda$ and λ is the zero function. The other two pieces are lines with slope 1 (so that the value of t is not expanded or contracted) that have been vertically shifted so that they are λ units closer to the t-axis.

Example 9.1 (Soft Threshold $s_\lambda(t)$). *Suppose that $\lambda = 2$ and we apply the soft threshold rule (9.4) to* $z = [1, 3.5, -1.25, 6, -3.6, 1.8, 2.4, -3.5]$. *The resulting vector is* $w = [0, 1.5, 0, 4, -1.6, 0, .4, -1.5]$ *and both z and w are plotted in Figure 9.4. Note that all the black dots between -2 and 2 are set to zero, while those black dots larger than 2 and smaller than -2 are shrunk so that they are 2 units closer to the t-axis.* □

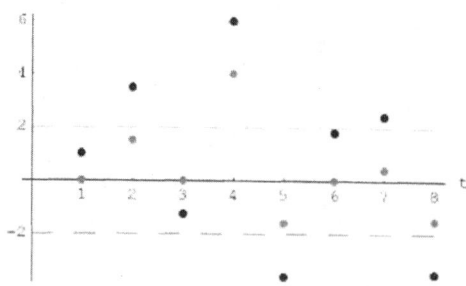

Figure 9.4   The vector z (black). Application of the soft threshold rule (9.4) to z gives w (gray).

Why Does Wavelet Shrinkage Work?

Before we describe two methods for choosing the tolerance λ, let's try to better understand why wavelet shrinkage works. Suppose that we perform one iteration of a wavelet transformation to a noisy vector $\mathbf{y} = \mathbf{v} + \mathbf{e}$. We obtain

$$\mathbf{z} = W_N \mathbf{y} = W_N(\mathbf{v} + \mathbf{e}) = W_N \mathbf{v} + W_N \mathbf{e} \tag{9.5}$$

Recall the basic structure of $W_N \mathbf{v}$. This vector contains the lowpass portion of the transformation in the first half of the vector and the highpass portion of the transformation in the second half of the vector. As we have seen throughout the book, most of the energy in $\mathbf{z}$ is stored in the lowpass (approximation) portion ℓ, while the highpass (details) portion $\mathbf{d}$ is sparse by comparison. What about $W_N \mathbf{e}$? You will show in Problem 9.1 that application of an orthogonal matrix W_N to a Gaussian white noise vector $\mathbf{e}$ returns a Gaussian white noise vector. In particular, the variance of the resultant is unchanged, so the noise level for $W_N \mathbf{e}$ is still σ! As Donoho points out in [28], since $\mathbf{d}$ is sparse, the highpass portion of $W_N \mathbf{y}$ is comprised primarily of noise. So application of the soft threshold rule (9.4) to the elements in the highpass portion of $W_N \mathbf{y}$ should serve to denoise the signal.

The final thing we need to discuss is how to choose the tolerance λ. There are several methods for choosing a tolerance for the soft threshold rule (9.4), and in the following sections, we discuss two methods for choosing λ.

PROBLEMS

9.1 Let $\mathbf{e} = [e_1, \ldots, e_N]^T$ be a Gaussian white noise vector. That is, for $k = 1, \ldots, N$, the entries e_k are independent and normally distributed with mean $E(e_k) = 0$ and variance $E(e_k^2) = \sigma^2$. In this problem, you will show that if W is an $N \times N$ orthogonal matrix, then the entries of y_j, $j = 1, \ldots, N$, of $\mathbf{y} = W\mathbf{e}$ will each have mean zero and variance σ^2 as well.

The following steps will help you organize your proof.

(a) Let w_{jk} denote the j,kth element of W, $j, k = 1, \ldots, N$. For a fixed $j = 1, \ldots, N$, write y_j in terms of w_{jk} and e_k, $k = 1, \ldots, N$. Now take the expectation of each side and utilize part (d) of Proposition A.3 from Appendix A.

(b) From part (a) we know that $\text{Var}(y_j) = E(y_j^2) = E\left(\left(\sum_{k=1}^{N} w_{jk} e_k\right)^2\right)$. Use the fact that for numbers $a_1, \ldots, a_N$ and $b_1, \ldots, b_N$, we have

$$\left(\sum_{k=1}^{N} a_k\right) \cdot \left(\sum_{m=1}^{N} b_m\right) = \sum_{k=1}^{N} \sum_{m=1}^{N} a_k b_m$$

to write this expectation as a double sum and then use part (d) of Proposition A.3 to pass the expectation through the double sum.

(c) Use the fact that e_k and e_m are independent for $k \neq m$ to reduce the double sum of part (b) to a single sum.

(d) Use what you know about orthogonal matrices and the fact that $E(e_k^2) = \sigma^2$ to show that $E(y_j^2) = \sigma^2$.

9.2   Consider the vector $\mathbf{v} = [-3, 2.6, .35, -5, -.3, 5, 3, 1.6, 6.2, .5, -.9, 0]^T$.

(a) Apply the soft threshold function (9.4) to $\mathbf{v}$ using $\lambda = 1.5$.

(b) Repeat part (a) but use $\lambda = 3.5$.

9.3   Compute $\lim\limits_{\lambda \to 0} s_\lambda(t)$ and $\lim\limits_{\lambda \to \infty} s_\lambda(t)$ and interpret the results in terms of the wavelet shrinkage process.

9.4   In this problem we develop an alternative formula for the soft threshold rule (9.4). Define the function

$$t_+ = \max\{0, t\}$$

(a) Graph t_+, $(t - \lambda)_+$, and $(|t| - \lambda)_+$ for some $\lambda > 0$.

(b) Observe that $(|t| - \lambda)_+ = s_\lambda(t)$ whenever $t \geq -\lambda$. What's wrong with the formula when $t < -\lambda$?

(c) Show that $s_\lambda(t) = \mathrm{sgn}(t)(|t| - \lambda)_+$.

The formula in (c) is useful for coding $s_\lambda(t)$ in a **CAS**.

9.5   Hard thresholding was discussed in this section. To apply hard thresholding with tolerance $\lambda > 0$ to elements of vector $\mathbf{z}$, we simply replace z_k with zero whenever $|z_k| \leq \lambda$ and leave z_k unchanged otherwise. There are instances (see Yoon and Vaidyanathan [87]) when hard thresholding yields desirable results.

(a) Using (9.4) as a model, write a piecewise-defined function $f_\lambda(t)$ that performs hard thresholding on input t.

(b) Write $f_\lambda(t)$ in terms of the shrinkage function $s_\lambda(t)$ and $\mathrm{sgn}(t)$.

(c) Sketch a graph of $f_\lambda(t)$ for $\lambda = 1, 1.5, 2$.

9.6   Repeat Problem 9.2 but use the hard thresholding function developed in Problem 9.5.

Computer Labs

◆ **9.1**   **Software Development: Wavelet Shrinkage.** From the text Web site, access the development package `waveletshrinkage`. In this development lab you will create a module that takes as input a vector or matrix, a wavelet filter **h**, the number of iterations i for the wavelet transform, and a tolerance λ, and returns a denoised version of the input vector or matrix. As part of the wavelet shrinkage algorithm, you will also write a module for the soft threshold function $s_\lambda(t)$ (9.4). Instructions are provided that allow you to add the code to the software package DiscreteWavelets.

9.2   Wavelet Shrinkage. From the text Web site, access the lab shrinkageintro. In this lab you will learn how to generate Gaussian white noise and add it to signals and images. You will then use the wavelet shrinkage module to experiment with denoising various test signals, audio signals, and digital images.

9.2 VISUSHRINK

VisuShrink utilizes the wavelet shrinkage algorithm (Algorithm 9.1) with a *universal threshold* λ^{univ}. To motivate the theorem of Donoho and Johnstone [29], suppose that we have observed

$$\mathbf{y} = \mathbf{v} + \mathbf{e}$$

Suppose further that we decide to estimate $\mathbf{v}$ with $\hat{\mathbf{w}} = (\hat{w}_1, \ldots, \hat{w}_N)$ by keeping certain values of y_k and setting other values $y_k = 0$. Let A be the set of indices for which we retain the values y_k. We have

$$\hat{w}_k = \begin{cases} y_k, & k \in A \\ 0, & k \notin A \end{cases}$$

We can compute the expected value $E(\|\hat{\mathbf{w}} - \mathbf{v}\|^2)$. We know (see Proposition A.3 in Section A.4 of Appendix A) that

$$E\left(\|\hat{\mathbf{w}} - \mathbf{v}\|^2\right) = E\left(\sum_{k=1}^{N}(\hat{w}_k - v_k)^2\right) = \sum_{k=1}^{N} E\left((\hat{w}_k - v_k)^2\right)$$

Now if $k \in A$, we have

$$E\left((\hat{w}_k - v_k)^2\right) = E\left((y_k - v_k)^2\right) = E\left((v_k + e_k - v_k)^2\right) = E(e_k^2)$$

Recall that the elements e_k are normally distributed with mean zero and variance σ^2. From Theorem A.1 we know that if we define random variable $Z = e_k/\sigma$, then Z^2 has a χ^2 distribution with 1 degree of freedom. Proposition A.7 then tells us (with $n = 1$) that

$$1 = E(Z^2) = E\left(\frac{e_k^2}{\sigma^2}\right) = \frac{1}{\sigma^2} E(e_k^2)$$

or

$$E(e_k^2) = \sigma^2 \tag{9.6}$$

Now if $k \notin A$, we have $\hat{w}_k = 0$, so that

$$E\big((\hat{w}_k - v_k)^2\big) = E(v_k^2) = v_k^2 \tag{9.7}$$

We can use (9.6) and (9.7) to define the function

$$f(k) = \begin{cases} \sigma^2, & k \in A \\ v_k^2, & k \notin A \end{cases}$$

and write the expected value

$$E\big(\|\hat{w} - v\|^2\big) = \sum_{k=1}^{N} f(k)$$

Now ideally, we would form A by choosing all the indices k so that $v_k^2 > \sigma^2$. The reason we would do this is that if $v_k^2 \leq \sigma^2$, then the true value v_k is below (in absolute value) the noise level and we aren't going to be able to recover it. Of course such an ideal rule is impossible to implement since we don't know the values v_k, but it serves as a good theoretical measure for our threshold rule. If we do form A in this manner, then our *ideal mean squared error* is

$$E(\|\hat{w} - v\|^2) = \sum_{k=1}^{N} \min(v_k^2, \sigma^2) \tag{9.8}$$

To better understand the ideal mean squared error, let's think about how we will use it. We will apply our threshold rule to the highpass portion of one iteration of the wavelet transform of y. We have

$$z = W_N y = W_N(v + e) = W_N v + W_N e$$

Let h denote the highpass portion of $W_N v$. Remember that $W_N e$ is again Gaussian white noise, so let's denote the components of e corresponding to h as e_h. Then we will apply our threshold rule to the vector $h + e_h$.

We expect that most of the components of h are zero so that most of the components of $h + e_h$ will be at or below the noise level σ. In such a case, a large number of the summands of the ideal mean squared error (9.8) will be σ^2.

An Accuracy Result for Wavelet Shrinkage

The entire point of the discussion thus far is to provide motivation for the result we now state.

Theorem 9.1 (Measuring the Error: Donoho and Johnstone [29]). *Suppose that* u *is any vector in* $\mathbb{R}^N$ *and that we are given the data* $y = u + e$ *where* e *is Gaussian* *white noise with noise level σ. Let $s_\lambda(t)$ be the soft threshold rule (9.4) with $\lambda = \sigma\sqrt{2\ln(N)}$. Let $\hat{u}$ be the vector formed by applying $s_\lambda(t)$ to $\mathbf{u}$. Then*

$$E\big(\|\hat{u} - \mathbf{u}\|^2\big) \le (2\ln(N) + 1)\left(\sigma^2 + \sum_{k=1}^{N} \min(u_k^2, \sigma^2)\right)$$

$\square$

Proof of Theorem 9.1. The proof requires some advanced ideas from analysis and statistics and is thus omitted. The interested reader is encouraged to see the proof given by Vidakovic ([81], Theorem 6.3.1). $\square$

Definition 9.1 (Universal Threshold). *The tolerance*

$$\boxed{\lambda = \sigma\sqrt{2\ln(N)}} \tag{9.9}$$

given in Theorem 9.1 is called the universal threshold *and hereafter denoted by* λ^{univ}. $\square$

It is important to understand what Theorem 9.1 is telling us. The choice of $\lambda^{univ} = \sigma\sqrt{2\ln(N)}$ gives us something of a *minimax* solution to the problem of minimizing the ideal mean squared error. For this λ^{univ} and *any* choice of $\mathbf{u}$, the soft threshold rule (9.4) produces a mean squared error that is always smaller than a constant times the noise level squared plus the ideal mean squared error. It is worth noting that the constant $2\ln(N) + 1$ is quite small relative to N (see Table 9.1).

Table 9.1 The values of $2\ln(N) + 1$ for various values of N.

N	$2\ln(N) + 1$
100	10.21
1,000	14.82
10,000	19.42
100,000	24.03

Estimating the Noise Level

Note that the universal threshold λ^{univ} given in Definition (9.1) depends on the size of the vector N and the noise level σ. In practice, we typically do not know σ so we have to estimate it.

Hampel [42], has shown that the *median absolute deviation* (*MAD*) converges to 0.6745σ as the sample size goes to infinity. MAD is defined in Definition A.4 in Section A.1 in Appendix A.

Donoho and Johnstone [29] suggest that we should estimate σ using the highpass elements in the first iteration of the wavelet transform, since this portion of the transform is predominantly noise. If we let $\mathbf{h}^1$ denote the highpass elements of the first iteration of the wavelet transformation, then we have the following estimator for σ:

$$\hat{\sigma} = MAD(\mathbf{h}^{(1)})/0.6745 \tag{9.10}$$

VisuShrink Examples

Next we look at three examples that implement VisuShrink. The first example illustrates the method on the test signal plotted in Figure 9.1. This example is somewhat artificial since we set the noise level when we form the noisy vector. We do, however, use (9.10) to estimate σ. The second example illustrates the application of VisuShrink to denoise a real signal. The final example shows how VisuShrink works when we use it to denoise a digital image.

Example 9.2 (1D VisuShrink). *We consider the one-dimensional signal and the noisy version plotted in Figure 9.1. To obtain the original signal, we set $N = 2048$ and uniformly sample the function $f(t) = 4\sin(4\pi t) - sgn(t - 0.3) - sgn(0.72 - t)$ on the interval $[0, 1)$. Thus, the components of our signal $\mathbf{v}$ are $v_k = f(k/2048)$, $k = 0, \ldots, 2047$. We set the noise level to $\sigma = .5$ and then create the Gaussian white noise vector $\mathbf{e}$ by generating 2048 random normal samples. We add this noise vector to the original to get the noisy signal $\mathbf{y}$.*

We use the D6 wavelet filter and perform 5 iterations of the wavelet transform on the noisy signal $\mathbf{y}$. After computing five iterations of the wavelet transform, we construct the estimate $\hat{\sigma}$ using (9.10). Recall that we take the highpass portion from the first iteration of the transformation and compute its median absolute deviation. We then divide this number by 0.6745 to obtain

$$\hat{\sigma} = 0.474412$$

Now the length of the lowpass portion of the transform is $2048/2^5 = 64$, so the length of the highpass portion of the transform is $2048 - 64 = 1984$. Using the formula (9.9) for λ^{univ} from Definition 9.1, we find our VisuShrink tolerance to be

$$\lambda^{univ} = 0.474412\sqrt{2\ln(1984)} \approx 1.84873$$

We have plotted the highpass portions of the D6 wavelet transform $\mathbf{z}$ in Figure 9.5(a). The horizontal lines in the figure are at heights $\pm\lambda^{univ}$. Note that most of the highpass elements fall between these two lines, so we are shrinking only a few of these values.

We apply the soft threshold rule (9.4) with $\lambda^{univ} = 1.84873$ to obtain a modified highpass portion of the wavelet transform. We join this new highpass portion to the original lowpass portion of **z** *to obtain the modified wavelet transformation* $\hat{\mathbf{z}}$. *We have plotted (a very sparse!) the highpass portion of* $\hat{\mathbf{z}}$ *in Figure 9.5(b).*

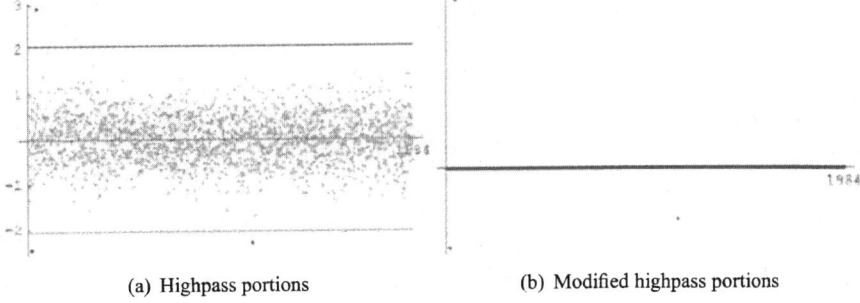

(a) Highpass portions   (b) Modified highpass portions

Figure 9.5 The highpass portions of five iterations of the wavelet transformation of **y**. The light gray values are those that will be set to zero by the soft threshold rule. The black values will be shrunk so that they are λ units closer to zero. The highpass portions of the modified wavelet transformation $\hat{\mathbf{z}}$ after application of the soft threshold rule (9.4) with $\lambda = 1.84873$.

Finally, we compute five iterations of the inverse wavelet transformation to arrive at our denoised signal. It is plotted in Figure 9.6.

It is interesting to observe what happens if we change filters. If we redo the example using the D4 filter, we obtain $\sigma = 0.470484$ *and* $\lambda^{univ} = 1.83342$.

We apply the soft threshold rule (9.4) to the wavelet transform **z** *(again performing five iterations) to obtain the modified wavelet transform* $\hat{\mathbf{z}}$. *Finally, we perform 5 iterations of the inverse wavelet transform to obtain the denoised signal* $\hat{\mathbf{v}}$. *We have plotted* $\hat{\mathbf{v}}$ *in Figure 9.6[1]. In Computer Lab 9.4, you will continue to investigate VisuShrink using different filters and different test signals.*

$\square$

Our next example utilizes a "real" data set.

Example 9.3 (Denoising the Speed Readings of a Sailboat). *The real data in this example come from research done in the College of Engineering at the University of St. Thomas. Faculty members Christopher Greene, Michael Hennessey, and Jeff Jalkio and student Colin Sullivan were interested in determining the optimal path of a sailboat given factors such as changing wind speed, boat type, sail configuration,*

---

[1]Compare the denoised signals in Figure 9.6. The signal on the left appears a bit smoother than the one on the right. It is no coincidence that the longer filter produces a smoother approximation. In the classical derivation of wavelet theory (see, e. g., Walnut [83]), these filters are derived from special functions called *scaling functions*. The smoothness of these functions is directly proportional to the filter length associated with them.

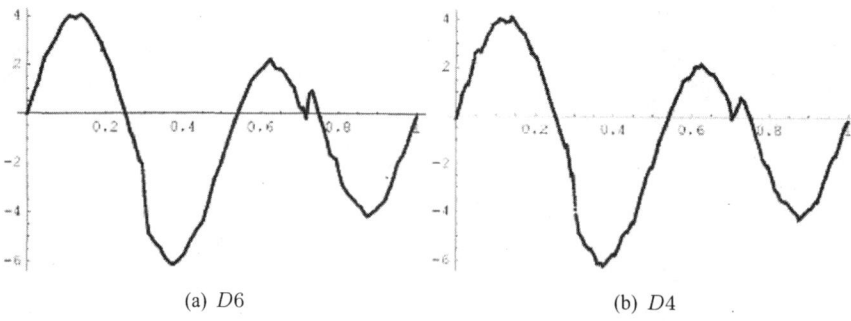

(a) $D6$ (b) $D4$

Figure 9.6 The denoised signal using D6 and D4 filters.

and current. They collected data from June 28 through July 1, 2004 by sailing preset courses on Lake Superior near Bayfield, Wisconsin. They gathered data from multiple sensors and then used these data in conjunction with ideas from calculus of variations to model the problem of determining the optimal course. Their results may be found in Hennessey et. al. [44].

The data plotted in Figure 9.7 are boat speeds measured during one of their course runs. The speed was measured in knots (1 knot ≈ 1.15 mph) once per second. The total number of observations is 4304, so the experiment was conducted for 71 minutes and 44 seconds. The maximum speed measured during the experiment was 7.7 knots and the minimum speed was 1.1 knots.

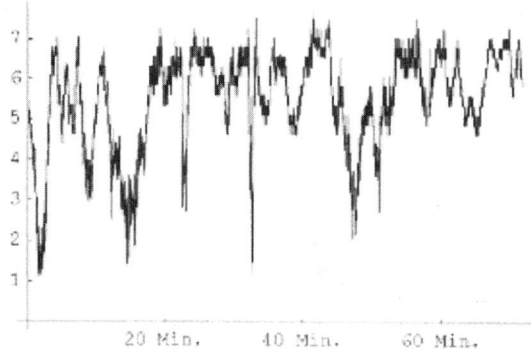

Figure 9.7 The boat speeds (recorded in knots at one second intervals).

Several factors can contribute to noise when attempting to record the boat speed. Factors such as sail trim, wind speed, steering, sail configuration, waves, and current can all affect the sensor's ability to record the boat speed accurately. It is important to point out that the noise we observe in this experiment cannot be classified as Gaussian white noise. Thus,, we cannot quantify how well wavelet shrinkage works on this example.

We first compute four iterations (as many iterations as the data allow) of the wavelet transform using the Coiflet C6 filter. We then take the highpass portion of the first iteration of the transform and use (9.10) to find an estimate for the noise level σ. We have

$$\hat{\sigma} = 0.169374 \qquad (9.11)$$

We can use (9.11) and the highpass portions of the wavelet transform to find the universal threshold λ^{univ}. We know that the size of the lowpass part of the transform is $4304/2^4 = 269$, so the number of highpass elements is $4304 - 269 = 4035$. We have

$$\lambda^{univ} = 0.169374\sqrt{2\ln(4035)} \approx 0.690197$$

Now we apply the soft threshold rule (9.4) with $\lambda^{univ} = 0.717013$ to the highpass portions of the wavelet transform to produce a modified highpass portion. The wavelet transform and the modified wavelet transform are plotted in Figure 9.8.

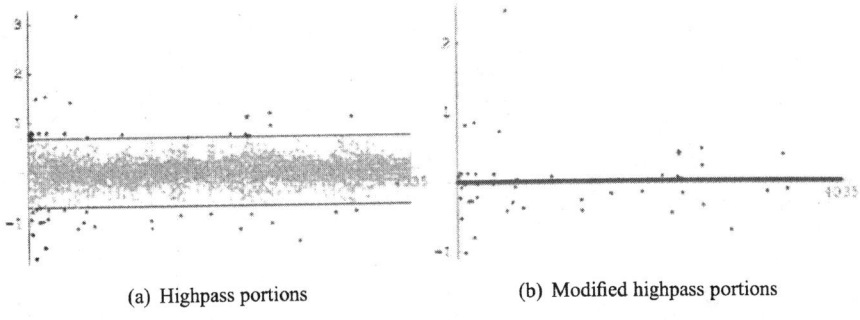

(a) Highpass portions   (b) Modified highpass portions

Figure 9.8 The highpass portions of the wavelet and modified wavelet transforms after application of the soft threshold rule.

Finally, we apply four iterations of the inverse wavelet transform to arrive at the denoised boat speeds. These values are plotted in Figure 9.9. □

Our final example deals with image denoising and illustrates a shortcoming of the VisuShrink method.

Example 9.4 (2D VisuShrink). *Let's consider the 256×384 image from Figure 9.2. The figure shows the original image Figure 9.2(a) and the noisy image Figure 9.2(b) with noise level $\sigma = 18$. We first perform two iterations of the wavelet transform using the D6 filter. To estimate σ, we take the horizontal, vertical, and diagonal components from the first iteration of the transformation (each 128×192) and concatenate all the rows to form vector $\mathbf{h}^1$ of size $3 \cdot 128 \cdot 192 = 294{,}912$. We have*

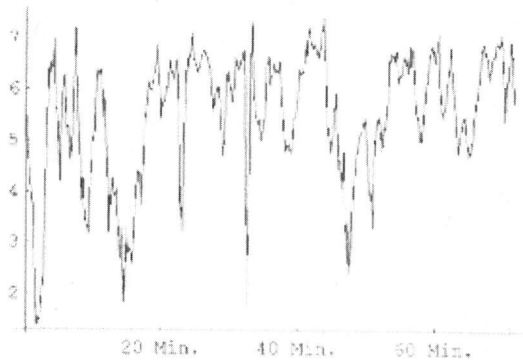

Figure 9.9   The denoised boat speeds.

$$\hat{\sigma} = MAD(\mathbf{h}^1)/0.6745 = 21.3035$$

We can use this value of $\hat{\sigma}$ to compute λ. Since we compute two iterations of the wavelet transform, the blur submatrix is of size $64 \cdot 96 = 6144$. Thus, the total number of components in the highpass portions of the transform is $N = 294{,}912 - 6144 = 288{,}768$. The value for λ^{univ} is

$$\lambda^{univ} = 21.3035\sqrt{2\ln(49{,}152)} \approx 101.862$$

We can now apply the soft threshold rule (9.4) with $\lambda = 101.862$ to the highpass portions of the wavelet transform. In Figure 9.10 we have plotted both the wavelet transform and the modified wavelet transform.

Finally, we perform two iterations of the inverse wavelet transform to obtain the denoised image. We plot the denoised image in Figure 9.11.   □

The denoised image in Figure 9.11 is not a good estimate of the original image shown in Figure 9.2. Several researchers (see, e. g. Fan et al. [34]) have pointed out that the VisuShrink technique tends to produce a large value for λ. We can easily ascertain from (9.9) that λ grows as the size of the data grows. Certainly, a large value for λ will remove the noise from the input signal or image, but it also sets to zero important parts of the highpass portion of the transform.

For Example 9.4, the value $\lambda = 101.862$ is quite high considering that the pixel values of the image range from 0 to 255. We can also see in Figure 9.10 that almost all of the highpass features have been deleted in the modified transform.

The next method we present for choosing λ is not data dependent. The procedure, called *SureShrink*, was developed by Donoho and Johnstone [30].

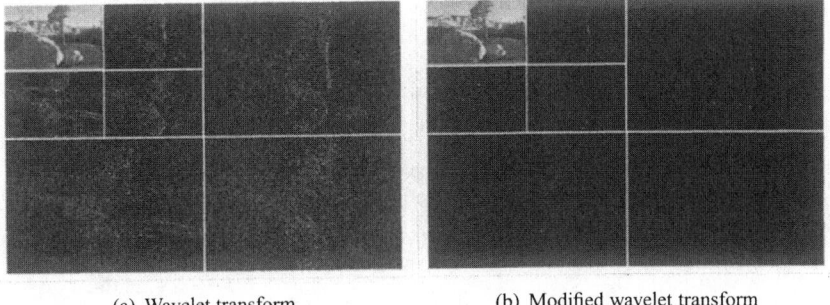

(a) Wavelet transform	(b) Modified wavelet transform

Figure 9.10 Two iterations of the wavelet transform using the D6 filter applied to the original image. The image in part (a) is the modified wavelet transform after application of the soft threshold rule (9.4) with $\lambda = 101.862$.

Figure 9.11 The denoised image using VisuShrink. Compare this image to those plotted in Figure 9.2.

PROBLEMS

9.7 Suppose that we observe $\mathbf{y} = \mathbf{v} + \mathbf{e}$, where the length of $\mathbf{y}$ is N, $\mathbf{v}$ is the true signal, and $\mathbf{e}$ is Gaussian white noise with noise level $\sigma > 0$. Further, suppose that $\|\mathbf{v}\| = 1$. Let n_σ denote the number of elements in $\mathbf{v}$ such that $|v_k| > \sigma$, $k = 1, \ldots, N$. Show that the ideal mean squared error given in (9.8) satisfies

$$\sum_{k=1}^{N} \min(v_k^2, \sigma^2) = \sigma^2 n_\sigma + (1 - C_e(\mathbf{v})_{n_\sigma})$$

where $C_e(\mathbf{v})$ is the cumulative energy vector for $\mathbf{v}$ (see Definition 3.2). In this way we see, as Donoho points out [28], that the ideal mean squared error "is explicitly a measure of the extent to which the energy is compressed into a few big coefficients" of $\mathbf{v}$.

9.8   Let $\mathbf{v} = [0, 0, 0, 1, 1, 4, 8, 1, 1, 0, 1, 2]^T$. (You can think of $\mathbf{v}$ as the true high-pass portion of wavelet-transformed data.) We use a **CAS** to add Gaussian white noise with $\sigma = 1.5$ and obtain the noisy data

$$\mathbf{y} \;=\; [-0.340713, 0.0903748, -1.65198, -1.00401, 0.270618, 2.21331,$$
$$9.88094, 0.689211, 0.995415, -0.642826, 1.83169, 1.20736]^T$$

(a) Using (9.10) and a **CAS** or a calculator, compute the estimated noise level $\hat{\sigma}$.

(b) Use $\hat{\sigma}$ from part (a) to compute λ^{univ} for this data set.

9.9   The universal threshold λ^{univ} is data dependent and it is possible that for large enough vector length N, the choice of λ^{univ} will result in all highpass values being shrunk to zero.

Suppose that we are given a noisy vector $\mathbf{y} = (y_1, \ldots, y_N)$ with $|y_k| \leq M$ for some $M > 0$ and each $k = 1, \ldots, N$. Suppose that $\mathbf{z}$ is the result of applying one iteration of the wavelet tranform using the Haar filter $\mathbf{h} = (h_0, h_1) = (\frac{\sqrt{2}}{2}, \frac{\sqrt{2}}{2})$. Let $\mathbf{d} = (d_1, \ldots, d_{N/2})$ denote the highpass portion of $\mathbf{z}$.

(a) Find a value L so that $|d_k| \leq L, k = 1, \ldots, N/2$.

(b) Suppose that we have computed the noise level estimate $\hat{\sigma}$. How large must N be so that $\lambda^{univ} \geq L$? In this case, we literally discard the entire highpass portion of the transformation!

(c) Find N in the case where $M = 10$ and $\hat{\sigma} = 3$ so that λ^{univ} uses only lowpass values to denoise the signal.

(d) Suppose that an $N \times N$ noisy image A satisfies $0 \leq a_{jk} \leq 300$, $j, k = 1, \ldots, N$, with $\hat{\sigma} = 15$. Find N so that that λ^{univ} uses only lowpass values to denoise the image.

Computer Labs

◆ **9.3**   **Software Development: VisuShrink.** From the text Web site, access the development package `visushrink`. In this development lab you will write a module that takes as input a wavelet-transformed vector or matrix and the number of iterations performed, and returns the universal threshold λ^{univ}. Instructions are provided that allow you to add the code to the software package `DiscreteWavelets`.

9.4   VisuShrink. From the text Web site, access the lab `visushrink`. In this lab you will use the wavelet shrinkage module with λ^{univ} to apply the VisuShrink method to various test signals, audio signals, and digital images using different orthogonal filters. You will also write a module to compute the NormalShrink tolerance and compare the results with those you obtain using the VisuShrink tolerance λ^{univ}.

9.3   SURESHRINK

The SureShrink method utilizes Stein's unbiased risk estimator (SURE) [68]. Donoho and Johnstone [30] suggest minimizing this estimator to find what they call λ^{sure}. Unlike λ^{univ}, the tolerance λ^{sure} does not depend on the size of the input vector. The SureShrink method also calls for a tolerance to be computed for *each* highpass portion of the wavelet transform. A test is performed on each highpass portion to determine its sparseness. If a highpass portion is deemed sparse, we compute λ^{univ} for that portion and use it in conjunction with the soft threshold rule (9.4). If the highpass portion is determined not to be sparse, we compute λ^{sure} for the portion and use it with (9.4).

Before we develop the SureShrink method, we need to discuss some ideas about vector-valued functions that we'll need to develop λ^{sure}. If you have had a course in multivariable calculus, then you are familiar with functions that map vectors to vectors. We will only be concerned with functions that map $\mathbb{R}^N$ to $\mathbb{R}^N$. Let $\mathbf{x} \in \mathbb{R}^N$. We denote such functions by

$$\mathbf{f}(\mathbf{x}) = (f_1(\mathbf{x}), \dots, f_N(\mathbf{x}))$$

where the functions $f_1, \dots, f_N$ that map $\mathbb{R}^N$ to $\mathbb{R}$ are called *coordinate functions*.

We have actually been using vector-valued functions throughout the book without actually calling them by name. In Problem 9.10 you will show that one iteration of the HWT applied to $\mathbf{x} \in \mathbb{R}^N$ can be expressed as a vector-valued function.

The soft threshold rule (9.4) we have utilized in this section can be expressed as a vector-valued function. Suppose that $\mathbf{x} \in \mathbb{R}^N$. Define

$$\mathbf{s}_\lambda(\mathbf{x}) = (s_1(\mathbf{x}), \dots, s_N(\mathbf{x})) \tag{9.12}$$

where all the coordinate functions are the same,

$$s_k(\mathbf{x}) = s_\lambda(x_k) \tag{9.13}$$

and $s_\lambda(t)$ is the soft threshold rule given in (9.4).

Review of Partial Differentiation

We will have occasion to differentiate coordinate functions. To perform this differentiation, we need the concept of a *partial derivative*.

Definition 9.2. *Let* $g(\mathbf{x}) = g(x_1, x_2, \ldots, x_N)$. *We define the* partial derivative *of* g with respect to x_k *as*

$$\frac{\partial}{\partial x_k} g(x_1, \ldots, x_N) = \lim_{h \to 0} \frac{g(x_1, \ldots, x_k + h, \ldots, x_N) - g(x_1, \ldots, x_N)}{h}$$

$\square$

Thus to compute a partial derivative $\frac{\partial g}{\partial x_k}$, we simply hold all other variables constant and differentiate with respect to x_k. We can often employ differentiation rules from elementary calculus to compute partial derivatives. For example, if

$$g(x_1, x_2, x_3) = 2x_1 x_2^2 - \sin(x_1) + e^{x_3}$$

then

$$\frac{\partial}{\partial x_1} g(x_1, x_2, x_3) = 2x_2^2 - \cos(x_1)$$

$$\frac{\partial}{\partial x_2} g(x_1, x_2, x_3) = 4x_1 x_2$$

$$\frac{\partial}{\partial x_3} g(x_1, x_2, x_3) = e^{x_3}$$

Consult a calculus book (e. g., Stewart [69]) for more information regarding partial derivatives.

Multivariable Soft Threshold Function

Our primary purpose for introducing the partial derivative is so that we can compute the partial derivatives of the coordinate functions $s_k(\mathbf{x})$, $k = 1, \ldots, N$ given by (9.13). We have

$$s_k(\mathbf{x}) = s_\lambda(x_k) = \begin{cases} x_k - \lambda, & x_k > \lambda \\ 0, & -\lambda \le x_k \le \lambda \\ x_k + \lambda, & x_k < -\lambda \end{cases}$$

so that we need to simply differentiate s_λ with respect to x_k:

$$\frac{\partial}{\partial x_k} s_k(\mathbf{x}) = s_\lambda'(x_k) = \begin{cases} 1, & x_k > \lambda \\ 0, & -\lambda < x_k < \lambda \\ 1, & x_k < -\lambda \end{cases} = \begin{cases} 1, & |x_k| > \lambda \\ 0, & |x_k| < \lambda \end{cases} \quad (9.14)$$

Note that the partial derivative in (9.14) is undefined at $\pm\lambda$.

Stein's Unbiased Risk Estimator

We are now ready to begin our development of the SureShrink method. For completeness, we first state Stein's theorem without proof.

Theorem 9.2 (SURE: Stein's Unbiased Risk Estimator [68]). *Suppose that the N-vector* $\mathbf{w}$ *is formed by adding N-vectors* $\mathbf{z}$ *and* ϵ. *That is,* $\mathbf{w} = \mathbf{z} + \epsilon$, *where* $\epsilon = (\epsilon_1, \ldots, \epsilon_N)$ *and each* ϵ_k *is normally distributed with mean* θ_k *and variance* 1. *Let* $\hat{\mathbf{z}}$ *be the estimator formed by*

$$\hat{\mathbf{z}} = \mathbf{w} + \mathbf{g}(\mathbf{w}) \qquad (9.15)$$

where the coordinate functions $g_k \colon \mathbb{R}^N \to \mathbb{R}$ *of the vector-valued function* $\mathbf{g} \colon \mathbb{R}^N \to \mathbb{R}^N$ *are differentiable except at a finite number of points.*[2] *Then*

$$E\big(\|\hat{\mathbf{z}} - \mathbf{z}\|^2\big) = E\left(N + \|\mathbf{g}(\mathbf{w})\|^2 + 2 \sum_{k=1}^{N} \frac{\partial}{\partial w_k} g_k(\mathbf{w}) \right) \qquad (9.16)$$

$\square$

Application of SURE to Wavelet Shrinkage

We want to apply Theorem 9.2 to the wavelet shrinkage method. We can view $\mathbf{w}$ as the highpass portion of the wavelet transform of our observed signal $\mathbf{y} = \mathbf{v} + \mathbf{e}$ so that $\mathbf{z}$ is the highpass portion of the wavelet transform of the true signal $\mathbf{v}$. The vector ϵ is the highpass portion of the wavelet transform of the noise $\mathbf{e}$ (see the discussion following (9.5)). Then $\hat{\mathbf{z}}$ is simply the threshold function (9.12). We just need to define $\mathbf{g}$ in (9.15) so that $\hat{\mathbf{z}} = \mathbf{s}_\lambda(\mathbf{w})$. This is straightforward. We take

$$\mathbf{g}(\mathbf{w}) = \mathbf{s}_\lambda(\mathbf{w}) - \mathbf{w} \qquad (9.17)$$

so that $\hat{\mathbf{z}} = \mathbf{w} + \mathbf{g}(\mathbf{w}) = \mathbf{w} + \mathbf{s}_\lambda(\mathbf{w}) - \mathbf{w} = \mathbf{s}_\lambda(\mathbf{w})$.

What Theorem 9.2 gives us is a an alternative way to write the mean squared error $E\big(\|\hat{\mathbf{z}} - \mathbf{z}\|^2\big)$. We can certainly simplify the right-hand side of (9.16) to obtain an equation with only λ unknown. If we want to minimize the mean squared error, then we must first simplify

$$f(\lambda) = N + \|\mathbf{g}(\mathbf{w})\|^2 + 2 \sum_{k=1}^{N} \frac{\partial}{\partial w_k} g_k(\mathbf{w}) \qquad (9.18)$$

---

[2]The theorem that appears in Stein [68] calls for g_k to be only *weakly differentiable*. The concept of weakly differentiable functions is beyond the scope of this book and we do not require a statement of the theorem in its weakest form.

and then find a value of λ that minimizes it. This minimum value will be our λ^{sure}. Let's simplify $f(\lambda)$. We have

$$\|\mathbf{g}(\mathbf{w})\|^2 = \sum_{k=1}^{N} g_k(\mathbf{w})^2$$

Now

$$g_k(\mathbf{w}) = s_\lambda(w_k) - w_k = \left\{ \begin{array}{ll} w_k - \lambda, & w_k > \lambda \\ 0, & |w_k| \leq \lambda \\ w_k + \lambda, & w_k < -\lambda \end{array} \right\} - w_k = \left\{ \begin{array}{ll} -\lambda, & w_k > \lambda \\ -w_k, & |w_k| \leq \lambda \\ \lambda, & w_k < -\lambda \end{array} \right.$$

so that

$$g_k(\mathbf{w})^2 = \left\{ \begin{array}{ll} \lambda^2, & |w_k| > \lambda \\ w_k^2, & |w_k| \leq \lambda \end{array} \right. = \min(\lambda^2, w_k^2)$$

Thus

$$\|\mathbf{g}(\mathbf{w})\|^2 = \sum_{k=1}^{N} \min(\lambda^2, w_k^2) \tag{9.19}$$

From (9.17) we know that $g_k(\mathbf{w}) = s_\lambda(w_k) - w_k$. Using (9.14), we have

$$\frac{\partial}{\partial w_k} g_k(\mathbf{w}) = s_\lambda'(w_k) - 1 = \left\{ \begin{array}{ll} 1, & |w_k| > \lambda \\ 0, & |w_k| < \lambda \end{array} \right\} - 1 = \left\{ \begin{array}{ll} 0, & |w_k| > \lambda \\ -1, & |w_k| < \lambda \end{array} \right.$$

Now we need to sum up all the partial derivatives given by (9.20). We have

$$2 \sum_{k=1}^{N} \frac{\partial}{\partial w_k} g_k(\mathbf{w}) = -2 \sum_{k=1}^{N} \left\{ \begin{array}{ll} 0, & |w_k| > \lambda \\ 1, & |w_k| < \lambda \end{array} \right. \tag{9.20}$$

The last sum is simply producing a count of the number of w_k's that are smaller (in absolute value) than λ. We will write this count as

$$\#\{k: |w_k| < \lambda\} \tag{9.21}$$

Inserting (9.21) into (9.20) gives

$$2 \sum_{k=1}^{N} \frac{\partial}{\partial w_k} g_k(\mathbf{w}) = -2 \cdot \#\{k: |w_k| < \lambda\} \tag{9.22}$$

Combining (9.19) and (9.22), we see that (9.18) can be written as

$$f(\lambda) = N - 2 \cdot \#\{k: |w_k| < \lambda\} + \sum_{k=1}^{N} \min(w_k^2, \lambda^2) \tag{9.23}$$

Simplifying the Estimator Function f(λ)

We want to find, if possible, a λ that minimizes $f(\lambda)$. While the formula (9.23) might look complicated, it is a straightforward process to minimize it.

The value of the function $f(\lambda)$ in (9.23) depends on the elements in **w**. Note that neither the factor $\#\{k\colon |w_k| < \lambda\}$ nor the term $\sum_{k=1}^{N} \min(w_k^2, \lambda^2)$ depend on the *ordering* of the elements of **w**. Thus, we assume in our subsequent analysis of $f(\lambda)$ that the elements of **w** are ordered so that

$$|w_1| \leq |w_2| \leq \cdots \leq |w_N|$$

Let's first consider $0 \leq \lambda \leq |w_1|$. In this case,

$$\#\{k\colon |w_k| < \lambda\} = 0 \quad \text{and} \quad \sum_{k=1}^{N} \min(w_k^2, \lambda^2) = \sum_{k=1}^{N} \lambda^2 = N\lambda^2$$

so that

$$f(\lambda) = N + N\lambda^2 \tag{9.24}$$

Now if $\lambda > |w_N|$, we have

$$\#\{k\colon |w_k| < \lambda\} = N \quad \text{and} \quad \sum_{k=1}^{N} \min(w_k^2, \lambda^2) = \sum_{k=1}^{N} w_k^2$$

so that

$$f(\lambda) = N - 2N + \sum_{k=1}^{N} w_k^2 = -N + \sum_{k=1}^{N} w_k^2 \tag{9.25}$$

What about $|w_1| < \lambda \leq |w_N|$? Note that the term $N - 2 \cdot \#\{k\colon |w_k| < \lambda\}$ is piecewise constant and will change only whenever λ moves past a $|w_k|$ that is bigger than the preceding $|w_{k-1}|$. We can write the second term as

$$\sum_{k=1}^{N} \min(w_k^2, \lambda^2) = \sum_{|w_k| < \lambda} w_k^2 + \#\{k\colon |w_k| \geq \lambda\}\lambda^2 \tag{9.26}$$

and it is easy to see that this term is piecewise quadratic and it will also change whenever λ moves past a $|w_k|$ that is bigger than the preceding $|w_{k-1}|$. So $f(\lambda)$ is a piecewise quadratic function with breaks occurring at *distinct* values of $|w_k|$. For $\lambda > |w_N|$, $f(\lambda)$ is the constant value given by (9.25). We can use (9.24) and (9.26) to note that whenever λ is between distinct values of $|w_k|$, $f'(\lambda) > 0$, so that $f(\lambda)$ is increasing between these values.

Examples of the Estimator Function f(λ)

Let's look at a simple example that summarizes our analysis of $f(\lambda)$ thus far.

Example 9.5 (Plotting f(λ)). *Let* $\mathbf{w} = [1, 1.1, 2, 2, 2.4]^T$. *Note that the elements of* $\mathbf{w}$ *are nonnegative and ordered smallest to largest. For* $0 < \lambda \leq 1 = w_1$, *we can use (9.24) and write* $f(\lambda) = 5 + 5\lambda^2$. *We can also use (9.25) and write* $f(\lambda) = -5 + (1^2 + 1.1^2 + 2^2 + 2^2 + 2.4^2) = 10.97$ *for* $\lambda > 2.4 = w_5$.

For $1 < \lambda \leq 1.1$, $\#\{k : |w_k| < \lambda\} = 1$ *and we can use (9.23) and write*

$$f(\lambda) = 5 - 2 \cdot 1 + 1^2 + 4\lambda^2 = 4 + 4\lambda^2$$

Note that the endpoint $|w_2| = 1.1$ *is included in this interval — the function doesn't change definitions again until* λ *moves past 1.1. We can use (9.23) for the remaining two intervals and describe* $f(\lambda)$:

$$f(\lambda) = \begin{cases} 5 + 5\lambda^2, & 0 < \lambda \leq 1 \\ 4 + 4\lambda^2, & 1 < \lambda \leq 1.1 \\ 3.21 + 3\lambda^2, & 1.1 < \lambda \leq 2 \\ 7.21 + \lambda^2, & 2 < \lambda \leq 2.4 \\ 10.97, & 2.4 < \lambda \end{cases}$$

Thus we see that $f(\lambda)$ *is a left continuous function. The function is plotted in Figure 9.12.* □

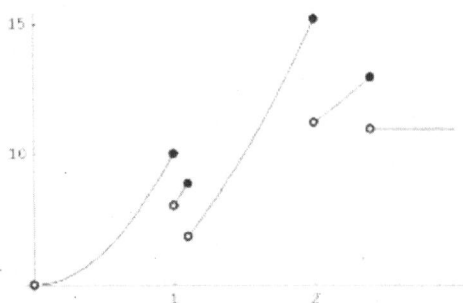

Figure 9.12 A plot of $f(\lambda)$. The function is left continuous at the breakpoints (noted by the black points) but is not right continuous at the breakpoints (noted by the open circle points).

In Figure 9.13 we have plotted $f(\lambda)$ for a different vector $\mathbf{w}$. From the graphs in Figures 9.12 and 9.13, it would seem that the desired minimum value of $f(\lambda)$ would either be at the left endpoint of some interval or the constant value $\sum_{k=1}^{N} w_k^2 - N$, but $f(\lambda)$ is left continuous at each breakpoint $|w_\ell|$. Theoretically, we could compute

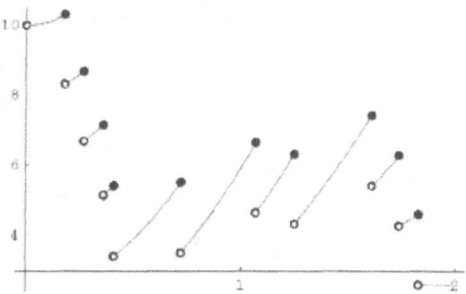

Figure 9.13 $f(\lambda)$ for $\mathbf{w} = [\,0.18, 0.27, 0.36, 0.41, 0.72, 1.07, 1.25, 1.61, 1.74, 1.83\,]^T$.

$$m_k = \lim_{\lambda \to |w_k|^+} f(\lambda)$$

for each of $|w_1|, \ldots, |w_N|$ and then pick the $|w_k|$ that corresponds to the minimum value of $\{m_1, m_2, \ldots, m_N, \sum_{k=1}^{N} w_k^2 - N\}$, but this is hardly tractable on a computer.

Minimizing $f(\lambda)$ on a Computer

A more feasible solution would be simply to change our function $f(\lambda)$ so that the quadratic pieces are defined on intervals $|w_\ell| \le \lambda < |w_{\ell+1}|$, where $|w_\ell| < |w_{\ell+1}|$. The easiest way to do this is simply to change our counting function to include $|w_\ell|$ in the count but exclude $|w_{\ell+1}|$ from the count. That is, we consider minimizing the function[3]

$$f(\lambda) = N - 2 \cdot \#\{k \colon |w_k| \le \lambda\} + \sum_{k=1}^{N} \min(w_k^2, \lambda^2) \qquad (9.27)$$

Note: We use (9.27) in all subsequent examples, problem sets, and computer labs.

In Figure 9.14 we have used (9.27) and plotted $f(\lambda)$ using the vectors from Example 9.5 and Figure 9.13. Note that the new formulation sets the constant value $\sum_{k=1}^{N} w_k^2 - N$ as $f(|w_N|)$.

[3]The reader with some background in analysis may realize that if the integrand of the expected value in (9.16) is viewed as an $L^2(\mathbb{R})$ function, we could have written (9.27) immediately instead of (9.18).

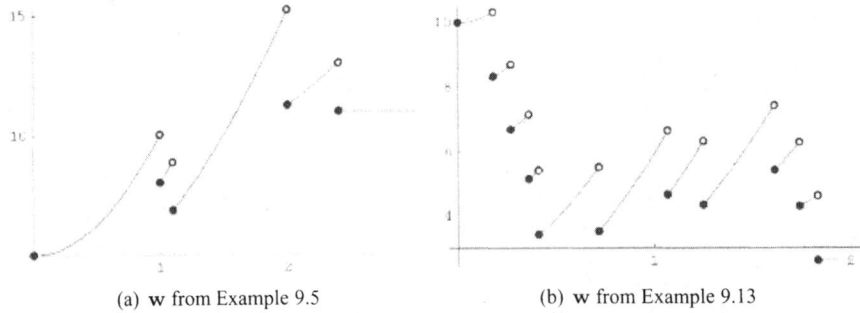

(a) **w** from Example 9.5   (b) **w** from Example 9.13

Figure 9.14   Two plots of $f(\lambda)$ using (9.27).   The graph in part (a) uses vector **w** $= [1, 1.1, 2, 2, 2.4]^T$ from Example 9.5, while the graph in part (b) uses vector **w** $= [0.18, 0.27, 0.36, 0.41, 0.72, 1.07, 1.25, 1.61, 1.74, 1.83]^T$ from Figure 9.13. The functions are now right continous at the breakpoints.

Choosing the SUREShrink Tolerance λ^{sure}

Now our minimum can simply be chosen to be $\lambda^{sure} = |w_\ell|$ where $f(|w_\ell|) \leq f(|w_k|)$, $k = 1, \ldots, N$ and $\lambda^{sure} \leq \lambda^{univ}$. Computationally, λ^{sure} is quite straightforward. In Problem 9.14 you will derive a recursive formula for quickly computing $f(|w_k|)$, and in Computer Lab 9.5 you will develop the routine for computing λ^{sure} as given in the software package WaveLab [31].

The reason we insist that λ^{sure} is no larger than λ^{univ} is that the SureShrink method has some problems when the input vector **y** is sparse. If $\mathbf{y} = \mathbf{v} + \mathbf{e}$, where **e** is Gaussian white noise, then **y** is considered to be sparse if only a small number of elements of **v** are nonzero. As Donoho and Johnstone point out [30], when the input vector is sparse, the coefficients that are essentially noise dominate the relatively small number of remaining terms. You will repeat the experiment that led them to this conclusion in Computer Lab 9.7.

We need some way to determine if an input vector **y** is sparse. Donoho and Johnstone [30] suggest computing the values

$$s = \frac{1}{N} \sum_{k=1}^{N} (y_k^2 - 1) \quad \text{and} \quad u = \frac{3}{2\sqrt{N}} \log_2(N) \tag{9.28}$$

If $s \leq u$, then **y** is sparse. In this case, we use λ^{univ} with the soft threshold rule (9.4). Otherwise, we use λ^{sure} with the soft threshold function.

An Example of SURESHrink

We illustrate the SureShrink method by returning to Example 9.4.

Example 9.6 (2D SureShrink). *In Example 9.4 we were given a noisy digital image of dimension* 256×384. *We performed two iterations of the wavelet transform using the D6 filter. Recall that the VisuShrink method returned an undesirable result. In this example we use the SureShrink method instead of the VisuShrink method.*

Each iteration of the wavelet transform produces three highpass detail matrices (see Figure 9.15), so we'll apply the SureShrink method to each of these six matrices.

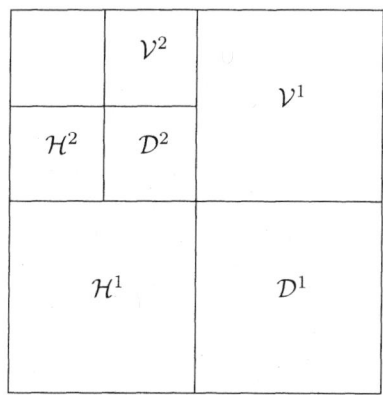

Figure 9.15 The detail matrices of the wavelet transform.

We add subscripts to our highpass matrices to indicate the iteration of the wavelet tranform. We thus have highpass matrices $\mathcal{V}_j, \mathcal{H}_j,$ *and* $\mathcal{D}_j$ *for* $j = 1, 2$.

Recall that the noise level is $\sigma = 18$. *To use Theorem 9.2, we need these highpass matrices to have noise level* $\sigma = 1$. *We can easily transform our highpass matrices to satisfy Theorem 9.2 by dividing each of them by* σ. *In practice, we typically don't know* σ, *so we'll use our estimate* $\hat{\sigma} = 21.3035$ *from Example 9.4. We set* $\hat{\mathcal{V}}_j = \mathcal{V}_j / \hat{\sigma}, \hat{\mathcal{H}}_j = \mathcal{H}_j / \hat{\sigma},$ *and* $\hat{\mathcal{D}}_j = \mathcal{D}_j / \hat{\sigma}, j = 1, 2$. *In a similar manner, we denote the six tolerances we compute as* $\lambda_j^{\mathcal{V}}, \lambda_j^{\mathcal{H}},$ *and* $\lambda_j^{\mathcal{D}}, $ *for* $j = 1, 2$.

We use $\hat{\mathcal{V}}_j, \hat{\mathcal{H}}_j,$ *and* $\hat{\mathcal{D}}_j$ *to find the tolerances* $\lambda_j^{\mathcal{V}}, \lambda_j^{\mathcal{H}},$ *and* $\lambda_j^{\mathcal{D}}$ *and then use these tolerances with (9.4) to shrink* $\hat{\mathcal{V}}_j, \hat{\mathcal{H}}_j,$ *and* $\hat{\mathcal{D}}_j$ *and obtain the modified highpass matrices* $\hat{\mathcal{V}}_j^s, \hat{\mathcal{H}}_j^s,$ *and* $\hat{\mathcal{D}}_j^s$. *Before we compute the inverse transform, we multiply the modified highpass matrices by* $\hat{\sigma}$ *to put them back to the original noise level.*

To determine whether to use λ^{sure} *or* λ^{univ}, *we use (9.28) to compute* $s_j^{\mathcal{V}}, s_j^{\mathcal{H}}, s_j^{\mathcal{D}}$ *and* $u_j^{\mathcal{V}}, u_j^{\mathcal{H}}, u_j^{\mathcal{D}}$ *for each of* $\hat{\mathcal{V}}_j, \hat{\mathcal{H}}_j, \hat{\mathcal{D}}_j$, *respectively. In each case we found the values of* $s_j^{\mathcal{V}}, s_j^{\mathcal{H}}, s_j^{\mathcal{D}}$ *less than or equal to their respective values* $u_j^{\mathcal{V}}, u_j^{\mathcal{H}}, u_j^{\mathcal{D}}$. *Thus, we choose* $\lambda_j^{\mathcal{V}}, \lambda_j^{\mathcal{H}},$ *and* $\lambda_j^{\mathcal{D}}$ *by minimizing (9.27).*

To apply the soft threshold function to $\hat{\mathcal{V}}_j$, we concatenate the rows of $\hat{\mathcal{V}}_j$ to obtain a vector. In the case where $j = 1$, the vector length is $128 \cdot 192 = 24{,}576$, and for the second iteration, the vector length is $64 \cdot 96 = 6144$. We repeat the process for $\hat{\mathcal{H}}_j$ and $\hat{\mathcal{D}}_j$. Table 9.2 gives the various tolerances.

Table 9.2   The various tolerances for use in the SureShrink method.

Iteration	Vertical	Horizontal	Diagonal
1	$\lambda_1^{\mathcal{V}} = 1.65607$	$\lambda_1^{\mathcal{H}} = 1.51588$	$\lambda_1^{\mathcal{D}} = 2.02219$
2	$\lambda_2^{\mathcal{V}} = 1.08117$	$\lambda_2^{\mathcal{H}} = 0.90912$	$\lambda_2^{\mathcal{D}} = 1.34043$

For comparative purposes, we can easily compute the universal thresholds for each level. For the first iteration, we have $\lambda^{univ} = \hat{\sigma}\sqrt{2 \cdot 24{,}576} = 67.7354$, and for the second iteration $\lambda^{univ} = \hat{\sigma}\sqrt{2 \cdot 6144} = 62.92$. Certainly, the λ^{sure} values given in Table 9.2 will do a better job of discerning noise from a true signal.

In Figure 9.16 we plot the original wavelet transform and the modified wavelet transform using the SureShrink method. We can see the differences in the VisuShrink and SureShrink methods by comparing the graphs in Figure 9.16 with the modified wavelet transform plotted in Figure 9.10.

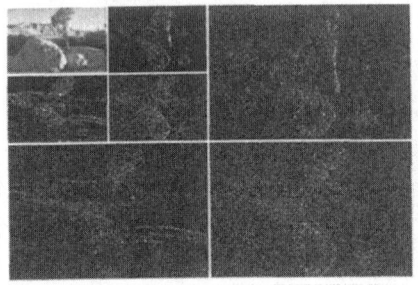
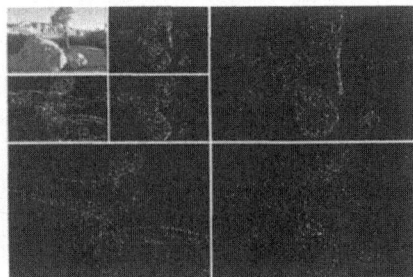

(a) Wavelet transform   (b) Modified wavelet transform

Figure 9.16   Two iterations of the wavelet transform using the D6 filter are applied to the original image. The modified wavelet transform is obtained by applying the soft threshold rule (9.4) to each $\hat{H}_{jk}$ using the corresponding λ_{jk} from Table 9.2.

To obtain the denoised image, we first multiply each of the $\hat{\mathcal{V}}_j$, $\hat{\mathcal{H}}_j$, and $\hat{\mathcal{D}}_j$, $j = 1, 2$, by $\hat{\sigma}$ and then compute two iterations of the inverse wavelet transform. The denoised image and the original image are plotted in Figure 9.17. For comparative purposes, we have included the plot of the denoised image from Example 9.4 using the VisuShrink method. ☐

(a) The original image

(b) Image with noise

(c) VisuShrink

(d) SureShrink

Figure 9.17   Comparison of the VisuShrink and SureShrink denoising methods.

PROBLEMS

9.10   Let $x \in \mathbb{R}^N$. Write one iteration of the HWT applied to x as a vector-valued function $\mathbf{f}(x)$. Identify the coordinate functions $f_1, \ldots, f_N$.

9.11   Let $w = [.2, -.5, .3, -.1, .2, -.1]^T$. Find an expression for $f(\lambda)$ (see Example 9.5) and plot the result.

9.12   In this problem you will investigate the location of λ^{sure} for vectors of different lengths whose *distinct* elements are the same.

  (a) Let $v = [\frac{1}{2}, \frac{1}{2}]^T$. Find an expression for $f(\lambda)$ (see Example 9.5) and plot the result. Where does the minimum of $f(\lambda)$ occur?

  (b) Repeat part (a) with $w = [\frac{1}{2}, 1, 1, 2]^T$.

  (c) Can you create a vector x whose elements are selected from $1/2, 1, 2$ so that the minimum of $f(\lambda)$ occurs at $\lambda = 0$? (*Hint:* A tolerance of $\lambda = 0$ means that we are keeping all coefficients — that is, there is no noise to remove. What must a vector look like in this case?)

  (d) Can you create a vector x whose elements are selected from $1/2, 1, 2$ so that the minimum of $f(\lambda)$ occurs at $\lambda = 2$?

9.13   Let v be a vector of length N and let us denote (9.27) by $f_v(\lambda)$ to indicate that it was constructed using v. Further suppose that $|v_m|$ minimizes (9.27) where $1 \le m \le N$.

  (a) Construct the N-vector w by permuting the elements of v. Show that v_m minimizes $f_w(\lambda)$.

  (b) Suppose that x is an N-vector whose entries are $\pm v_k$. Show that $|v_m|$ minimizes $f_x(\lambda)$.

  (c) *True or False:* $|v_m|$ minimizes $f_y(\lambda)$ where $y_k = cv_k$ for $c \ne 0$. If the statement is true, prove it. If the statement is false, find a counterexample.

  (d) *True or False:* $|v_m|$ minimizes f_z where $z_k = v_k + c$ for $c \ne 0$. If the statement is true, prove it. If the statement is false, find a counterexample.

9.14   In this problem you will derive a recursive formula that will allow you quickly to compute the values $f(|w_\ell|)$, $\ell = 2, \ldots, N$. For this problem, assume that $|w_1| \le |w_2| \le \cdots \le |w_N|$.

  (a) Using (9.27), show that

$$f(|w_\ell|) = N - 2\ell + \sum_{k=1}^{\ell} w_k^2 + (N - \ell)w_\ell^2$$

  (b) Replace ℓ by $\ell + 1$ in part (a) to find an analogous expression for $f(|w_{\ell+1}|)$.

(c) Using parts (a) and (b), show that

$$f(|w_{\ell+1}|) = f(|w_\ell|) - 2 + (N - \ell)(w_{\ell+1}^2 - w_\ell^2)$$

To employ this recursion formula, we must first use (9.27) and compute $f(|w_1|) = N(1 + |w_1|^2)$. Then we can compute all the values $f(|w_\ell|)$, $\ell = 2, \ldots, N$ using the formula in part (c). To find λ^{sure}, we simply pick the minimum of these values. The algorithm for computing λ^{sure} in the WaveLab software [31] is based on the formula in part (a).

9.15   We create the vector $\mathbf{w}$ by adding Gaussian white noise with $\sigma = 1$ to the vector $\mathbf{v} = [0, 0.5, 1, 1.5, 2, 2.5, 3, 3.5]^T$. We obtain

$$\mathbf{w} = [-0.159354, -1.62378, 1.22804, 1.57117, 3.15511,$$
$$2.67804, 2.38617, 3.98616]^T$$

Use (9.28) and a **CAS** or a calculator to determine if $\mathbf{w}$ is sparse.

9.16   Repeat Problem 9.15 using

$$\mathbf{w} = [0.0885254, 0.689566, -0.707012, 0.33514, 1.22295,$$
$$2.68978, 0.831615, 1.16285]^T$$

In this case, $\mathbf{w}$ was created by adding Gaussian white noise with $\sigma = 1$ to the vector $\mathbf{v} = [0, 0.25, 0.5, 0.75, 1, 1.25, 1.5, 1.75]^T$.

9.17   Yoon and Vaidyanathan [87] note that hard thresholding can "yield abrupt artifacts in the denoised signal". This is caused by the jump discontinuity in the threshold function $f_\lambda(t)$ (see Problem 9.5 in Section 9.1). Moreover, Bruce and Gao [10] argue that "soft shrink tends to have a bigger bias" because the large elements in the highpass portion are scaled toward zero. Some researchers have developed customized threshold functions for the purpose of alleviating the problems with hard and soft thresholding. In this problem you will study such a function.

(a) Bruce and Gao [37], introduced a *firm threshold function*. Let λ_1 and λ_2 be two positive numbers with $\lambda_1 < \lambda_2$. Define the firm threshold function by

$$f_{\lambda_1, \lambda_2}(t) = \begin{cases} 0, & |t| \leq \lambda_1 \\ \text{sgn}(t)\frac{\lambda_2(|t|-\lambda_1)}{(\lambda_2-\lambda_1)}, & \lambda_1 < |t| \leq \lambda_2 \\ t, & |t| > \lambda_2 \end{cases}$$

Sketch $f_{\lambda_1, \lambda_2}(t)$ for the following threshold pairs: $\lambda_1 = 1/2, \lambda_2 = 1, \lambda_1 = 3/4, \lambda_2 = 1, \lambda_1 = .9, \lambda_2 = 1$, and $\lambda_1 = 1, \lambda_2 = 100$.

(b) Write $f_{\lambda_1, \lambda_2}(t)$ as a piecewise function that does not utilize $|\cdot|$ or sgn. How many pieces are there? What type of function is each piece?

(c) For which values of λ_1, λ_2 is $f_{\lambda_1, \lambda_2}(t)$ continuous? Continuously differentiable?

(d) What happens to $f_{\lambda_1, \lambda_2}(t)$ as λ_2 approaches infinity? (*Hint:* To see the result geometrically, use part (b) to draw $f_{\lambda_1, \lambda_2}(t)$ with a fixed λ_1 and progressively larger values of λ_2. To find the result analytically, view $f_{\lambda_1, \lambda_2}(t)$ as a function of λ_2 and use L'Hospital's rule.)

(e) What happens to $f_{\lambda_1, \lambda_2}(t)$ as λ_1 approaches λ_2? (*Hint:* Use the hint from part (c) and also consider what happens to $|t|$ as $\lambda_1 \to \lambda_2$.)

9.18 The *customized soft threshold function* developed in Yoon and Vaidyanathan [87] is related to the firm threshold function defined in Problem 9.17. Let λ, γ, and α be real numbers such that $0 < \gamma < \lambda$, and $0 \leq \alpha \leq 1$. The customized shrinkage function is then defined by

$$f_{\lambda, \gamma, \alpha}(t) = \begin{cases} t - \mathrm{sgn}(t)(1 - \alpha)\lambda, & |t| \geq \lambda \\ 0, & |t| \leq \gamma \\ \mathrm{sgn}(t)\, \alpha \left(\frac{|t|-\gamma}{\lambda-\gamma} \right)^2 \left((\alpha - 3) \left(\frac{|t|-\gamma}{\lambda-\gamma} \right) + 4 - \alpha \right), & \text{otherwise} \end{cases}$$

(9.29)

(a) Plot $f_{1, \frac{1}{2}, \alpha}(t)$ for $\alpha = 0, \frac{1}{4}, \frac{1}{2}, \frac{3}{4}, 1$.

(b) Write $f_{\lambda, \gamma, \alpha}(t)$ as a piecewise-defined function that does not utilize $|\cdot|$ and sgn. How many pieces are there? What type of function is each piece?

(c) Show that $\lim\limits_{\alpha \to 0} f_{\lambda, \gamma, \alpha}(t)$ is the soft threshold function (9.4).

(d) What is $\lim\limits_{\gamma \to \lambda} f_{\lambda, \gamma, 1}(t)$? The hints in Problem 9.17 will be useful for computing this limit.

(e) For which values of λ, γ, and α is $f_{\lambda, \gamma, \alpha}(t)$ continuous? Continuously differentiable? Compare your results to part (c) of Problem 9.17.

9.19 Consider the soft threshold function $s_\lambda(t)$ defined in (9.4) and the hard threshold function $f_\lambda(t)$ given in Problem 9.5 from Section 9.1.

(a) Plot the following linear combinations

   (i) $\frac{1}{3}f_\lambda(t) + \frac{2}{3}s_\lambda(t)$.

   (ii) $\frac{1}{4}f_\lambda(t) + \frac{3}{4}s_\lambda(t)$.

   (iii) $\frac{3}{4}f_\lambda(t) + \frac{1}{4}s_\lambda(t)$.

(b) Find a piecewise representation of the function

$$\alpha f_\lambda(t) + (1 - \alpha)s_\lambda(t)$$

(c) Use part (b) to show that for $|t| > \lambda$ and $|t| \leq \gamma$, $f_{\lambda, \gamma, \alpha}(t) = \alpha f_\lambda(t) + (1 - \alpha)s_\lambda(t)$.

In this way we see that the customized threshold function is a linear combination[4] of hard and soft threshold functions except on the interval from $(\gamma, \lambda]$. On this interval, the quadratic piece given in (9.29) serves to connect the other two pieces smoothly.

Computer Labs

◆ **9.5   Software Development: Constructing λ^{sure}.** From the text Web site, access the development package `lambdasure`. In this development lab you will write a module that implements Donoho's scheme (see Problem 9.14) for computing λ^{sure}.

◆ **9.6   Software Development: Wavelet Shrinkage.** From the text Web site, access the development package `waveletshrinkage`. To proceed, you must have completed Computer Lab 9.1. In this development lab you will modify the wavelet shrinkage module so that it accepts a list of tolerances. The module will apply elements of the list to the appropriate portions of the wavelet transform to perform SureShrink. The module will also be altered so that it accepts a threshold function as an argument. In this way the module can be used in conjunction with the soft threshold functions described in Problems 9.17 and 9.18. Instructions are provided that allow you to add the code to the software package `DiscreteWavelets`.

9.7   Sparseness of λ^{sure}. From the text Web site, access the lab `sparsesure`. In this lab, you will reproduce the experiment performed by Donoho and Johnstone [30] that shows that SureShrink has problems when the highpass portion of the wavelet transform (of the original input) contains only a few nonzero elements.

9.8   SureShrink. From the text Web site, access the lab `sureshrink`. In this lab you will use the wavelet shrinkage module with λ^{sure} to apply the SureShrink method to various audio signals and digital images. You will also experiment with the customized threshold functions described in Problems 9.17 and 9.18.

9.9   Color Image Denoising. From the text Web site, access the lab `colordenoise`. In this lab you will use the wavelet shrinkage module in conjunction with the RGB to HSI conversion lab (Computer Lab 3.6) to perform color image denoising.

---

[4]The linear combination here is special; it is also known as a *convex combination* of the functions $f_\lambda(t)$ and $s_\lambda(t)$ since the multipliers α and $1 - \alpha$ are nonnegative and sum to 1. You may have seen convex combinations in analytic geometry — given two points **a** and **b**, the line segment connecting the points is given by $t\mathbf{a} + (1 - t)\mathbf{b}$ for $0 \leq t \leq 1$.

CHAPTER 10

BIORTHOGONAL FILTERS

Researchers have learned that in many cases, *symmetric filters* work best in applications such as image compression. For now, think of a symmetric filter as one consisting of a finite number of elements whose values are mirrored across a central axis through the filter. For example, the Haar filter is symmetric since $h_0 = h_1 = \sqrt{2}/2$. Here the central axis is between h_0 and h_1. For our purposes, symmetric filters **h** of odd length will reflect about a central axis that goes through h_0. $(h_{-2}, h_{-1}, h_0, h_1, h_2) = (1, 2, 3, 2, 1)$ is an example of such a filter.

Unfortunately, the Haar filter is the only finite-length, symmetric, orthogonal filter whose Fourier series $H(\omega)$ satisfies the zero derivative conditions at $\omega = \pi$ (see Daubechies [24]). Thus, if we wish to construct symmetric filters, we need to prioritize the desired properties obeyed by our filters. We want to consider only finite-length filters **h**, and for **h** to be considered a lowpass filter, its Fourier series $H(\omega)$ must satisfy $H(\pi) = 0$. That leaves orthogonality. What exactly does orthogonality give us? First and foremost, an orthogonal filter **h** gives us an orthogonal transform matrix W_N whose inverse W_N^{-1} is simply the transpose W_N^T. Remember, though, that once we know the inverse of a nonsingular matrix, we never have to recompute

Discrete Wavelet Transformations: An Elementary Approach With Applications. By P. J. Van Fleet **351**
Copyright © 2008 John Wiley & Sons, Inc.

it. So why not simply construct symmetric lowpass and highpass filters **h** and **g**, use them to construct the transform matrix, and then compute and store its inverse? One problem is that we can't be sure that the inverse would have the nice lowpass/highpass structure possessed by the transform matrix. Remember, this structure leads to fast algorithms for computing the transform. If the inverse does not possess a regular pattern, then we have no choice but to use regular matrix multiplication to compute the inverse transform, and that could result in a slower process. Moreover, we used W_N^T not only to invert the transform but also to process the rows of matrix A in the two-dimensional wavelet transform! So the inverse W_N^{-1} is actually used in the two-dimensional transform and we see that in the nonorthogonal case, we need the inverse of our transform matrix to be the transpose of another wavelet transform matrix! The natural solution to this problem is, if possible, to construct *two* sets of filter pairs $\tilde{\mathbf{h}}, \tilde{\mathbf{g}}$ and **h**, **g** and use them to form wavelet transform matrices $\tilde{W}_N$ and W_N, respectively, so that $\tilde{W}_N^{-1} = W_N^T$. The filters $\tilde{\mathbf{h}}$ and **h** are called a *biorthogonal filter pair*.

We could attempt our construction by mimicking the ideas of Chapter 7. That is, we could set filter lengths on **h** and $\tilde{\mathbf{h}}$, write down some lowpass conditions that must be satisfied by the Fourier series $H(\omega)$ and $\tilde{H}(\omega)$, and then add to our list the equations that result from insisting that $\tilde{W}_N W_N^T$ is the $N \times N$ identity matrix I_N. There are many problems with this approach. How do we choose the lengths of the two filters? How many derivative conditions do we need to impose on $H(\omega)$ and $\tilde{H}(\omega)$ at $\omega = \pi$? The quadratic system that results from $\tilde{W}_N W_N^T = I_N$ needs to be solved for *two* filters **h** and $\tilde{\mathbf{h}}$. There are also the associated highpass filters **g** and $\tilde{\mathbf{g}}$. Can we construct them from **h** and $\tilde{\mathbf{h}}$ in a manner similar to that used in Chapter 7?

We have reached a critical juncture in the book. We could proceed with the somewhat ad hoc method for filter development described in the preceding paragraph or we could try another method. In this chapter we describe an alternative approach. In so doing, we analyze the problem in a way most common to mathematicians — we develop theoretical results (theorems and propositions) that will allow us to perform the filter pair construction desired.

In Section 10.1 we generalize the construction of Section 8.1 and learn how to build general Fourier series $\tilde{H}(\omega)$ and $H(\omega)$ that yield the biorthogonal filters $\tilde{\mathbf{h}}$ and **h**. We will then see how to construct $\tilde{G}(\omega)$ and $G(\omega)$ from the Fourier series $H(\omega)$ and $\tilde{H}(\omega)$ and thus obtain highpass filters $\tilde{\mathbf{g}}$ and **g**. In Section 10.2 we learn how to use the results of Section 10.1 to construct *biorthogonal spline filter* pairs. The construction allows us easily to impose the constraint that the filters be symmetric.

The committee for the development of the JPEG2000 image compression standard decided to use the biorthogonal wavelet transform as part of the compression scheme. For lossy compression, they wanted to use an odd-length symmetric biorthogonal filter pair whose lengths are similar. Filter lengths 9 and 7 are long enough to give a good approximation of the input data and at the same time are not so long as to slow down computations. The (9,7) biorthogonal spline filter pair $\tilde{\mathbf{h}}$, **h** was not adopted because the number of derivatives of $H(\omega)$ and $\tilde{H}(\omega)$ equal to zero at $\omega = \pi$ are disparate. A biorthogonal filter pair that does not have this problem is *Cohen–*

Daubechies–Feauveau 9/7 filter. This symmetric filter was adopted by the JPEG2000 committee and we describe its construction in Section 10.3.

10.1 CONSTRUCTING BIORTHOGONAL FILTERS

In this section we generalize the results of Sections 8.1 and 8.2. The goal is to construct *two* lowpass filters filters h and $\tilde{\text{h}}$ and their associated highpass filters g and $\tilde{\text{g}}$. These filters can then be used to form wavelet transform matrices W_N and $\tilde{W}_N$ so that the inverse of one is the transpose of the other. The lowpass filters h and $\tilde{\text{h}}$ that result from this construction are called *biorthogonal filters*.

We first develop the properties necessary to construct W_N and $\tilde{W}_N$ and conclude the section by considering symmetric biorthogonal filter pairs.

Creating Two Wavelet Transform Matrices

As we mentioned in the introductory remarks for this chapter, Daubechies showed [24] that the only symmetric, finite-length, orthogonal filter is the Haar filter. We talked about the limitations of the Haar wavelet transform at the beginning of Chapter 7 — the short filter length sometimes fails to detect large changes in the input data. So we are interested in ultimately constructing symmetric filters of length greater than 2. To do so, we must give up orthogonality as a filter constraint.

Thus, we have arrived at the fundamental question of this section. Is it possible to construct a wavelet transform W_N built from a symmetric lowpass filter h and highpass filter g so that the transpose of W_N^{-1} is also a wavelet transform matrix? Since W_N is not orthogonal, we need to construct a wavelet transform matrix $\tilde{W}_N$ so that $\tilde{W}_N^T = W_N^{-1}$. Since $\tilde{W}_N$ is a wavelet transform matrix, we will need to build a symmetric lowpass filter $\tilde{\text{h}}$ and a highpass filter $\tilde{\text{g}}$.

Let H and G be $\frac{N}{2} \times N$ matrices constructed from the lowpass and highpass filters h and g, respectively. Then our wavelet matrix can be written in block form:

$$W = \left[\frac{H}{G} \right]$$

In a similar fashion we can construct

$$\tilde{W} = \left[\frac{\tilde{H}}{\tilde{G}} \right]$$

from lowpass and highpass filters $\tilde{\text{h}}$ and $\tilde{\text{g}}$. Since we want $W_N \tilde{W}_N^T = I_N$, where I_N is the $N \times N$ identity matrix, we have in block form

$$W_N \tilde{W}_N^T = \left[\frac{H}{G} \right] \cdot \left[\tilde{H}^T \middle| \tilde{G}^T \right] = \left[\begin{array}{c|c} H\tilde{H}^T & H\tilde{G}^T \\ \hline G\tilde{H}^T & G\tilde{G}^T \end{array} \right] = I_N$$

Now $H\tilde{H}^T$ is an $\frac{N}{2} \times \frac{N}{2}$ matrix and this product must be the $\frac{N}{2} \times \frac{N}{2}$ identity matrix. In a similar manner, $G\tilde{G}^T = I_{N/2}$, and if $0_{N/2}$ is the $\frac{N}{2} \times \frac{N}{2}$ zero matrix, we have

$$H\tilde{G}^T = G\tilde{H}^T = 0_{N/2}$$

An Example of the General Transform Matrix

Let's have a closer look at $H\tilde{H}^T$ for some particular filters $\mathbf{h}$ and $\tilde{\mathbf{h}}$ and dimension N.

Example 10.1 ($\mathbf{W_N}$ and $\mathbf{\tilde{W}_N}$ for Two Generic Filters). *Consider the lowpass filter* $\mathbf{h} = (h_{-2}, h_{-1}, h_0, h_1, h_2)$. *Let's suppose that we wish to apply a wavelet transform matrix to vectors of length $N = 10$. Then the lowpass portion of the transform matrix might look like*

$$H = \begin{bmatrix} h_0 & h_1 & h_2 & 0 & 0 & 0 & 0 & 0 & h_{-2} & h_{-1} \\ h_{-2} & h_{-1} & h_0 & h_1 & h_2 & 0 & 0 & 0 & 0 & 0 \\ 0 & 0 & h_{-2} & h_{-1} & h_0 & h_1 & h_2 & 0 & 0 & 0 \\ 0 & 0 & 0 & 0 & h_{-2} & h_{-1} & h_0 & h_1 & h_2 & 0 \\ h_2 & 0 & 0 & 0 & 0 & 0 & h_{-2} & h_{-1} & h_0 & h_1 \end{bmatrix} \quad (10.1)$$

Note that we have formed H by placing h_0 in the first column of row 1 and wrapping the two filter coefficients with negative subscripts at the end of the first row. Each subsequent row of H is formed by rotating the elements of the preceding row 2 units right. In Section 11.3 we will see how the formulation (10.1) with symmetric filters is useful in reducing boundary effects caused by wrapping rows of the transform matrix. Unlike the wavelet transform matrices from Chapter 7, the filter coefficients do not appear in reversed order. There are a couple of reasons for this. The first is that we have seen in the past that we can use reflections of filters to build wavelet transform matrices, and the second is that when the filter coefficients are symmetric, reversing the order of the coefficients has no effect on the construction.

Now let's build the matrix $\tilde{H}$ from lowpass filter $\tilde{\mathbf{h}} = (\tilde{h}_{-1}, \tilde{h}_0, \tilde{h}_1)$. We have

$$\tilde{H} = \begin{bmatrix} \tilde{h}_0 & \tilde{h}_1 & 0 & 0 & 0 & 0 & 0 & 0 & 0 & \tilde{h}_{-1} \\ 0 & \tilde{h}_{-1} & \tilde{h}_0 & \tilde{h}_1 & 0 & 0 & 0 & 0 & 0 & 0 \\ 0 & 0 & 0 & \tilde{h}_{-1} & \tilde{h}_0 & \tilde{h}_1 & 0 & 0 & 0 & 0 \\ 0 & 0 & 0 & 0 & 0 & \tilde{h}_{-1} & \tilde{h}_0 & \tilde{h}_1 & 0 & 0 \\ 0 & 0 & 0 & 0 & 0 & 0 & 0 & \tilde{h}_{-1} & \tilde{h}_0 & \tilde{h}_1 \end{bmatrix} \quad (10.2)$$

We use (10.1) and (10.2) to compute $H\tilde{H}^T = I_5$. The result of the multiplication is the 5×5 matrix

$$HH̃^T = \begin{bmatrix} a & b & 0 & 0 & c \\ c & a & b & 0 & 0 \\ 0 & c & a & b & 0 \\ 0 & 0 & c & a & b \\ b & 0 & 0 & c & a \end{bmatrix}$$

where

$$a = h_{-1}\tilde{h}_{-1} + h_0\tilde{h}_0 + h_1\tilde{h}_1$$
$$b = h_1\tilde{h}_{-1} + h_2\tilde{h}_0$$
$$c = h_{-2}\tilde{h}_0 + h_{-1}\tilde{h}_1$$

Since $HH̃^T = I_5$, $a = 1$ and $b = c = 0$, so we have the following conditions:

$$1 = h_{-1}\tilde{h}_{-1} + h_0\tilde{h}_0 + h_1\tilde{h}_1$$
$$0 = h_1\tilde{h}_{-1} + h_2\tilde{h}_0 \qquad (10.3)$$
$$0 = h_{-2}\tilde{h}_0 + h_{-1}\tilde{h}_1$$

Another way to view the orthogonality conditions (10.3) is to display $\tilde{\mathbf{h}}$ in one row and the 2-shifts of $\mathbf{h}$ that overlap it in the subsequent rows:

$$
\begin{array}{ccccccc}
 & & & \tilde{h}_1 & \tilde{h}_0 & \tilde{h}_{-1} & \\
h_2 & h_1 & h_0 & h_{-1} & h_{-2} & & \\
 & h_2 & h_1 & h_0 & h_{-1} & h_{-2} & \\
 & & h_2 & h_1 & h_0 & h_{-1} & h_{-2}
\end{array} \qquad (10.4)
$$

Using (10.4), we see that dotting the top row with rows 2 and 4 must result in zero while dotting the top row with row three must be 1. We can rewrite (10.4) algebraically and see that the orthogonality conditions that must be satisfied by the elements of H and $\tilde{H}$ are

$$\sum_{k=-1}^{1} \tilde{h}_k h_k = 1 \qquad and \qquad \sum_{k=-1}^{1} \tilde{h}_k h_{k-2m} = 0, \qquad m = -1,1$$

This system should look familiar to you. Other than the summation limits on the second sum and the fact that we have $\tilde{h}_k$ instead of h_k for one of the factors in each sum, the system could be the same as that given in Corollary 8.1 from Section 8.1! The fact that the system resembles that of Corollary 8.1 shouldn't surprise you — the orthogonality conditions (8.6) and (8.7) resulted when we multiplied HH^T to get the identity matrix. In this example, we are computing $HH̃^T$ to get the identity matrix.

It should be clear that the elements of G and $\tilde{G}$ must satisfy a similar system.[1] *That is, $G\tilde{G}^T = I_5$ implies that*

$$\sum_{k=-1}^{1} \tilde{g}_k g_k = 1 \quad and \quad \sum_{k=-1}^{1} \tilde{g}_k g_{k-2m} = 0, \quad m = -1, 1$$

For the identity $H\tilde{G}^T = 0_5$, it is a straightforward computation to see that the elements of these matrices must satisfy

$$\sum_{k=-1}^{1} \tilde{g}_k h_{k-2m} = 0, \quad m = -1, 0, 1 \tag{10.5}$$

and for $G\tilde{H}^T = 0$, we have

$$\sum_{k=-1}^{1} \tilde{h}_k g_{k-2m} = 0, \quad m = -1, 0, 1 \tag{10.6}$$

In Problem 10.2 you will verify the results of (10.5) and (10.6).

The identities from (10.5) and (10.6) match those from (8.14) in Theorem 8.2 with $\mathbf{a} = \tilde{\mathbf{g}}, \mathbf{b} = \mathbf{h}$ in (10.5) and $\mathbf{a} = \tilde{\mathbf{h}}, \mathbf{b} = \mathbf{g}$ in (10.6).   □

Note: A formulation similar to (10.1) and (10.2) is used for all subsequent wavelet transform matrices.

Orthogonality for Two Filters

Example 10.1 was provided in part to motivate the following result. Theorem 10.1 is a generalization of Theorem 8.1. The proof is almost identical to that of Theorem 8.1 so it is left as Problem 10.3.

Theorem 10.1 (Orthogonality Conditions for Two Filters). *Suppose that $\mathbf{a}$ and $\mathbf{b}$ are bi-infinite sequences whose Fourier series are given by*

$$A(\omega) = \sum_{k=-\infty}^{\infty} a_k e^{ik\omega} \quad and \quad B(\omega) = \sum_{k=-\infty}^{\infty} b_k e^{ik\omega}$$

respectively. Then

$$A(\omega)\overline{B(\omega)} + A(\omega + \pi)\overline{B(\omega + \pi)} = 2 \tag{10.7}$$

---

[1]For now, the matrices G and $\tilde{G}$ look just like the matrices in (10.1) and (10.2), with h and $\tilde{h}$ replaced by g and $\tilde{g}$, respectively.

if and only if

$$\sum_{k=-\infty}^{\infty} a_k b_k = 1 \tag{10.8}$$

and for $m \in \mathbb{Z}$ *with* $m \neq 0$,

$$\sum_{k=-\infty}^{\infty} a_k b_{k-2m} = 0 \tag{10.9}$$

$\square$

Proof of Theorem 10.1. The proof is left as Problem 10.3. $\square$

Orthogonality in the Fourier Domain

We can use Theorem 10.1 to identify the orthogonality conditions that the filter pairs **h** and $\tilde{\mathbf{h}}$, and **g** and $\tilde{\mathbf{g}}$ must satisfy to ensure $H_N \tilde{H}_N^T = I_{N/2}, G_N \tilde{G}_N^T = I_{N/2}$, respectively. We take $A(\omega) = \tilde{H}(\omega)$ and $B(\omega) = H(\omega)$ in Theorem 10.1 to write

$$\boxed{\tilde{H}(\omega)\overline{H(\omega)} + \tilde{H}(\omega + \pi)\overline{H(\omega + \pi)} = 2} \tag{10.10}$$

If $H(\omega)$ and $\tilde{H}(\omega)$ satisfy (10.10), then we have

$$\sum_{k \in \mathbb{Z}} \tilde{h}_k h_k = 1 \tag{10.11}$$

and for $m \in \mathbb{Z}, m \neq 0$,

$$\sum_{k \in \mathbb{Z}} \tilde{h}_k h_{k-2m} = 0 \tag{10.12}$$

A similar substitution in Theorem 10.1 using $\tilde{G}(\omega)$ and $G(\omega)$ gives

$$\boxed{\tilde{G}(\omega)\overline{G(\omega)} + \tilde{G}(\omega + \pi)\overline{G(\omega + \pi)} = 2} \tag{10.13}$$

so that

$$\sum_{k \in \mathbb{Z}} \tilde{g}_k g_k = 1 \tag{10.14}$$

and for $m \in \mathbb{Z}, m \neq 0$,

$$\sum_{k \in \mathbb{Z}} \tilde{g}_k g_{k-2m} = 0 \tag{10.15}$$

To ensure that $H\tilde{G}^T = 0$ and $G\tilde{H}^T = 0$, we utilize Theorem 8.2. If we take $A(\omega) = \tilde{G}$ and $B(\omega) = H(\omega)$ in Theorem 8.2, we have

$$\boxed{\tilde{G}(\omega)\overline{H(\omega)} + \tilde{G}(\omega + \pi)\overline{H(\omega + \pi)} = 0} \tag{10.16}$$

so that for all $m \in \mathbb{Z}$,

$$\sum_{k \in \mathbb{Z}} \tilde{g}_k h_{k-2m} = 0 \qquad (10.17)$$

and if we take $A(\omega) = \tilde{H}(\omega)$ and $B(\omega) = G(\omega)$ in Theorem 8.2, we have

$$\boxed{\tilde{H}(\omega)\overline{G(\omega)} + \tilde{H}(\omega + \pi)\overline{G(\omega + \pi)} = 0} \qquad (10.18)$$

so that for all $m \in \mathbb{Z}$,

$$\sum_{k \in \mathbb{Z}} \tilde{h}_k g_{k-2m} = 0 \qquad (10.19)$$

Thus, conditions (10.10), (10.13), (10.16), and (10.18) are exactly the conditions we need the Fourier series associated with $\mathbf{h}$, $\tilde{\mathbf{h}}$, $\mathbf{g}$, and $\tilde{\mathbf{g}}$ to satisfy in order to construct wavelet transform matrices W_N and $\tilde{W}_N$ that satisfy $W_N \tilde{W}_N^T = I_N$.

Biorthogonal Filters Defined

We are now ready to define biorthogonal filters.

Definition 10.1 (Biorthogonal Filters). *Suppose that the Fourier series $H(\omega)$ and $\tilde{H}(\omega)$ of filters $\mathbf{h}$ and $\tilde{\mathbf{h}}$, respectively, satisfy*

$$\boxed{\tilde{H}(\omega)\overline{H(\omega)} + \tilde{H}(\omega + \pi)\overline{H(\omega + \pi)} = 2} \qquad (10.20)$$

Then $\mathbf{h}$ and $\tilde{\mathbf{h}}$ are called a biorthogonal filter pair. *The identity (10.20) is called the* biorthogonality condition.

□

Building the Four Filters

Our task is to find Fourier series $H(\omega)$, $\tilde{H}(\omega)$, $G(\omega)$, and $\tilde{G}(\omega)$ that satisfy (10.10), (10.13), (10.16), and (10.18), respectively. From these Fourier series we can extract the filters $\mathbf{h}$, $\tilde{\mathbf{h}}$, $\mathbf{g}$, and $\tilde{\mathbf{g}}$.

In all of our constructions to date, we have created the highpass filter $\mathbf{g}$ from the lowpass filter $\mathbf{h}$, or equivalently, as we saw in Section 8.2, the Fourier series $G(\omega)$ from the Fourier series $H(\omega)$. In fact, Proposition 8.2 tells us exactly how to do this. It seems natural that a similar choice for $G(\omega)$ and $\tilde{G}(\omega)$ might work for the filters in this section as well.

The following proposition summarizes the process needed to construct biorthogonal filter pairs.

Proposition 10.1 (Constructing $G(\omega)$ and $\tilde{G}(\omega)$ From $H(\omega)$ and $\tilde{H}(\omega)$). *Suppose that* h *and* $\tilde{h}$ *are a biorthogonal filter pair. That is, the Fourier series $H(\omega)$ and $\tilde{H}(\omega)$ satisfy the biorthogonality condition (10.20).*

If

$$\boxed{G(\omega) = e^{i(n\omega+b)}\overline{\tilde{H}(\omega+\pi)}} \tag{10.21}$$

and

$$\boxed{\tilde{G}(\omega) = e^{i(n\omega+b)}\overline{H(\omega+\pi)}} \tag{10.22}$$

for odd integer n *and real number* b, *then*

$$\tilde{G}(\omega)\overline{G(\omega)} + \tilde{G}(\omega+\pi)\overline{G(\omega+\pi)} = 2 \tag{10.23}$$

$$\tilde{H}(\omega)\overline{G(\omega)} + \tilde{H}(\omega+\pi)\overline{G(\omega+\pi)} = 0 \tag{10.24}$$

$$\tilde{G}(\omega)\overline{H(\omega)} + \tilde{G}(\omega+\pi)\overline{H(\omega+\pi)} = 0 \tag{10.25}$$

$\square$

Proof of Proposition 10.1. The proof is quite straightforward in that we assume that (10.20) holds and simply plug (10.21) and (10.22) into the left sides of (10.23), (10.24), and (10.25) and verify that the identities hold. For (10.23), we start by using (10.22) and (10.21) to write

$$\tilde{G}(\omega)\overline{G(\omega)} = e^{i(n\omega+b)}\overline{H(\omega+\pi)}\overline{e^{i(n\omega+b)}\overline{\tilde{H}(\omega+\pi)}}$$
$$= e^{i(n\omega+b)}\overline{H(\omega+\pi)}e^{-i(n\omega+b)}\tilde{H}(\omega+\pi)$$
$$= \tilde{H}(\omega+\pi)\overline{H(\omega+\pi)} \tag{10.26}$$

If we replace ω by $\omega+\pi$ in (10.26) and recall that $H(\omega)$ and $\tilde{H}(\omega)$ are 2π-periodic functions, we have
$$\tilde{G}(\omega+\pi)\overline{G(\omega+\pi)} = \tilde{H}(\omega)\overline{H(\omega)} \tag{10.27}$$

Adding (10.26) and (10.27) gives

$$\tilde{G}(\omega)\overline{G(\omega)} + \tilde{G}(\omega+\pi)\overline{G(\omega+\pi)} = \tilde{H}(\omega)\overline{H(\omega)} + \tilde{H}(\omega+\pi)\overline{H(\omega+\pi)} = 2$$

and establishes (10.23). To prove (10.24) we use (10.22) to write

$$\tilde{H}(\omega)\overline{G(\omega)} = \tilde{H}(\omega)\overline{e^{i(n\omega+b)}\overline{\tilde{H}(\omega+\pi)}} = \tilde{H}(\omega)e^{-i(n\omega+b)}\tilde{H}(\omega+\pi) \tag{10.28}$$

If we replace ω by $\omega+\pi$ in (10.28) and simplify, we have

$$\tilde{H}(\omega+\pi)\overline{G(\omega+\pi)} = \tilde{H}(\omega+\pi)e^{-i(n\omega+n\pi+b)}\tilde{H}(\omega+2\pi)$$
$$= \tilde{H}(\omega+\pi)e^{-in\pi}e^{-i(n\omega+b)}\tilde{H})\omega)$$
$$= (-1)^n\tilde{H}(\omega+\pi)e^{-i(n\omega+b)}\tilde{H}(\omega) \tag{10.29}$$

But n is an odd integer so $(-1)^n = -1$ and we see that (10.29) is the opposite of (10.28). Adding (10.29) and (10.28) gives zero and proves (10.24).

The proof of (10.25) is very similar to the proof for (10.24) and we leave it as Problem 10.5. ☐

We have the following corollary that tells us how to write the filter elements of g and g̃.

Corollary 10.1 (The Filters g and g̃ in Terms of h̃ and h). *Let n be an odd integer and b be any real number. If*

$$G(\omega) = e^{i(n\omega+b)}\overline{\tilde{H}(\omega + \pi)} \qquad and \qquad \tilde{G}(\omega) = e^{i(n\omega+b)}\overline{H(\omega + \pi)}$$

then the elements of g *and* g̃ *are*

$$g_k = -e^{ib}(-1)^k\tilde{h}_{n-k} \qquad and \qquad \tilde{g}_k = -e^{ib}(-1)^k h_{n-k} \qquad (10.30)$$

respectively. ☐

Proof of Corollary 10.1. The proof is left as Problem 10.6. ☐

Proposition 10.1 reduces our construction to the task of finding two Fourier series $\tilde{H}(\omega)$ and $H(\omega)$ that satisfy the biorthogonality condition (10.20). Once we have $\tilde{H}(\omega)$ and $H(\omega)$, Proposition 10.1 tells us how to construct $\tilde{G}(\omega)$ and $G(\omega)$ so that the orthogonality conditions (10.13), (10.16), and (10.18) are satisfied. We can extract h̃ and h from $\tilde{H}(\omega)$ and $H(\omega)$, respectively, and Corollary 10.1 tells us how to build g̃ and g.

Note: Since we consider only real-valued filters in this book, the only values that b can assume in (10.30) are 0 and π. It is easy to see that the filters resulting from $b = \pi$ are the negatives of the filters obtained by setting $b = 0$. Moreover, the role of the odd integer n in (10.30) is to shift the filter elements n units right (left) if n is positive (negative). Throughout the remainder of the book, we use $n = 1$. Thus, we will use the rules

$$\boxed{g_k = (-1)^k\tilde{h}_{1-k} \qquad and \qquad \tilde{g}_k = (-1)^k h_{1-k}} \qquad (10.31)$$

for our filter pair g and g̃.

A Biorthogonal Filter Pair

Let's look at an example.

Example 10.2 (A Biorthogonal Filter Pair). *Consider the filters*

$$\tilde{\mathbf{h}} = \left(\tilde{h}_{-1}, \tilde{h}_0, \tilde{h}_1 \right) = (\frac{\sqrt{2}}{4}, \frac{\sqrt{2}}{2}, \frac{\sqrt{2}}{4})$$

and

$$\mathbf{h} = (h_{-2}, h_{-1}, h_0, h_1, h_2) = \left(-\frac{\sqrt{2}}{2}, \sqrt{2}, \frac{3\sqrt{2}}{4}, -\frac{\sqrt{2}}{2}, \frac{\sqrt{2}}{4} \right)$$

Verify the biorthogonality condition (10.20), find the wavelet filters, and write down the wavelet matrices W_8 and $\tilde{W}_8$.

Solution. *The Fourier series are*

$$\tilde{H}(\omega) = \frac{\sqrt{2}}{4} e^{-i\omega} + \frac{\sqrt{2}}{2} + \frac{\sqrt{2}}{4} e^{i\omega}$$

and

$$H(\omega) = -\frac{\sqrt{2}}{2} e^{-2i\omega} + \sqrt{2}e^{-i\omega} + \frac{3\sqrt{2}}{4} - \frac{\sqrt{2}}{2} e^{i\omega} + \frac{\sqrt{2}}{4} e^{2i\omega}$$

It is easy to verify that $\tilde{H}(0) = H(0) = \sqrt{2}$ and $\tilde{H}(\pi) = H(\pi) = 0$.

Conjugating $H(\omega)$ gives

$$\overline{H(\omega)} = \frac{\sqrt{2}}{4} e^{-2i\omega} - \frac{\sqrt{2}}{2} e^{-i\omega} + \frac{3\sqrt{2}}{4} + \sqrt{2}e^{i\omega} - \frac{\sqrt{2}}{2} e^{2i\omega}$$

We first compute $\tilde{H}(\omega)\overline{H(\omega)}$. After some simplification, we obtain

$$\tilde{H}(\omega)\overline{H(\omega)} = \frac{1}{8} e^{-3i\omega} + 1 + \frac{9}{8} e^{i\omega} - \frac{1}{4} e^{3i\omega} \qquad (10.32)$$

If we replace ω by $\omega + \pi$ in (10.32), and use the fact that $e^{\pm 3i\omega} = e^{i\pi} = -1$, we obtain

$$\tilde{H}(\omega + \pi)\overline{H(\omega + \pi)} = \frac{1}{8} e^{-3i(\omega+\pi)} + 1 + \frac{9}{8} e^{i(\omega+\pi)} - \frac{1}{4} e^{3i(\omega+\pi)}$$

$$= \frac{1}{8} e^{-3i\pi} e^{-3i\omega} + 1 + \frac{9}{8} e^{i\pi} e^{i\omega} - \frac{1}{4} e^{3i\pi} e^{3i\omega}$$

$$= -\frac{1}{8} e^{-3i\omega} + 1 - \frac{9}{8} e^{i\omega} + \frac{1}{4} e^{3i\omega} \qquad (10.33)$$

If we add (10.32) and (10.33), we obtain the orthogonality condition (10.20).

To find the wavelet filter $\tilde{\mathbf{g}}$, we use (10.31). For $k = -1$ we have

$$\tilde{g}_{-1} = (-1)^{-1}h_{1-(-1)} = -h_2 = -\frac{\sqrt{2}}{4}$$

The other nonzero values result when $k = 0, 1, 2, 3$. We have

$$\tilde{g}_0 = h_1 = -\frac{\sqrt{2}}{2} \qquad \tilde{g}_1 = -h_0 = -\frac{3\sqrt{2}}{4}$$

$$\tilde{g}_2 = h_{-1} = \sqrt{2} \qquad \tilde{g}_3 = -h_{-2} = \frac{\sqrt{2}}{2}$$

Again using (10.31), we see that the nonzero elements of $\mathbf{g}$ are

$$g_0 = \tilde{h}_1 = \frac{\sqrt{2}}{4} \qquad g_1 = -\tilde{h}_0 = -\frac{\sqrt{2}}{2} \qquad g_2 = \tilde{h}_{-1} = \frac{\sqrt{2}}{4}$$

We can now write down the wavelet matrices W_8 and $\tilde{W}_8$. It should be easy to generalize the formulation of these matrices for any even dimension N.

$$W_8 = \left[\begin{array}{cccccccc}
h_0 & h_1 & h_2 & 0 & 0 & 0 & h_{-2} & h_{-1} \\
h_{-2} & h_{-1} & h_0 & h_1 & h_2 & 0 & 0 & 0 \\
0 & 0 & h_{-2} & h_{-1} & h_0 & h_1 & h_2 & 0 \\
h_2 & 0 & 0 & 0 & h_{-2} & h_{-1} & h_0 & h_1 \\
\hline
g_0 & g_1 & g_2 & 0 & 0 & 0 & 0 & 0 \\
0 & 0 & g_0 & g_1 & g_2 & 0 & 0 & 0 \\
0 & 0 & 0 & 0 & g_0 & g_1 & g_2 & 0 \\
g_2 & 0 & 0 & 0 & 0 & 0 & g_0 & g_1
\end{array}\right]$$

$$= \left[\begin{array}{cccccccc}
\frac{3\sqrt{2}}{4} & -\frac{\sqrt{2}}{2} & \frac{\sqrt{2}}{4} & 0 & 0 & 0 & -\frac{\sqrt{2}}{2} & \sqrt{2} \\
-\frac{\sqrt{2}}{2} & \sqrt{2} & \frac{3\sqrt{2}}{4} & -\frac{\sqrt{2}}{2} & \frac{\sqrt{2}}{4} & 0 & 0 & 0 \\
0 & 0 & -\frac{\sqrt{2}}{2} & \sqrt{2} & \frac{3\sqrt{2}}{4} & -\frac{\sqrt{2}}{2} & \frac{\sqrt{2}}{4} & 0 \\
\frac{\sqrt{2}}{4} & 0 & 0 & 0 & -\frac{\sqrt{2}}{2} & \sqrt{2} & \frac{3\sqrt{2}}{4} & -\frac{\sqrt{2}}{2} \\
\hline
\frac{\sqrt{2}}{4} & -\frac{\sqrt{2}}{2} & \frac{\sqrt{2}}{4} & 0 & 0 & 0 & 0 & 0 \\
0 & 0 & \frac{\sqrt{2}}{4} & -\frac{\sqrt{2}}{2} & \frac{\sqrt{2}}{4} & 0 & 0 & 0 \\
0 & 0 & 0 & 0 & \frac{\sqrt{2}}{4} & -\frac{\sqrt{2}}{2} & \frac{\sqrt{2}}{4} & 0 \\
\frac{\sqrt{2}}{4} & 0 & 0 & 0 & 0 & 0 & \frac{\sqrt{2}}{4} & -\frac{\sqrt{2}}{2}
\end{array}\right]$$

and

$$\tilde{W}_8 = \begin{bmatrix} \tilde{h}_0 & \tilde{h}_1 & 0 & 0 & 0 & 0 & 0 & \tilde{h}_{-1} \\ 0 & \tilde{h}_{-1} & \tilde{h}_0 & \tilde{h}_1 & 0 & 0 & 0 & 0 \\ 0 & 0 & 0 & \tilde{h}_{-1} & \tilde{h}_0 & \tilde{h}_1 & 0 & 0 \\ 0 & 0 & 0 & 0 & 0 & \tilde{h}_{-1} & \tilde{h}_0 & \tilde{h}_1 \\ \tilde{g}_0 & \tilde{g}_1 & \tilde{g}_2 & \tilde{g}_3 & 0 & 0 & 0 & \tilde{g}_{-1} \\ 0 & \tilde{g}_{-1} & \tilde{g}_0 & \tilde{g}_1 & \tilde{g}_2 & \tilde{g}_3 & 0 & 0 \\ 0 & 0 & 0 & \tilde{g}_{-1} & \tilde{g}_0 & \tilde{g}_1 & \tilde{g}_2 & \tilde{g}_3 \\ \tilde{g}_2 & \tilde{g}_3 & 0 & 0 & 0 & \tilde{g}_{-1} & \tilde{g}_0 & \tilde{g}_1 \end{bmatrix}$$

$$= \begin{bmatrix} \frac{\sqrt{2}}{2} & \frac{\sqrt{2}}{4} & 0 & 0 & 0 & 0 & 0 & \frac{\sqrt{2}}{4} \\ 0 & \frac{\sqrt{2}}{4} & \frac{\sqrt{2}}{2} & \frac{\sqrt{2}}{4} & 0 & 0 & 0 & 0 \\ 0 & 0 & 0 & \frac{\sqrt{2}}{4} & \frac{\sqrt{2}}{2} & \frac{\sqrt{2}}{4} & 0 & 0 \\ 0 & 0 & 0 & 0 & 0 & \frac{\sqrt{2}}{4} & \frac{\sqrt{2}}{2} & \frac{\sqrt{2}}{4} \\ -\frac{\sqrt{2}}{2} & -\frac{3\sqrt{2}}{4} & \sqrt{2} & \frac{\sqrt{2}}{2} & 0 & 0 & 0 & -\frac{\sqrt{2}}{4} \\ 0 & -\frac{\sqrt{2}}{4} & -\frac{\sqrt{2}}{2} & -\frac{3\sqrt{2}}{4} & \sqrt{2} & \frac{\sqrt{2}}{2} & 0 & 0 \\ 0 & 0 & 0 & -\frac{\sqrt{2}}{4} & -\frac{\sqrt{2}}{2} & -\frac{3\sqrt{2}}{4} & \sqrt{2} & \frac{\sqrt{2}}{2} \\ \sqrt{2} & \frac{\sqrt{2}}{2} & 0 & 0 & 0 & -\frac{\sqrt{2}}{4} & -\frac{\sqrt{2}}{2} & -\frac{3\sqrt{2}}{4} \end{bmatrix}$$

*You should use a **CAS** to verify that $\tilde{W}_8^{-1} = W_8^T$.* ◻

Finite-Length Filters

The filters in Example 10.2 were given to us — no information was provided about how they were obtained. To formulate methods for using Proposition 10.1 to construct biorthogonal filter pairs, it is worthwhile to have a closer look at our results thus far for finite-length filters.

As was the case in Section 8.1 (see Proposition 8.1), there are some constraints on the starting and stopping indices of finite-length filters. Let's look at some examples.

Example 10.3 (Finite Length Filters and Orthogonality). *Suppose that each of the following filter pairs* h *and* $\tilde{\text{h}}$ *satisfy (10.10). Write down and explore the orthogonality conditions for the elements of* h *and* h.

(a) $\text{h} = (h_{-2}, h_{-1}, h_0, h_1, h_2, h_3)$ *and* $\tilde{\text{h}} = (\tilde{h}_{-1}, \tilde{h}_0, \tilde{h}_1)$.

(b) $\text{h} = (h_{-2}, h_{-1}, h_0, h_1, h_2, h_3)$ *and* $\tilde{\text{h}} = (\tilde{h}_0, \tilde{h}_1)$.

(c) $\text{h} = (h_{-2}, h_{-1}, h_0, h_1, h_2, h_3)$ *and* $\tilde{\text{h}} = (\tilde{h}_{-1}, \tilde{h}_0, \tilde{h}_1, \tilde{h}_2)$.

(d) $\text{h} = (h_{-3}, \ldots, h_3)$ *and* $\tilde{\text{h}} = (\tilde{h}_{-1}, \tilde{h}_0, \tilde{h}_1)$

Solution. *For each filter pair, we will write down* $\tilde{\mathbf{h}}$ *in the top row of a table and then all the 2-translates of* $\mathbf{h}$ *that overlap* $\tilde{\mathbf{h}}$ *in subsequent rows. The table for the filters in (a) is*

				$\tilde{\mathbf{h}}_1$	$\tilde{\mathbf{h}}_0$	$\tilde{\mathbf{h}}_{-1}$			
h_3	h_2	h_1	h_0	h_{-1}	h_{-2}				
		h_3	h_2	h_1	h_0	h_{-1}	h_{-2}		
			h_3	h_2	h_1	h_0	h_{-1}	h_{-2}	
				$\mathbf{h_3}$	h_2	h_1	h_0	h_{-1}	h_{-2}

Dotting row 1 with row 3 is (10.11), while row 1 dotted with the remaining rows gives the nontrivial identities in (10.12). Note that dotting row 1 with row 5 gives $\tilde{h}_{-1}h_3 = 0$, *so that either* $\tilde{h}_{-1} = 0$ *or* $h_3 = 0$. *In the former case we have contradicted the fact that* $\tilde{h}_{-1} \neq 0$ *(guaranteed by the fact that* $\tilde{\mathbf{h}}$ *is a finite-length filter with starting index* -1), *and in the case of the latter, we have contradicted the fact that* $\mathbf{h}$ *is a finite-length filter with stopping index 3.*

We have the same problem in the general case if the lengths of the filters are of different parity. There will always be a 2-translate of one filter that overlaps the other filter for only one term.

Let's consider the filters in part (b). In this case, the lengths of both filters are even and in table form we have

				$\tilde{\mathbf{h}}_1$	$\tilde{\mathbf{h}}_0$		
h_3	h_2	h_1	h_0	h_{-1}	h_{-2}		
		h_3	h_2	h_1	h_0	h_{-1}	h_{-2}
			h_3	h_2	h_1	h_0	h_{-1}	h_{-2}

The orthogonality conditions are

$$\tilde{h}_1 h_1 + \tilde{h}_0 h_0 = 1$$
$$\tilde{h}_1 h_{-1} + \tilde{h}_0 h_{-2} = 0$$
$$\tilde{h}_1 h_3 + \tilde{h}_0 h_2 = 0$$

For part (c), both filters are again even and the table is

				$\tilde{\mathbf{h}}_2$	$\tilde{\mathbf{h}}_1$	$\tilde{\mathbf{h}}_0$	$\tilde{\mathbf{h}}_{-1}$			
h_3	h_2	h_1	h_0	h_{-1}	$\mathbf{h_{-2}}$					
	h_3	h_2	h_1	h_0	h_{-1}	h_{-2}				
		h_3	h_2	h_1	h_0	h_{-1}	h_{-2}			
			h_3	h_2	h_1	h_0	h_{-1}	h_{-2}		
				$\mathbf{h_3}$	h_2	h_1	h_0	h_{-1}	h_{-2}	

The orthogonality conditions are

$$\tilde{h}_{-1}h_{-1} + \tilde{h}_0 h_0 + \tilde{h}_1 h_1 + \tilde{h}_2 h_2 = 1$$
$$\tilde{h}_{-1}h_1 + \tilde{h}_0 h_2 + \tilde{h}_1 h_3 = 0$$
$$\mathbf{\tilde{h}_{-1}h_3 = 0}$$
$$\tilde{h}_0 h_{-2} + \tilde{h}_1 h_{-1} + \tilde{h}_2 h_0 = 0$$
$$\mathbf{\tilde{h}_2 h_{-2} = 0}$$

Note that the first bold identity implies that either $\tilde{h}_{-1} = 0$, or $h_3 = 0$ while the second bold identity says that either $h_{-2} = 0$ or $\tilde{h}_2 = 0$. In any case, we have some starting or stopping index of a filter equal to zero and this contradicts the definition of a finite-length filter.

For part (d), the table is

						$\tilde{h}_1$		$\tilde{h}_0$		$\tilde{h}_{-1}$				
h_3	h_2	h_1	h_0	h_{-1}	h_{-2}	$\mathbf{h_{-3}}$								
		h_3	h_2	h_1	h_0	h_{-1}	h_{-2}	h_{-3}						
				h_3	h_2	h_1	h_0	h_{-1}	h_{-2}	h_{-3}				
						h_3	h_2	h_1	h_0	h_{-1}	h_{-2}	h_{-3}		
								$\mathbf{h_3}$	h_2	h_1	h_0	h_{-1}	h_{-2}	h_{-3}

and the orthogonality conditions are

$$\tilde{h}_{-1}h_{-1} + \tilde{h}_0 h_0 + \tilde{h}_1 h_1 = 1$$
$$\mathbf{\tilde{h}_1 h_{-3} = 0}$$
$$\tilde{h}_{-1}h_{-3} + \tilde{h}_0 h_{-2} + \tilde{h}_1 h_{-1} = 0$$
$$\tilde{h}_{-1}h_1 + \tilde{h}_0 h_2 + \tilde{h}_1 h_3 = 0$$
$$\mathbf{\tilde{h}_{-1}h_3 = 0}$$

Note that the second equation says that either $\tilde{h}_1 = 0$, or $h_{-3} = 0$ while the last equation implies that either $\tilde{h}_{-1} = 0$ or $h_3 = 0$. In either case we have contradicted the assumption that $\tilde{h}$ and h are finite-length filters. □

Lengths of Filters and Orthogonality

What went "wrong" with parts (a), (c), and (d) of Example 10.3? Part (b) contained two even-length filters and we were able to write down the orthogonality conditions. The problem is that in the case of even-length filters, we need to make sure that every possible 2-shift of the longer filter overlaps at least two terms of the shorter filter. See Problem 10.4 for more examples of finite-length filters that satisfy orthogonality conditions.

In the case of finite filters of even length, the best way to ensure that we do not have problems like those encountered in part (c) of Example 10.3 is to insist that the stopping indices for the filters either both be even or both be odd, and for part (d) we require one stopping index to be odd and the other even. Proposition 10.2 tells us the exact orthogonality conditions that are satisfied by finite-length filters.

Proposition 10.2 (Orthogonality for Finite-Length Filters). *Consider the finite length filters* $\mathbf{a} = (a_\ell, \ldots, a_L)$ *and* $\mathbf{b} = (b_{\ell'}, \ldots, b_{L'})$ *with* $L < L'$ *whose Fourier series* $A(\omega)$ *and* $B(\omega)$ *satisfy (10.7). Denote by* N *and* N' *the lengths of filters* $\mathbf{a}$ *and* $\mathbf{b}$, *respectively. Then* N *and* N' *have the same parity (see Definition 8.2). In the case where* N *and* N' *are even,* L *and* L' *have the same parity and in the case where* N *and* N' *are odd,* L *and* L' *have different parity.* □

Proof of Proposition 10.2. The proof is left as Problem 10.7. □

Symmetric Filters

We can provide more information about the system of equations that must be satisfied by $\tilde{\mathbf{h}}$ and $\mathbf{h}$ if we insist that the filters are symmetric. We now give a precise definition of what we mean by symmetric filters.

Since all filters we consider are of finite length, we need to clarify what we mean by a symmetric filter. In the case when the filter length is odd, we require our filter to be symmetric about zero and when the filter length is even, we ask that our filters be symmetric about $\frac{1}{2}$. We have the following definition.

Definition 10.2 (Symmetric Filters). *Let* $\mathbf{h} = (h_\ell, \ldots, h_L)$ *be a finite-length filter with length* $N = L - \ell + 1$. *We say that* $\mathbf{h}$ *is symmetric if*

(a) $h_k = h_{-k}$ *for all* $k \in \mathbb{Z}$ *whenever* N *is odd.*

(b) $h_k = h_{1-k}$ *for all* $k \in \mathbb{Z}$ *whenever* N *is even.*

□

Note that the Haar filter $\mathbf{h} = (h_0, h_1) = (\frac{\sqrt{2}}{2}, \frac{\sqrt{2}}{2})$ is a symmetric filter satisfying $h_k = h_{1-k}$ for $k = 0,1$. An example of a symmetric filter of odd length is $\mathbf{h} = (h_{-1}, h_0, h_1) = (1, 2, 1)$. The filter $\tilde{\mathbf{h}}$ from Example 10.2 is symmetric, whereas, the filter $\mathbf{h}$ from that example is not.

It is clear that symmetry imposes some restrictions on ℓ and L in the case where $\mathbf{h} = (h_\ell, \ldots, h_L)$. We have the following proposition.

Proposition 10.3 (Starting and Stopping Indices for Symmetric Finite-Length Filters). *Suppose that* $\mathbf{a} = (a_\ell, \ldots, a_L)$ *is a symmetric finite-length filter with* $L > 0$. *Let* $N = L - \ell + 1$ *denote the length of* $\mathbf{a}$.
If N *is odd, then* $\ell = -L$. *If* N *is even, then* $\ell = -L + 1$. □

Proof of Proposition 10.3. The proof is left as Problem 10.8. □

A System of Equations for Constructing Symmetric Biorthogonal Filter Pairs

The following corollary gives us some more information on the orthogonality conditions satisfied by symmetric filters in Proposition 10.2.

Corollary 10.2 (Orthogonality: A System of Equations). *Suppose that* $\mathbf{a} = (a_{\tilde{\ell}}, \ldots, a_{\tilde{L}})$ *and* $\mathbf{b} = (b_{\ell}, \ldots, b_L)$ *are symmetric finite-length filters with* $\tilde{L} < L$ *that satisfy the conditions of Proposition 10.2. Let* $\tilde{N}$ *and* N *be the lengths of* $\mathbf{a}$ *and* $\mathbf{b}$*, respectively. If* $\tilde{N}$ *and* N *are both odd (even) and we define* $p = -\tilde{L}$ *(* $p = -\tilde{L} + 1$ *), then*

$$\sum_{k=p}^{\tilde{L}} a_k b_k = 1 \tag{10.34}$$

$$\sum_{k=p}^{\tilde{L}} a_k b_{k-2m} = 0 \qquad m = 1, \ldots, \tilde{L} \tag{10.35}$$

☐

Proof of Corollary 10.2. The proof is left as Problem 10.9.

☐

Constructing a Biorthogonal Filter Pair

The filters that appear in Example 10.4 can be constructed using the techniques developed in Section 10.2.

Example 10.4 (A Biorthogonal Filter Pair). *Construct a symmetric biorthogonal filter pair* $\tilde{\mathbf{h}}$ *and* $\mathbf{h}$ *whose lengths are 2 and 6, respectively.*
Solution. We can use Proposition 10.3 to find the starting and stopping indices of the filters. We have $\tilde{\mathbf{h}} = (\tilde{h}_0, \tilde{h}_1)$ *and* $\mathbf{h} = (h_{-2}, h_{-1}, h_0, h_1, h_2, h_3)$*. Since we are to build symmetric filters, we see that*

$$\tilde{h}_0 = \tilde{h}_1 \qquad h_0 = h_1 \qquad h_{-1} = h_2 \qquad h_{-2} = h_3$$

so that the terminal index for $\tilde{\mathbf{h}}$ *is* $\tilde{L} = 1$ *and the terminal index for* $\mathbf{h}$ *is* $L = 3$.
The Fourier series for each filter are

$$\tilde{H}(\omega) = \tilde{h}_0 + \tilde{h}_1 e^{i\omega} = \tilde{h}_0 + \tilde{h}_0 e^{i\omega} \tag{10.36}$$

and

$$\begin{aligned} H(\omega) &= h_{-2}\, e^{-2i\omega} + h_{-1}\, e^{-i\omega} + h_0 + h_1 e^{i\omega} + h_2 e^{2i\omega} + h_3 e^{3i\omega} \\ &= h_3\, e^{-2i\omega} + h_2 e^{-i\omega} + h_1 + h_1 e^{i\omega} + h_2 e^{2i\omega} + h_3 e^{3i\omega} \end{aligned} \tag{10.37}$$

$\tilde{h}$ *and* h *must be lowpass filters. In the case of* $\tilde{h}$, *we plug zero into (10.36) to obtain*

$$\tilde{H}(0) = 2h_0 = \sqrt{2}$$

so that $h_0 = h_1 = \sqrt{2}/2$. *Thus we see that the only symmetric, length 2, lowpass filter is the Haar filter. For* h *we plug* $\omega = 0$ *and* $\omega = \pi$ *into (10.37) and write*

$$H(0) = 2h_1 + 2h_2 + 2h_3 = \sqrt{2} \quad and \quad H(\pi) = h_3 - h_2 + h_1 - h_1 + h_2 - h_3 = 0$$

Note that the second equation is always true and the first reduces to

$$h_1 + h_2 + h_3 = \sqrt{2}/2$$

We now use Corollary 10.2 with $\tilde{L} = 1$, $a = \tilde{h}$ *and* $b = h$. *Using the fact that* $\tilde{h}_0 = \tilde{h}_1 = \sqrt{2}/2$ *and* $h_0 = h_1$, *the first equation (10.34) reduces to*

$$1 = \tilde{h}_0 h_0 + \tilde{h}_1 h_1 = \frac{\sqrt{2}}{2} h_1 + \frac{\sqrt{2}}{2} h_1 = \sqrt{2} h_1$$

or $h_0 = h_1 = \frac{\sqrt{2}}{2}$.

Using the fact that $h_{-2} = h_3$ *and* $h_{-1} = h_2$, *the second equation (10.35) in Corollary 10.2 becomes*

$$0 = \tilde{h}_0 h_{-2} + \tilde{h}_1 h_{-1} = \frac{\sqrt{2}}{2} h_3 + \frac{\sqrt{2}}{2} h_2$$

or $h_2 = -h_3$. *If we let* $h_3 = a$, *where* $a \in \mathbb{R}$, *then we can write* h *completely in terms of* a. *We have*

$$h = (h_{-2}, h_{-1}, h_0, h_1, h_2, h_3) = (h_3, h_2, h_1, h_1, h_2, h_3)$$

$$= \left(a, -a, \frac{\sqrt{2}}{2}, \frac{\sqrt{2}}{2}, -a, a \right)$$

$\square$

PROBLEMS

10.1 Find filters (if possible) that satisfy the following constraints.

(a) h is a symmetric, finite-length filter with length 6.

(b) g is a symmetric, finite-length filter with length 7.

(c) h is a symmetric, finite length-filter with length 6 whose Fourier series satisfies $H(0) = \sqrt{2}$ and $H(\pi) = 0$.

(d) g is a symmetric, finite length-filter with length 7 whose Fourier series satisfies $G(0) = 0$ and $G(\pi) = \sqrt{2}$.

(d) g is a symmetric, finite-length filter with length 4 whose Fourier series satisfies $G(0) = 0$ and $G(\pi) = \sqrt{2}$.

10.2   Verify the identities (10.5) and (10.6) in Example 10.1.

10.3   Prove Theorem 10.1. (*Hint:* The proof is nearly identical to that of Theorem 8.1 — replace the $\overline{A(\omega)}$ and $\overline{A(\omega + \pi)}$ in the first identity of that proof with $\overline{B(\omega)}$ and $\overline{B(\omega + \pi)}$, respectively, and continue from there.)

10.4   Assume that each of the following filter pairs h and $\tilde{\text{h}}$ satisfy (10.10). Write down the orthogonality conditions for each filter pair, and identity those pairs that violate the definition of a finite-length filter.

(a) $\mathbf{h} = (h_{-3}, \ldots, h_3)$, $\tilde{\mathbf{h}} = (\tilde{h}_{-1}, \tilde{h}_0, \tilde{h}_1)$.

(b) $\mathbf{h} = (h_{-4}, \ldots, h_4)$, $\tilde{\mathbf{h}} = (\tilde{h}_{-1}, \tilde{h}_0, \tilde{h}_1)$.

(c) $\mathbf{h} = (h_{-4}, \ldots, h_5)$, $\tilde{\mathbf{h}} = (\tilde{h}_{-2}, \ldots, \tilde{h}_3)$.

(d) $\mathbf{h} = (h_{-4}, \ldots, h_5)$, $\tilde{\mathbf{h}} = (\tilde{h}_{-3}, \ldots, \tilde{h}_4)$.

10.5   Complete the proof of Proposition 10.1 by establishing (10.25).

10.6   Prove Corollary 10.1.

10.7   Prove Proposition 10.2.

10.8   Prove Proposition 10.3.

10.9   Prove Corollary 10.2.

10.10   Show that if h is a symmetric finite-length filter of even length, then the Fourier series $H(\omega)$ satisfies $H(\pi) = 0$.

10.11   Suppose that h is a lowpass symmetric filter with $h_k = h_{-k}$ for $k = 1, \ldots, L$. Further assume that the terminal index L is an even integer. Show that

$$h_0 + 2(h_2 + h_4 + \cdots + h_L) = \frac{\sqrt{2}}{2}$$

and

$$h_1 + h_3 + \cdots + h_{L-1} = \frac{\sqrt{2}}{4}$$

What can you infer if the terminal index L is an odd integer?

10.12   In this problem we investigate the symmetry of filters g built using the rule $g_k = (-1)^k h_{1-k}$.

(a) Suppose that **h** is an odd-length symmetric filter. Show that $g_k = g_{2-k}$.

(b) Suppose that **h** is an even-length symmetric filter. Show that $g_k = -g_{1-k}$. We say that this filter is *antisymmetric*.

10.13 Suppose that

$$\mathbf{h} = (h_{-1}, h_0, h_1, h_2) = \left(-\frac{\sqrt{2}}{4}, \frac{3\sqrt{2}}{4}, \frac{3\sqrt{2}}{4}, -\frac{\sqrt{2}}{4}\right)$$

and

$$\tilde{\mathbf{h}} = (\tilde{h}_{-1}, \tilde{h}_0, \tilde{h}_1, \tilde{h}_2) = \left(\frac{\sqrt{2}}{8}, \frac{3\sqrt{2}}{8}, \frac{3\sqrt{2}}{8}, \frac{\sqrt{2}}{8}\right)$$

(This filter pair can be constructed using ideas from Section 10.2.)

You might want to use a **CAS** for parts (b) and (d), but you should do the remaining parts by hand.

(a) Show that both **h** and **h̃** are lowpass filters.

(b) Show that **h** and **h̃** are a biorthogonal filter pair. That is, show that the filters satisfy (10.10).

(c) Use Proposition 10.1 to build $G(\omega)$ and $\tilde{G}(\omega)$ (take $n = 1$ and $b = \pi$).

(d) Verify (10.23)-(10.25) in Proposition 10.1.

(e) Use Corollary 10.1 to find the filters **g** and **g̃**.

(f) Find the wavelet matrices W_{10} and $\tilde{W}_{10}$ (see Example 10.1) and then verify that $\tilde{W}_{10} W_{10}^T$ is the identity matrix.

10.14 Repeat parts (a), (e), and (f) of Problem 10.13 for filters [2] **h** and **h̃** where

$$\mathbf{h} = (h_{-2}, h_{-1}, h_0, h_1, h_2) = (-\frac{\sqrt{2}}{20}, \frac{\sqrt{2}}{4}, \frac{3\sqrt{2}}{5}, \frac{\sqrt{2}}{4}, -\frac{\sqrt{2}}{20})$$

and

$$\begin{aligned}
\tilde{\mathbf{h}} &= (h_{-3}, h_{-2}, h_{-1}, h_0, h_1, h_2, h_3) \\
&= (-\frac{3\sqrt{2}}{280}, -\frac{3\sqrt{2}}{56}, \frac{73\sqrt{2}}{280}, \frac{17\sqrt{2}}{28}, \frac{73\sqrt{2}}{280}, -\frac{3\sqrt{2}}{56}, -\frac{3\sqrt{2}}{280})
\end{aligned}$$

You will probably want to use a **CAS** for this problem.

---

[2] These filters are constructed in Dauchechies [26]. The length 5 filter is the well-known filter of Burt and Adelson [12] used in image encoding.

10.15   Write down the highpass filters $\tilde{g}$ and g for the filters given in Example 10.4.

10.16   Find all length 3 symmetric lowpass filters.

10.17   Find all six-term symmetric filters that along with the Haar filter form a biorthogonal filter pair.

10.18   In Example 10.4, suppose that we wanted symmetric biorthogonal filters of length 2 and 10. The length 2 filter is the Haar filter. Can you use the solution for h in Example 10.4 to quickly write down a length 10 filter? To write down a symmetric length $4M - 2$ filter (where $M = 1, 2, \ldots$) that forms a biorthogonal filter pair with the Haar filter? Are these solutions unique?

10.2   BIORTHOGONAL SPLINE FILTERS

Proposition 10.1 reduces our task of finding four filters h, $\tilde{h}$, g, and $\tilde{g}$ to finding a symmetric finite-length biorthogonal filter pair h and $\tilde{h}$. Solving for the filters in Example 10.4 was tractable since the length of $\tilde{h}$ was only 2. In general, finding two filters that satisfy (10.10) is a daunting task. Requiring the filters to be symmetric certainly reduces the number of variables in the quadratic system that we would obtain by expanding the terms in (10.10). Since we want the filters to be lowpass, we insist that $H(0) = \tilde{H}(0) = \sqrt{2}$ and $H(\pi) = \tilde{H}(\pi) = 0$.

Daubechies' [24] analysis of this system led her to a conclusion that greatly simplifies the task at hand: Why not simply *pick* a symmetric finite-length lowpass filter for $\tilde{h}$?

Choosing $\tilde{h}$ makes $\tilde{H}(\omega)$ known, and if $\tilde{H}(\omega)$ is fixed in (10.10), then the only unknowns in that identity are the filter elements of h, and the resulting equation is linear in the elements of h. We know that the lowpass conditions $H(0) = \sqrt{2}$ and $H(0) = 0$ are linear as well, so by choosing $\tilde{h}$, we arrive at a linear system! This is exactly what happened in Example 10.4 — once we determined $\tilde{h}$, the system we solved to find h was linear. As you learned in a linear algebra class, we can always write down the solutions (if they exist) of a linear system.

Thus we have reduced our task to simply solving a linear system. The next question should be obvious. How do we find a symmetric finite-length lowpass filter $\tilde{h}$?

The Filter $\tilde{H}(\omega) = \sqrt{2}\cos^{\tilde{N}}\left(\frac{\omega}{2}\right)$

The answer to the question posed above is that we have already found $\tilde{h}$! If you worked Problem 4.41 in Section 4.3, then you learned that if $\tilde{N}$ is an even integer,

$$\cos^{\tilde{N}}\left(\frac{\omega}{2}\right) = \sum_{k=-\frac{\tilde{N}}{2}}^{\frac{\tilde{N}}{2}} a_k e^{ik\omega} \tag{10.38}$$

where

$$a_k = \frac{1}{2^{\tilde{N}}} \binom{\tilde{N}}{\frac{\tilde{N}}{2} - k} = \frac{\tilde{N}!}{2^{\tilde{N}} (\frac{\tilde{N}}{2} - k)! \, (\frac{\tilde{N}}{2} + k)!} \tag{10.39}$$

Let's take

$$\tilde{H}(\omega) = \sqrt{2} \cos^{\tilde{N}} \left(\frac{\omega}{2} \right) = \sqrt{2} \sum_{k=-\frac{\tilde{N}}{2}}^{\frac{\tilde{N}}{2}} \frac{1}{2^{\tilde{N}}} \binom{\tilde{N}}{\frac{\tilde{N}}{2} - k} e^{ik\omega} \tag{10.40}$$

In this form, we see immediately that $\tilde{H}(0) = \sqrt{2} \cos^{\tilde{N}}(0) = \sqrt{2}$ and $\tilde{H}(\pi) = \sqrt{2} \cos^{\tilde{N}}(\frac{\pi}{2}) = 0$. Moreover, we can use (10.39) to write $\tilde{h}_k = \sqrt{2} a_k$, and it is readily evident from (10.39) that $a_k = a_{-k}$, so that $\tilde{h}_k = \tilde{h}_{-k}$. Thus, $\tilde{\mathbf{h}}$ is a symmetric filter with odd length $\tilde{N} + 1$. Let's look at some examples.

Example 10.5 (Even Powers of $\cos\left(\frac{\omega}{2}\right)$). *Write down the Fourier series for*

(a) $\tilde{H}(\omega) = \sqrt{2} \cos^2 \left(\frac{\omega}{2} \right)$.

(b) $\tilde{H}(\omega) = \sqrt{2} \cos^4 \left(\frac{\omega}{2} \right)$.

(c) $\tilde{H}(\omega) = \sqrt{2} \cos^6 \left(\frac{\omega}{2} \right)$.

Solution. *For part (a) we have* $\tilde{N} = 2$ *and*

$$a_k = \frac{2!}{2^2 (1-k)! \, (1+k)!}, \qquad k = -1, 0, 1$$

so that $a_0 = \frac{1}{2}$ *and* $a_1 = a_{-1} = \frac{1}{4}$. *We have* $\tilde{\mathbf{h}} = (\tilde{h}_{-1}, \tilde{h}_0, \tilde{h}_1) = \left(\frac{\sqrt{2}}{4}, \frac{\sqrt{2}}{2}, \frac{\sqrt{2}}{4} \right)$ *and the Fourier series is*

$$\tilde{H}(\omega) = \frac{\sqrt{2}}{4} e^{-i\omega} + \frac{\sqrt{2}}{2} + \frac{\sqrt{2}}{4} e^{i\omega}$$

For part (b) we have $\tilde{N} = 4$ *and*

$$a_k = \frac{4!}{2^4 (2-k)! \, (2+k)!}, \qquad k = -2, -1, 0, 1, 2$$

so that $a_0 = \frac{3}{8}$, $a_1 = a_{-1} = \frac{1}{4}$, *and* $a_2 = a_{-2} = \frac{1}{16}$. *Thus, the Fourier series for part (b) is*

$$\tilde{H}(\omega) = \frac{\sqrt{2}}{16} e^{-2i\omega} + \frac{\sqrt{2}}{4} e^{-i\omega} + \frac{3\sqrt{2}}{8} + \frac{\sqrt{2}}{4} e^{i\omega} + \frac{\sqrt{2}}{16} e^{2i\omega}$$

Finally, for part (c), $\tilde{N} = 6$ and

$$a_k = \frac{6!}{2^6(3-k)!\,(3+k)!}, \qquad k = 0, \pm 1, \pm 2, \pm 3$$

We have $a_0 = \frac{5}{16}$, $a_1 = a_{-1} = \frac{15}{64}$, $a_2 = a_{-2} = \frac{3}{32}$, and $a_3 = a_{-3} = \frac{1}{64}$. The Fourier series for part (c) is

$$\tilde{H}(\omega) = \frac{\sqrt{2}}{64}\, e^{-3i\omega} + \frac{3\sqrt{2}}{32}\, e^{-2i\omega} + \frac{15\sqrt{2}}{64}\, e^{-i\omega} + \frac{5\sqrt{2}}{16}$$

$$+ \frac{15\sqrt{2}}{64}\, e^{i\omega} + \frac{3\sqrt{2}}{32}\, e^{2i\omega} + \frac{\sqrt{2}}{64}\, e^{3i\omega}$$

$\square$

When $\tilde{N}$ is even, we have seen that $\tilde{H}(\omega) = \sqrt{2}\cos^{\tilde{N}}\left(\frac{\omega}{2}\right)$ produces a symmetric filter $\tilde{\mathbf{h}} = (\tilde{h}_{-\tilde{N}/2}, \ldots, \tilde{h}_{\tilde{N}/2})$ of odd length $\tilde{N} + 1$. How can we generate even-length symmetric filters? Let's try raising $\cos\left(\frac{\omega}{2}\right)$ to an odd power. For example, for $\tilde{N} = 3$, we have

$$\begin{aligned}
\cos^3\left(\frac{\omega}{2}\right) &= \left(\frac{e^{i\omega/2} + e^{-i\omega/2}}{2}\right)^3 \\
&= \frac{1}{8}\left(e^{-i\omega/2}(e^{i\omega}+1)\right)^3 \\
&= \frac{e^{-3i\omega/2}}{8}\left(1 + 3\,e^{i\omega} + 3\,e^{2i\omega} + e^{3i\omega}\right) \\
&= \frac{1}{8}\,e^{-3i\omega/2} + \frac{3}{8}\,e^{-i\omega/2} + \frac{3}{8}\,e^{i\omega/2} + \frac{1}{8}\,e^{3i\omega/2}
\end{aligned}$$

Unfortunately, this is not a Fourier series, due to the division by 2 in each of the complex exponentials. It is promising, though, since the length of the series is even and the coefficients are symmetric. Perhaps we can modify the expression so that it would work. In particular, if we multiply the entire expression by $e^{i\omega/2}$, we would shift the complex exponentials so that they are of the proper form. We have

$$e^{i\omega/2}\cos^3\left(\frac{\omega}{2}\right) = \frac{1}{8}\,e^{-i\omega} + \frac{3}{8} + \frac{3}{8}\,e^{i\omega} + \frac{1}{8}\,e^{2i\omega}$$

Thus, if we take

$$\tilde{H}(\omega) = \sqrt{2}\,e^{i\omega/2}\cos^3\left(\frac{\omega}{2}\right) = \frac{\sqrt{2}}{8}\,e^{-i\omega} + \frac{3\sqrt{2}}{8} + \frac{3\sqrt{2}}{8}\,e^{i\omega} + \frac{\sqrt{2}}{8}\,e^{2i\omega}$$

we see immediately that

$$\tilde{H}(0) = \sqrt{2}\,e^0\cos^3(0) = \sqrt{2} \qquad \text{and} \qquad \tilde{H}(\pi) = \sqrt{2}\,e^{i\pi/2}\cos^3(\pi/2) = 0$$

so that $\tilde{H}(\omega)$ produces a lowpass filter. Moreover, the length of the filter is 4 and the filter $\tilde{\mathbf{h}} = (\tilde{h}_{-1}, \tilde{h}_0, \tilde{h}_1, \tilde{h}_2) = \left(\frac{\sqrt{2}}{8}, \frac{3\sqrt{2}}{8}, \frac{3\sqrt{2}}{8}, \frac{\sqrt{2}}{8} \right)$ satisfies $\tilde{h}_k = \tilde{h}_{1-k}$, so that it is symmetric about $k = \frac{1}{2}$. In Problem 10.20 you will show that for $\tilde{N}$ odd,

$$\tilde{H}(\omega) = \sqrt{2}\, e^{i\omega/2} \cos^{\tilde{N}} \left(\frac{\omega}{2} \right) = \sqrt{2} \sum_{k=-\frac{\tilde{N}-1}{2}}^{\frac{\tilde{N}+1}{2}} \frac{1}{2^{\tilde{N}}} \binom{\tilde{N}}{\frac{\tilde{N}-1}{2} + k} e^{ik\omega} \qquad (10.41)$$

is an even-length filter that is symmetric about $k = \frac{1}{2}$.

Spline Filters

We can combine (10.40) and (10.41) to form the following proposition.

Definition 10.3 (Spline Filter of Length $\tilde{N} + 1$). *Let $\tilde{N}$ be a positive integer. Then $\tilde{\mathbf{h}}$, given elementwise by*

$$\tilde{h}_k = \frac{\sqrt{2}}{2^{\tilde{N}}} \binom{\tilde{N}}{\frac{\tilde{N}}{2} - k}, \qquad k = 0, \pm 1, \ldots, \pm \tilde{N}/2 \qquad (10.42)$$

when $\tilde{N}$ is even, and

$$\tilde{h}_k = \frac{\sqrt{2}}{2^{\tilde{N}}} \binom{\tilde{N}}{\frac{\tilde{N}-1}{2} + k}, \qquad k = -\frac{\tilde{N}-1}{2}, \ldots, \frac{\tilde{N}+1}{2} \qquad (10.43)$$

when $\tilde{N}$ is odd, is called the spline filter of length $\tilde{N} + 1$. *We will call*

$$\tilde{H}(\omega) = \begin{cases} \sqrt{2} \cos^{\tilde{N}} \left(\frac{\omega}{2} \right), & \tilde{N} = 2\tilde{\ell} \\ \sqrt{2} e^{i\omega/2} \cos^{\tilde{N}} \left(\frac{\omega}{2} \right), & \tilde{N} = 2\tilde{\ell} + 1 \end{cases} \qquad (10.44)$$

the Fourier series *associated with the spline filter. Here $\tilde{\ell}$ is any positive integer.* ☐

In Problem 10.25 you will show that for a given positive integer $\tilde{N}$, we can write the spline filter given in (10.42) – (10.43) as

$$\tilde{h}_{k - \tilde{N}/2} = \frac{\sqrt{2}}{2^{\tilde{N}}} \binom{\tilde{N}}{k}, \qquad k = 0, \ldots, \tilde{N}$$

Those readers familiar with Pascal's triangle will undoubtedly recognize the spline filter of length $\tilde{N} + 1$ as the $\tilde{N}$th row of Pascal's triangle multiplied by $\sqrt{2}/2^{\tilde{N}}$!

Why the Name "Spline Filters"?

Let's digress for just a moment and discuss the term spline filter. In mathematics, *spline functions*[3] are piecewise polynomials that are used in many areas of applied mathematics. The fundamental (B-spline) piecewise constant spline is the *box function*

$$B_0(t) = \begin{cases} 1, & 0 \le t < 1 \\ 0, & \text{otherwise} \end{cases}$$

Higher-degree splines $B_{\tilde{N}}(t)$ are defined recursively by the formula[4]

$$B_{\tilde{N}+1}(t) = \int_0^1 B_{\tilde{N}}(t - u)\, du \tag{10.45}$$

The spline $B_{\tilde{N}}(t)$ is a piecewise polynomial of degree $\tilde{N}$. It is easy to show that

$$B_1(t) = \begin{cases} t, & 0 \le t < 1 \\ 2 - t, & 1 \le t < 2 \\ 0, & \text{otherwise} \end{cases} \tag{10.46}$$

is the *triangle function* (see Problem 10.26). The splines $B_0(t)$ and $B_1(t)$ are plotted in Figure 10.1.

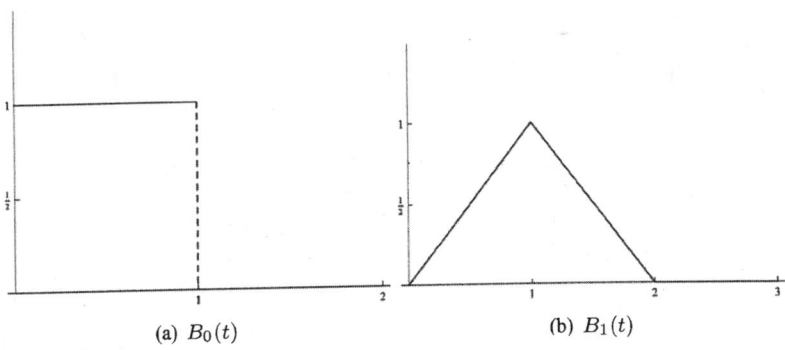

(a) $B_0(t)$ (b) $B_1(t)$

Figure 10.1   Plots of the piecewise constant spline $B_0(t)$ and the piecewise linear spline $B_1(t)$.

[3] The word *spline* comes from architecture. A spline is a flexible piece of hard rubber that is used in mechanical drawing to render curves.
[4] Those readers with some background in analysis will recognize that this definition is simply the $\tilde{N}$-fold convolution product of $B_0(t)$ with itself.

One important property of splines is that they satisfy a *dilation equation*. A function $f(t)$ satisfies a dilation equation if it can be written in the form

$$f(t) = \sum_{k \in \mathbb{Z}} h_k f(2t - k) \tag{10.47}$$

So $f(t)$ is formed by a linear combination of 2-dilates and translates of itself. The dilation equation is of fundamental importance in the classical derivation of wavelet theory. Indeed, the filters used for the discrete transforms we developed are the coefficients h_k in (10.47) for some function $f(t)$.

It is easy to check that both $B_0(t)$ and $B_1(t)$ satisfy dilation equations. We have

$$B_0(t) = 1 \cdot B_0(2t) + 1 \cdot B_0(2t - 1) \tag{10.48}$$

and (see Figure 10.2)

$$B_1(t) = \frac{1}{2} B_1(2t) + 1 \cdot B_1(2t - 1) + \frac{1}{2} B_1(2t - 2) \tag{10.49}$$

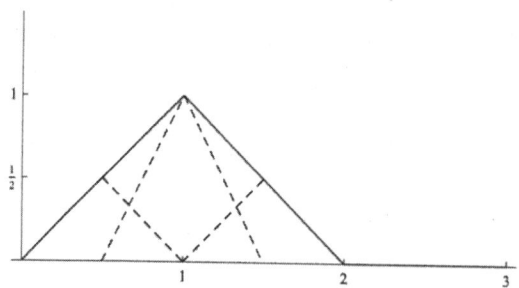

Figure 10.2   The two-dilates and translates $\frac{1}{2} B_1(2t)$, $B_1(2t - 1)$, and $\frac{1}{2} B_1(2t - 2)$ of $B_1(t)$ used to construct $B_1(t)$.

Note that from Definition 10.3, the spline filter of length 2 is the Haar filter $(\frac{\sqrt{2}}{2}, \frac{\sqrt{2}}{2}) = \frac{\sqrt{2}}{2}(1,1)$ and the spline filter of length 3 is $(\frac{\sqrt{2}}{4}, \frac{\sqrt{2}}{2}, \frac{\sqrt{2}}{4}) = \frac{\sqrt{2}}{2}(\frac{1}{2}, 1, \frac{1}{2})$. These filters are $\frac{\sqrt{2}}{2}$ times the coefficients $(1, 1)$ and $(\frac{1}{2}, 1, \frac{1}{2})$ that appear in the dilation equations (10.48) and (10.49). In fact, it can be shown (see Chui [18]) that

$$B_{\tilde{N}}(t) = \sum_{k=0}^{\tilde{N}+1} 2^{-\tilde{N}} \binom{\tilde{N} + 1}{k} B_{\tilde{N}}(2t - k) \tag{10.50}$$

The important thing to take from (10.50) is that the filter elements $2^{-\tilde{N}} \binom{\tilde{N} + 1}{k}$ are very similar to the spline filters defined in Definition 10.3.

Traditionally, the approach for obtaining wavelet filters consists of a search for a *scaling function* (such as $B_{\tilde{N}}(t)$) that satisfies a dilation equation. This is a fascinating topic but not the approach we adopt in this book. The interested reader is strongly encouraged to view the books by Boggess and Narcowich [7], Frazier [35], and Walnut [83]. The advanced reader would also enjoy the development of Daubechies [26].

Constructing Biorthogonal Filter Pairs

We are ready to actually construct some biorthogonal filter pairs. The best way to illustrate the process is via examples.

Example 10.6 (Biorthogonal Filter Pair with Lengths 3 and 5). *Let start by taking* $\tilde{N} = 2$ *so that the length of the spline filter is* $\tilde{N} + 1 = 3$. *We have*

$$\tilde{H}(\omega) = \sqrt{2}\cos^2\left(\frac{\omega}{2}\right) = \frac{\sqrt{2}}{4}e^{-i\omega} + \frac{\sqrt{2}}{2} + \frac{\sqrt{2}}{4}e^{i\omega} \qquad (10.51)$$

so that

$$\tilde{\mathbf{h}} = (\tilde{h}_{-1}, \tilde{h}_0, \tilde{h}_1) = (\tilde{h}_1, \tilde{h}_0, \tilde{h}_1) = \left(\frac{\sqrt{2}}{4}, \frac{\sqrt{2}}{2}, \frac{\sqrt{2}}{4}\right) \qquad (10.52)$$

We must also choose an odd length for $\mathbf{h}$, *so the filter must have the form* $\mathbf{h} = (h_{-L}, \ldots, h_L)$ *for some* $L > 0$. *Proposition 10.2 tells us that since the lengths of the filters are odd, the stopping indices must have different parity. Since the stopping index of* $\tilde{\mathbf{h}}$ *is 1, we need to choose L even. Let's take* $L = 2$ *so that the filter we seek is*

$$\mathbf{h} = (h_{-2}, h_{-1}, h_0, h_1, h_2) = (h_2, h_1, h_0, h_1, h_2)$$

Understand that there are infinitely many choices for L — we pick two since a length 5 filter is easy to work with by hand.

We use Corollary 10.2 with $\mathbf{a} = \tilde{\mathbf{h}}$ *(so that* $L = 1$ *) and* $\mathbf{b} = \mathbf{h}$ *to write down the system*

$$\sum_{k=-1}^{1} \tilde{h}_k h_k = 1 \qquad and \qquad \sum_{k=-1}^{1} \tilde{h}_k h_{k-2} = 0$$

Before we plug in the values in (10.52), we use symmetry (i.e., $\tilde{h}_{-1} = \tilde{h}_1$, $h_{-1} = h_1$, *and* $h_{-2} = h_2$) *to rewrite these three equations:*

$$\tilde{h}_0 h_0 + 2\tilde{h}_1 h_1 = 1 \qquad and \qquad \tilde{h}_0 h_2 + \tilde{h}_1 h_1 = 0$$

We next insert the values from (10.52) to obtain

$$\frac{\sqrt{2}}{2}h_0 + 2\frac{\sqrt{2}}{4}h_1 \qquad = 1$$

$$\frac{\sqrt{2}}{4}h_1 + \frac{\sqrt{2}}{2}h_2 = 0$$

We next insert the values from (10.52) to obtain

$$\frac{\sqrt{2}}{2}h_0 + 2\frac{\sqrt{2}}{4}h_1 \qquad = 1$$

$$\frac{\sqrt{2}}{4}h_1 + \frac{\sqrt{2}}{2}h_2 = 0$$

Simplifying gives

$$h_0 + h_1 \qquad = \sqrt{2}$$

$$h_1 + 2h_2 = 0$$

This linear system has an infinite number of solutions since there are more variables than equations. This gives us the freedom to add other constraints. We could add the constraint

$$H(0) = h_0 + 2h_1 + 2h_2 = \sqrt{2}$$

but that constraint is redundant since that is what we obtain if we add the two equations of our system together. (Actually, in Problem 10.19 you will show that whenever $\tilde{H}(0) = \sqrt{2}$, $\tilde{H}(\pi) = 0$, and (10.10) holds, we have $H(0) = \sqrt{2}$.)

We certainly need to add the condition

$$H(\pi) = h_0 - 2h_1 + 2h_2 = 0$$

*to ensure that **h** is a lowpass filter. Thus, our system is*

$$h_0 + \ h_1 \qquad\qquad = \sqrt{2}$$

$$h_1 + 2h_2 = 0 \qquad\qquad (10.53)$$

$$h_0 - 2h_1 + 2h_2 = 0$$

This system is easily solved by hand. If we subtract the second equation from the third, we have

$$h_0 - 3h_1 = 0$$

Insert $h_0 = 3h_1$ into the first equation and simplify to obtain $h_1 = \frac{\sqrt{2}}{4}$, so that $h_0 = \frac{3\sqrt{2}}{4}$. Inserting $h_1 = \frac{\sqrt{2}}{4}$ into the second equation and simplifying gives $h_2 = -\frac{\sqrt{2}}{8}$. Our filter is

$$\boxed{\mathbf{h} = (h_2, h_1, h_0, h_1, h_2) = \left(-\frac{\sqrt{2}}{8}, \frac{\sqrt{2}}{4}, \frac{3\sqrt{2}}{4}, \frac{\sqrt{2}}{4}, -\frac{\sqrt{2}}{8}\right)}$$

Since this filter pair was formed by starting with the spline filter $\tilde{\mathbf{h}}$, we will refer to this filter pair as the $(5, 3)$ biorthogonal spline filter pair.   □

As you can see, finding biorthogonal filters in this manner is almost easier than finding orthogonal filters from Chapter 7 or Section 8.3! The fact that the systems we must solve are linear really simplifies the problem.

The Biorthogonal Wavelet Transform Matrix

Before constructing more biorthogonal filter pairs, let's see how we would actually implement the filter pair from Example 10.6.

Example 10.7 (Constructing and Implementing the Biorthogonal Wavelet Transform Matrix). *Let's consider the* $(5,3)$ *biorthogonal spline filter pair from Example 10.6. We have*

$$\tilde{\mathbf{h}} = (\tilde{h}_{-1}, \tilde{h}_0, \tilde{h}_1) = (\tilde{h}_1, \tilde{h}_0, \tilde{h}_1) = \left(\frac{\sqrt{2}}{4}, \frac{\sqrt{2}}{2}, \frac{\sqrt{2}}{4} \right)$$

and

$$\mathbf{h} = (h_{-2}, h_{-1}, h_0, h_1, h_2) = (h_2, h_1, h_0, h_1, h_2)$$

$$= \left(-\frac{\sqrt{2}}{8}, \frac{\sqrt{2}}{4}, \frac{3\sqrt{2}}{4}, \frac{\sqrt{2}}{4}, -\frac{\sqrt{2}}{8} \right)$$

To form the highpass filters, we will use Corollary 10.1 with $n = 1$. *We have*

$$g_k = (-1)^k \tilde{h}_{1-k} \quad \text{and} \quad \tilde{g}_k = (-1)^k h_{1-k}$$

so that we arrive at Table 10.1.

Table 10.1 The highpass filter pair $\mathbf{g}$ and $\tilde{\mathbf{g}}$ for the $(5, 3)$ biorthogonal spline filter pair.

$\tilde{g}_k$	$(-1)^k h_{1-k}$		g_k	$(-1)^k \tilde{h}_{1-k}$	
$\tilde{g}_{-1}$	$-h_2$	$\sqrt{2}/8$			
$\tilde{g}_0$	h_1	$\sqrt{2}/4$	g_0	$\tilde{h}_1$	$\sqrt{2}/4$
$\tilde{g}_1$	$-h_0$	$-3\sqrt{2}/4$	g_1	$-\tilde{h}_0$	$-\sqrt{2}/2$
$\tilde{g}_2$	h_1	$\sqrt{2}/4$	g_2	$\tilde{h}_1$	$\sqrt{2}/4$
$\tilde{g}_3$	$-h_2$	$\sqrt{2}/8$			

If we process a vector with an orthogonal filter $\mathbf{p} = (p_0, \ldots, p_L)$ *from Chapter 7, then we reverse the elements of* $\mathbf{p}$ *and form the first row of the lowpass portion of the wavelet transform matrix by writing*

$$p_L \quad p_{L-1} \quad \cdots \quad p_1 \quad p_0 \quad 0 \quad 0 \quad \cdots \quad 0$$

Since we are using symmetric filters, we will alter our wavelet transform matrix. We will learn in Section 11.3 that centering the symmetric filters $\mathbf{h}$ *and* $\tilde{\mathbf{h}}$ *in the first rows of* W_N *and* $\tilde{W}_N$, *respectively, reduces the edge effects that result when we apply a wavelet transform to a vector or image. Thus we will build our wavelet transform matrices using centered versions of our symmetric filters.*

When we developed orthogonal wavelet transformations in Chapters 6 and 7, we reversed the order of the filter coefficients in each row. This step was taken to make our transform matrices behave like the convolution matrices developed in Chapter 5. Due to symmetry, there is no need to reverse the order of $(5,3)$ biorthogonal spline filter pair filter coefficients in our transform matrix.

For the sake of illustration, let's suppose that we wish to process a vector of length 10. Then we note that $\tilde{h}_{-1} = \tilde{h}_1$ and form the first row of $\tilde{W}_{10}$ as

$$\tilde{h}_0 \quad \tilde{h}_1 \quad 0 \quad 0 \quad 0 \quad 0 \quad 0 \quad 0 \quad 0 \quad \tilde{h}_1$$

We can construct subsequent rows of the lowpass portion of $\tilde{W}_{10}$ simply by rotating this row 2 units right. For the bottom half of W_{10}, we use the filter $\tilde{\mathbf{g}} = (\tilde{g}_{-1}, \tilde{g}_0, \tilde{g}_1, \tilde{g}_2, \tilde{g}_3)$ and place $\tilde{g}_0$ in the first column of the first row of the highpass portion of $\tilde{W}_{10}$. We have

$$\tilde{W}_{10} = \left[\begin{array}{cccccccccc}
\tilde{h}_0 & \tilde{h}_1 & 0 & 0 & 0 & 0 & 0 & 0 & 0 & \tilde{h}_1 \\
0 & \tilde{h}_1 & \tilde{h}_0 & \tilde{h}_1 & 0 & 0 & 0 & 0 & 0 & 0 \\
0 & 0 & 0 & \tilde{h}_1 & \tilde{h}_0 & \tilde{h}_1 & 0 & 0 & 0 & 0 \\
0 & 0 & 0 & 0 & 0 & \tilde{h}_1 & \tilde{h}_0 & \tilde{h}_1 & 0 & 0 \\
0 & 0 & 0 & 0 & 0 & 0 & 0 & \tilde{h}_1 & \tilde{h}_0 & \tilde{h}_1 \\
\hline
\tilde{g}_0 & \tilde{g}_1 & \tilde{g}_2 & \tilde{g}_3 & 0 & 0 & 0 & 0 & 0 & \tilde{g}_{-1} \\
0 & \tilde{g}_{-1} & \tilde{g}_0 & \tilde{g}_1 & \tilde{g}_2 & \tilde{g}_3 & 0 & 0 & 0 & 0 \\
0 & 0 & 0 & \tilde{g}_{-1} & \tilde{g}_0 & \tilde{g}_1 & \tilde{g}_2 & \tilde{g}_3 & 0 & 0 \\
0 & 0 & 0 & 0 & 0 & \tilde{g}_{-1} & \tilde{g}_0 & \tilde{g}_1 & \tilde{g}_2 & \tilde{g}_3 \\
\tilde{g}_2 & \tilde{g}_3 & 0 & 0 & 0 & 0 & 0 & \tilde{g}_{-1} & \tilde{g}_0 & \tilde{g}_1
\end{array}\right] \qquad (10.54)$$

$$= \left[\begin{array}{cccccccccc}
\frac{\sqrt{2}}{2} & \frac{\sqrt{2}}{4} & 0 & 0 & 0 & 0 & 0 & 0 & 0 & \frac{\sqrt{2}}{4} \\
0 & \frac{\sqrt{2}}{4} & \frac{\sqrt{2}}{2} & \frac{\sqrt{2}}{4} & 0 & 0 & 0 & 0 & 0 & 0 \\
0 & 0 & 0 & \frac{\sqrt{2}}{4} & \frac{\sqrt{2}}{2} & \frac{\sqrt{2}}{4} & 0 & 0 & 0 & 0 \\
0 & 0 & 0 & 0 & 0 & \frac{\sqrt{2}}{4} & \frac{\sqrt{2}}{2} & \frac{\sqrt{2}}{4} & 0 & 0 \\
0 & 0 & 0 & 0 & 0 & 0 & 0 & \frac{\sqrt{2}}{4} & \frac{\sqrt{2}}{2} & \frac{\sqrt{2}}{4} \\
\hline
\frac{\sqrt{2}}{4} & -\frac{3\sqrt{2}}{4} & \frac{\sqrt{2}}{4} & \frac{\sqrt{2}}{4} & 0 & 0 & 0 & 0 & 0 & \frac{\sqrt{2}}{8} \\
0 & \frac{\sqrt{2}}{8} & \frac{\sqrt{2}}{4} & -\frac{3\sqrt{2}}{4} & \frac{\sqrt{2}}{4} & \frac{\sqrt{2}}{8} & 0 & 0 & 0 & 0 \\
0 & 0 & 0 & \frac{\sqrt{2}}{8} & \frac{\sqrt{2}}{4} & -\frac{3\sqrt{2}}{4} & \frac{\sqrt{2}}{4} & \frac{\sqrt{2}}{8} & 0 & 0 \\
0 & 0 & 0 & 0 & 0 & \frac{\sqrt{2}}{8} & \frac{\sqrt{2}}{4} & -\frac{3\sqrt{2}}{4} & \frac{\sqrt{2}}{4} & \frac{\sqrt{2}}{8} \\
\frac{\sqrt{2}}{4} & \frac{\sqrt{2}}{8} & 0 & 0 & 0 & 0 & 0 & \frac{\sqrt{2}}{8} & \frac{\sqrt{2}}{4} & -\frac{3\sqrt{2}}{4}
\end{array}\right]$$

In a similar manner, we can form W_{10}:

$$W_{10} = \left[\begin{array}{cccccccccc}
h_0 & h_1 & h_2 & 0 & 0 & 0 & 0 & 0 & h_2 & h_1 \\
h_2 & h_1 & h_0 & h_1 & h_2 & 0 & 0 & 0 & 0 & 0 \\
0 & 0 & h_2 & h_1 & h_0 & h_1 & h_2 & 0 & 0 & 0 \\
0 & 0 & 0 & 0 & h_2 & h_1 & h_0 & h_1 & h_2 & 0 \\
h_2 & 0 & 0 & 0 & 0 & 0 & h_2 & h_1 & h_0 & h_1 \\
\hline
g_0 & g_1 & g_2 & 0 & 0 & 0 & 0 & 0 & 0 & 0 \\
0 & 0 & g_0 & g_1 & g_2 & 0 & 0 & 0 & 0 & 0 \\
0 & 0 & 0 & 0 & g_0 & g_1 & g_2 & 0 & 0 & 0 \\
0 & 0 & 0 & 0 & 0 & 0 & g_0 & g_1 & g_2 & 0 \\
g_2 & 0 & 0 & 0 & 0 & 0 & 0 & 0 & g_0 & g_1
\end{array}\right]$$

$$= \left[\begin{array}{cccccccccc}
\frac{3\sqrt{2}}{4} & \frac{\sqrt{2}}{4} & -\frac{\sqrt{2}}{8} & 0 & 0 & 0 & 0 & 0 & -\frac{\sqrt{2}}{8} & \frac{\sqrt{2}}{4} \\
-\frac{\sqrt{2}}{8} & \frac{\sqrt{2}}{4} & \frac{3\sqrt{2}}{4} & \frac{\sqrt{2}}{4} & -\frac{\sqrt{2}}{8} & 0 & 0 & 0 & 0 & 0 \\
0 & 0 & -\frac{\sqrt{2}}{8} & \frac{\sqrt{2}}{4} & \frac{3\sqrt{2}}{4} & \frac{\sqrt{2}}{4} & -\frac{\sqrt{2}}{8} & 0 & 0 & 0 \\
0 & 0 & 0 & 0 & -\frac{\sqrt{2}}{8} & \frac{\sqrt{2}}{4} & \frac{3\sqrt{2}}{4} & \frac{\sqrt{2}}{4} & -\frac{\sqrt{2}}{8} & 0 \\
-\frac{\sqrt{2}}{8} & 0 & 0 & 0 & 0 & 0 & -\frac{\sqrt{2}}{8} & \frac{\sqrt{2}}{4} & \frac{3\sqrt{2}}{4} & \frac{\sqrt{2}}{4} \\
\hline
\frac{\sqrt{2}}{4} & -\frac{\sqrt{2}}{2} & \frac{\sqrt{2}}{4} & 0 & 0 & 0 & 0 & 0 & 0 & 0 \\
0 & 0 & \frac{\sqrt{2}}{4} & -\frac{\sqrt{2}}{2} & \frac{\sqrt{2}}{4} & 0 & 0 & 0 & 0 & 0 \\
0 & 0 & 0 & 0 & \frac{\sqrt{2}}{4} & -\frac{\sqrt{2}}{2} & \frac{\sqrt{2}}{4} & 0 & 0 & 0 \\
0 & 0 & 0 & 0 & 0 & 0 & \frac{\sqrt{2}}{4} & -\frac{\sqrt{2}}{2} & \frac{\sqrt{2}}{4} & 0 \\
\frac{\sqrt{2}}{4} & 0 & 0 & 0 & 0 & 0 & 0 & 0 & \frac{\sqrt{2}}{4} & -\frac{\sqrt{2}}{2}
\end{array}\right]$$

*You should verify using a **CAS** that $W_{10}^{-1} = \tilde{W}_{10}^T$.*

If we wish to apply these filter pairs to a digital image A of size $M \times N$, then we use $\tilde{W}_M$ to process the columns of A and W_N^T to process the rows of A. We have

$$B = \tilde{W}_M A W_N^T \tag{10.55}$$

The inverse is easy as well — we just need to get the dimensions right on the wavelet matrices. Remember that W_M^T is the inverse of $\tilde{W}_M$ and $\tilde{W}_N$ is the inverse of W_N^T. Thus, we multiply (10.55) on the left by W_M and on the right by $\tilde{W}_N$ to obtain

$$W_M^T B \tilde{W}_N = (W_M^T \tilde{W}_M) A (W_N^T \tilde{W}_N) = I_M A I_N = A$$

In Figure 10.3 we have applied the two-dimensional biorthogonal wavelet transform to a 160×240 digital image. In Section 11.1 you will learn how to write code that compute one- and two-dimensional biorthogonal wavelet transforms. □

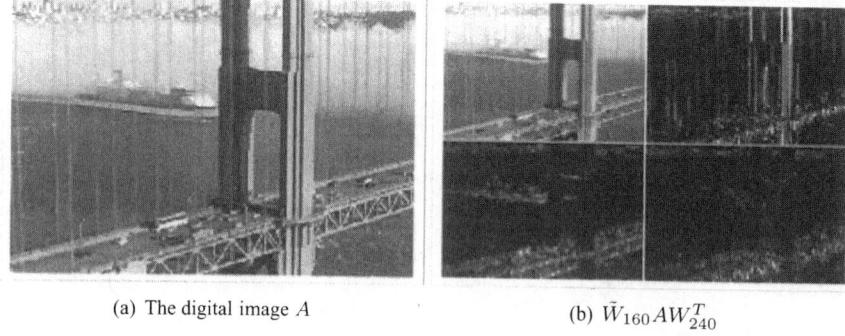

(a) The digital image A (b) $\tilde{W}_{160} A W_{240}^T$

Figure 10.3 An image processed with the $(5, 3)$ biorthogonal spline filter pair.

The $(8, 4)$ Biorthogonal Filter Pair

In Example 10.6 we constructed a biorthogonal filter pair where the lengths of both filters are odd. In the next example we construct a pair of even-length biorthogonal filters.

Example 10.8 (Biorthogonal Filter Pair with Lengths 8 and 4). *For this example let's take $\tilde{N} = 3$, so that we obtain the length 4 spline filter from the Fourier series*

$$\tilde{H}(\omega) = \sqrt{2}\, e^{i\omega/2} \cos^3\left(\frac{\omega}{2}\right)$$

The corresponding filter is

$$\tilde{\mathbf{h}} = (\tilde{h}_{-1}, \tilde{h}_0, \tilde{h}_1, \tilde{h}_2) = (\tilde{h}_2, \tilde{h}_1, \tilde{h}_1, \tilde{h}_2) = \left(\frac{\sqrt{2}}{8}, \frac{3\sqrt{2}}{8}, \frac{3\sqrt{2}}{8}, \frac{\sqrt{2}}{8}\right) \quad (10.56)$$

From Proposition 10.2 we know that the stopping index of any even-length filter $\mathbf{h}$ must be even. For this example we take the stopping index to be 4, so the length 8 filter $\mathbf{h}$ we seek is

$$\mathbf{h} = (h_{-3}, h_{-2}, h_{-1}, h_0, h_1, h_2, h_3, h_4) = (h_4, h_3, h_2, h_1, h_1, h_2, h_3, h_4) \quad (10.57)$$

From Corollary 10.2 we have the following three orthogonality conditions:

$$2\tilde{h}_1 h_1 + 2\tilde{h}_2 h_2 = 1$$
$$\tilde{h}_2 h_4 + \tilde{h}_1 h_3 + \tilde{h}_1 h_2 + \tilde{h}_2 h_1 = 0$$
$$\tilde{h}_1 h_4 + \tilde{h}_2 h_3 = 0$$

If we insert the values from (10.56) into this system and simplify, we obtain the following system:

$$
\begin{aligned}
3h_1 + h_2 &= 2\sqrt{2} \\
h_1 + 3h_2 + 3h_3 + h_4 &= 0 \\
h_3 + 3h_4 &= 0
\end{aligned}
$$

Since there are more unknowns than equations, this system has an infinite number of solutions. Thus, we have the freedom to add another constraint. From Problem 10.19 we know that $H(0) = \sqrt{2}$, and from Problem 10.10 in Section 10.1 we know that $H(\pi) = 0$. Let's mimic the ideas of Chapter 7 and add the derivative condition $H'(\pi) = 0$. We have

$$
H(\omega) = h_4 e^{-3i\omega} + h_3 e^{-2i\omega} + h_2 e^{-i\omega} + h_1 + h_1 e^{i\omega} + h_2 e^{2i\omega} + h_3 e^{3i\omega} + h_4 e^{4i\omega}
$$

so that

$$
\begin{aligned}
H'(\omega) = &-3ih_4 e^{-3i\omega} - 2ih_3 e^{-2i\omega} - ih_2 e^{-i\omega} \\
&+ ih_1 e^{i\omega} + 2ih_2 e^{2i\omega} + 3ih_3 e^{3i\omega} + 4ih_4 e^{4i\omega}
\end{aligned}
\tag{10.58}
$$

If we replace ω by π in (10.58), set the result to zero and simplify, we obtain

$$
0 = -h_1 + 3h_2 - 5h_3 + 7h_4
$$

Adding this equation to our linear system gives us the new system

$$
\begin{aligned}
3h_1 + h_2 &= 2\sqrt{2} \\
h_1 + 3h_2 + 3h_3 + h_4 &= 0 \\
h_3 + 3h_4 &= 0 \\
-h_1 + 3h_2 - 5h_3 + 7h_4 &= 0
\end{aligned}
\tag{10.59}
$$

Solving this system gives

$$
h_1 = \frac{45\sqrt{2}}{64} \qquad h_2 = -\frac{7\sqrt{2}}{64} \qquad h_3 = -\frac{9\sqrt{2}}{64} \qquad h_4 = \frac{3\sqrt{2}}{64}
\tag{10.60}
$$

Since this filter pair was formed by starting with the spline filter $\tilde{\mathbf{h}}$, we refer to this filter pair as the $(8, 4)$ biorthogonal spline filter pair.

$\square$

The Transform Matrix for the $(8, 4)$ Biorthogonal Spline Filter Pair

Suppose that we wish to use the $(8, 4)$ biorthogonal spline filter pair to transform a vector $\mathbf{v} = (v_1, \ldots, v_{12})^T$. Then we use the lowpass filter $\tilde{\mathbf{h}}$ given by (10.56) and the highpass filter

$$
\begin{aligned}
\tilde{\mathbf{g}} &= (\tilde{g}_{-3}, \tilde{g}_{-2}, \ldots, \tilde{g}_3, \tilde{g}_4) \\
&= (-h_4, h_3, -h_2, h_1, -h_0, h_{-1}, -h_{-2}, h_{-3}) \\
&= (-h_4, h_3, -h_2, h_1, -h_1, h_2, -h_3, h_4)
\end{aligned}
$$

where $h_1, \ldots, h_4$ are given by (10.60). We write the transform matrix $\tilde{W}_{12}$ as

$$
\tilde{W}_{12} =
\begin{bmatrix}
\tilde{h}_0 & \tilde{h}_1 & \tilde{h}_2 & 0 & 0 & 0 & 0 & 0 & 0 & 0 & 0 & \tilde{h}_{-1} \\
0 & \tilde{h}_{-1} & \tilde{h}_0 & \tilde{h}_1 & \tilde{h}_2 & 0 & 0 & 0 & 0 & 0 & 0 & 0 \\
0 & 0 & 0 & \tilde{h}_{-1} & \tilde{h}_0 & \tilde{h}_1 & \tilde{h}_2 & 0 & 0 & 0 & 0 & 0 \\
0 & 0 & 0 & 0 & 0 & \tilde{h}_{-1} & \tilde{h}_0 & \tilde{h}_1 & \tilde{h}_2 & 0 & 0 & 0 \\
0 & 0 & 0 & 0 & 0 & 0 & 0 & \tilde{h}_{-1} & \tilde{h}_0 & \tilde{h}_1 & \tilde{h}_2 & 0 \\
\tilde{h}_2 & 0 & 0 & 0 & 0 & 0 & 0 & 0 & 0 & \tilde{h}_{-1} & \tilde{h}_0 & \tilde{h}_1 \\
\tilde{g}_0 & \tilde{g}_1 & \tilde{g}_2 & \tilde{g}_3 & \tilde{g}_4 & 0 & 0 & 0 & 0 & \tilde{g}_{-3} & \tilde{g}_{-2} & \tilde{g}_{-1} \\
\tilde{g}_{-2} & \tilde{g}_{-1} & \tilde{g}_0 & \tilde{g}_1 & \tilde{g}_2 & \tilde{g}_3 & \tilde{g}_4 & 0 & 0 & 0 & 0 & \tilde{g}_{-3} \\
0 & \tilde{g}_{-3} & \tilde{g}_{-2} & \tilde{g}_{-1} & \tilde{g}_0 & \tilde{g}_1 & \tilde{g}_2 & \tilde{g}_3 & \tilde{g}_4 & 0 & 0 & 0 \\
0 & 0 & 0 & \tilde{g}_{-3} & \tilde{g}_{-2} & \tilde{g}_{-1} & \tilde{g}_0 & \tilde{g}_1 & \tilde{g}_2 & \tilde{g}_3 & \tilde{g}_4 & 0 \\
\tilde{g}_4 & 0 & 0 & 0 & 0 & \tilde{g}_{-3} & \tilde{g}_{-2} & \tilde{g}_{-1} & \tilde{g}_0 & \tilde{g}_1 & \tilde{g}_2 & \tilde{g}_3 \\
\tilde{g}_2 & \tilde{g}_3 & \tilde{g}_4 & 0 & 0 & 0 & 0 & \tilde{g}_{-3} & \tilde{g}_{-2} & \tilde{g}_{-1} & \tilde{g}_0 & \tilde{g}_1
\end{bmatrix}
$$

$$
=
\begin{bmatrix}
\tilde{h}_1 & \tilde{h}_1 & \tilde{h}_2 & 0 & 0 & 0 & 0 & 0 & 0 & 0 & 0 & \tilde{h}_2 \\
0 & \tilde{h}_2 & \tilde{h}_1 & \tilde{h}_1 & \tilde{h}_2 & 0 & 0 & 0 & 0 & 0 & 0 & 0 \\
0 & 0 & 0 & \tilde{h}_2 & \tilde{h}_1 & \tilde{h}_1 & \tilde{h}_2 & 0 & 0 & 0 & 0 & 0 \\
0 & 0 & 0 & 0 & 0 & \tilde{h}_2 & \tilde{h}_1 & \tilde{h}_1 & \tilde{h}_2 & 0 & 0 & 0 \\
0 & 0 & 0 & 0 & 0 & 0 & 0 & \tilde{h}_2 & \tilde{h}_1 & \tilde{h}_1 & \tilde{h}_2 & 0 \\
\tilde{h}_2 & 0 & 0 & 0 & 0 & 0 & 0 & 0 & 0 & \tilde{h}_2 & \tilde{h}_1 & \tilde{h}_1 \\
h_1 & -h_0 & h_{-1} & -h_{-2} & h_{-3} & 0 & 0 & 0 & 0 & -h_4 & h_3 & -h_2 \\
h_3 & -h_2 & h_1 & -h_0 & h_{-1} & -h_{-2} & h_{-3} & 0 & 0 & 0 & 0 & -h_4 \\
0 & -h_4 & h_3 & -h_2 & h_1 & -h_0 & h_{-1} & -h_{-2} & h_{-3} & 0 & 0 & 0 \\
0 & 0 & 0 & -h_4 & h_3 & -h_2 & h_1 & -h_0 & h_{-1} & -h_{-2} & h_{-3} & 0 \\
h_{-3} & 0 & 0 & 0 & 0 & -h_4 & h_3 & -h_2 & h_1 & -h_0 & h_{-1} & -h_{-2} \\
h_{-1} & -h_{-2} & h_{-3} & 0 & 0 & 0 & 0 & -h_4 & h_3 & -h_2 & h_1 & -h_0
\end{bmatrix}
$$

If we insert the values from (10.56) and (10.60), $\tilde{W}_{12}$ becomes

$$
\tilde{W}_{12} = \frac{\sqrt{2}}{64}
\begin{bmatrix}
24 & 24 & 8 & 0 & 0 & 0 & 0 & 0 & 0 & 0 & 0 & 8 \\
0 & 8 & 24 & 24 & 8 & 0 & 0 & 0 & 0 & 0 & 0 & 0 \\
0 & 0 & 0 & 8 & 24 & 24 & 8 & 0 & 0 & 0 & 0 & 0 \\
0 & 0 & 0 & 0 & 0 & 8 & 24 & 24 & 8 & 0 & 0 & 0 \\
0 & 0 & 0 & 0 & 0 & 0 & 0 & 8 & 24 & 24 & 8 & 0 \\
8 & 0 & 0 & 0 & 0 & 0 & 0 & 0 & 0 & 8 & 24 & 24 \\
45 & -45 & -7 & 9 & 3 & 0 & 0 & 0 & 0 & -3 & -9 & 7 \\
-9 & 7 & 45 & -45 & -7 & 9 & 3 & 0 & 0 & 0 & 0 & -3 \\
0 & -3 & -9 & 7 & 45 & -45 & -7 & 9 & 3 & 0 & 0 & 0 \\
0 & 0 & 0 & -3 & -9 & 7 & 45 & -45 & -7 & 9 & 3 & 0 \\
3 & 0 & 0 & 0 & 0 & -3 & -9 & 7 & 45 & -45 & -7 & 9 \\
-7 & 9 & 3 & 0 & 0 & 0 & 0 & -3 & -9 & -7 & 45 & -45
\end{bmatrix}
$$

For coding purposes, we have formed $\tilde{W}_{12}$ in a manner consistent to that used to form the transform matrix for the $(5, 3)$ biorthogonal filter pair from Example 10.7. That is, we do not reverse the order of the filter coefficients in each row, and the first element in the first row is $\tilde{h}_0$.

For the inverse transformation we need W_{12}. We have

$$
W_{12} =
\begin{bmatrix}
h_0 & h_1 & h_2 & h_3 & h_4 & 0 & 0 & 0 & 0 & h_{-3} & h_{-2} & h_{-1} \\
h_{-2} & h_{-1} & h_0 & h_1 & h_2 & h_3 & h_4 & 0 & 0 & 0 & 0 & h_{-3} \\
0 & h_{-3} & h_{-2} & h_{-1} & h_0 & h_1 & h_2 & h_3 & h_4 & 0 & 0 & 0 \\
0 & 0 & 0 & h_{-3} & h_{-2} & h_{-1} & h_0 & h_1 & h_2 & h_3 & h_4 & 0 \\
h_4 & 0 & 0 & 0 & 0 & h_{-3} & h_{-2} & h_{-1} & h_0 & h_1 & h_2 & h_3 \\
h_2 & h_3 & h_4 & 0 & 0 & 0 & 0 & h_{-3} & h_{-2} & h_{-1} & h_0 & h_1 \\
g_0 & g_1 & g_2 & 0 & 0 & 0 & 0 & 0 & 0 & 0 & 0 & g_{-1} \\
0 & g_{-1} & g_0 & g_1 & g_2 & 0 & 0 & 0 & 0 & 0 & 0 & 0 \\
0 & 0 & 0 & g_{-1} & g_0 & g_1 & g_2 & 0 & 0 & 0 & 0 & 0 \\
0 & 0 & 0 & 0 & 0 & g_{-1} & g_0 & g_1 & g_2 & 0 & 0 & 0 \\
0 & 0 & 0 & 0 & 0 & 0 & 0 & g_{-1} & g_0 & g_1 & g_2 & 0 \\
g_2 & 0 & 0 & 0 & 0 & 0 & 0 & 0 & 0 & g_{-1} & g_0 & g_1
\end{bmatrix}
$$

$$
=
\begin{bmatrix}
h_1 & h_1 & h_2 & h_3 & h_4 & 0 & 0 & 0 & 0 & h_4 & h_3 & h_2 \\
h_3 & h_2 & h_1 & h_1 & h_2 & h_3 & h_4 & 0 & 0 & 0 & 0 & h_4 \\
0 & h_4 & h_3 & h_2 & h_1 & h_1 & h_2 & h_3 & h_4 & 0 & 0 & 0 \\
0 & 0 & 0 & h_4 & h_3 & h_2 & h_1 & h_1 & h_2 & h_3 & h_4 & 0 \\
h_4 & 0 & 0 & 0 & 0 & h_4 & h_3 & h_2 & h_1 & h_1 & h_2 & h_3 \\
h_2 & h_3 & h_4 & 0 & 0 & 0 & 0 & h_4 & h_3 & h_2 & h_1 & h_1 \\
\tilde{h}_1 & -\tilde{h}_1 & \tilde{h}_2 & 0 & 0 & 0 & 0 & 0 & 0 & 0 & 0 & -\tilde{h}_2 \\
0 & -\tilde{h}_2 & \tilde{h}_1 & -\tilde{h}_1 & \tilde{h}_2 & 0 & 0 & 0 & 0 & 0 & 0 & 0 \\
0 & 0 & 0 & -\tilde{h}_2 & \tilde{h}_1 & -\tilde{h}_1 & \tilde{h}_2 & 0 & 0 & 0 & 0 & 0 \\
0 & 0 & 0 & 0 & 0 & -\tilde{h}_2 & \tilde{h}_1 & -\tilde{h}_1 & \tilde{h}_2 & 0 & 0 & 0 \\
0 & 0 & 0 & 0 & 0 & 0 & 0 & -\tilde{h}_2 & \tilde{h}_1 & -\tilde{h}_1 & \tilde{h}_2 & 0 \\
\tilde{h}_2 & 0 & 0 & 0 & 0 & 0 & 0 & 0 & 0 & -\tilde{h}_2 & \tilde{h}_1 & -\tilde{h}_1
\end{bmatrix}
$$

If we insert the values from (10.56) and (10.60), W_{12} becomes

$$
W_{12} = \tfrac{\sqrt{2}}{64}
\begin{bmatrix}
45 & 45 & -7 & -9 & 3 & 0 & 0 & 0 & 0 & 3 & -9 & -7 \\
-9 & -7 & 45 & 45 & -7 & -9 & 3 & 0 & 0 & 0 & 0 & 3 \\
0 & 3 & -9 & -7 & 45 & 45 & -7 & -9 & 3 & 0 & 0 & 0 \\
0 & 0 & 0 & 3 & -9 & -7 & 45 & 45 & -7 & -9 & 3 & 0 \\
3 & 0 & 0 & 0 & 0 & 3 & -9 & -7 & 45 & 45 & -7 & -9 \\
-7 & -9 & 3 & 0 & 0 & 0 & 0 & 3 & -9 & -7 & 45 & 45 \\
24 & -24 & 8 & 0 & 0 & 0 & 0 & 0 & 0 & 0 & 0 & -8 \\
0 & -8 & 24 & -24 & 8 & 0 & 0 & 0 & 0 & 0 & 0 & 0 \\
0 & 0 & 0 & -8 & 24 & -24 & 8 & 0 & 0 & 0 & 0 & 0 \\
0 & 0 & 0 & 0 & 0 & -8 & 24 & -24 & 8 & 0 & 0 & 0 \\
0 & 0 & 0 & 0 & 0 & 0 & 0 & -8 & 24 & -24 & 8 & 0 \\
8 & 0 & 0 & 0 & 0 & 0 & 0 & 0 & 0 & -8 & 24 & -24
\end{bmatrix}
$$

Daubechies Formulation of $H(\omega)$

As you can see from Examples 10.6 and 10.8, sometimes we have to add extra conditions to the orthogonality constraints in order to obtain our solution for $\mathbf{h}$. In Problem 10.23 you will consider other systems and see that we cannot always solve them. Rather than trying to write down derivative conditions for $H(\omega)$ at $\omega = \pi$ for every possible pair of filter lengths, or trying to determine conditions on the lengths of the filters which ensure that a solution for $\mathbf{h}$ exists, we instead appeal to a result due to Ingrid Daubechies [24]. This result tells us exactly when, given a spline filter $\tilde{\mathbf{h}}$, we can find the symmetric filter $\mathbf{h}$. Moreover, her result gives an explicit form for $H(\omega)$.

Theorem 10.2 (Daubechies Solution for $H(\omega)$). *Suppose that $\tilde{\mathbf{h}}$ is the spline filter whose Fourier series is given by (10.44) in Definition 10.3. If N and $\tilde{N}$ have the same parity and $\mathbf{h}$ is the filter whose Fourier series is*

$$
H(\omega) =
\begin{cases}
\sqrt{2}\cos^N\left(\frac{\omega}{2}\right) \displaystyle\sum_{j=0}^{\ell+\tilde{\ell}-1} \binom{\ell+\tilde{\ell}-1+j}{j} \sin^{2j}\left(\frac{\omega}{2}\right), & N = 2\ell \\[2ex]
\sqrt{2}e^{i\omega/2}\cos^N\left(\frac{\omega}{2}\right) \displaystyle\sum_{j=0}^{\ell+\tilde{\ell}} \binom{\ell+\tilde{\ell}+j}{j} \sin^{2j}\left(\frac{\omega}{2}\right), & N = 2\ell + 1
\end{cases}
\tag{10.61}
$$

then $\mathbf{h}$ is a symmetric finite-length filter (with $h_k = h_{-k}$ when N is even and $h_k = h_{1-k}$ when N is odd) with

$$
H(0) = \sqrt{2} \quad \text{and} \quad H(\pi) = 0
$$

and $\tilde{H}(\omega)$ and $H(\omega)$ satisfy the orthogonality condition (10.10). Here $\tilde{N} = 2\tilde{\ell}$ when N is even and $\tilde{N} = 2\tilde{\ell} + 1$ when N is odd. $\quad\square$

Proof of Theorem 10.2. Since $H(\omega)$ is a finite sum, it is clear that $\mathbf{h}$ will be a finite length filter. It is also clear from (10.61) that $H(0) = \sqrt{2}$ and $H(\pi) = 0$.

To show that **h** is a symmetric filter, consider the case first where N is even. Since $\cos\left(\frac{\omega}{2}\right) = \cos(-\frac{\omega}{2})$ and $\sin^{2j}(-\frac{\omega}{2}) = (-\sin\left(\frac{\omega}{2}\right))^{2j} = \sin^{2j}\left(\frac{\omega}{2}\right)$, we have $H(\omega) = H(-\omega)$. We can thus use Problem 4.24 from Section 4.3 to infer that $h_k = h_{-k}$. Thus when N is even, **h** must be an odd length filter symmetric about zero.

In the case when N is odd, note that the function

$$P(\omega) = \sqrt{2}\cos^N\left(\frac{\omega}{2}\right) \sum_{j=0}^{\ell+\tilde{\ell}} \binom{\ell + \tilde{\ell} + j}{j} \sin^{2j}\left(\frac{\omega}{2}\right)$$

will also be even, so that if we expand $P(\omega)$ as a linear combination of complex exponentials, the coefficients will be symmetric about zero. However, since N is odd, the complex exponentials coming from $\cos^N\left(\frac{\omega}{2}\right)$ will be of the form $e^{(2m+1)i\omega/2}$ for integer m (see the discussion on page 373) so that multiplying $P(\omega)$ by $e^{i\omega/2}$ not only gives $H(\omega) = e^{i\omega/2}P(\omega)$, but it shifts the complex exponentials $\frac{1}{2}$ unit right. Thus we see that $H(\omega)$ is a Fourier series with the coefficients of $P(\omega)$ shifted 1 unit right. This gives us $h_k = h_{1-k}$, so that **h** is an even-length filter that is symmetric about $k = \frac{1}{2}$.

To complete the proof, we need to show that $\tilde{H}(\omega)$ and $H(\omega)$ satisfy the orthogonality relation (10.10). We will give the proof for the case when $\tilde{N}$ and N are both even and leave the odd case as Problem 10.27.

Suppose that $\tilde{N}$ and N are even. Then both $\tilde{H}(\omega)$ and $H(\omega)$ are real-valued, so

$$\tilde{H}(\omega)\overline{H(\omega)} = \tilde{H}(\omega)H(\omega)$$

$$= \sqrt{2}\cos^{\tilde{N}}\left(\frac{\omega}{2}\right)\sqrt{2}\cos^N\left(\frac{\omega}{2}\right) \sum_{j=0}^{\ell+\tilde{\ell}-1} \binom{\ell + \tilde{\ell} - 1 + j}{j} \sin^{2j}\left(\frac{\omega}{2}\right)$$

$$= 2\cos^{\tilde{N}+N}\left(\frac{\omega}{2}\right) \sum_{j=0}^{\ell+\tilde{\ell}-1} \binom{\ell + \tilde{\ell} - 1 + j}{j} \sin^{2j}\left(\frac{\omega}{2}\right) \qquad (10.62)$$

Using the trigonometric identities $\cos(t + \frac{\pi}{2}) = -\sin t$ and $\sin(t + \frac{\pi}{2}) = \cos t$, we replace ω by $\omega + \pi$ in (10.62) to write

$$\tilde{H}(\omega + \pi)\overline{H(\omega + \pi)} = \tilde{H}(\omega + \pi)H(\omega + \pi)$$

$$= 2\cos^{\tilde{N}+N}(\frac{\omega}{2} + \pi/2)$$

$$\sum_{j=0}^{\ell+\tilde{\ell}-1} \binom{\ell + \tilde{\ell} - 1 + j}{j} \sin^{2j}(\frac{\omega}{2} + \pi/2)$$

$$= 2(-\sin\left(\frac{\omega}{2}\right))^{\tilde{N}+N} \sum_{j=0}^{\ell+\tilde{\ell}-1} \binom{\ell + \tilde{\ell} - 1 + j}{j} \cos^{2j}\left(\frac{\omega}{2}\right)$$

$$= 2\sin^{\tilde{N}+N}\left(\frac{\omega}{2}\right) \sum_{j=0}^{\ell+\tilde{\ell}-1} \binom{\ell + \tilde{\ell} - 1 + j}{j} \cos^{2j}\left(\frac{\omega}{2}\right)$$

$$(10.63)$$

where the last identity in (10.63) results from the fact that $\tilde{N} + N$ is an even integer. Combining (10.62) and (10.63) yields

$$\tilde{H}(\omega)\overline{H(\omega)} + \tilde{H}(\omega + \pi)\overline{H(\omega + \pi)}$$

$$= 2\left(\cos^{\tilde{N}+N}\left(\frac{\omega}{2}\right) \sum_{j=0}^{\ell+\tilde{\ell}-1} \binom{\ell + \tilde{\ell} - 1 + j}{j} \sin^{2j}\left(\frac{\omega}{2}\right)\right.$$

$$\left. + \sin^{\tilde{N}+N}\left(\frac{\omega}{2}\right) \sum_{j=0}^{\ell+\tilde{\ell}-1} \binom{\ell + \tilde{\ell} - 1 + j}{j} \cos^{2j}\left(\frac{\omega}{2}\right)\right)$$

Now let $K = \ell + \tilde{\ell}$ so that $\tilde{N} + N = 2(\tilde{\ell} + \ell) = 2K$. Rewriting the last identity gives

$$\tilde{H}(\omega)\overline{H(\omega)} + \tilde{H}(\omega + \pi)\overline{H(\omega + \pi)}$$

$$= 2\left(\cos^{2K}\left(\frac{\omega}{2}\right) \sum_{j=0}^{K-1} \binom{K - 1 + j}{j} \sin^{2j}\left(\frac{\omega}{2}\right)\right.$$

$$\left. + \sin^{2K}\left(\frac{\omega}{2}\right) \sum_{j=0}^{K-1} \binom{K - 1 + j}{j} \cos^{2j}\left(\frac{\omega}{2}\right)\right)$$

But using Lemma 8.1, we see that the right side of this identity is 2 and we have proved (10.10). $\square$

In Problem 10.33 you will show that the length of h in Theorem 10.2 is $2N + \tilde{N} - 1$. We now use Theorem 10.2 to state the following definition:

Definition 10.4 (Biorthogonal Spline Filter Pair). *Suppose that N and $\tilde{N}$ have the same parity. Any biorthogonal filter pair formed using Theorem 10.2 will be called the $(2N + \tilde{N} - 1, \tilde{N})$ biorthogonal spline filter pair or the biorthogonal spline filter pair.* □

Using the Daubechies Theorem

Let's look at an example of how to use Theorem 10.2.

Example 10.9 (Creating Biorthogonal Spline Filter Pairs Using Theorem 10.2). *Let's create a biorthogonal spline filter pair using Theorem 10.2 for $\tilde{N} = N = 2$. Thus, $\tilde{\ell} = \ell = 1$. For $\mathbf{h}$ we have*

$$\tilde{H}(\omega) = \sqrt{2}\cos^2\left(\frac{\omega}{2}\right) = \frac{\sqrt{2}}{4}e^{-i\omega} + \frac{\sqrt{2}}{2} + \frac{\sqrt{2}}{4}e^{i\omega}$$

We now use (10.61) to write

$$H(\omega) = \sqrt{2}\cos^2\left(\frac{\omega}{2}\right)\sum_{j=0}^{1}\binom{1+j}{j}\sin^{2j}\left(\frac{\omega}{2}\right)$$

$$= \sqrt{2}\cos^2\left(\frac{\omega}{2}\right)\left(1 + 2\sin^2\left(\frac{\omega}{2}\right)\right)$$

$$= \sqrt{2}\left(\frac{e^{i\omega/2} + e^{-i\omega/2}}{2}\right)^2\left(1 + 2\left(\frac{e^{i\omega/2} - e^{-i\omega/2}}{2i}\right)^2\right)$$

$$= -\frac{\sqrt{2}}{8}e^{-2i\omega} + \frac{\sqrt{2}}{4}e^{-i\omega} + \frac{3\sqrt{2}}{4} + \frac{\sqrt{2}}{4}e^{i\omega} - \frac{\sqrt{2}}{8}e^{2i\omega}$$

Thus, we can read the symmetric filter $\mathbf{h}$ from the Fourier series. We have

$$\mathbf{h} = (h_{-2}, h_{-1}, h_0, h_1, h_2) = \left(-\frac{\sqrt{2}}{8}, \frac{\sqrt{2}}{4}, \frac{3\sqrt{2}}{4}, \frac{\sqrt{2}}{4}, -\frac{\sqrt{2}}{8}\right)$$

Compare the results of this example with that of Example 10.6. In Problem 10.29, you will show that the biorthogonal filter pair obtained in Example 10.8 can be constructed using Theorem 10.2 with $\tilde{N} = N = 3$. □

Example: Signal Compression

We conclude this section by returning to Example 8.2.

Example 10.10 (Signal Compression Using the $(5, 3)$ Biorthogonal Spline Filter Pair). *We perform six iterations of the wavelet transform using the $(5, 3)$ biorthogonal* *spline filter pair on the 7,744-sample audio file from Example 7.1. (The algorithms used for this computation are discussed in Chapter 11.) In Figure 10.4 we show the cumulative energy of this transform, and for comparative purposes we plot the cumulative energy of the transform using the D4 filter.*

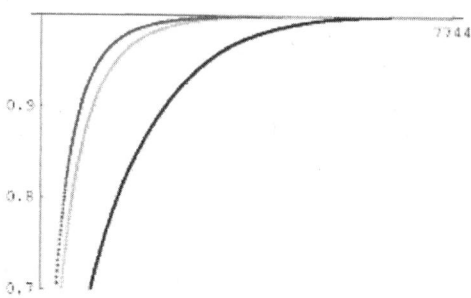

Figure 10.4   The cumulative energies of the original audio file (black), the D4 filter (gray), and the $(5, 3)$ biorthogonal spline filter pair (light gray).

Note that the D4 filter slightly outperformed the (5,3) filter pair in this example. □

Which Biorthogonal Filter Pair to Use?

Using Definition 10.3 and Theorem 10.2, we can easily generate many biorthogonal filter pairs for use in applications. How do we know which filter pair to choose?

As a rule of thumb, longer filters $\mathbf{h}$ whose Fourier series $\tilde{H}(\omega)$ satisfies $\tilde{H}^{(m)}(\pi) = 0$ for a relatively large m return a more accurate approximation of the original data in the lowpass portion of the transform. The classical treatment of wavelet theory develops the lowpass filters by creating *scaling functions* that are building blocks used to represent the space of square-integrable functions. Here, the designer has control of features such as orthogonality, symmetry, and smoothness. The length of the resulting lowpass filter is directly proportional to the smoothness of the scaling functions. We touched on this fact in our earlier discussion of the spline scaling functions $B_N(t)$ on page 374. Note that as N increases, the spline function $B_N(t)$ becomes smoother and the resulting spline filter grows in length.

One disadvantage of using a long filter is that the computation speed is slower. Another problem is that biorthogonal spline filter pairs often have lengths where one filter is quite a bit longer than the other (see Problem 10.33). This can be problematic since $\tilde{\mathbf{g}}$ is constructed from $\mathbf{h}$, and if $\mathbf{h}$ is substantially longer say than $\tilde{\mathbf{h}}$, then the resulting wavelet transform matrix $\tilde{W}$ is built with filters where the approximation ability of the lowpass portion of the transform is not as good as that of the highpass portion of the transform. In applications such as image compression, we prefer filter pairs whose lengths are similar. In Section 10.3 we address this problem by constructing a symmetric biorthogonal filter pair whose lengths differ by 2.

PROBLEMS

10.19   Suppose that $\mathbf{h}$ and $\tilde{\mathbf{h}}$ are filters whose Fourier series $H(\omega)$ and $\tilde{H}(\omega)$, respectively satisfy $\tilde{H}(0) = \sqrt{2}$, $\tilde{H}(\pi) = 0$, and (10.10). Show that $H(0) = \sqrt{2}$.

10.20   Show that the filter given by

$$\tilde{H}(\omega) = \sqrt{2}\, e^{i\omega/2} \cos^{\tilde{N}}\left(\frac{\omega}{2}\right)$$

has the form given by (10.41) and show that $\tilde{h}_k = \tilde{h}_{1-k}$ for $k = -\frac{\tilde{N}-1}{2}, \dots, \frac{\tilde{N}+1}{2}$.

10.21   Write down the spline filter of length $\tilde{N} + 1$ where

(a) $\tilde{N} = 1$.

(b) $\tilde{N} = 3$.

(c) $\tilde{N} = 5$.

10.22   Verify that the solution to the system (10.59) is indeed given by (10.60).

10.23   Using Examples 10.6 and 10.8 as guides, find (if possible) the following biorthogonal filter pairs.

(a) $\tilde{\mathbf{h}}$ is the spline filter of length 5 and $\mathbf{h}$ is a symmetric filter of length 7.

(b) $\tilde{\mathbf{h}}$ is the spline filter of length 4 and $\mathbf{h}$ is also a symmetric filter of length 4.

(c) $\tilde{\mathbf{h}}$ is the spline filter of length 6 and $\mathbf{h}$ is a symmetric filter of length 10. (A **CAS** will be useful for this problem.)

(d) $\tilde{\mathbf{h}}$ is the spline filter of length 7 and $\mathbf{h}$ is a symmetric filter of length 11.

10.24   Write down the wavelet matrices W_{12} and $\tilde{W}_{12}$ that you would use to process a vector $\mathbf{v} = (v_1, \dots, v_{12})^T$ using the filters from

(a) part (a) of Problem 10.23.

(b) part (b) of Problem 10.23.

10.25   Show that the elements $\tilde{h}_k$ of the spline filter $\tilde{\mathbf{h}}$ given in Definition 10.3 can be written as

$$\tilde{h}_{k-\frac{\tilde{N}}{2}} = \frac{\sqrt{2}}{2^{\tilde{N}}}\binom{\tilde{N}}{k}, \qquad k = 0, \dots, \tilde{N}$$

Thus, the spline filter of length $\tilde{N} + 1$ can be written using the $\tilde{N} + 1$ row of Pascal's triangle!

10.26 Use the defining relation (10.45) to verify the formula for the triangle function (10.46).

10.27 Complete the proof that $\tilde{H}(\omega)$ and $H(\omega)$ satisfy (10.10) in Theorem 10.2 in the case where both $\tilde{N}$ and N are odd. (*Hint:* Take $K = \ell + \tilde{\ell} + 1$ and mimic the proof when $\tilde{N}$ N are both even.)

10.28 What biorthogonal spline filter pair do you get if you take $N = \tilde{N} = 1$ in Theorem 10.2 and Definition 10.3?

10.29 Consider the spline filter $\tilde{\mathbf{h}}$ given by (10.43) for $\tilde{N} = 3$. Show that if we take this filter and $N = 3$, Theorem 10.2 produces the eight-term filter $\mathbf{h}$ constructed in Example 10.8.

10.30 Use Theorem 10.2 and Definition 10.3 to find the biorthogonal spline filter pairs for the given values of $\tilde{N}$ and N. You might wish to use a **CAS** for some of the computations.

(a) $\tilde{N} = 3$, $N = 1$.

(b) $\tilde{N} = 1$, $N = 3$.

(c) $\tilde{N} = 4$, $N = 2$.

10.31 For each of the filter pairs you found in Problem 10.30, find m and $\tilde{m}$ such that

$$H^{(k)}(\pi) = 0, \ k = 0, \ldots, m - 1 \qquad \text{and} \qquad \tilde{H}^{(k)}(\pi) = 0, \ k = 0, \ldots, \tilde{m} - 1$$

but $H^{(m)}(\pi) \neq 0$ and $\tilde{H}^{(\tilde{m})}(\pi) \neq 0$.

10.32 Let $\tilde{H}(\omega)$ be the Fourier series given by (10.40). Show that $\tilde{H}^{(k)}(\pi) = 0$ for $k = 0, \ldots, \tilde{N} - 1$ and $\tilde{H}^{(\tilde{N})}(\pi) \neq 0$.

10.33 Suppose that N and $\tilde{N}$ have the same parity. In this problem you will show that the length of the filter $\mathbf{h}$ from Theorem 10.2 is $2N + \tilde{N} - 1$. The following steps will help you organize your work.

(a) Consider the case where both N and $\tilde{N}$ are even. Write $N = 2\ell$ and $\tilde{N} = 2\tilde{\ell}$. Argue that

$$\cos\left(\frac{\omega}{2}\right)^N = \cos\left(\frac{\omega}{2}\right)^{2\ell} = \sum_{k=-\ell}^{\ell} a_k e^{ik\omega}$$

and

$$\sum_{j=0}^{\ell+\tilde{\ell}-1} \binom{\ell + \tilde{\ell} - 1 + j}{j} \sin^{2j}\left(\frac{\omega}{2}\right) = \sum_{m=-\ell-\tilde{\ell}+1}^{\ell+\tilde{\ell}-1} b_k e^{im\omega}$$

for numbers a_k, $k = -\ell, \ldots, \ell$ and b_m, $m = -\ell - \tilde{\ell} + 1, \ldots, \ell + \tilde{\ell} - 1$. Use Problem 4.18 from Section 4.2 to expand $\cos\left(\frac{\omega}{2}\right)$ and $\sin\left(\frac{\omega}{2}\right)$. We do not want to find a_k and b_m explicitly — we just want to know where these series start and stop.

(b) Multiply the results from part (a) to show that $H(\omega)$ can be expanded as a Fourier series of the form

$$H(\omega) = \sum_{k=-2\ell+\tilde{\ell}-1}^{2\ell+\tilde{\ell}-1} c_k e^{ik\omega}$$

for some numbers c_k.

(c) Use part (b) to show that the length of the Fourier series $H(\omega)$ is $4\ell + 2\tilde{\ell} - 1 = 2N + \tilde{N} - 1$, as desired.

(d) Repeat steps (a) – (c) for the case when both N and $\tilde{N}$ are odd.

10.34   Suppose that $\tilde{h}$ and h are a biorthogonal spline filter pair constructed using Theorem 10.2. Then the length of $\tilde{h}$ is $\tilde{N} + 1$, and from Problem 10.33 we know that the length of h is $2N + \tilde{N} - 1$.

(a) Show that the length of $\tilde{h}$ is the same as the length of h if and only if $N = 1$.

(b) From part (a) we see that odd-length biorthogonal filter pairs cannot have the same length. Is it possible to construct an odd-length biorthogonal filter pair using Theorem 10.2 so that the difference of the lengths of the filters is 2? If so, explain how you would choose $\tilde{N}$ and N.

10.35   Suppose that N and $\tilde{N}$ are both even. Then from Problem 10.33 we know that the length of $h = (h_{-L}, \ldots, h_L)$ is $2N + \tilde{N} - 1$. We also know from Definition 10.3 that the length of $\tilde{h} = (\tilde{h}_{-L'}, \ldots, \tilde{h}_{L'})$ is $\tilde{N} + 1$. Find L and L' in terms of N and $\tilde{N}$. Repeat the problem in the case where N and $\tilde{N}$ are both odd.

Computer Lab

◆ **10.1 Software Development - Generating Biorthogonal Spline Filter Pairs.** From the text Web site, access the package `biorthsplinefilters`. In this development lab you will generate code that will produce biorthogonal spline filter pairs using Theorem 10.2. Your code will utilize the formula given in Definition 10.3 to construct $\tilde{h}$ and Theorem 10.2 to construct h. Instructions are provided that allow you to add the code to the software package `DiscreteWavelets`.

10.3   THE COHEN–DAUBECHIES–FEAUVEAU 9/7 FILTER

The CDF 9/7 Filter Pair

The JPEG2000 committee chose the Cohen–Daubechies–Feauveau 9/7 biorthogonal filter pair (hereafter the CDF97 filter pair — the lengths of the filters are 9 and 7) developed by Cohen et al. [20] for the lossy version of the JPEG2000 compression standard. Although this filter pair is not a member of the biorthogonal spline filter pair family that we developed in Section 10.2, the filters are symmetric and their Fourier transforms satisfy the conditions necessary to classify them as lowpass filters.

The filter lengths of the CDF97 filter pair are relatively short and similar in length. Although it is possible to develop a biorthogonal spline filter pair with lengths 9 and 7 (see Problem 10.36), Unser and Blu [79] cite several reasons why the CDF97 filter pair is preferred over the $(9,7)$ biorthogonal spline filter pair. Unser and Blu report that the CDF97 filter pair produces a wavelet transform matrix that is closer to orthogonal than that generated by the $(9,7)$ biorthogonal spline filter pair. [5]

Smoothness of CDF97 Fourier Series at $\omega = \pi$

Another key advantage of the CDF97 over the $(9,7)$ biorthogonal spline pair is recognized when we analyze the Fourier series of each filter. Suppose that $\tilde{\mathbf{h}}$ and $\mathbf{h}$ are the biorthogonal spline filter pair whose lengths are 7 and 9, respectively. Let

$$\tilde{H}(\omega) = \sum_{k=-3}^{3} \tilde{h}_k e^{ik\omega} \qquad \text{and} \qquad H(\omega) = \sum_{k=-4}^{4} h_k e^{ik\omega}$$

denote the Fourier series associated with each filter. Then we can show (see Problem 10.37) that

$$H(\pi) = H'(\pi) = 0 \qquad \text{and} \qquad \tilde{H}^{(m)}(\pi) = 0, \qquad m = 0,\ldots,5 \qquad (10.64)$$

The moduli of the Fourier series $\tilde{H}(\omega)$ and $H(\omega)$ are plotted in Figure 10.5.

It turns out that the Fourier series $H(\omega)$ and $\tilde{H}(\omega)$ for the CDF97 filters $\mathbf{h}$ and $\tilde{\mathbf{h}}$, respectively, satisfy

$$H^{(m)}(\pi) = \tilde{H}^{(m)}(\pi) = 0, \qquad m = 0,1,2,3 \qquad (10.65)$$

Thus, we see that while the combined number of zeros at $\omega = \pi$ of $\tilde{H}(\omega)$ and $H(\omega)$ for each filter pair is eight, the zeros at $\omega = \pi$ of $\tilde{H}(\omega)$ and $H(\omega)$ for the CDF97 filter pair are equally balanced.

As we work through the construction of the CDF97 filter pair, we will have a better understanding of why the zeros are equally balanced. The process for constructing

---

[5]The mathematics behind this argument is beyond the scope of this book.

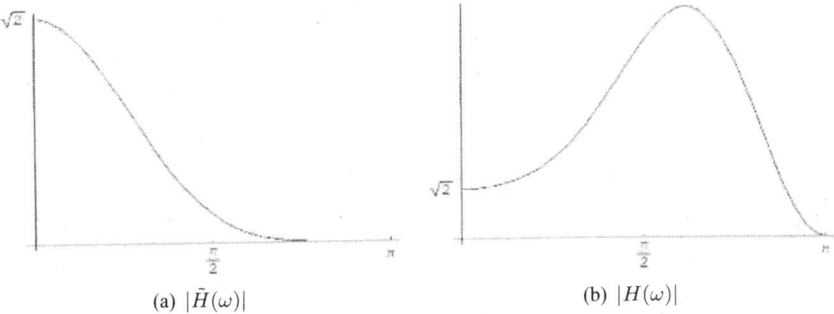

(a) $|\tilde{H}(\omega)|$ (b) $|H(\omega)|$

Figure 10.5   Plots of $|\tilde{H}(\omega)|$ and $|H(\omega)|$ for the $(9, 7)$ biorthogonal spline filter pair. Note that, as indicated by (10.64), $H(\omega)$ is not very "flat" at $\omega = \pi$.

the CDF97 filter pair requires us to better understand the form of the solutions to the biorthogonality condition

$$\tilde{H}(\omega)\overline{H(\omega)} + \tilde{H}(\omega + \pi)\overline{H(\omega + \pi)} = 2 \tag{10.66}$$

that is fundamental for constructing biorthogonal filter pairs.

Note: The results that follow are from Cohen et al. [20]. Whereas we consider only odd-length symmetric filters in this section, the results given in Cohen et al. [20] consider even-length symmetric filters as well.

$H(\omega)$ for Symmetric Odd-Length Filters

Suppose that $\mathbf{h} = (h_{-L}, \ldots, h_L)$ is an odd-length symmetric filter. In Problem 4.25 from Section 4.3 we showed that if $h_k = h_{-k}$, then the Fourier series $H(\omega) = \sum_{k=-L}^{L} h_k e^{ik\omega}$ is an even function. We can then use part (c) of Problem 4.24 in Section 4.3 to write $H(\omega)$ as a linear combinations of cosine functions:

$$H(\omega) = h_0 + 2 \sum_{k=1}^{L} h_k \cos(k\omega) \tag{10.67}$$

It turns out that we can write $H(\omega)$ in (10.67) as a linear combination of powers of $\cos(\omega)$. In other words, when we start with a symmetric, odd-length filter $\mathbf{h}$, $H(\omega)$ can be expressed as a polynomial in $\cos(\omega)$. In Problem 10.38, you will show that for $k \in \mathbf{Z}$, the function $\cos(k\omega)$ can be written as a polynomial in $\cos(\omega)$. For example,

$\cos(3\omega) = 4\cos^3(\omega) - 3\cos(\omega)$ (you will verify this in Problem 10.39). Here, we have written $\cos(3\omega)$ by composing $\cos(\omega)$ and the polynomial $P(t) = 4t^3 - 3t$.

Using Problem 10.38, we can rewrite (10.67) as a polynomial in $\cos(\omega)$. Let's look at an example.

Example 10.11 (Writing Even H(ω) as a Polynomial in Cosine). *Consider the length 7 symmetric filter*

$$\mathbf{h} = (h_{-3}, \ldots, h_3) = \left(\frac{3\sqrt{2}}{32}, -\frac{3\sqrt{2}}{8}, \frac{5\sqrt{2}}{32}, \frac{5\sqrt{2}}{4}, \frac{5\sqrt{2}}{32}, -\frac{3\sqrt{2}}{8}, \frac{3\sqrt{2}}{32}\right)$$

Using part (c) of Problem 4.24 in Section 4.3, we can write the associated Fourier series $H(\omega)$ as

$$H(\omega) = \frac{5\sqrt{2}}{4} + 2\left(\frac{5\sqrt{2}}{32}\cos(\omega) - \frac{3\sqrt{2}}{8}\cos(2\omega) + \frac{3\sqrt{2}}{32}\cos(3\omega)\right)$$

Since $\cos(2\omega) = 2\cos^2(\omega) - 1$ and $\cos(3\omega) = 4\cos^3(\omega) - 3\cos(\omega)$, we have

$$H(\omega) = \frac{5\sqrt{2}}{4} + \frac{5\sqrt{2}}{16}\cos(\omega) - \frac{3\sqrt{2}}{4}(2\cos^2(\omega) - 1) + \frac{3\sqrt{2}}{16}(4\cos^3(\omega) - 3\cos(\omega))$$

$$= 2\sqrt{2} - \frac{\sqrt{2}}{4}\cos(\omega) - \frac{3\sqrt{2}}{2}\cos^2(\omega) + \frac{3\sqrt{2}}{4}\cos^3(\omega)$$

Thus, we see $H(\omega)$ can be written as a cubic polynomial P in $\cos(\omega)$ where

$$P(t) = 2\sqrt{2} - \frac{\sqrt{2}}{4}t - \frac{3\sqrt{2}}{2}t^2 + \frac{3\sqrt{2}}{4}t^3 \qquad (10.68)$$

□

Lowpass Conditions and H(ω)

For **h** to be a lowpass filter, we require that $H(0) = \sqrt{2}$ and $H(\pi) = 0$. Moreover, if $H(\omega)$ must also satisfy $H^{(\ell)}(\pi) = 0$ for some $\ell = 0, 1, \ldots$, then $(1 + \cos(\omega))^\ell$ must be a factor of $H(\omega)$. Thus, we can write

$$H(\omega) = \sqrt{2}\,(1 + \cos(\omega))^\ell\, q(\cos(\omega))$$

for some polynomial q where $q(\cos(\pi)) = q(-1) \neq 0$. Since

$$\sqrt{2} = H(0) = \sqrt{2}\,(1 + \cos(0))^\ell\, q(\cos(0)) = \sqrt{2} \cdot 2^\ell \cdot q(1)$$

we also must have $q(1) = 2^{-\ell}$. Using the half-angle formula

$$\cos^2\left(\frac{\omega}{2}\right) = \frac{1}{2}(1 + \cos(\omega)) \qquad \text{or} \qquad 2\cos^2\left(\frac{\omega}{2}\right) = 1 + \cos(\omega)$$

we can write

$$H(\omega) = \sqrt{2} \, \cos^{2\ell}\left(\frac{\omega}{2}\right) 2^{\ell} \, q(\cos(\omega))$$

If we let $p(\omega) = 2^{\ell} q(\omega)$, we have $p(-1) = 2^{\ell} q(-1) \neq 0$ and $p(1) = 2^{\ell} q(1) = 2^{\ell} \cdot 2^{-\ell} = 1$, so that

$$H(0) = \sqrt{2} \, \cos^{2\ell}(0) \, p(\cos(0)) = \sqrt{2}$$

We summarize this discussion in Proposition 10.4.

Proposition 10.4 ($H(\omega)$ for Odd-Length Symmetric Filters). *Consider the filter* $\mathbf{h} = (h_{-L}, \ldots, h_L)$ *with* $h_k = h_{-k}$ *for* $k = 1, \ldots L$ *whose Fourier series* $H(\omega)$ *satisfies* $H(0) = \sqrt{2}$. *Then*

$$H(\omega) = \sum_{k=-L}^{L} h_k \, e^{ik\omega}$$

is an even function (see Problem 4.25 from Section 4.3) and $H(\omega)$ *can be written as a power of* $\cos\left(\frac{\omega}{2}\right)$ *times a polynomial in* $\cos(\omega)$. *That is,*

$$H(\omega) = \sqrt{2} \, \cos^{2\ell}\left(\frac{\omega}{2}\right) p(\cos(\omega)) \tag{10.69}$$

where $p(1) = 1$ *and* $p(-1) \neq 0$.

$\square$

Let's apply Proposition 10.4 to the filter from Example 10.11.

Example 10.12 (Expressing $H(\omega)$ Using Proposition 10.4). *For the filter* $\mathbf{h}$ *in Example 10.11, we found that* $H(\omega)$ *could be expressed in terms of* $P(\cos(\omega))$ *where* P *is given by (10.68). We factor out* $\sqrt{2}/4(1+t)$ *to obtain*

$$P(t) = \frac{\sqrt{2}}{4}(1+t)(3t^2 - 9t + 8)$$

Our goal is to write $H(\omega)$ *in the form given by (10.69). Toward this end, we substitute* $t = \cos(w)$ *and use the identity* $1 + \cos(\omega) = 2\cos^2\left(\frac{\omega}{2}\right)$ *to write*

$$H(\omega) = P(\cos(\omega))$$
$$= \frac{\sqrt{2}}{4}(1 + \cos(\omega))(3\cos^2(\omega) - 9\cos(\omega) + 8)$$
$$= \frac{\sqrt{2}}{2}\cos^2\left(\frac{\omega}{2}\right)(3\cos^2(\omega) - 9\cos(\omega) + 8)$$
$$= \sqrt{2}\cos^2\left(\frac{\omega}{2}\right)\left(\frac{3}{2}\cos^2(\omega) - \frac{9}{2}\cos(\omega) + 4\right)$$

so we see that $\ell = 1$ and $p(t)$ from Proposition 10.4 is

$$p(t) = \frac{3}{2}t^2 - \frac{9}{2}t + 4$$

It is easy to check that $p(-1) \neq 0$ and $p(1) = 1$. $\square$

Existence of Even Solutions to (10.66)

The next result says that for a given *even* $H(\omega)$, if we find a solution $\tilde{H}(\omega)$ to the orthogonality condition (10.66), then we can construct an even solution. This result was stated and proved in Cohen et al. [20].

Proposition 10.5 (Even Solutions to (10.66)). *Suppose that $H(\omega)$ is given and $H(\omega) = H(-\omega)$. If there exists a solution $\tilde{H}(\omega)$ to (10.66), then there exists an even solution.*

$\square$

Proof of Proposition 10.5. If $\tilde{H}(\omega) = \tilde{H}(-\omega)$, then the proof is complete. Otherwise, consider the Fourier series

$$E(\omega) = \frac{1}{2}\left(\tilde{H}(\omega) + \tilde{H}(-\omega)\right)$$

It is easy to check that $E(\omega) = E(-\omega)$. We need to verify that $E(\omega)$ satisfies (10.66). Toward this end, we first compute

$$E(\omega)\overline{H(\omega)} = \frac{1}{2}(\tilde{H}(\omega) + \tilde{H}(-\omega))\overline{H(\omega)} = \frac{1}{2}\tilde{H}(\omega)\overline{H(\omega)} + \frac{1}{2}\tilde{H}(-\omega)\overline{H(\omega)}$$

$$(10.70)$$

and next compute

$$E(\omega + \pi)\overline{H(\omega + \pi)} = \frac{1}{2}(\tilde{H}(\omega + \pi) + \tilde{H}(-(\omega + \pi)))\overline{H(\omega + \pi)}$$

$$= \frac{1}{2}\tilde{H}(\omega + \pi)\overline{H(\omega + \pi)} + \frac{1}{2}\tilde{H}(-(\omega + \pi))\overline{H(\omega + \pi)}$$

$$(10.71)$$

We next add $E(\omega)\overline{H}(\omega)$ and $E(\omega + \pi)\overline{H(\omega + \pi)}$. Adding the first terms of (10.70) and (10.71) gives

$$\frac{1}{2}\tilde{H}(\omega)\overline{H(\omega)} + \frac{1}{2}\tilde{H}(\omega + \pi)\overline{H(\omega + \pi)} = \frac{1}{2}(\tilde{H}(\omega)\overline{H(\omega)} + \tilde{H}(\omega + \pi)\overline{H(\omega + \pi)})$$

But $\tilde{H}(\omega)$ is a solution to (10.66) so this sum is 1. Thus we have

$$E(\omega)\overline{H(\omega)} + E(\omega + \pi)\overline{H(\omega + \pi)}$$

$$= 1 + \frac{1}{2}\left(\tilde{H}(-\omega)\overline{H(\omega)} + \tilde{H}(-\omega - \pi)\overline{H(\omega + \pi)}\right) \quad (10.72)$$

We need to replace the $-\omega$ that appears as an argument in the functions in (10.72). In Problem 10.40 you will verify that replacing ω by $-\omega$ leaves the left side of (10.72) unchanged. Let's work on the right side of (10.72). In particular, let

$$g(\omega) = \tilde{H}(-\omega)\overline{H(\omega)} + \tilde{H}(-\omega - \pi)\overline{H(\omega + \pi)}$$

We replace ω by $-\omega$ and use the fact that $H(\omega)$ is even and both $H(\omega)$ and $\tilde{H}(\omega)$ are 2π–periodic functions in the second term to write

$$\begin{aligned} g(-\omega) &= \tilde{H}(\omega)\overline{H(-\omega)} + \tilde{H}(\omega - \pi)\overline{H(-(\omega - \pi))} \\ &= \tilde{H}(\omega)\overline{H(\omega)} + \tilde{H}(\omega - \pi)\overline{H(\omega - \pi)} \\ &= \tilde{H}(\omega)\overline{H(\omega)} + \tilde{H}(\omega - \pi + 2\pi)\overline{H(\omega - \pi + 2\pi)} \\ &= \tilde{H}(\omega)\overline{H(\omega)} + \tilde{H}(\omega + \pi)\overline{H(\omega + \pi)} = 2 \end{aligned}$$

since $\tilde{H}(\omega)$ is a solution to (10.66). Inserting this result into the second term on the right side of (10.72) gives the desired result.   □

To summarize, if h is a symmetric, odd-length filter, then we can write its Fourier series in the form

$$H(\omega) = \sqrt{2}\,\cos^{2\ell}\left(\frac{\omega}{2}\right) p(\cos(\omega)) \tag{10.73}$$

where ℓ is a nonnegative integer and p is a polynomial with $p(-1) \neq 0$ and $p(1) = 1$.

Moreover, Proposition 10.5 tells us that for this even $H(\omega)$, if we can find a solution to (10.66), then we can find an even solution to (10.66). Combining this result with Proposition 10.4 tells us that the even solution $\tilde{H}(\omega)$ to (10.66) can be expressed as

$$\tilde{H}(\omega) = \sqrt{2}\,\cos^{2\tilde{\ell}}\left(\frac{\omega}{2}\right) \tilde{p}(\cos(\omega)) \tag{10.74}$$

where $\tilde{\ell}$ is a nonnegative integer and $\tilde{p}$ is a polynomial with $\tilde{p}(-1) \neq 0$ and $\tilde{p}(1) = 1$.

Characterizing Solutions to (10.66)

We are now ready for the main result of this section. Theorem 10.3, due to Cohen et al. [20],[6] allows us to characterize the polynomials p and $\tilde{p}$ given in (10.73) and (10.74), respectively, and thus develop many different solutions to the orthogonality condition (10.66).

---

[6] We only state and prove the result for odd-length symmetric filters. The result is valid for even-length symmetric filters as well. Moreover, we have only given a special case of the result. The result, Proposition 6.4 in [20], gives an even more general constraint that the product of p and $\tilde{p}$ must satisfy. This general constraint leads to even more possible formulations of h and h.

Theorem 10.3 (Characterizing Solutions to (10.66)). *Suppose that* $\tilde{\mathbf{h}}$ *and* $\mathbf{h}$ *are odd-length, symmetric filters whose Fourier series given by (10.74) and (10.73), respectively, solve (10.66). Then the polynomials* p *and* $\tilde{p}$ *must satisfy*

$$p(\cos(\omega))\,\tilde{p}(\cos(\omega)) = \sum_{j=0}^{K-1} \binom{K-1+j}{j} \sin^{2j}\left(\frac{\omega}{2}\right) \qquad (10.75)$$

where $K = \ell + \tilde{\ell}$ *and the degree of* $p(t)\tilde{p}(t)$ *is less than* K. $\qquad\square$

Proof of Theorem 10.3. The proof of Theorem 10.3 is quite technical. For this reason, we leave it as (challenging!) Problem 10.41. $\qquad\square$

You should recognize the right-hand side of (10.75) — it appeared as part of $H(\omega)$ for odd-length biorthogonal spline filter pairs in Theorem 10.2![7] That is, the Fourier series for the biorthogonal spline filter pair $\tilde{\mathbf{h}}$ and $\mathbf{h}$ are

$$\tilde{H}(\omega) = \sqrt{2}\cos^{\tilde{N}}\left(\frac{\omega}{2}\right)$$

and

$$H(\omega) = \sqrt{2}\cos^{N}\left(\frac{\omega}{2}\right) \sum_{j=0}^{K-1} \binom{K-1+j}{j}\sin^{2j}\left(\frac{\omega}{2}\right)$$

where $\tilde{N} = 2\tilde{\ell}$ and $N = 2\ell$ are nonnegative, even integers. So we see now that this filter pair is simply a special case of Theorem 10.3 with $\tilde{p}(t) = 1$ and $p(t) = \sum_{j=0}^{K-1}\binom{K-j+1}{j}t^j$!

The Construction of Cohen, Daubechies, and Feauveau

In developing the CDF97 filter pair, Cohen et al. decided to factor the right-hand side of (10.75) and spread the factors as equally as possibly between $\tilde{p}$ and p. This differs from the spline case where all the factors of the right-hand side of (10.75) are assigned to $p(t)$. Thus, the idea is to choose a K and then factor

$$P(t) = \sum_{j=0}^{K-1} \binom{K-1+j}{j}t^j$$

Recall that the roots of any polynomial with real coefficients are either real or occur in complex conjugate pairs. Cohen et al. sought to build one of the polynomials (say, p) by multiplying the linear factors built from the complex roots of $P(t)$ and

---

[7]The right side of (10.75) also appears in the Fourier series for the Coiflet filter of Theorem 8.3!

the other (say, $\tilde{p}$) as the product of the linear factors from the real roots of $P(t)$. In Problem 10.42 you will find the roots of $P(t)$ where $K = 2$ or 3. When $K = 1$ there are no roots, and in the other cases there are only real roots or only complex roots. The first case, where there is a combination of roots, is when $K = 4$. The CDF97 filter pair is built from this case. When $K = 4$, we take $\ell = \tilde{\ell} = 2$ and

$$P(t) = \sum_{j=0}^{3} \binom{3+j}{j} t^j = 1 + 4t + 10t^2 + 20t^3$$

Using a **CAS**, we can find that the roots (to six digits) of $P(t)$ are

$$r_1 = -0.342484 \qquad r_2 = -0.078808 - .373391i \qquad r_3 = -0.078808 + .373391i$$

Since the leading coefficient of $P(t)$ is 20, we have

$$P(t) = 20(t - r_1)(t - r_2)(t - r_3)$$

We build $\tilde{p}$ using r_1 and p from r_2 and r_3. The question is how to distribute the leading coefficient 20 of $P(t)$. For now, we write

$$\tilde{p}(t) = a(t - r_1) \qquad \text{and} \qquad p(t) = \frac{20}{a}(t - r_2)(t - r_3)$$

We now use (10.73) and (10.74) with $\tilde{\ell} = \ell = 2$ and $\tilde{p}(t) = a(t + 0.342484)$. If we make the substitution $t = \sin^2\left(\frac{\omega}{2}\right)$, we can write $\tilde{H}(\omega)$ as

$$\tilde{H}(\omega) = \sqrt{2}\cos^4\left(\frac{\omega}{2}\right) a\left(\sin^2\left(\frac{\omega}{2}\right) + 0.342484\right) \tag{10.76}$$

and

$$H(\omega) = \sqrt{2}\cos^4\left(\frac{\omega}{2}\right) \frac{20}{a}\left(\sin^2\left(\frac{\omega}{2}\right) + 0.078808 + 0.373391i\right)$$
$$\cdot \left(\sin^2\left(\frac{\omega}{2}\right) + 0.078808 - 0.373391i\right) \tag{10.77}$$

We can find a by remembering that both $\tilde{H}(\omega)$ and $H(\omega)$ must satisfy the lowpass condition $\tilde{H}(0) = H(0) = \sqrt{2}$. So we can plug $\omega = 0$ into, say, (10.76), set the result equal to $\sqrt{2}$, and solve for a. We can then plug the value for a into (10.77). We have

$$\sqrt{2} = \tilde{H}(0) = \sqrt{2}\cos^4(0)a\left(\sin^2(0) + 0.342484\right) = \sqrt{2}\,(0.342484)a$$

so that $a = 2.920696$.

The last step in the construction is easy but tedious. We first insert a into (10.76) and (10.77). Next we must expand each of (10.76) and (10.77) into Fourier series. The coefficients of these Fourier series will be our filter coefficients.

To expand (10.76) and (10.77) into Fourier series, we replace $\cos\left(\frac{\omega}{2}\right)$ and $\sin\left(\frac{\omega}{2}\right)$ by $\frac{e^{i\frac{\omega}{2}}+e^{-i\omega/2}}{2}$, $\frac{e^{i\omega/2}-e^{-i\omega/2}}{2i}$, respectively, and then use a **CAS** to expand both series. In so doing (see Problem 10.44), we see that $\tilde{H}(\omega)$ consists of seven terms and $H(\omega)$ has nine terms. That is, to six digits,

$$\tilde{H}(\omega) = \sum_{k=-3}^{3} \tilde{h}_k e^{ik\omega}$$

where

$$
\begin{array}{rl}
\tilde{h}_0 = & 0.788486 \\
\tilde{h}_{-1} = \tilde{h}_1 = & 0.418092 \\
\tilde{h}_{-2} = \tilde{h}_2 = & -0.040689 \\
\tilde{h}_{-3} = \tilde{h}_3 = & -0.064539
\end{array}
\qquad (10.78)
$$

and

$$H(\omega) = \sum_{k=-4}^{4} h_k e^{ik\omega}$$

where

$$
\begin{array}{rl}
h_0 = & 0.852699 \\
h_{-1} = h_1 = & 0.377403 \\
h_{-2} = h_2 = & -0.110624 \\
h_{-3} = h_3 = & -0.023850 \\
h_{-4} = h_4 = & 0.037829
\end{array}
\qquad (10.79)
$$

We have the following definition.

Definition 10.5 (Cohen–Daubechies–Feauveau 9/7 Filter). *We define the* Cohen–Daubechies–Feauveau 9/7 biorthogonal filter pair *by the values given in (10.78) and (10.79). We call this filter pair the* CDF97 filter pair. ☐

In Problem 10.45 you will show that the Fourier series $\tilde{H}(\omega)$, $H(\omega)$ for the CDF97 filters $\tilde{\mathbf{h}}$ and $\mathbf{h}$, respectively, satisfy

$$\tilde{H}^{(m)}(\pi) = H^{(m)}(\pi) = 0, \qquad m = 0, 1, 2, 3$$

In Figure 10.6 we have plotted $|\tilde{H}(\omega)|$ and $|H(\omega)|$ for the CDF97 filter pair. You can see both filters are equally "flat" at $\omega = \pi$. Compare these functions to those plotted in Figure 10.5.

The Wavelet Filters $\tilde{\mathbf{g}}$ and $\mathbf{g}$

We now use (10.31) to compute the associated wavelet filters $\tilde{\mathbf{g}}$ and $\mathbf{g}$ for the CDF97 filter pair. and obtain the results shown in Table 10.2

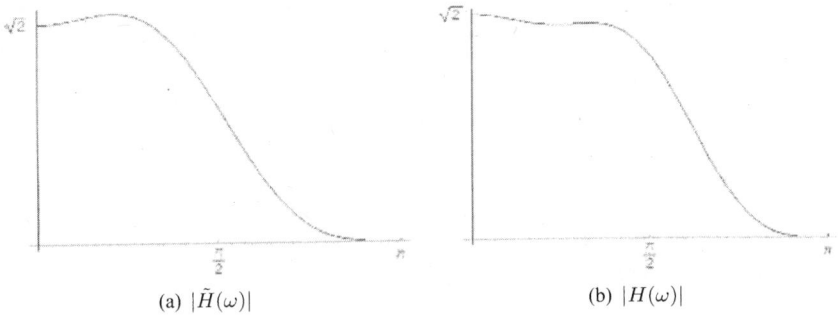

(a) $|\tilde{H}(\omega)|$ (b) $|H(\omega)|$

Figure 10.6   Plots of $|\tilde{H}(\omega)|$ and $|H(\omega)|$ for the CDF97 filter.

Table 10.2   The highpass filter pair **g** and $\tilde{\mathbf{g}}$ for the CDF97 filter pair.

$\tilde{g}_k$	$(-1)^k h_{1-k}$		g_k	$(-1)^k \tilde{h}_{1-k}$	
$\tilde{g}_{-3}$	$-h_4$	-0.037829			
$\tilde{g}_{-2}$	h_3	-0.023850	g_{-2}	$\tilde{h}_3$	-0.064539
$\tilde{g}_{-1}$	$-h_2$	0.110624	g_{-1}	$-\tilde{h}_2$	0.040689
$\tilde{g}_0$	h_1	0.377403	g_0	$\tilde{h}_1$	0.418092
$\tilde{g}_1$	$-h_0$	-0.852699	g_1	$-\tilde{h}_0$	-0.788486
$\tilde{g}_2$	h_1	0.377403	g_2	$\tilde{h}_1$	0.418092
$\tilde{g}_3$	$-h_2$	0.110624	g_3	$-\tilde{h}_2$	0.040689
$\tilde{g}_4$	h_3	-0.023850	g_4	$\tilde{h}_3$	-0.064539
$\tilde{g}_5$	$-h_4$	-0.037829			

PROBLEMS

10.36   What values of $\tilde{N}$ and N will produce the $(9, 7)$ biorthogonal spline filter pair?

10.37   Use the software you wrote in Computer Lab 10.1 from Section 10.2 to find the $(9, 7)$ biorthogonal spline filter pair. Form the Fourier series $H(\omega)$ and $\tilde{H}(\omega)$ for each filter and show that the formulas (10.64) hold. Show also that $H''(\pi) \neq 0$ and $\tilde{H}^{(6)}(\pi) \neq 0$. (*Hint:* A **CAS** is useful for this problem.)

10.38   In this problem you will show that for any nonnegative integer k, $\cos(k\omega)$ can be written as a polynomial in $\cos(\omega)$.

(a) By DeMoivre's theorem (Problem 4.20 from Section 4.2), we know that

$$(\cos(\omega) + i \sin(\omega))^k = \cos(k\omega) + i \sin(k\omega)$$

Use the binomial theorem (Problem 4.41 from Section 4.3) to show that for any nonnegative integer k, we have

$$\cos(k\omega) + i\sin(k\omega) = \sum_{j=0}^{k} \binom{k}{j} \cos^{k-j}(\omega) i^j \sin^j(\omega)$$

(b) Identify the real portion of the right-hand side of the identity in part (a). Can you give an explicit formula for this real part?

(c) Use part (b) to equate the real parts of the identity in part (a) to complete the proof.

10.39   Use the ideas from Problem 10.38 to verify the following trigonometric identities:

(a) $\cos(2\omega) = 2\cos^2(\omega) - 1$.

(b) $\cos(3\omega) = 4\cos^3(\omega) - 3\cos(\omega)$.

(c) $\cos(4\omega) = 8\cos^4(\omega) - 8\cos^2(\omega) + 1$.

10.40   Let $g(\omega) = E(\omega)\overline{H(\omega)} + E(\omega + \pi)\overline{H(\omega + \pi)}$ and assume that $E(\omega)$ and $H(\omega)$ are even, 2π-periodic functions. Show that $g(-\omega) = g(\omega)$.

10.41   In this problem you will prove Theorem 10.3. This is a challenging problem, so we have divided the proof into the following steps.

(a) Insert (10.74) and (10.73) into (10.66) and use trigonometric identities for $\cos(t + \frac{\pi}{2})$ and $\cos(t + \pi)$ to show that

$$\cos^{2K}\left(\frac{\omega}{2}\right) p(\cos(\omega))\tilde{p}(\cos(\omega)) + \sin^{2K}\left(\frac{\omega}{2}\right) p(-\cos(\omega))\tilde{p}(-\cos(\omega)) = 1$$
(10.80)

where $K = \ell + \tilde{\ell}$.

(b) Explain why we can always rewrite $p(\cos(\omega))$ and $\tilde{p}(\cos(\omega))$ as polynomials whose arguments are $\frac{1-\cos(\omega)}{2}$.

(c) Let $P(t) = p(t)\tilde{p}(t)$. Use the trigonometric identities $\sin^2\left(\frac{\omega}{2}\right) = \frac{1-\cos(\omega)}{2}$ and $\cos^2\left(\frac{\omega}{2}\right) = \frac{1+\cos(\omega)}{2}$ to rewrite (10.80) as

$$\cos^{2K}\left(\frac{\omega}{2}\right) P\left(\sin^2\left(\frac{\omega}{2}\right)\right) + \sin^{2K}\left(\frac{\omega}{2}\right) P\left(\cos^2\left(\frac{\omega}{2}\right)\right) = 1$$

or

$$(1-t)^K P(t) + t^K P(1-t) = 1$$
(10.81)

where $t = \sin^2\left(\frac{\omega}{2}\right)$.

(d) The proof now relies on a result known as *Bezout's Theorem*. We state below the version of this theorem given in [26].

Bezout's Theorem: If p_1 and p_2 are polynomials of degree n_1 and n_2, respectively, and if p_1 and p_2 have no common zeros, then there exist unique polynomials q_1 and q_2 of degree at most $n_2 - 1$ and $n_1 - 1$, respectively, so that

$$p_1(t)q_1(t) + p_2(t)q_2(t) = 1 \qquad (10.82)$$

Note that the polynomials $p_1(t) = (1-t)^K$ and $p_2(t) = t^K$ share no common zeros and both are degree K polynomials. Thus, by Bezout's theorem, there exist *unique* polynomials $q_1(t)$ and $q_2(t)$, both of degree at most $K - 1$, such that

$$(1-t)^K q_1(t) + t^K q_2(t) = 1$$

Replace t by $1 - t$ and use Bezout's theorem to explain why $q_2(t) = q_1(1-t)$.

(e) Use part (d) to show that

$$(1-t)^K q_1(t) + t^K q_1(1-t) = 1$$

so that $P(t) = q_1(t)$ solves (10.81) and must be a polynomial of at most $K - 1$.

(f) Use (10.81) to show that

$$P(t) = (1-t)^{-K} - t^K(1-t)^{-K}P(1-t)$$

(g) Show that the MacLaurin series for $f(t) = (1-t)^{-K}$ is

$$f(t) = \sum_{j=0}^{\infty} \binom{K+j-1}{j} t^j$$

Find the interval of convergence for this series.

(h) Insert the series you found in part (g) into the identity from part (f). Explain why

$$P(t) = \sum_{j=0}^{K-1} \binom{K+j-1}{j} t^j \qquad (10.83)$$

(*Hint:* What is the degree of $P(t)$?) Replacing t by $\sin^2\left(\frac{\omega}{2}\right)$ completes the proof.

10.42   Consider the polynomial given in (10.83). Write down the polynomial $P(t)$ for $K = 1$, $K = 2$, and $K = 3$. For $K = 2, 3$ find the roots of $P(t)$.

10.43   The proof of Bezout's theorem is broken into two parts. The first part of the proof establishes the existence of a solution to the identity (10.82). Existence is established using the Euclidean algorithm and proof by induction. Although this part of the proof is a bit advanced for the scope of this book (the interested reader is referred to Daubechies [26]), the second part of proof, showing that the solution is unique, is straightforward. In this exercise you will establish the uniqueness of solutions to (10.82) in Bezout's theorem. The following steps will help you organize your work.

(a) To establish the uniqueness of solution to (10.82), assume that there are two solution pairs, $q_1(t), q_2(t)$ and $r_1(t), r_2(t)$, with degrees at most $n_2 - 1$ and $n_1 - 1$, respectively. Insert these solutions into (10.82) and subtract to show that

$$p_1(t)(q_1(t) - r_1(t)) = p_2(t)(r_2(t) - q_2(t))$$

(b) Why must the set of zeros of $p_2(t)$ be contained in the set of zeros for $q_1(t) - r_1(t)$?

(c) What is the only polynomial $q_1(t) - r_1(t)$ for which the set of zeros of $p_2(t)$ is contained in the set of zeros for $q_1(t) - r_1(t)$? Explain your answer and why this establishes the uniqueness of the solution to (10.82).

10.44   Use a **CAS** with $\tilde{\ell} = 2$, $\tilde{p}(\cos(\omega)) = 2.920696(\sin^2\left(\frac{\omega}{2}\right) + 0.342484)$ and the identities for $\cos\left(\frac{\omega}{2}\right)$ and $\sin\left(\frac{\omega}{2}\right)$ from Problem 4.18 in Section 4.2 to verify that the coefficients of the Fourier series for (10.76) are listed in (10.78). Repeat the problem for the Fourier series given by (10.77) and verify that the Fourier coefficients for that series are given by (10.79).

10.45   Show that (10.65) holds and that neither $\tilde{H}^{(4)}(\pi)$ nor $H^{(4)}(\pi)$ equals zero.

Computer Lab

◆ **10.2 Software Development: Generating the CDF97 Filters.** From the text Web site, access the package cdf97filters. In this development lab you will generate code that will produce the CDF97 filter pair. Your code will expand the Fourier series given by (10.76) and (10.77) to find the filters. Instructions are provided that allow you to add the code to the software package DiscreteWavelets.

CHAPTER 11

COMPUTING BIORTHOGONAL
WAVELET TRANSFORMATIONS

The biorthogonal wavelet transformation built from the symmetric biorthogonal filter pairs developed in Chapter 10 often exhibit better results in image compression applications than those obtained using the orthogonal filters developed in Chapters 7 and 8. In this chapter we develop the computational tools necessary to realize this fact.

In Section 11.1 we develop an algorithm for computing one iteration of the biorthogonal wavelet transformation. The algorithm for computing the inverse biorthogonal wavelet transform is described in Section 11.2. Both algorithms are designed to work with symmetric biorthogonal filter pairs. Algorithms for computing multiple iterations or two-dimensional biorthogonal transformations are omitted since they are very similar to the iterative and two-dimensional transform algorithms presented in Sections 6.2 and 6.3. You are asked to write code for all algorithms in the computer labs that follow Sections 11.1 and 11.2.

The most important reason for developing symmetric biorthogonal filters is that we can exploit the symmetry obeyed by the filters to better handle transforming data at the end of a vector or on the right and bottom edges of an image. In the final

Discrete Wavelet Transformations: An Elementary Approach With Applications. By P. J. Van Fleet   **407**
Copyright © 2008 John Wiley & Sons, Inc.

section of the chapter we modify the algorithms from Sections 11.1 and 11.2 to take advantage of symmetric biorthogonal filters.

11.1 COMPUTING THE BIORTHOGONAL WAVELET TRANSFORMATION

We are now ready to develop an algorithm for the implementation of the biorthogonal wavelet transform. The only new algorithm we develop in this section is a routine for computing one iteration of the one-dimensional biorthogonal wavelet transform. Once we have constructed this algorithm, the algorithms to do multiple iterations or two-dimensional transformations are almost identical to those developed in Sections 6.2 and 6.3 for orthogonal wavelet transforms.

Note: Although it is possible to develop biorthogonal filter pairs that are not symmetric, the algorithms we develop in this section will assume that the input filters are either symmetric about $k = 0$ or $k = \frac{1}{2}$.

The Lowpass Portion of the Transform for Odd-Length Filters

Unlike the orthogonal wavelet transformations developed in Chapter 7, the starting and stopping indices for lowpass and highpass filters for the biorthogonal wavelet transform are not always the same. For example, the wavelet matrix (10.54) for the $(5,3)$ biorthogonal spline filter pair used $\tilde{\mathbf{h}} = (\tilde{h}_{-1}, \tilde{h}_0, \tilde{h}_1)$ and $\tilde{\mathbf{g}} = (\tilde{g}_{-1}, \tilde{g}_0, \tilde{g}_1, \tilde{g}_2, \tilde{g}_3)$. Thus, we will consider algorithms for computing the lowpass and highpass portions of the transform as separate cases.

To get an idea of how the algorithm works, let's look at the lowpass portion $\tilde{H}_6$ of $\tilde{W}_{12}$ formed from some odd-length filters. Remember that we will apply this matrix to $\mathbf{v} = (v_1, \ldots, v_{12})^T$ to obtain the top half of the wavelet transform. For the sake of illustration, we choose *not* to exploit the symmetry of our filters — that is, we will not replace $\tilde{h}_{-1}$ with $\tilde{h}_1$, and so on. Here are the $\tilde{H}_6$ matrices for filters of length 3, 5, 7, and 9.

Length 3 filter:

$$
\tilde{H}_6 = \begin{bmatrix}
\tilde{h}_0 & \tilde{h}_1 & 0 & 0 & 0 & 0 & 0 & 0 & 0 & 0 & 0 & \tilde{h}_{-1} \\
0 & \tilde{h}_{-1} & \tilde{h}_0 & \tilde{h}_1 & 0 & 0 & 0 & 0 & 0 & 0 & 0 & 0 \\
0 & 0 & 0 & \tilde{h}_{-1} & \tilde{h}_0 & \tilde{h}_1 & 0 & 0 & 0 & 0 & 0 & 0 \\
0 & 0 & 0 & 0 & 0 & \tilde{h}_{-1} & \tilde{h}_0 & \tilde{h}_1 & 0 & 0 & 0 & 0 \\
0 & 0 & 0 & 0 & 0 & 0 & 0 & \tilde{h}_{-1} & \tilde{h}_0 & \tilde{h}_1 & 0 & 0 \\
0 & 0 & 0 & 0 & 0 & 0 & 0 & 0 & 0 & \tilde{h}_{-1} & \tilde{h}_0 & \tilde{h}_1
\end{bmatrix}
$$

Length 5 filter:

$$
\tilde{H}_6 = \begin{bmatrix}
\tilde{h}_0 & \tilde{h}_1 & \tilde{h}_2 & 0 & 0 & 0 & 0 & 0 & 0 & 0 & \tilde{h}_{-2} & \tilde{h}_{-1} \\
\tilde{h}_{-2} & \tilde{h}_{-1} & \tilde{h}_0 & \tilde{h}_1 & \tilde{h}_2 & 0 & 0 & 0 & 0 & 0 & 0 & 0 \\
0 & 0 & \tilde{h}_{-2} & \tilde{h}_{-1} & \tilde{h}_0 & \tilde{h}_1 & \tilde{h}_2 & 0 & 0 & 0 & 0 & 0 \\
0 & 0 & 0 & 0 & \tilde{h}_{-2} & \tilde{h}_{-1} & \tilde{h}_0 & \tilde{h}_1 & \tilde{h}_2 & 0 & 0 & 0 \\
0 & 0 & 0 & 0 & 0 & 0 & \tilde{h}_{-2} & \tilde{h}_{-1} & \tilde{h}_0 & \tilde{h}_1 & \tilde{h}_2 & 0 \\
\tilde{h}_2 & 0 & 0 & 0 & 0 & 0 & 0 & 0 & \tilde{h}_{-2} & \tilde{h}_{-1} & \tilde{h}_0 & \tilde{h}_1
\end{bmatrix}
$$

Length 7 filter:

$$
\tilde{H}_6 = \begin{bmatrix}
\tilde{h}_0 & \tilde{h}_1 & \tilde{h}_2 & \tilde{h}_3 & 0 & 0 & 0 & 0 & 0 & \tilde{h}_{-3} & \tilde{h}_{-2} & \tilde{h}_{-1} \\
\tilde{h}_{-2} & \tilde{h}_{-1} & \tilde{h}_0 & \tilde{h}_1 & \tilde{h}_2 & \tilde{h}_3 & 0 & 0 & 0 & 0 & 0 & \tilde{h}_{-3} \\
0 & \tilde{h}_{-3} & \tilde{h}_{-2} & \tilde{h}_{-1} & \tilde{h}_0 & \tilde{h}_1 & \tilde{h}_2 & \tilde{h}_3 & 0 & 0 & 0 & 0 \\
0 & 0 & 0 & \tilde{h}_{-3} & \tilde{h}_{-2} & \tilde{h}_{-1} & \tilde{h}_0 & \tilde{h}_1 & \tilde{h}_2 & \tilde{h}_3 & 0 & 0 \\
0 & 0 & 0 & 0 & 0 & \tilde{h}_{-3} & \tilde{h}_{-2} & \tilde{h}_{-1} & \tilde{h}_0 & \tilde{h}_1 & \tilde{h}_2 & \tilde{h}_3 \\
\tilde{h}_2 & \tilde{h}_3 & 0 & 0 & 0 & 0 & 0 & \tilde{h}_{-3} & \tilde{h}_{-2} & \tilde{h}_{-1} & \tilde{h}_0 & \tilde{h}_1
\end{bmatrix}
$$

Length 9 filter:

$$
\tilde{H}_6 = \begin{bmatrix}
\tilde{h}_0 & \tilde{h}_1 & \tilde{h}_2 & \tilde{h}_3 & \tilde{h}_4 & 0 & 0 & 0 & \tilde{h}_{-4} & \tilde{h}_{-3} & \tilde{h}_{-2} & \tilde{h}_{-1} \\
\tilde{h}_{-2} & \tilde{h}_{-1} & \tilde{h}_0 & \tilde{h}_1 & \tilde{h}_2 & \tilde{h}_3 & \tilde{h}_4 & 0 & 0 & 0 & \tilde{h}_{-4} & \tilde{h}_{-3} \\
\tilde{h}_{-4} & \tilde{h}_{-3} & \tilde{h}_{-2} & \tilde{h}_{-1} & \tilde{h}_0 & \tilde{h}_1 & \tilde{h}_2 & \tilde{h}_3 & \tilde{h}_4 & 0 & 0 & 0 \\
0 & 0 & \tilde{h}_{-4} & \tilde{h}_{-3} & \tilde{h}_{-2} & \tilde{h}_{-1} & \tilde{h}_0 & \tilde{h}_1 & \tilde{h}_2 & \tilde{h}_3 & \tilde{h}_4 & 0 \\
\tilde{h}_4 & 0 & 0 & 0 & \tilde{h}_{-4} & \tilde{h}_{-3} & \tilde{h}_{-2} & \tilde{h}_{-1} & \tilde{h}_0 & \tilde{h}_1 & \tilde{h}_2 & \tilde{h}_3 \\
\tilde{h}_2 & \tilde{h}_3 & \tilde{h}_4 & 0 & 0 & 0 & \tilde{h}_{-4} & \tilde{h}_{-3} & \tilde{h}_{-2} & \tilde{h}_{-1} & \tilde{h}_0 & \tilde{h}_1
\end{bmatrix}
$$

When we compute $\mathbf{y} = \tilde{H}_6 \mathbf{v}$ for each of the matrices above, the pattern is basically the same — to obtain an element y_k in $\mathbf{y}$, we must dot $(\tilde{h}_{-\tilde{L}}, \ldots, \tilde{h}_{\tilde{L}})$ with some portion of $\mathbf{v}$ that is of length $2\tilde{L} + 1$. Unfortunately, some of the rows of $\tilde{H}_6$ "wrap" — that is, the elements of $\tilde{\mathbf{h}}$ occur at both the beginning and the end of some rows. Thus, we must be careful when choosing the $2\tilde{L} + 1$ elements of $\mathbf{v}$ to dot with $\tilde{\mathbf{h}}$.

Note that there are wrapped rows at both the top and bottom of $\tilde{H}$. Upon closer inspection, we see that there is a pattern to the number of elements at the end of the wrapped rows on top. The first row always has $\tilde{h}_{-\tilde{L}}, \ldots, \tilde{h}_{-1}$ (a total of $\tilde{L}$ elements) in the last $\tilde{L}$ columns. The second row ends with $\tilde{h}_{-\tilde{L}}, \ldots, \tilde{h}_{-3}$, and so on. Note also that the bottom row in each of the $\tilde{H}_6$ begins with $\tilde{h}_2, \ldots, \tilde{h}_{\tilde{L}}$ (a total of $\tilde{L} - 1$ elements) and the row before it begins with $\tilde{h}_4, \ldots, \tilde{h}_{\tilde{L}}$, and so on.

Perhaps the easiest way to write the algorithm to compute $\mathbf{y} = \tilde{H}_6 \mathbf{v}$ would be to create a *new* vector $\mathbf{w}$ where we prepend the last $\tilde{L}$ elements of $\mathbf{v}$ to $\mathbf{v}$ and append the first $\tilde{L} - 1$ elements of $\mathbf{v}$ to the result.

For example, for filters of length 7 ($\tilde{L} = 3$), we form $\mathbf{w}$ as

$$
\mathbf{w} = [v_{10}, v_{11}, v_{12} \mid \underbrace{v_1, \ldots, v_{12}}_{\mathbf{v}} \mid v_1, v_2]^T
$$

Now if we take the first seven elements $[v_{10}, v_{11}, v_{12}, v_1, v_2, v_3, v_4]^T$ of $\mathbf{w}$ and dot them with $\tilde{\mathbf{h}}$, we obtain

$$y_1 = [v_{10}, v_{11}, v_{12}, v_1, v_2, v_3, v_4] \cdot \tilde{\mathbf{h}}$$
$$= v_{10}\tilde{h}_{-3} + v_{11}\tilde{h}_{-2} + v_{12}\tilde{h}_{-1} + v_1\tilde{h}_0 + v_2\tilde{h}_1 + v_3\tilde{h}_2 + v_4\tilde{h}_3$$

If we shift right by two and take $(v_{12}, v_1, v_2, v_3, v_4, v_5, v_6)$ from $\mathbf{w}$ and dot it with $\tilde{\mathbf{h}}$, we can compute y_2. We continue this pattern until we reach

$$y_6 = [v_8, v_9, v_{10}, v_{11}, v_{12}, v_1, v_2]^T \cdot \tilde{\mathbf{h}}$$

Undoubtedly, we could simplify the computations by exploiting the symmetry of $\tilde{\mathbf{h}}$. You will consider such simplifications in Problem 11.1.

For the filters of length 3, 5, and 9, we build

$$\mathbf{w} = [\underbrace{v_{12-(\tilde{L}-1)}, \ldots, v_{12}}_{\tilde{L} \text{ terms}} \mid \underbrace{v_1, v_2, \ldots, v_{12}}_{\mathbf{v}} \mid \underbrace{v_1, \ldots, v_{\tilde{L}-1}}_{\tilde{L}-1 \text{ terms}}]^T \qquad (11.1)$$

Of course, in the case when $\tilde{\mathbf{h}} = (\tilde{h}_{-1}, \tilde{h}_0, \tilde{h}_1)$, there is no wrapping row at the bottom of $\tilde{H}_6$.

We could then compute y_k, $k = 1, \ldots, 6$, as follows:

$$y_1 = [\tilde{h}_{-\tilde{L}}, \ldots, \tilde{h}_{\tilde{L}}] \cdot [w_1, \ldots, w_{2\tilde{L}+1}]^T$$
$$y_2 = [\tilde{h}_{-\tilde{L}}, \ldots, \tilde{h}_{\tilde{L}}] \cdot [w_3, \ldots, w_{2\tilde{L}+3}]^T$$
$$y_3 = [\tilde{h}_{-\tilde{L}}, \ldots, \tilde{h}_{\tilde{L}}] \cdot [w_5, \ldots, w_{2\tilde{L}+5}]^T$$
$$y_4 = [\tilde{h}_{-\tilde{L}}, \ldots, \tilde{h}_{\tilde{L}}] \cdot [w_7, \ldots, w_{2\tilde{L}+7}]^T$$
$$y_5 = [\tilde{h}_{-\tilde{L}}, \ldots, \tilde{h}_{\tilde{L}}] \cdot [w_9, \ldots, w_{2\tilde{L}+9}]^T$$
$$y_6 = [\tilde{h}_{-\tilde{L}}, \ldots, \tilde{h}_{\tilde{L}}] \cdot [w_{11}, \ldots, w_{2\tilde{L}+11}]^T$$

This algorithm generalizes regardless of the length of $\mathbf{v}$. That is, for N an even positive integer and $\mathbf{v} \in \mathbf{R}^N$, we have

$$y_k = [\tilde{h}_{-\tilde{L}}, \ldots, \tilde{h}_{\tilde{L}}] \cdot [w_{2k-1}, \ldots, w_{2\tilde{L}+2k-1}]^T, \quad k = 1, \ldots, \frac{N}{2}$$
$$= \sum_{j=0}^{2\tilde{L}} \tilde{h}_{j-\tilde{L}} w_{2k-1+j}, \quad k = 1, \ldots, \frac{N}{2} \qquad (11.2)$$

where $\mathbf{w}$ is built from $\mathbf{v}$ using (11.1):

$$\mathbf{w} = [v_{N-(\tilde{L}-1)}, \ldots, v_N \mid v_1, v_2, \ldots, v_N \mid v_1, \ldots, v_{\tilde{L}-1}]^T \qquad (11.3)$$

The Lowpass Portion of the Transform for Even-Length Filters

We now consider even-length filters for the lowpass portion $\tilde{H}_6$ of $\tilde{W}_{12}$. Recall (see page 384) that for the symmetric, even-length filter

$$\tilde{\mathbf{h}} = (\tilde{h}_{-\tilde{L}+1}, \ldots, \tilde{h}_0, \tilde{h}_1, \ldots, \tilde{h}_{\tilde{L}}) = (\tilde{h}_{\tilde{L}}, \ldots, \tilde{h}_1, \tilde{h}_1, \ldots, \tilde{h}_{\tilde{L}})$$

we place the last $\tilde{L} + 1$ elements $\tilde{h}_0, \ldots, \tilde{h}_{\tilde{L}}$ of $\tilde{\mathbf{h}}$ in the first $\tilde{L} + 1$ columns of the first row of $\tilde{H}_6$ and the first $\tilde{L} - 1$ elements $\tilde{h}_{-\tilde{L}+1}, \ldots, \tilde{h}_{-1}$ of $\tilde{\mathbf{h}}$ in the last $\tilde{L} - 1$ columns of the first row of $\tilde{H}_6$. For the sake of illustration, it is easier to describe the computation $\mathbf{y} = \tilde{H}_6 \mathbf{v}$ if we *do not* exploit the symmetry of the filter. Here are the $\tilde{H}_6$ matrices for filters of length 4, 6, 8, and 10.

Length 4 filter:

$$\tilde{H}_6 = \begin{bmatrix} \tilde{h}_0 & \tilde{h}_1 & \tilde{h}_2 & 0 & 0 & 0 & 0 & 0 & 0 & 0 & 0 & \tilde{h}_{-1} \\ 0 & \tilde{h}_{-1} & \tilde{h}_0 & \tilde{h}_1 & \tilde{h}_2 & 0 & 0 & 0 & 0 & 0 & 0 & 0 \\ 0 & 0 & 0 & \tilde{h}_{-1} & \tilde{h}_0 & \tilde{h}_1 & \tilde{h}_2 & 0 & 0 & 0 & 0 & 0 \\ 0 & 0 & 0 & 0 & 0 & \tilde{h}_{-1} & \tilde{h}_0 & \tilde{h}_1 & \tilde{h}_2 & 0 & 0 & 0 \\ 0 & 0 & 0 & 0 & 0 & 0 & 0 & \tilde{h}_{-1} & \tilde{h}_0 & \tilde{h}_1 & \tilde{h}_2 & 0 \\ \tilde{h}_2 & 0 & 0 & 0 & 0 & 0 & 0 & 0 & 0 & \tilde{h}_{-1} & \tilde{h}_0 & \tilde{h}_1 \end{bmatrix}$$

Length 6 filter:

$$\tilde{H}_6 = \begin{bmatrix} \tilde{h}_0 & \tilde{h}_1 & \tilde{h}_2 & \tilde{h}_3 & 0 & 0 & 0 & 0 & 0 & 0 & \tilde{h}_{-2} & \tilde{h}_{-1} \\ \tilde{h}_{-2} & \tilde{h}_{-1} & \tilde{h}_0 & \tilde{h}_1 & \tilde{h}_2 & \tilde{h}_3 & 0 & 0 & 0 & 0 & 0 & 0 \\ 0 & 0 & \tilde{h}_{-2} & \tilde{h}_{-1} & \tilde{h}_0 & \tilde{h}_1 & \tilde{h}_2 & \tilde{h}_3 & 0 & 0 & 0 & 0 \\ 0 & 0 & 0 & 0 & \tilde{h}_{-2} & \tilde{h}_{-1} & \tilde{h}_0 & \tilde{h}_1 & \tilde{h}_2 & \tilde{h}_3 & 0 & 0 \\ 0 & 0 & 0 & 0 & 0 & 0 & \tilde{h}_{-2} & \tilde{h}_{-1} & \tilde{h}_0 & \tilde{h}_1 & \tilde{h}_2 & \tilde{h}_3 \\ \tilde{h}_2 & \tilde{h}_3 & 0 & 0 & 0 & 0 & 0 & 0 & \tilde{h}_{-2} & \tilde{h}_{-1} & \tilde{h}_0 & \tilde{h}_1 \end{bmatrix}$$

Length 8 filter:

$$\tilde{H}_6 = \begin{bmatrix} \tilde{h}_0 & \tilde{h}_1 & \tilde{h}_2 & \tilde{h}_3 & \tilde{h}_4 & 0 & 0 & 0 & 0 & \tilde{h}_{-3} & \tilde{h}_{-2} & \tilde{h}_{-1} \\ \tilde{h}_{-2} & \tilde{h}_{-1} & \tilde{h}_0 & \tilde{h}_1 & \tilde{h}_2 & \tilde{h}_3 & \tilde{h}_4 & 0 & 0 & 0 & 0 & \tilde{h}_{-3} \\ 0 & \tilde{h}_{-3} & \tilde{h}_{-2} & \tilde{h}_{-1} & \tilde{h}_0 & \tilde{h}_1 & \tilde{h}_2 & \tilde{h}_3 & \tilde{h}_4 & 0 & 0 & 0 \\ 0 & 0 & 0 & \tilde{h}_{-3} & \tilde{h}_{-2} & \tilde{h}_{-1} & \tilde{h}_0 & \tilde{h}_1 & \tilde{h}_2 & \tilde{h}_3 & \tilde{h}_4 & 0 \\ \tilde{h}_4 & 0 & 0 & 0 & 0 & \tilde{h}_{-3} & \tilde{h}_{-2} & \tilde{h}_{-1} & \tilde{h}_0 & \tilde{h}_1 & \tilde{h}_2 & \tilde{h}_3 \\ \tilde{h}_2 & \tilde{h}_3 & \tilde{h}_4 & 0 & 0 & 0 & 0 & \tilde{h}_{-3} & \tilde{h}_{-2} & \tilde{h}_{-1} & \tilde{h}_0 & \tilde{h}_1 \end{bmatrix}$$

Length 10 filter:

$$\tilde{H}_6 = \begin{bmatrix} \tilde{h}_0 & \tilde{h}_1 & \tilde{h}_2 & \tilde{h}_3 & \tilde{h}_4 & \tilde{h}_5 & 0 & 0 & \tilde{h}_{-4} & \tilde{h}_{-3} & \tilde{h}_{-2} & \tilde{h}_{-1} \\ \tilde{h}_{-2} & \tilde{h}_{-1} & \tilde{h}_0 & \tilde{h}_1 & \tilde{h}_2 & \tilde{h}_3 & \tilde{h}_4 & \tilde{h}_5 & 0 & 0 & \tilde{h}_{-4} & \tilde{h}_{-3} \\ \tilde{h}_{-4} & \tilde{h}_{-3} & \tilde{h}_{-2} & \tilde{h}_{-1} & \tilde{h}_0 & \tilde{h}_1 & \tilde{h}_2 & \tilde{h}_3 & \tilde{h}_4 & \tilde{h}_5 & 0 & 0 \\ 0 & 0 & \tilde{h}_{-4} & \tilde{h}_{-3} & \tilde{h}_{-2} & \tilde{h}_{-1} & \tilde{h}_0 & \tilde{h}_1 & \tilde{h}_2 & \tilde{h}_3 & \tilde{h}_4 & \tilde{h}_5 \\ \tilde{h}_4 & \tilde{h}_5 & 0 & 0 & \tilde{h}_{-4} & \tilde{h}_{-3} & \tilde{h}_{-2} & \tilde{h}_{-1} & \tilde{h}_0 & \tilde{h}_1 & \tilde{h}_2 & \tilde{h}_3 \\ \tilde{h}_2 & \tilde{h}_3 & \tilde{h}_4 & \tilde{h}_5 & 0 & 0 & \tilde{h}_{-4} & \tilde{h}_{-3} & \tilde{h}_{-2} & \tilde{h}_{-1} & \tilde{h}_0 & \tilde{h}_1 \end{bmatrix}$$

We will again construct a vector $\mathbf{w}$ from $\mathbf{v}$ in order to compute y_k, $k = 1, \ldots, 6$. The vector $\mathbf{w}$ for the even-length filter is quite similar to (11.1) formed for odd-length filters. Note that for the even-length filters, there are $\tilde{L} - 1$ elements at the end of the first row of $\tilde{H}_6$. The number of terms at the end of subsequent rows decreases by 2.

There are $\tilde{L} - 1$ (just as the case for odd-length filters) elements at the beginning of the last row of $\tilde{H}_6$, and this total decreases by 2 as we move up to preceding rows. Thus, we form $\mathbf{w}$ by first prepending the last $\tilde{L} - 1$ elements of $\mathbf{v}$ to $\mathbf{v}$ and then appending the last $\tilde{L} - 1$ elements of $\mathbf{v}$ to the result. We have

$$\mathbf{w} = [\underbrace{v_{12-(\tilde{L}-2)}, \ldots, v_{12}}_{\tilde{L}-1 \text{ terms}} \mid \underbrace{v_1, v_2, \ldots, v_{12}}_{\mathbf{v}} \mid \underbrace{v_1, \ldots, v_{\tilde{L}-1}}_{\tilde{L}-1 \text{ terms}}]^T \tag{11.4}$$

We then compute y_k, $k = 1, \ldots, 6$ as follows:

$$y_1 = [\tilde{h}_{-\tilde{L}+1}, \ldots, \tilde{h}_{\tilde{L}}] \cdot [w_1, \ldots, w_{2\tilde{L}}]^T$$

$$y_2 = [\tilde{h}_{-\tilde{L}+1}, \ldots, \tilde{h}_{\tilde{L}}] \cdot [w_3, \ldots, w_{2\tilde{L}+2}]^T$$

$$y_3 = [\tilde{h}_{-\tilde{L}+1}, \ldots, \tilde{h}_{\tilde{L}}] \cdot [w_5, \ldots, w_{2\tilde{L}+4}]^T$$

$$y_4 = [\tilde{h}_{-\tilde{L}+1}, \ldots, \tilde{h}_{\tilde{L}}] \cdot [w_7, \ldots, w_{2\tilde{L}+6}]^T$$

$$y_5 = [\tilde{h}_{-\tilde{L}+1}, \ldots, \tilde{h}_{\tilde{L}}] \cdot [w_9, \ldots, w_{2\tilde{L}+8}]^T$$

$$y_6 = [\tilde{h}_{-\tilde{L}+1}, \ldots, \tilde{h}_{\tilde{L}}] \cdot [w_{11}, \ldots, w_{2\tilde{L}+10}]^T$$

This algorithm generalizes no matter the length of $\mathbf{v}$. For N an even positive integer and $\mathbf{v} \in \mathbf{R}^N$, we have

$$y_k = [\tilde{h}_{-\tilde{L}+1}, \ldots, \tilde{h}_{\tilde{L}})] \cdot [w_{2k-1}, \ldots, w_{2\tilde{L}+2k-2}]^T, \quad k = 1, \ldots, \frac{N}{2}$$

$$= \sum_{j=1}^{2\tilde{L}} \tilde{h}_{-\tilde{L}+j}\, w_{2k-2+j}, \quad k = 1, \ldots, \frac{N}{2} \tag{11.5}$$

where $\mathbf{w}$ is built from $\mathbf{v}$ using (11.4):

$$\mathbf{w} = [v_{N-(\tilde{L}-2)}, \ldots, v_N \mid \underbrace{v_1, v_2, \ldots, v_N}_{\mathbf{v}} \mid v_1, \ldots, v_{\tilde{L}-1}]^T \tag{11.6}$$

In Problem 11.3 you will be asked to explain why the formation of $\mathbf{w}$ works for a symmetric lowpass filter of arbitrary length.

The Highpass Portion of the 1D Biorthogonal Wavelet Transform

We now turn our attention to writing formulas for computing the highpass portion of the wavelet transform $\tilde{W}_N$. That is, given $\mathbf{v} \in \mathbb{R}^N$, where N is an even integer, we seek formulas for the components $z_1, \ldots, z_{N/2}$ of $\mathbf{z} = \tilde{G}_{N/2}\mathbf{v}$. Here $\tilde{G}_{N/2}$ represents the highpass (bottom) block of $\tilde{W}_N$.

We follow the same procedure that we used to find the components of the lowpass portion of the transform. We again consider separate cases for even and odd filter lengths.

Odd-Length Highpass Filters

For odd-length highpass filters $\tilde{g}$, recall that $\tilde{g}_k = (-1)^k h_{1-k}$, where $\mathbf{h} = (h_{-L}, \ldots, h_L)$. In this case, $\tilde{g}$ has length $2L + 1$ and the nonzero elements of $\tilde{g}$ are

$$\tilde{\mathbf{g}} = (\tilde{g}_{-L+1}, \ldots, \tilde{g}_0, \ldots, \tilde{g}_{L+1})$$

As was the case with lowpass filters, we now form vector $\mathbf{w}$ from $\mathbf{v}$ by prepending and appending to $\mathbf{v}$ certain elements of $\mathbf{v}$. We can then use $\mathbf{w}$ to find a formula for the elements z_k, $k = 1, \ldots, \frac{N}{2}$ of $\mathbf{z} = \tilde{G}_{N/2}\mathbf{v}$.

In Problem 11.4, you will be asked to write down $\tilde{G}_{N/2}$ for highpass filters of length 3, 5, 7, and 9 and use them to show that

$$\mathbf{w} = [v_{N-(L-2)}, \ldots, v_N \mid v_1, \ldots, v_N \mid v_1, \ldots, v_L]^T \tag{11.7}$$

We can use $\mathbf{w}$ to compute

$$z_1 = [\tilde{g}_{-L+1}, \ldots, \tilde{g}_{L+1}] \cdot [w_1, \ldots, w_{2L+1}]^T$$
$$z_2 = [\tilde{g}_{-L+1}, \ldots, \tilde{g}_{L+1}] \cdot [w_3, \ldots, w_{2L+3}]^T$$
$$z_3 = [\tilde{g}_{-L+1}, \ldots, \tilde{g}_{L+1}] \cdot [w_5, \ldots, w_{2L+5}]^T$$
$$\vdots$$
$$z_{N/2} = [\tilde{g}_{-L+1}, \ldots, \tilde{g}_{L+1}] \cdot [w_{2(N/2)-1}, \ldots, w_{2L+2(N/2)-1}]^T$$

or

$$z_k = \sum_{j=0}^{2L} \tilde{g}_{j+1-L} \, w_{2k-1+j}, \qquad k = 1, \ldots, \frac{N}{2} \tag{11.8}$$

Even-Length Highpass Filters

In the case of even-length filters, we use $\mathbf{h} = (h_{-L+1}, \ldots, h_0, \ldots, h_L)$ to construct $\tilde{g}$. Again using the formula $\tilde{g}_k = (-1)^k h_{1-k}$, we see that

$$\tilde{\mathbf{g}} = (\tilde{g}_{-L+1}, \ldots, \tilde{g}_0, \ldots, \tilde{g}_L)$$

In Problem 11.5 you will be asked to write down $\tilde{G}_{N/2}$ for highpass filters of length $4, 6, 8$, and 10 and use them to show that $\mathbf{w}$ in this case is the same as (11.7). We can thus write the elements $z_k, k = 1, \ldots, \frac{N}{2}$, of $\mathbf{z} = \tilde{G}_{N/2}\mathbf{v}$ as

$$z_1 = [\tilde{g}_{-L+1}, \ldots, \tilde{g}_L] \cdot [w_1, \ldots, w_{2L}]^T$$

$$z_2 = [\tilde{g}_{-L+1}, \ldots, \tilde{g}_L] \cdot [w_3, \ldots, w_{2L+2}]^T$$

$$z_3 = [\tilde{g}_{-L+1}, \ldots, \tilde{g}_L] \cdot [w_5, \ldots, w_{2L+4}]^T$$

$$\vdots$$

$$z_{N/2} = [\tilde{g}_{-L+1}, \ldots, \tilde{g}_L] \cdot [w_{2(N/2)-1}, \ldots, w_{2L+2(N/2)-2}]^T$$

or

$$z_k = \sum_{j=1}^{2L} \tilde{g}_{-L+j}\, w_{2k-2+j}, \qquad k = 1, \ldots, \frac{N}{2} \qquad (11.9)$$

An Algorithm for One Iteration of the Biorthogonal Wavelet Transform

We can now use (11.2), (11.5), (11.8), and (11.9) to create an algorithm to perform one iteration of the one-dimensional biorthogonal wavelet transformation.

Algorithm 11.1 (Biorthogonal Wavelet Transform: 1D). *We are given a vector* $\mathbf{v} \in \mathbb{R}^N$ *(N even) and a pair of biorthogonal symmetric lowpass filters* $\tilde{\mathbf{h}}$ *and* $\mathbf{h}$ *and we return one iteration of the wavelet transform.*

*This algorithm uses the **Mod** function, that takes two integers* i *and* j *and returns the remainder of* i *divided by* j. ***Mod** is useful for determining whether an integer is odd or even. If we compute **Mod**$[i,2]$, then we obtain the remainder of* i *divided by* 2. *If* i *is odd, the remainder will be* 1, *and if* i *is even, then the remainder is* 0.

The first step in the algorithm is to compute $\tilde{\mathbf{g}}$. *We then compute the lowpass portion of the transform. Toward that end, we create the vector* $\mathbf{w}$ *given by either (11.3) or (11.6) (depending on whether the length of* $\tilde{\mathbf{h}}$ *is odd or even). We then use a loop to compute* $y_1, \ldots, y_{N/2}$. *The loop is built using either (11.2) or (11.5) (again depending on whether the length of* $\tilde{\mathbf{h}}$ *is odd or even). The algorithm then moves on to the computation of the highpass portion of the transform. We create the vector* $\mathbf{w}$ *using (11.7). We then use either (11.8) or (11.9) to create a loop that gives us the highpass components* $y_k, k = \frac{N}{2} + 1, \ldots, N$ *of the transform.*

Algorithm: BWT1D1

Input: Vector $\mathbf{v}$ of even length N and symmetric biorthogonal filter pair $\tilde{\mathbf{h}}$, $\mathbf{h}$. It is assumed that N is at least as large as the length of the longer filter.

Output: Vector $\mathbf{y}$ where $\mathbf{y} = \bar{W}_N \mathbf{v}$.

// Create the highpass filter --- assume filter length is odd.

$p = L + 1$

// Change the stop variable if the filter length is even.

If [**Mod** [**Length** [$\tilde{h}$] , 2] = 0,
$\qquad p = p - 1$
]

For [$k = -L + 1$, $k \leq p$, $k{+}{+}$,
$\qquad \tilde{g}_k = (-1)^k h_{1-k}$
]

// Build the vector $\mathbf{w}$ using (11.3) and (11.6).
// Use the index variable i to handle odd/even filter lengths.

$i = 0$

If [**Mod** [**Length** [$\tilde{h}$] , 2] = 0,
$\qquad i = 1$
]

$\mathbf{w} = $ **Join** [**Join** [$(v_{N-(\tilde{L}-1)}, \ldots, v_N), \mathbf{v}$] , $(v_1, \ldots, v_{\tilde{L}-1-i})$]

// Compute the lowpass portion of the transform.

For [$k = 1$, $k \leq \frac{N}{2}$, $k{+}{+}$,
$\qquad y_k = 0$
$\qquad$ **For** [$j = i$, $j \leq 2\tilde{L}$, $j{+}{+}$,
$\qquad\qquad y_k = y_k + h_{j-\tilde{L}} \, w_{2k-1+j-i}$
$\qquad$]
]

// Build the vector $\mathbf{w}$ using (11.7).

$\mathbf{w} = $ **Join** [**Join** [$(v_{N-(L-2)}, \ldots, v_N), \mathbf{v}$] , $(v_1, \ldots, v_L)$]

// Compute the highpass portion of the transform.

For $[k = 1,\ k \leq \frac{N}{2},\ k++,$

     $y_{k+N/2} = 0$

     For $[j = i,\ j \leq 2L,\ j++,$

         $y_{k+N/2} = y_{k+N/2} + g_{j-L+1-i}\, w_{2k-1+j-i}$

     ]

]

Return [y]

$\square$

PROBLEMS

11.1   In (11.2) we constructed a formula for computing the values $y_1, \ldots, y_{N/2}$ of the lowpass portion of a wavelet transformation using an odd-length symmetric filter of length $2\tilde{L} + 1$. This formula does not utilize the fact that $\tilde{h}_k = \tilde{h}_{-k}, k = 1, \ldots, \tilde{L}$. Rewrite (11.2) as $\tilde{h}_0$ times the appropriate value of **w** plus a sum of $\tilde{L}$ terms.

11.2   Repeat Problem 11.1 for (11.5). In this case, your formula for y_k, $k = 1, \ldots, \frac{N}{2}$ should be a sum of $\tilde{L}$ terms.

11.3   Explain why the formation of **w** (11.1) works for any lowpass symmetric filter of odd length. Repeat the problem for arbitrary even length symmetric lowpass filters.

11.4   Suppose that $\mathbf{f} = (h_{-L}, \ldots, h_L)$ is a symmetric, odd-length filter and set $N = 24$. Write down $\tilde{G}_{12}$ for $L = 3, 5, 7, 9$ and use these matrices to establish (11.7).

11.5   Suppose that $\mathbf{f} = (h_{-L+1}, \ldots, h_L)$ is a symmetric, even-length filter and set $N = 24$. Write down $\tilde{G}_{12}$ for $L = 4, 6, 8, 10$ and use these matrices to establish (11.7).

11.6   Let $\mathbf{v} = [1, 8, 0, 4, 9, 2, 8, 4]^T$. Using the $(5, 3)$ biorthogonal spline filter pair, use Algorithm 11.1 to compute one iteration of the biorthogonal wavelet transform of **v**. Do this calculation by hand.

Computer Labs

♦ **11.1 Software Development: 1D Biorthogonal Wavelet Transforms** From the text Web site, access the development package bwt1d. In this development lab you will create a function that will compute one iteration of the one-dimensional biorthogonal wavelet transform using biorthogonal spline filter pairs. Construction of iterated transforms are included in this lab. Instructions are provided that allow you to add all of these functions to the software package DiscreteWavelets.

♦ **11.2 Software Development: 2D Biorthogonal Wavelet Transforms.** From the text Web site, access the development package bwt2d. In this development lab

biorthogonal wavelet transform using biorthogonal spline filter pairs. Instructions are provided that allow you to add all of these functions to the software package `DiscreteWavelets`.

11.2   COMPUTING THE INVERSE BIORTHOGONAL WAVELET TRANSFORMATION

We now turn our attention to the development of an algorithm for computing one iteration of the one-dimensional inverse biorthogonal wavelet transform. We begin by looking at the inverse transform in block form. That is, for N an even integer and $\mathbf{v} \in \mathbb{R}^N$, we compute

$$\mathbf{y} = W_N^T \mathbf{v} = \left[H_{N/2}^T \middle| G_{N/2}^T \right] \mathbf{v}$$

If we partition $\mathbf{v}$ into two blocks $\mathbf{s}$ and $\mathbf{t}$, each of length $N/2$, we have

$$\mathbf{y} = W_N^T \mathbf{v} = \left[H_{N/2}^T \middle| G_{N/2}^T \right] \left[\frac{\mathbf{s}}{\mathbf{t}} \right] = H_{N/2}^T \mathbf{s} + G_{N/2}^T \mathbf{t} \tag{11.10}$$

We will first develop an auxiliary algorithm to compute $H_{N/2}^T \mathbf{s}$ and $G_{N/2}^T \mathbf{t}$. The algorithm for computing the inverse is then straightforward — we build g from $\tilde{h}$, form $\mathbf{s}$ and $\mathbf{t}$ from $\mathbf{v}$, call the auxiliary algorithm to compute vectors $\mathbf{q} = H_{N/2}^T \mathbf{s}$ and $\mathbf{r} = G_{N/2}^T \mathbf{t}$, and add the results.

Computing $H_{N/2}^T \mathbf{s}$ for Odd-Length Filters

Perhaps the best way to develop an algorithm for computing $H_{N/2}^T \mathbf{s}$ is to write out matrices $H_{N/2}^T$ for various filters and then look for patterns. We first consider the case where the filter length of $\mathbf{h}$ is odd and use $N = 16$. We have $H_8^T \mathbf{s}$ for the following filters:

Length 3 filter:

$$H_8^T \mathbf{s} = \begin{bmatrix} h_0 & 0 & 0 & 0 & 0 & 0 & 0 & 0 \\ h_1 & h_{-1} & 0 & 0 & 0 & 0 & 0 & 0 \\ 0 & h_0 & 0 & 0 & 0 & 0 & 0 & 0 \\ 0 & h_1 & h_{-1} & 0 & 0 & 0 & 0 & 0 \\ 0 & 0 & h_0 & 0 & 0 & 0 & 0 & 0 \\ 0 & 0 & h_1 & h_{-1} & 0 & 0 & 0 & 0 \\ 0 & 0 & 0 & h_0 & 0 & 0 & 0 & 0 \\ 0 & 0 & 0 & h_1 & h_{-1} & 0 & 0 & 0 \\ 0 & 0 & 0 & 0 & h_0 & 0 & 0 & 0 \\ 0 & 0 & 0 & 0 & h_1 & h_{-1} & 0 & 0 \\ 0 & 0 & 0 & 0 & 0 & h_0 & 0 & 0 \\ 0 & 0 & 0 & 0 & 0 & h_1 & h_{-1} & 0 \\ 0 & 0 & 0 & 0 & 0 & 0 & h_0 & 0 \\ 0 & 0 & 0 & 0 & 0 & 0 & h_1 & h_{-1} \\ 0 & 0 & 0 & 0 & 0 & 0 & 0 & h_0 \\ h_{-1} & 0 & 0 & 0 & 0 & 0 & 0 & h_1 \end{bmatrix} \cdot \begin{bmatrix} s_1 \\ s_2 \\ s_3 \\ s_4 \\ s_5 \\ s_6 \\ s_7 \\ s_8 \end{bmatrix}$$

Length 5 filter:

$$H_8^T \mathbf{s} = \begin{bmatrix} h_0 & h_{-2} & 0 & 0 & 0 & 0 & 0 & h_2 \\ h_1 & h_{-1} & 0 & 0 & 0 & 0 & 0 & 0 \\ h_2 & h_0 & h_{-2} & 0 & 0 & 0 & 0 & 0 \\ 0 & h_1 & h_{-1} & 0 & 0 & 0 & 0 & 0 \\ 0 & h_2 & h_0 & h_{-2} & 0 & 0 & 0 & 0 \\ 0 & 0 & h_1 & h_{-1} & 0 & 0 & 0 & 0 \\ 0 & 0 & h_2 & h_0 & h_{-2} & 0 & 0 & 0 \\ 0 & 0 & 0 & h_1 & h_{-1} & 0 & 0 & 0 \\ 0 & 0 & 0 & h_2 & h_0 & h_{-2} & 0 & 0 \\ 0 & 0 & 0 & 0 & h_1 & h_{-1} & 0 & 0 \\ 0 & 0 & 0 & 0 & h_2 & h_0 & h_{-2} & 0 \\ 0 & 0 & 0 & 0 & 0 & h_1 & h_{-1} & 0 \\ 0 & 0 & 0 & 0 & 0 & h_2 & h_0 & h_{-2} \\ 0 & 0 & 0 & 0 & 0 & 0 & h_1 & h_{-1} \\ h_{-2} & 0 & 0 & 0 & 0 & 0 & h_2 & h_0 \\ h_{-1} & 0 & 0 & 0 & 0 & 0 & 0 & h_1 \end{bmatrix} \cdot \begin{bmatrix} s_1 \\ s_2 \\ s_3 \\ s_4 \\ s_5 \\ s_6 \\ s_7 \\ s_8 \end{bmatrix}$$

Length 7 filter:

$$
H_8^T \mathbf{s} =
\begin{bmatrix}
h_0 & h_{-2} & 0 & 0 & 0 & 0 & 0 & h_2 \\
h_1 & h_{-1} & h_{-3} & 0 & 0 & 0 & 0 & h_3 \\
h_2 & h_0 & h_{-2} & 0 & 0 & 0 & 0 & 0 \\
h_3 & h_1 & h_{-1} & h_{-3} & 0 & 0 & 0 & 0 \\
0 & h_2 & h_0 & h_{-2} & 0 & 0 & 0 & 0 \\
0 & h_3 & h_1 & h_{-1} & h_{-3} & 0 & 0 & 0 \\
0 & 0 & h_2 & h_0 & h_{-2} & 0 & 0 & 0 \\
0 & 0 & h_3 & h_1 & h_{-1} & h_{-3} & 0 & 0 \\
0 & 0 & 0 & h_2 & h_0 & h_{-2} & 0 & 0 \\
0 & 0 & 0 & h_3 & h_1 & h_{-1} & h_{-3} & 0 \\
0 & 0 & 0 & 0 & h_2 & h_0 & h_{-2} & 0 \\
0 & 0 & 0 & 0 & h_3 & h_1 & h_{-1} & 0 \\
0 & 0 & 0 & 0 & 0 & h_2 & h_0 & h_{-2} \\
h_{-3} & 0 & 0 & 0 & 0 & h_3 & h_1 & h_{-1} \\
h_{-2} & 0 & 0 & 0 & 0 & 0 & h_2 & h_0 \\
h_{-1} & h_{-3} & 0 & 0 & 0 & 0 & h_3 & h_1
\end{bmatrix}
\cdot
\begin{bmatrix}
s_1 \\ s_2 \\ s_3 \\ s_4 \\ s_5 \\ s_6 \\ s_7 \\ s_8
\end{bmatrix}
$$

Length 9 filter:

$$
H_8^T \mathbf{s} =
\begin{bmatrix}
h_0 & h_{-2} & h_{-4} & 0 & 0 & 0 & h_4 & h_2 \\
h_1 & h_{-1} & h_{-3} & 0 & 0 & 0 & 0 & h_3 \\
h_2 & h_0 & h_{-2} & h_{-4} & 0 & 0 & 0 & h_4 \\
h_3 & h_1 & h_{-1} & h_{-3} & 0 & 0 & 0 & 0 \\
h_4 & h_2 & h_0 & h_{-2} & h_{-4} & 0 & 0 & 0 \\
0 & h_3 & h_1 & h_{-1} & h_{-3} & 0 & 0 & 0 \\
0 & h_4 & h_2 & h_0 & h_{-2} & h_{-4} & 0 & 0 \\
0 & 0 & h_3 & h_1 & h_{-1} & h_{-3} & 0 & 0 \\
0 & 0 & h_4 & h_2 & h_0 & h_{-2} & h_{-4} & 0 \\
0 & 0 & 0 & h_3 & h_1 & h_{-1} & h_{-3} & 0 \\
0 & 0 & 0 & h_4 & h_2 & h_0 & h_{-2} & h_{-4} \\
0 & 0 & 0 & 0 & h_3 & h_1 & h_{-1} & h_{-3} \\
h_{-4} & 0 & 0 & 0 & h_4 & h_2 & h_0 & h_{-2} \\
h_{-3} & 0 & 0 & 0 & 0 & h_3 & h_1 & h_{-1} \\
h_{-2} & h_{-4} & 0 & 0 & 0 & h_4 & h_2 & h_0 \\
h_{-1} & h_{-3} & 0 & 0 & 0 & 0 & h_3 & h_1
\end{bmatrix}
\cdot
\begin{bmatrix}
s_1 \\ s_2 \\ s_3 \\ s_4 \\ s_5 \\ s_6 \\ s_7 \\ s_8
\end{bmatrix}
$$

Length 11 filter:

$$H_8^T \mathbf{s} = \begin{bmatrix} h_0 & h_{-2} & h_{-4} & 0 & 0 & 0 & h_4 & h_2 \\ h_1 & h_{-1} & h_{-3} & h_{-5} & 0 & 0 & h_5 & h_3 \\ h_2 & h_0 & h_{-2} & h_{-4} & 0 & 0 & 0 & h_4 \\ h_3 & h_1 & h_{-1} & h_{-3} & h_{-5} & 0 & 0 & h_5 \\ h_4 & h_2 & h_0 & h_{-2} & h_{-4} & 0 & 0 & 0 \\ h_5 & h_3 & h_1 & h_{-1} & h_{-3} & h_{-5} & 0 & 0 \\ 0 & h_4 & h_2 & h_0 & h_{-2} & h_{-4} & 0 & 0 \\ 0 & h_5 & h_3 & h_1 & h_{-1} & h_{-3} & h_{-5} & 0 \\ 0 & 0 & h_4 & h_2 & h_0 & h_{-2} & h_{-4} & 0 \\ 0 & 0 & h_5 & h_3 & h_1 & h_{-1} & h_{-3} & h_{-5} \\ 0 & 0 & 0 & h_4 & h_2 & h_0 & h_{-2} & h_{-4} \\ h_{-5} & 0 & 0 & h_5 & h_3 & h_1 & h_{-1} & h_{-3} \\ h_{-4} & 0 & 0 & 0 & h_4 & h_2 & h_0 & h_{-2} \\ h_{-3} & h_{-5} & 0 & 0 & h_5 & h_3 & h_1 & h_{-1} \\ h_{-2} & h_{-4} & 0 & 0 & 0 & h_4 & h_2 & h_0 \\ h_{-1} & h_{-3} & h_{-5} & 0 & 0 & h_5 & h_3 & h_1 \end{bmatrix} \cdot \begin{bmatrix} s_1 \\ s_2 \\ s_3 \\ s_4 \\ s_5 \\ s_6 \\ s_7 \\ s_8 \end{bmatrix}$$

Length 13 filter:

$$H_8^T \mathbf{s} = \begin{bmatrix} h_0 & h_{-2} & h_{-4} & h_{-6} & 0 & h_6 & h_4 & h_2 \\ h_1 & h_{-1} & h_{-3} & h_{-5} & 0 & 0 & h_5 & h_3 \\ h_2 & h_0 & h_{-2} & h_{-4} & h_{-6} & 0 & h_6 & h_4 \\ h_3 & h_1 & h_{-1} & h_{-3} & h_{-5} & 0 & 0 & h_5 \\ h_4 & h_2 & h_0 & h_{-2} & h_{-4} & h_{-6} & 0 & h_6 \\ h_5 & h_3 & h_1 & h_{-1} & h_{-3} & h_{-5} & 0 & 0 \\ h_6 & h_4 & h_2 & h_0 & h_{-2} & h_{-4} & h_{-6} & 0 \\ 0 & h_5 & h_3 & h_1 & h_{-1} & h_{-3} & h_{-5} & 0 \\ 0 & h_6 & h_4 & h_2 & h_0 & h_{-2} & h_{-4} & h_{-6} \\ 0 & 0 & h_5 & h_3 & h_1 & h_{-1} & h_{-3} & h_{-5} \\ h_{-6} & 0 & h_6 & h_4 & h_2 & h_0 & h_{-2} & h_{-4} \\ h_{-5} & 0 & 0 & h_5 & h_3 & h_1 & h_{-1} & h_{-3} \\ h_{-4} & h_{-6} & 0 & h_6 & h_4 & h_2 & h_0 & h_{-2} \\ h_{-3} & h_{-5} & 0 & 0 & h_5 & h_3 & h_1 & h_{-1} \\ h_{-2} & h_{-4} & h_{-6} & 0 & h_6 & h_4 & h_2 & h_0 \\ h_{-1} & h_{-3} & h_{-5} & 0 & 0 & h_5 & h_3 & h_1 \end{bmatrix} \cdot \begin{bmatrix} s_1 \\ s_2 \\ s_3 \\ s_4 \\ s_5 \\ s_6 \\ s_7 \\ s_8 \end{bmatrix}$$

Let's look for patterns in the H_8^T above. We first note that the filter $\mathbf{h} = (h_{-L}, \ldots, h_L)$, $L = 1, 2, \ldots, 6$ is split into two parts. One part is comprised of the elements of $\mathbf{h}$ with even indices, and the second part is built from the odd components of $\mathbf{h}$. Note that the elements of each part run in reverse order, but due to the symmetric nature of $\mathbf{h}$, we can ignore this fact. Let's call the part of $\mathbf{h}$ with even indices $\mathbf{h}^e$ and the part of $\mathbf{h}$ with odd indices $\mathbf{h}^o$. For example, for $\mathbf{h} = (h_{-3}, \ldots, h_3)$, we have $\mathbf{h}^e = (h_{-2}, h_0, h_2)$ and $\mathbf{h}^o = (h_{-3}, h_{-1}, h_1, h_3)$.

We also note that the odd components $q_1, q_3, \ldots, q_{15}$ of $\mathbf{q} = H_8^T \mathbf{s}$ are built by dotting a portion of $\mathbf{s}$ with $\mathbf{h^e}$, while the even components $q_2, q_4, \ldots, q_{16}$ are built by dotting a portion of $\mathbf{s}$ with $\mathbf{h^o}$.

Let's consider the computation in the case where the filter length is 7 ($L = 3$). If we form the vectors

$$\mathbf{c} = [c_1, c_2, \ldots, c_9, c_{10}] = [s_8, s_1, s_2, \ldots, s_7, s_8, s_1]^T \in \mathbb{R}^{10}$$
$$\mathbf{d} = [d_1, d_2, \ldots, d_{10}, d_{11}] = [s_8, s_1, s_2, \ldots, s_7, s_8, s_1, s_2]^T \in \mathbb{R}^{11}$$

then

$$q_1 = \mathbf{h^e} \cdot [c_1, c_2, c_3] \qquad\qquad q_2 = \mathbf{h^o} \cdot [d_1, d_2, d_3, d_4]^T$$
$$q_3 = \mathbf{h^e} \cdot [c_2, c_3, c_4] \qquad\qquad q_4 = \mathbf{h^o} \cdot [d_2, d_3, d_4, d_5]^T$$
$$q_5 = \mathbf{h^e} \cdot [c_3, c_4, c_5] \qquad\qquad q_6 = \mathbf{h^o} \cdot [d_3, d_4, d_5, d_6]^T$$
$$q_7 = \mathbf{h^e} \cdot [c_4, c_5, c_6] \qquad\qquad q_8 = \mathbf{h^o} \cdot [d_4, d_5, d_6, d_7]^T$$
$$q_9 = \mathbf{h^e} \cdot [c_5, c_6, c_7] \qquad\qquad q_{10} = \mathbf{h^o} \cdot [d_5, d_6, d_7, d_8]^T$$
$$q_{11} = \mathbf{h^e} \cdot [c_6, c_7, c_8] \qquad\qquad q_{12} = \mathbf{h^o} \cdot [d_6, d_7, d_8, d_9]^T$$
$$q_{13} = \mathbf{h^e} \cdot [c_7, c_8, c_9] \qquad\qquad q_{14} = \mathbf{h^o} \cdot [d_7, d_8, d_9, d_{10}]^T$$
$$q_{15} = \mathbf{h^e} \cdot [c_8, c_9, c_{10}] \qquad\qquad q_{16} = \mathbf{h^o} \cdot [d_8, d_9, d_{10}, d_{11}]^T$$

Note that once $\mathbf{c}$ and $\mathbf{d}$ are determined, the computations are very straightforward. For the odd-indexed values, we simply dot $\mathbf{h^e}$ with 3-tuples from $\mathbf{c}$. For q_1, we start with the 3-tuple (c_1, c_2, c_3) and then shift 1 unit right in $\mathbf{c}$ to obtain the next 3-tuple. A similar procedure works for computing the even-indexed values of $\mathbf{q}$. In summation form, we have

$$q_{2k-1} = \sum_{\ell=1}^{3} h_\ell^e c_{k+\ell-1}, \quad k = 1, \ldots, 8$$

and

$$q_{2k} = \sum_{\ell=1}^{4} h_\ell^o d_{k+\ell-1}, \quad k = 1, \ldots, 8$$

These formulas will generalize for any even-length vector $\mathbf{v}$ and any odd-length filter. The question is: How do we form $\mathbf{c}$ and $\mathbf{d}$ for arbitrary $\mathbf{v}$ and odd-length filter?

Note that for the length 7 filter, $\mathbf{c}$ was formed by prepending s_8 to $\mathbf{s}$ and then appending s_1 to the result. We can look at rows 1 and 15 of each of the H_8^T above to get an idea of how many elements to prepend and append to $\mathbf{s}$. Table 11.1 summarizes our findings.

If we let p denote the number of elements at the end of $\mathbf{s}$ that we must prepend to $\mathbf{s}$ and a denote the number of elements at the beginning of $\mathbf{s}$ we must append to the result in order to form $\mathbf{c}$, we can write general formulas in terms of L. We have

$$p = a = \lfloor L/2 \rfloor \tag{11.11}$$

Table 11.1   The number of elements to prepend and append to $\mathbf{s}^a$.

Filter Length	L	Prepend from Row 1	Append from Row 15
3	1	0	0
5	2	1	1
7	3	1	1
9	4	2	2
11	5	2	2
13	6	3	3

[a]The first column is the filter length of $\mathbf{h}$. The second column is the index of the last element in $\mathbf{h}$. The number in the third column is the number of filter elements that appear at the *end* of the first row of H_8^T. This number represents the number of elements at the end of $\mathbf{s}$ that we must prepend to $\mathbf{s}$ when forming $\mathbf{c}$. The number in the fourth column is the number of filter elements at the *beginning* of row 15 in H_8^T. This is the number of elements at the beginning of $\mathbf{s}$ that we need to append to $\mathbf{s}$ to form $\mathbf{c}$.

where $\lfloor \cdot \rfloor$ is the floor function introduced in Problem 6.26(c) in Section 6.4. In this case, for even N and arbitrary odd-length filter, we have

$$\mathbf{c} = [s_{N/2-p+1}, \ldots, s_{N/2} \mid \underbrace{s_1, \ldots, s_{N/2}}_{\mathbf{s}} \mid s_1, \ldots, s_a]^T \qquad (11.12)$$

Now, we can easily write down a formula for the odd components q_{2k+1}, $k = 1, \ldots, \frac{N}{2}$. If we reindex the elements of $\mathbf{h}^e$ as $\mathbf{h}^e = (h_1^e, h_2^e, \ldots, h_\ell^e)$, we have

$$q_{2k-1} = \sum_{j=1}^{\ell} h_j^e c_{k+j-1}, \qquad k = 1, \ldots, \frac{N}{2} \qquad (11.13)$$

To determine the general form of $\mathbf{d}$, we do a similar analysis of rows 2 and 16 of each of the H_8^T above. The results are summarized in Table 11.2.

Again, we can find formulas for p and a. In this case, we have

$$p = \left\lfloor \frac{L-1}{2} \right\rfloor \quad \text{and} \quad a = \left\lfloor \frac{L+1}{2} \right\rfloor \qquad (11.14)$$

so that

$$\mathbf{d} = [s_{N/2-p+1}, \ldots, s_{N/2} \mid \underbrace{s_1, \ldots, s_{N/2}}_{\mathbf{s}} \mid s_1, \ldots s_a]^T \qquad (11.15)$$

The formula for the even components q_{2k}, $k = 1, \ldots, \frac{N}{2}$ is straightforward. Again, we reindex the elements of $\mathbf{h}^o$ as $\mathbf{h}^o = (h_1^o, h_2^o, \ldots, h_m^o)$ and write

$$q_{2k} = \sum_{j=1}^{m} h_j^o d_{k+j-1}, \qquad k = 1, \ldots, \frac{N}{2} \qquad (11.16)$$

Table 11.2 The number of elements to prepend and append to s^a.

Filter Length	L	Prepend from Row 2	Append from Row 16
3	1	0	1
5	2	0	1
7	3	1	2
9	4	1	2
11	5	2	3
13	6	2	3

^{a}The first column is the filter length of **h**. The second column is the index of the last element in **h**. The number in the third column is the number of filter elements that appear at the *end* of the second row of H_8^T. This number represents the number of elements at the end of **s** that we must prepend to **s** when forming **d**. The number in the fourth column is the number of filter elements at the *beginning* of row 16 in H_8^T. This is the number of elements at the beginning of **s** that we need to append to **s** to form **d**.

Computing $H_{N/2}^T s$ for Even-Length Filters

Let's now write down $H_8^T s$ using filters of length 2, 4, 6, 8, 10, 12. Our goal here is to find patterns we can use to create vectors **c** and **d** and build formulas like (11.13) and (11.16).

Length 2 filter:

$$H_8^T s = \begin{bmatrix} h_0 & 0 & 0 & 0 & 0 & 0 & 0 & 0 \\ h_1 & 0 & 0 & 0 & 0 & 0 & 0 & 0 \\ 0 & h_0 & 0 & 0 & 0 & 0 & 0 & 0 \\ 0 & h_1 & 0 & 0 & 0 & 0 & 0 & 0 \\ 0 & 0 & h_0 & 0 & 0 & 0 & 0 & 0 \\ 0 & 0 & h_1 & 0 & 0 & 0 & 0 & 0 \\ 0 & 0 & 0 & h_0 & 0 & 0 & 0 & 0 \\ 0 & 0 & 0 & h_1 & 0 & 0 & 0 & 0 \\ 0 & 0 & 0 & 0 & h_0 & 0 & 0 & 0 \\ 0 & 0 & 0 & 0 & h_1 & 0 & 0 & 0 \\ 0 & 0 & 0 & 0 & 0 & h_0 & 0 & 0 \\ 0 & 0 & 0 & 0 & 0 & h_1 & 0 & 0 \\ 0 & 0 & 0 & 0 & 0 & 0 & h_0 & 0 \\ 0 & 0 & 0 & 0 & 0 & 0 & h_1 & 0 \\ 0 & 0 & 0 & 0 & 0 & 0 & 0 & h_0 \\ 0 & 0 & 0 & 0 & 0 & 0 & 0 & h_1 \end{bmatrix} \cdot \begin{bmatrix} s_1 \\ s_2 \\ s_3 \\ s_4 \\ s_5 \\ s_6 \\ s_7 \\ s_8 \end{bmatrix} \qquad (11.17)$$

Length 4 filter:

$$
H_8^T \mathbf{s} =
\begin{bmatrix}
h_0 & 0 & 0 & 0 & 0 & 0 & 0 & h_2 \\
h_1 & h_{-1} & 0 & 0 & 0 & 0 & 0 & 0 \\
h_2 & h_0 & 0 & 0 & 0 & 0 & 0 & 0 \\
0 & h_1 & h_{-1} & 0 & 0 & 0 & 0 & 0 \\
0 & h_2 & h_0 & 0 & 0 & 0 & 0 & 0 \\
0 & 0 & h_1 & h_{-1} & 0 & 0 & 0 & 0 \\
0 & 0 & h_2 & h_0 & 0 & 0 & 0 & 0 \\
0 & 0 & 0 & h_1 & h_{-1} & 0 & 0 & 0 \\
0 & 0 & 0 & h_2 & h_0 & 0 & 0 & 0 \\
0 & 0 & 0 & 0 & h_1 & h_{-1} & 0 & 0 \\
0 & 0 & 0 & 0 & h_2 & h_0 & 0 & 0 \\
0 & 0 & 0 & 0 & 0 & h_1 & h_{-1} & 0 \\
0 & 0 & 0 & 0 & 0 & h_2 & h_0 & 0 \\
0 & 0 & 0 & 0 & 0 & 0 & h_1 & h_{-1} \\
0 & 0 & 0 & 0 & 0 & 0 & h_2 & h_0 \\
h_{-1} & 0 & 0 & 0 & 0 & 0 & 0 & h_1
\end{bmatrix}
\cdot
\begin{bmatrix}
s_1 \\ s_2 \\ s_3 \\ s_4 \\ s_5 \\ s_6 \\ s_7 \\ s_8
\end{bmatrix}
$$

Length 6 filter:

$$
H_8^T \mathbf{s} =
\begin{bmatrix}
h_0 & h_{-2} & 0 & 0 & 0 & 0 & 0 & h_2 \\
h_1 & h_{-1} & 0 & 0 & 0 & 0 & 0 & h_3 \\
h_2 & h_0 & h_{-2} & 0 & 0 & 0 & 0 & 0 \\
h_3 & h_1 & h_{-1} & 0 & 0 & 0 & 0 & 0 \\
0 & h_2 & h_0 & h_{-2} & 0 & 0 & 0 & 0 \\
0 & h_3 & h_1 & h_{-1} & 0 & 0 & 0 & 0 \\
0 & 0 & h_2 & h_0 & h_{-2} & 0 & 0 & 0 \\
0 & 0 & h_3 & h_1 & h_{-1} & 0 & 0 & 0 \\
0 & 0 & 0 & h_2 & h_0 & h_{-2} & 0 & 0 \\
0 & 0 & 0 & h_3 & h_1 & h_{-1} & 0 & 0 \\
0 & 0 & 0 & 0 & h_2 & h_0 & h_{-2} & 0 \\
0 & 0 & 0 & 0 & h_3 & h_1 & h_{-1} & 0 \\
0 & 0 & 0 & 0 & 0 & h_2 & h_0 & h_{-2} \\
0 & 0 & 0 & 0 & 0 & h_3 & h_1 & h_{-1} \\
h_{-2} & 0 & 0 & 0 & 0 & 0 & h_2 & h_0 \\
h_{-1} & 0 & 0 & 0 & 0 & 0 & h_3 & h_1
\end{bmatrix}
\cdot
\begin{bmatrix}
s_1 \\ s_2 \\ s_3 \\ s_4 \\ s_5 \\ s_6 \\ s_7 \\ s_8
\end{bmatrix}
$$

Length 8 filter:

$$H_8^T \mathbf{s} = \begin{bmatrix}
h_0 & h_{-2} & 0 & 0 & 0 & 0 & h_4 & h_2 \\
h_1 & h_{-1} & h_{-3} & 0 & 0 & 0 & 0 & h_3 \\
h_2 & h_0 & h_{-2} & 0 & 0 & 0 & 0 & h_4 \\
h_3 & h_1 & h_{-1} & h_{-3} & 0 & 0 & 0 & 0 \\
h_4 & h_2 & h_0 & h_{-2} & 0 & 0 & 0 & 0 \\
0 & h_3 & h_1 & h_{-1} & h_{-3} & 0 & 0 & 0 \\
0 & h_4 & h_2 & h_0 & h_{-2} & 0 & 0 & 0 \\
0 & 0 & h_3 & h_1 & h_{-1} & h_{-3} & 0 & 0 \\
0 & 0 & h_4 & h_2 & h_0 & h_{-2} & 0 & 0 \\
0 & 0 & 0 & h_3 & h_1 & h_{-1} & h_{-3} & 0 \\
0 & 0 & 0 & h_4 & h_2 & h_0 & h_{-2} & 0 \\
0 & 0 & 0 & 0 & h_3 & h_1 & h_{-1} & h_{-3} \\
0 & 0 & 0 & 0 & h_4 & h_2 & h_0 & h_{-2} \\
h_{-3} & 0 & 0 & 0 & 0 & h_3 & h_1 & h_{-1} \\
h_{-2} & 0 & 0 & 0 & 0 & h_4 & h_2 & h_0 \\
h_{-1} & h_{-3} & 0 & 0 & 0 & 0 & h_3 & h_1
\end{bmatrix} \cdot \begin{bmatrix}
s_1 \\ s_2 \\ s_3 \\ s_4 \\ s_5 \\ s_6 \\ s_7 \\ s_8
\end{bmatrix}$$

Length 10 filter:

$$H_8^T \mathbf{s} = \begin{bmatrix}
h_0 & h_{-2} & h_{-4} & 0 & 0 & 0 & h_4 & h_2 \\
h_1 & h_{-1} & h_{-3} & 0 & 0 & 0 & h_5 & h_3 \\
h_2 & h_0 & h_{-2} & h_{-4} & 0 & 0 & 0 & h_4 \\
h_3 & h_1 & h_{-1} & h_{-3} & 0 & 0 & 0 & h_5 \\
h_4 & h_2 & h_0 & h_{-2} & h_{-4} & 0 & 0 & 0 \\
h_5 & h_3 & h_1 & h_{-1} & h_{-3} & 0 & 0 & 0 \\
0 & h_4 & h_2 & h_0 & h_{-2} & h_{-4} & 0 & 0 \\
0 & h_5 & h_3 & h_1 & h_{-1} & h_{-3} & 0 & 0 \\
0 & 0 & h_4 & h_2 & h_0 & h_{-2} & h_{-4} & 0 \\
0 & 0 & h_5 & h_3 & h_1 & h_{-1} & h_{-3} & 0 \\
0 & 0 & 0 & h_4 & h_2 & h_0 & h_{-2} & h_{-4} \\
0 & 0 & 0 & h_5 & h_3 & h_1 & h_{-1} & h_{-3} \\
h_{-4} & 0 & 0 & 0 & h_4 & h_2 & h_0 & h_{-2} \\
h_{-3} & 0 & 0 & 0 & h_5 & h_3 & h_1 & h_{-1} \\
h_{-2} & h_{-4} & 0 & 0 & 0 & h_4 & h_2 & h_0 \\
h_{-1} & h_{-3} & 0 & 0 & 0 & h_5 & h_3 & h_1
\end{bmatrix} \cdot \begin{bmatrix}
s_1 \\ s_2 \\ s_3 \\ s_4 \\ s_5 \\ s_6 \\ s_7 \\ s_8
\end{bmatrix}$$

Length 12 filter:

$$
H_8^T \mathbf{s} =
\begin{bmatrix}
h_0 & h_{-2} & h_{-4} & 0 & 0 & h_6 & h_4 & h_2 \\
h_1 & h_{-1} & h_{-3} & h_{-5} & 0 & 0 & h_5 & h_3 \\
h_2 & h_0 & h_{-2} & h_{-4} & 0 & 0 & h_6 & h_4 \\
h_3 & h_1 & h_{-1} & h_{-3} & h_{-5} & 0 & 0 & h_5 \\
h_4 & h_2 & h_0 & h_{-2} & h_{-4} & 0 & 0 & h_6 \\
h_5 & h_3 & h_1 & h_{-1} & h_{-3} & h_{-5} & 0 & 0 \\
h_6 & h_4 & h_2 & h_0 & h_{-2} & h_{-4} & 0 & 0 \\
0 & h_5 & h_3 & h_1 & h_{-1} & h_{-3} & h_{-5} & 0 \\
0 & h_6 & h_4 & h_2 & h_0 & h_{-2} & h_{-4} & 0 \\
0 & 0 & h_5 & h_3 & h_1 & h_{-1} & h_{-3} & h_{-5} \\
0 & 0 & h_6 & h_4 & h_2 & h_0 & h_{-2} & h_{-4} \\
h_{-5} & 0 & 0 & h_5 & h_3 & h_1 & h_{-1} & h_{-3} \\
h_{-4} & 0 & 0 & h_6 & h_4 & h_2 & h_0 & h_{-2} \\
h_{-3} & h_{-5} & 0 & 0 & h_5 & h_3 & h_1 & h_{-1} \\
h_{-2} & h_{-4} & 0 & 0 & h_6 & h_4 & h_2 & h_0 \\
h_{-1} & h_{-3} & h_{-5} & 0 & 0 & h_5 & h_3 & h_1
\end{bmatrix}
\cdot
\begin{bmatrix}
s_1 \\ s_2 \\ s_3 \\ s_4 \\ s_5 \\ s_6 \\ s_7 \\ s_8
\end{bmatrix}
$$

As was the case with the odd filters, we will split the filter $\mathbf{h}$ into two parts — $\mathbf{h}^e$ will hold the components of $\mathbf{h}$ with even indices and $\mathbf{h}^o$ will hold the components of $\mathbf{h}$ with odd indices. For example, for a filter of length 6 ($L = 3$), we have

$$
\mathbf{h}^e = (h_2, h_0, h_{-2}) \qquad \text{and} \qquad \mathbf{h}^o = (h_3, h_1, h_{-1})
$$

The basic approach is the same as for the highpass portion of the odd-length filters — to compute $\mathbf{q}$ we need to build vectors $\mathbf{c}$ and $\mathbf{d}$ and dot portions of them with $\mathbf{h}^e$, $\mathbf{h}^o$, respectively. Let's first look at the computations for the odd components q_{2k-1}, $k = 1, \ldots, 8$. Table 11.3 summarizes our findings.

We again use the floor function $\lfloor \cdot \rfloor$ to describe the number of elements (in terms of L) we must prepend and append to $\mathbf{s}$ to form $\mathbf{c}$. In this case we have

$$
p = \left\lfloor \frac{L}{2} \right\rfloor \qquad \text{and} \qquad a = \left\lfloor \frac{L-1}{2} \right\rfloor \tag{11.18}
$$

We can use p and a in (11.12) to form $\mathbf{c}$ from vector $\mathbf{s}$ of length $\frac{N}{2}$. If we reindex the elements of $\mathbf{h}^e$ as $\mathbf{h}^e = (h_1^e, \ldots, h_\ell^e)$, we can compute the odd components of $\mathbf{q}$ as

$$
q_{2k-1} = \sum_{j=1}^{m} h_j^e c_{k+j-1}, \qquad k = 1, \ldots, \frac{N}{2} \tag{11.19}
$$

For the even components of $\mathbf{q}$, we can use Table 11.4 to help us find formulas for p and a.

Table 11.3   The number of elements to prepend and append to s^a.

Filter Length	L	Prepend from Row 1	Append from Row 15
2	1	0	0
4	2	1	0
6	3	1	1
8	4	2	1
10	5	2	2
12	6	3	2

^{a}The first column is the filter length of **h**. The second column is the index of the last element in **h**. The number in the third column is the number of filter elements that appear at the *end* of the first row of H_8^{lT}. This number represents the number of elements at the end of **s** that we must prepend to **s** when forming **c**. The number in the fourth column is the number of filter elements at the *beginning* of row 15 in H_8^{lT}. This is the number of elements at the beginning of **s** that we need to append to **s** to form **c**.

Table 11.4   The number of elements to prepend and append to s^a.

Filter Length	L	Prepend from Row 2	Append from Row 16
2	1	0	0
4	2	0	1
6	3	1	1
8	4	1	2
10	5	2	2
12	6	2	3

^{a}The first column is the filter length of **h**. The second column is the index of the last element in **h**. The number in the third column is the number of filter elements that appear at the *end* of the second row of H_8^{lT}. This number represents the number of elements at the end of **s** that we must prepend to **s** when forming **d**. The number in the fourth column is the number of filter elements at the *beginning* of row 16 in H_8^{lT}. This is the number of elements at the beginning of **s** that we need to append to **s** to form **d**.

Again we seek formulas for p and a that will help us form **d** (11.15). Note that the third and fourth columns of Table 11.4 are the same as the fourth and third columns of Table 11.3, so we can simply interchange the roles of p and a from (11.18) to obtain

$$p = \left\lfloor \frac{L-1}{2} \right\rfloor \quad \text{and} \quad a = \left\lfloor \frac{L}{2} \right\rfloor \tag{11.20}$$

We can now write formulas for the even components of **q** in the case where **s** is a vector of length $\frac{N}{2}$ and **h** is a symmetric even-length filter. We first reindex the elements of $\mathbf{h^e}$ as $\mathbf{h^o} = (h_1^o, \ldots, h_m^o)$ and use (11.20) with (11.15) to build **d**. Then the even components of $\mathbf{q}$ are

$$q_{2k} = \sum_{j=1}^{\ell} h_o^e d_{k+j-1}, \qquad k = 1, \ldots, \frac{N}{2} \qquad (11.21)$$

Computing $\mathbf{r} = G_{N/2}^T \mathbf{t}$ for Even-Length Filters

We now consider the problem of computing $\mathbf{r} = G_{N/2}^T \mathbf{t}$ (see (11.10), where $G_{N/2}$ is built from the highpass filter $\mathbf{g}$. Recall that $\mathbf{g}$ is constructed from the biorthogonal spline filter $\tilde{\mathbf{h}}$ using the componentwise rule $g_k = (-1)^k \tilde{h}_{1-k}$.

We first consider the case where the length of $\mathbf{g}$ is even. In this case we have $\tilde{\mathbf{h}} = (\tilde{h}_{-\tilde{L}+1}, \ldots, \tilde{h}_{\tilde{L}})$. If we use the rule $g_k = (-1)^k \tilde{h}_{1-k}$, we see that $\mathbf{g}$ is the finite-length filter $(g_{-\tilde{L}+1}, \ldots, g_{\tilde{L}})$. For example, when $\tilde{\mathbf{h}} = (\tilde{h}_{-2}, \tilde{h}_{-1}, \tilde{h}_0, \tilde{h}_1, \tilde{h}_2, \tilde{h}_3)$, we build $\mathbf{g}$ simply by reflecting the elements of $\mathbf{h}$ around 0, shift them one to the right, and alternate signs. The reflection and shift leaves the indices of $\mathbf{g}$ the same as those for $\tilde{\mathbf{h}}$, so we have

$$\mathbf{g} = (g_{-2}, g_{-1}, g_0, g_1, g_2, g_3) = (-\tilde{h}_3, \tilde{h}_2, -\tilde{h}_1, \tilde{h}_0, -\tilde{h}_{-1}, \tilde{h}_{-2})$$

From a computational standpoint, the even-length case here is no different than the even-length case for the lowpass filter! That is, for even-length high pass filters, we simply load $\mathbf{g}$ into vectors $\mathbf{g}^e$ and $\mathbf{g}^o$, determine $\mathbf{c}$ and $\mathbf{d}$ using (11.18) and (11.20), respectively, and then use (11.19) and (11.21) to compute the elements of $\mathbf{t}$.

Computing $\mathbf{r} = G_{N/2}^T \mathbf{t}$ for Odd-Length Filters

Unfortunately, odd-length, highpass filters are different from their lowpass counterparts. When we reflect the filter $\tilde{\mathbf{h}} = (\tilde{h}_{-\tilde{L}}, \ldots, \tilde{h}_{\tilde{L}})$ about the origin and shift 1 unit right, the starting and stopping indices of $\mathbf{g}$ are not the same as those for $\tilde{\mathbf{h}}$. We have

$$\mathbf{g} = (g_{-\tilde{L}+1}, \ldots, g_{\tilde{L}+1}) \qquad (11.22)$$

Since the computations for $H_{N/2}^T$ s depended on the stopping index L, let's remain consistent and use the terminal index for $\mathbf{g}$ in our computations. Using (11.22), we see that $L = \tilde{L} + 1$.

In Problem 11.7 you are asked to mimic the procedure we use to compute $H_{N/2}$ s and write out $G_8^T \mathbf{t}$ for odd highpass filters of length $3, 5, \ldots, 13$ and use these matrices to derive formulas for $\mathbf{g}^o$, $\mathbf{g}^e$, p, and a. In this problem you will show that

$$\begin{aligned} g_{k+1}^e &= g_{L-2k}, & k &= 0, \ldots, L-1 \\ g_{k+1}^o &= g_{L-1-2k}, & k &= 0, \ldots, L-2 \end{aligned} \qquad (11.23)$$

when L is even and

$$
\begin{array}{ll}
g^o_{k+1} = g_{L-2k}, & k = 0, \ldots, L-1 \\
g^e_{k+1} = g_{L-1-2k}, & k = 0, \ldots, L-2
\end{array}
\tag{11.24}
$$

when L is odd. Furthermore, upon completion of the problem, you will see that

$$
\mathbf{c} = [t_{N-p+1}, \ldots, t_{N/2} \mid \underbrace{t_1, \ldots, t_{N/2}}_{\mathbf{t}} \mid t_1, \ldots, t_a]^T
$$

where

$$
p = a = \left\lfloor \frac{L-1}{2} \right\rfloor
\tag{11.25}
$$

and

$$
\mathbf{d} = [t_{N-p+1}, \ldots, t_N \mid \underbrace{t_1, \ldots, t_N}_{\mathbf{t}} \mid t_1, \ldots, t_a]^T
$$

$$
p = \left\lfloor \frac{L}{2} \right\rfloor \quad \text{and} \quad a = \left\lfloor \frac{L-2}{2} \right\rfloor
\tag{11.26}
$$

An Auxiliary Algorithm for Computing $H^T_{N/2}\mathbf{s}$ and $G^T_{N/2}\mathbf{t}$

We are now ready to write the auxiliary algorithm to compute $H^T_{N/2}\mathbf{s}$ and $G^T_{N/2}\mathbf{t}$ (see (11.10)).

Algorithm 11.2 (Computing $\mathbf{H}^T_{N/2}\mathbf{s}$ and $\mathbf{G}^T_{N/2}\mathbf{t}$). *We are given a vector $\mathbf{u} \in \mathbb{R}^{\frac{N}{2}}$ (N even), a filter $\mathbf{f}$, and a lowpass/highpass flag i, and we return the N-vector $\mathbf{y} = H_{N/2}\mathbf{u}$ if $i = 0$ or $\mathbf{y} = G_{N/2}\mathbf{u}$ if $i = 1$.*

The algorithm is easy to read if we split it into two cases. The first case computes $\mathbf{y}$ in the case where the filter length is even, and the second case computes $\mathbf{y}$ when the filter length is odd.

When the length of $\mathbf{f}$ is even, it doesn't matter if the input filter is a lowpass filter $\mathbf{h}$ or a highpass filter $\mathbf{g}$. In either case, the elements of both filters are centered around $k = \frac{1}{2}$. So we have $\mathbf{f} = (f_{-L+1}, \ldots, f_L)$, and both $\mathbf{f}^e$ and $\mathbf{f}^o$ are of length L. Of course, how $\mathbf{f}^e$ and $\mathbf{f}^o$ are formed depends on whether L is even or odd. Thus, we have a branch that considers both cases. Once we have formed $\mathbf{f}^e$ and $\mathbf{f}^o$, we build $\mathbf{c}$ and $\mathbf{d}$ using (11.12) and (11.15), respectively. This case is concluded when we use (11.19) and (11.21) to obtain $\mathbf{y}$.

When the length of the input filter is odd, we have to change the limits on the loops that build $\mathbf{f}^e$ and $\mathbf{f}^o$ as well as adjusting the values for p and a. The flag value is quite helpful here. We can simply subtract i from the upper limits of these loops to build the correct $\mathbf{f}^e$ and $\mathbf{f}^o$. There are also simple functions in terms of i that we can use when computing p and a for the $\mathbf{c}$ and $\mathbf{d}$ vectors.

Unlike the even-length filter, the $\mathbf{f}^e$ and $\mathbf{f}^o$ vectors have different lengths. For this reason, we use variables ℓ and m to represent the lengths of $\mathbf{f}^e$ and $\mathbf{f}^o$, respectively. Once we have constructed $\mathbf{f}^e$ and $\mathbf{f}^o$, the algorithm proceeds much like the even-filter-length case. We use (11.12) and (11.15) to construct $\mathbf{c}$ and $\mathbf{d}$, respectively, and (11.13) and (11.16) to obtain $\mathbf{y}$.

Algorithm: `IBWTht`

Input: Vector $\mathbf{u}$ of length $\frac{N}{2}$, filter $\mathbf{f}$ with terminal index L, and lowpass/highpass flag i. It is assumed that N is at least as large as the length of $\mathbf{f}$.

Output: Vector $\mathbf{y}$ where $\mathbf{y} = H_{N/2}^T \mathbf{u}$ when $i = 0$ or $\mathbf{y} = G_{N/2}^T \mathbf{u}$ when $i = 1$.

// First consider the case where the length of $\mathbf{f}$ is even.

If [**Mod** [**Length** [**f**] , 2]= 0,

    // Build $\mathbf{f}^e$ and $\mathbf{f}^o$.

    If [**Mod** [L, 2] = 0,
        For [$k = 0, k \le L - 1, k + +$,
            $f_{k+1}^e = f_{L-2k}$
            $f_{k+1}^o = f_{L-1-2k}$
        ]
    Else
        For [$k = 0, k \le L - 1, k + +$,
            $f_{k+1}^o = f_{L-2k}$
            $f_{k+1}^e = f_{L-1-2k}$
        ]
    ]

    // Use (11.18) and (11.12) to construct $\mathbf{c}$.

    $p = \left\lfloor \frac{L}{2} \right\rfloor$

    $a = \left\lfloor \frac{L-1}{2} \right\rfloor$

    $\mathbf{c} = \mathbf{Join} \left[\mathbf{Join} \left[(u_{N/2-p+1}, \ldots, u_{N/2}), \mathbf{u} \right], (u_1, \ldots, u_a) \right]$

    // Use (11.20) and (11.15) to construct $\mathbf{d}$.

$$p = \left\lfloor \frac{L-1}{2} \right\rfloor$$

$$a = \left\lfloor \frac{L}{2} \right\rfloor$$

$\mathbf{d} = \mathbf{Join}\left[\mathbf{Join}\left[(u_{N/2-p+1}, \ldots, u_{N/2}), \mathbf{u}\right], (u_1, \ldots, u_a)\right]$

// Use (11.19) and (11.21) to compute the components of $\mathbf{y}$.

For $[k = 1, k \leq \frac{N}{2}, k++,$
$\quad y_{2k-1} = 0$
$\quad y_{2k} = 0$

$\quad$ **For** $[j = 1, j \leq L, j++,$
$\quad\quad y_{2k-1} = y_{2k-1} + f_j^e\, c_{k+j-1}$
$\quad\quad y_{2k} = y_{2k} + f_j^o\, d_{k+j-1}$
$\quad$]

]
Else

// Now handle the case where the length of $\mathbf{f}$ is odd.

// For the highpass $(i = 1)$ case, the terminal values in the loops
// to construct $\mathbf{f^e}$ and $\mathbf{f^o}$ are one less (see (11.23) and (11.24)).
// In these cases, we simply subtract i from the upper
// limits of the loops.

$\quad$ **If** $[\mathbf{Mod}[L, 2] = 0,$
$\quad\quad$ **For** $[k = 0, k \leq L - i, k++,$
$\quad\quad\quad f_{k+1}^e = f_{L-2k}$
$\quad\quad$]

$\quad\quad$ **For** $[k = 0, k \leq L - 1 - i, k++,$
$\quad\quad\quad f_{k+1}^o = f_{L-1-2k}$
$\quad\quad$]

$\quad\quad \ell = L + 1 - i$
$\quad\quad m = L - i$

$\quad$ **Else**
$\quad\quad$ **For** $[k = 0, k \leq L - i, k++,$
$\quad\quad\quad f_{k+1}^o = f_{L-2k}$

]

For $[k = 0, k \leq L - 1 - i, k++,$
$$f^e_{k+1} = f_{L-1-2k}$$
]

$\ell = L - i$
$m = L + 1 - i$

]

// Use (11.11) and (11.12) to construct **c**.
// When we are using highpass filters, we have to use (11.25) and
// (11.26) to to set the values for p and a. We can use the flag i
// to adjust these values.

$$p = \left\lfloor \frac{L-i}{2} \right\rfloor$$

$$a = \left\lfloor \frac{L-i}{2} \right\rfloor$$

$$\mathbf{c} = \mathbf{Join}\left[\mathbf{Join}\left[(u_{N/2-p+1}, \ldots, u_{N/2}), \mathbf{u}\right], (u_1, \ldots, u_a)\right]$$

// Use (11.14) and (11.15) to construct **d**

$$p = \left\lfloor \frac{L-1+i}{2} \right\rfloor$$

$$a = \left\lfloor \frac{L+1-3i}{2} \right\rfloor$$

$$\mathbf{d} = \mathbf{Join}\left[\mathbf{Join}\left[(u_{N/2-p+1}, \ldots, u_{N/2}), \mathbf{u}\right], (u_1, \ldots, u_a)\right]$$

// Use (11.13) and (11.16) to compute the components of **y**.

// Unlike the even-filter-length case, the number of terms
// in each summation is different and the role of $\mathbf{f^e}$ and
// $\mathbf{f^o}$ are reversed.

For $[k = 1, k \leq \frac{N}{2}, k++,$
$$y_{2k-1} = 0$$
$$y_{2k} = 0$$

    For $[j = 1, j \leq m, j++,$
$$y_{2k-1} = y_{2k-1} + f^e_j c_{k+j-1}$$

```
        ]

    For [j = 1, j ≤ ℓ, j ++,
            y_{2k} = y_{2k} + f_j^o d_{k+j-1}
        ]


    ]

]
```

Return [y]

□

An Algorithm for Computing the Inverse Biorthogonal Wavelet Transform

We are now ready to present an algorithm for computing one iteration of the one-dimensional inverse biorthogonal wavelet transform.

Algorithm 11.3 (Biorthogonal Inverse Wavelet Transform: 1D). *We are given a vector* $\mathbf{v} \in \mathbb{R}^N$ *(N even) and a symmetric biorthgonal filter pair* $\tilde{\mathbf{h}}$ *and* $\mathbf{h}$ *and we return one iteration of the inverse biorthogonal wavelet transform.*

The first step in the algorithm is to build $\mathbf{g}$ *from* $\tilde{\mathbf{h}}$. *This construction is identical to that used in Algorithm 11.1 to build* $\tilde{\mathbf{g}}$ *from* $\mathbf{h}$.

We next use (11.10) to first form $\mathbf{s}$ *from the top half of* $\mathbf{v}$ *and then form* $\mathbf{t}$ *from the bottom half of* $\mathbf{v}$. *Now that we have* $\mathbf{s}$ *and* $\mathbf{t}$, *we call Algorithm 11.2 twice — once with* $\mathbf{h}$, $\mathbf{s}$, *and* $i = 0$ *and then again with* $\mathbf{g}$, $\mathbf{t}$, *and* $i = 1$. *We add the results of these calls to complete the computation.*

Algorithm: IBWT1D1

Input:	Vector **v** of even length N and symmetric biorthogonal filter pair $\tilde{\mathbf{h}}$, $\mathbf{h}$. It is assumed that N is at least as large as the length of the longer filter.

Output: Vector **y** where $\mathbf{y} = W_N^T \mathbf{v}$.

// Create the highpass filter --- assume filter length is odd.

$$p = \tilde{L} + 1$$

// Change the stop variable if the filter length is even.

If [Mod [Length [h] , 2] = 0,
$$p = p - 1$$
]

For [$k = -\tilde{L} + 1$, $k \leq p$, $k{+}{+}$,
$$g_k = (-1)^k \tilde{h}_{1-k}$$
]

$$\mathbf{s} = [y_1, \ldots, y_{\frac{N}{2}}]^T$$
$$\mathbf{t} = [y_{\frac{N}{2}+1}, \ldots, y_N]^T$$

q = IBWTht [s, h, 0]
r = IBWTht [t, g, 1]

Return [q+r]

□

PROBLEMS

11.7   In this problem you will verify the construction of $\mathbf{g}^e$ and $\mathbf{g}^o$ (11.23) and (11.24) and the values for p and a given by (11.25) and (11.26) for odd-length filters.

(a) For $\tilde{L} = 1, \ldots, 6$ and $\tilde{\mathbf{h}} = (\tilde{h}_{-\tilde{L}}, \ldots, \tilde{h}_{\tilde{L}})$, write down $\mathbf{g}$ using the rule $g_k = (-1)^k \tilde{h}_{1-k}$ and verify that the index for the last element in $\mathbf{g}$ is $L = \tilde{L} + 1$. Thus the filter $\mathbf{g} = (g_{-L+1}, \ldots, g_{L+1})$.

(b) Use your result in part (a) for $L = 1, 3, 5$ to verify (11.24). Explain why the result holds in general.

(c) Use your result in part (a) for $L = 2, 4, 6$ to verify (11.23). Explain why the result holds in general.

(d) For $L = 1, \ldots, 6$ and $\mathbf{t} \in \mathbb{R}^{16}$, write out $G_8^T \mathbf{t}$. Make a chart similar to Table 11.1 and use it to verify the values listed for p and a in (11.25). These values are then used to construct $\mathbf{c}$.

(e) For $L = 1, \ldots, 6$ and $\mathbf{t} \in \mathbb{R}^{16}$, write out $G_8^T \mathbf{t}$. Make a chart similar to Table 11.2 and use it to verify the values listed for p and a in (11.26). These values are then used to construct $\mathbf{d}$.

11.8   Use Algorithm 11.3 with the $(5, 3)$ biorthogonal spline filter pair and compute one iteration of the inverse biorthogonal wavelet transformation of the vector you obtained in Problem 11.6 from Section 11.1.

Computer Labs

♦ **11.3 Software Development: 1D Biorthogonal Wavelet Transforms** From the text Web site, access the development package bwt1d. In this development lab you will create a function that will compute one iteration of the one-dimensional biorthogonal wavelet transform using biorthogonal spline filter pairs. Additionally, you will construct a function that computes one iteration of the inverse biorthogonal wavelet transform. Construction of iterated transforms and their inverses are also included in this lab. Instructions are provided that allow you to add all of these functions to the software package DiscreteWavelets.

♦ **11.4 Software Development: 2D Biorthogonal Wavelet Transforms.** From the text Web site, access the development package bwt2d. In this development lab you will create a routine that will compute one iteration of the inverse two-dimensional biorthogonal wavelet transform using biorthogonal spline filter pairs. Additionally, you will construct routines that compute one iteration of the inverse biorthogonal wavelet transform. Instructions are provided that allow you to add all of these functions to the software package DiscreteWavelets.

11.5 Data Compression. From the text Web site, access the lab biorthcomp. In this lab you will use biorthogonal spline filter pairs to perform basic data compression. You will also compare your results with those you obtained using the Daubechies filters in Computer Lab 7.8 from Section 7.3 and Coiflet filters in Computer Lab 8.3 from Section 8.3.

11.3 SYMMETRY AND BOUNDARY EFFECTS

In our construction of orthogonal W_N in Chapter 7, we wrapped rows near the bottom parts of the lowpass and highpass portions of the transform matrix. For example, if we want to process vector $\mathbf{v} = (v_1, \ldots, v_{10})^T$ using the D6 filter, the lowpass portion of W_{10} is

$$H_5 = \begin{bmatrix} h_5 & h_4 & h_3 & h_2 & h_1 & h_0 & 0 & 0 & 0 & 0 \\ 0 & 0 & h_5 & h_4 & h_3 & h_2 & h_1 & h_0 & 0 & 0 \\ 0 & 0 & 0 & 0 & h_5 & h_4 & h_3 & h_2 & h_1 & h_0 \\ h_1 & h_0 & 0 & 0 & 0 & 0 & h_5 & h_4 & h_3 & h_2 \\ h_3 & h_2 & h_1 & h_0 & 0 & 0 & 0 & 0 & h_5 & h_4 \end{bmatrix} \qquad (11.27)$$

As we move our filter two to the right in each subsequent row, our filter finally reaches the last column in row 3. In rows 4 and 5, we wrap the filter around the first two and four columns, respectively, of the row. The reason for wrapping the rows is that it is easier to create an orthogonal matrix — if we truncate these rows, we would have many more conditions to satisfy in order to make W_N orthogonal. The major drawback of this construction is that we implicitly assume that our data are periodic. Note that dotting row 4 of H_5 in (11.27) with $\mathbf{v}$ uses components $v_7, v_8, v_9, v_{10}, v_1,$ and v_2. Now if $\mathbf{v}$ can be viewed as a repeating or periodic signal, say $v_9 = v_1, v_8 = v_2, \ldots$ this computation makes perfect sense. But in practice, digital images or audio signals can hardly be considered periodic. And even if the data are periodic, there is no special correlation of the filter coefficients $h_0, \ldots, h_5$ to exploit this.

In Section 11.3 we develop techniques utilizing symmetric filters that minimize the problems caused by wrapping rows in the transform. The material that appears in this section is outlined in Walnut [83].

Using Symmetry with a Biorthogonal Filter Pair

To motivate the process for handling the wrapping rows, let's consider an example.

Example 11.1 (Modifying the Biorthogonal Wavelet Transform). *Suppose that we wish to transform the vector* $\mathbf{v} = (v_1, v_2, \ldots, v_8)^T$ *using a biorthogonal filter pair* $\mathbf{h}$ *and* $\tilde{\mathbf{h}}$ *of lengths* 5 *and* 3, *respectively.*[1] *The top half of the transform* $\tilde{W}_8$ *uses the symmetric filter* $\tilde{\mathbf{h}} = (\tilde{h}_{-1}, \tilde{h}_0, \tilde{h}_1) = (\tilde{h}_1, \tilde{h}_0, \tilde{h}_1)$. *If we let* $\mathbf{y}$ *denote the lowpass portion of the biorthogonal wavelet transform, we have*

$$\mathbf{y} = \tilde{H}_4 \mathbf{v} = \begin{bmatrix} \tilde{h}_0 & \tilde{h}_1 & 0 & 0 & 0 & 0 & 0 & \tilde{h}_1 \\ 0 & \tilde{h}_1 & \tilde{h}_0 & \tilde{h}_1 & 0 & 0 & 0 & 0 \\ 0 & 0 & 0 & \tilde{h}_1 & \tilde{h}_0 & \tilde{h}_1 & 0 & 0 \\ 0 & 0 & 0 & 0 & 0 & \tilde{h}_1 & \tilde{h}_0 & \tilde{h}_1 \end{bmatrix} \begin{bmatrix} v_1 \\ v_2 \\ v_3 \\ v_4 \\ v_5 \\ v_6 \\ v_7 \\ v_8 \end{bmatrix} = \begin{bmatrix} \tilde{h}_1 v_8 + \tilde{h}_0 v_1 + \tilde{h}_1 v_2 \\ \tilde{h}_1 v_2 + \tilde{h}_0 v_3 + \tilde{h}_1 v_4 \\ \tilde{h}_1 v_4 + \tilde{h}_0 v_5 + \tilde{h}_1 v_6 \\ \tilde{h}_1 v_6 + \tilde{h}_0 v_7 + \tilde{h}_1 v_8 \end{bmatrix}$$

Notice that the first element of $\mathbf{y}$ *is* $\tilde{h}_0 v_1 + \tilde{h}_1 v_2 + \tilde{h}_1 v_8$. *This computation uses elements from the top and bottom of* $\mathbf{v}$, *but unlike the filter in the orthogonal case (11.27), the filter coefficients are symmetric. A more natural combination of elements for* y_1 *should use only the first few elements from* $\mathbf{v}$.

Since we can never assume that our input would satisfy such a symmetry condition, we will try instead to periodize the input. Let's create the 14-vector

$$\mathbf{v}^p = [v_1, v_2, \ldots, v_7, v_8, v_7, v_6, v_5, v_4, v_3, v_2]^T \tag{11.28}$$

Let's look at the structure of $\mathbf{v}^p$. *We have basically removed the first and last elements of* $\mathbf{v}$, *reversed the result, and appended it to* $\mathbf{v}$. *We are guaranteed an even length vector with this process and also note that the last element of* $\mathbf{v}^p$ *is* v_2. *Now consider*

---

[1] You could use, for example, the $(5, 3)$ biorthogonal spline filter pair developed in Section 10.2.

the computation

$$\mathbf{y}^p = \tilde{H}_7 \mathbf{v}^p$$

$$= \begin{bmatrix} \tilde{h}_0 & \tilde{h}_1 & 0 & 0 & 0 & 0 & 0 & 0 & 0 & 0 & 0 & 0 & \tilde{h}_1 \\ 0 & \tilde{h}_1 & \tilde{h}_0 & \tilde{h}_1 & 0 & 0 & 0 & 0 & 0 & 0 & 0 & 0 & 0 \\ 0 & 0 & 0 & \tilde{h}_1 & \tilde{h}_0 & \tilde{h}_1 & 0 & 0 & 0 & 0 & 0 & 0 & 0 \\ 0 & 0 & 0 & 0 & 0 & \tilde{h}_1 & \tilde{h}_0 & \tilde{h}_1 & 0 & 0 & 0 & 0 & 0 \\ 0 & 0 & 0 & 0 & 0 & 0 & 0 & \tilde{h}_1 & \tilde{h}_0 & \tilde{h}_1 & 0 & 0 & 0 \\ 0 & 0 & 0 & 0 & 0 & 0 & 0 & 0 & 0 & \tilde{h}_1 & \tilde{h}_0 & \tilde{h}_1 & 0 & 0 \\ 0 & 0 & 0 & 0 & 0 & 0 & 0 & 0 & 0 & 0 & 0 & \tilde{h}_1 & \tilde{h}_0 & \tilde{h}_1 \end{bmatrix} \cdot \begin{bmatrix} v_1 \\ v_2 \\ v_3 \\ v_4 \\ v_5 \\ v_6 \\ v_7 \\ v_8 \\ v_7 \\ v_6 \\ v_5 \\ v_4 \\ v_3 \\ v_2 \end{bmatrix}$$

$$= \begin{bmatrix} \tilde{h}_0 v_1 + 2\tilde{h}_1 v_2 \\ \tilde{h}_1 v_2 + \tilde{h}_0 v_3 + \tilde{h}_1 v_4 \\ \tilde{h}_1 v_4 + \tilde{h}_0 v_5 + \tilde{h}_1 v_6 \\ \tilde{h}_1 v_6 + \tilde{h}_0 v_7 + \tilde{h}_1 v_8 \\ \hline \tilde{h}_1 v_8 + \tilde{h}_0 v_7 + \tilde{h}_1 v_6 \\ \tilde{h}_1 v_6 + \tilde{h}_0 v_5 + \tilde{h}_1 v_4 \\ \tilde{h}_1 v_4 + \tilde{h}_0 v_3 + \tilde{h}_1 v_2 \end{bmatrix} \tag{11.29}$$

Note that the elements of $\mathbf{y}^p$ exhibit some symmetry: in particular, $y_2^p = y_7^p$, $y_3^p = y_6^p$, and $y_4^p = y_5^p$. So $\mathbf{y}_p$ has the form

$$\mathbf{y}^p = \left[y_1^p,\ y_2^p,\ y_3^p,\ y_4^p \,\middle|\, y_4^p,\ y_3^p,\ y_2^p \right]^T \tag{11.30}$$

More important, the first element $y_1^p = \tilde{h}_0 v_1 + 2\tilde{h}_1 v_2$ uses a combination of only the first two elements of $\mathbf{v}$.

The value y_1^p combines only the first few elements of $\mathbf{v}$, where y_1 uses v_8. So y_1^p provides a more accurate representation of the top portion of $\mathbf{v}$ than y_1 does. Also note that $y_2^p = y_2$, $y_3^p = y_3$, and $y_4^p = y_4$. Thus, it makes sense to use $(y_1^p, y_1^p, y_3^p, y_4^p)^T$ as the lowpass portion of our transform.

We now consider the highpass portion for both $\tilde{G}_4 \mathbf{v}$ and $\tilde{G}_7 \mathbf{v}^p$. The highpass portion of the transform uses the five-term filter $\tilde{\mathbf{g}}$, where $\tilde{g}_k = (-1)^k h_{1-k}$, $k = -1, \ldots, 3$. Here $\mathbf{h}$ is the symmetric filter

$$\mathbf{h} = (h_{-2}, h_{-1}, h_0, h_1, h_2) = (h_2, h_1, h_0, h_1, h_2)$$

so that $g_{-1} = -h_2$, $g_0 = h_1$, $g_1 = -h_0$, $g_2 = h_1$, and $g_3 = -h_2$. We have

$$\mathbf{z} = \tilde{G}_4 \mathbf{v} = \begin{bmatrix} g_0 & g_1 & g_2 & g_3 & 0 & 0 & 0 & g_{-1} \\ 0 & g_{-1} & g_0 & g_1 & g_2 & g_3 & 0 & 0 \\ 0 & 0 & 0 & g_{-1} & g_0 & g_1 & g_2 & g_3 \\ g_2 & g_3 & 0 & 0 & 0 & g_{-1} & g_0 & g_1 \end{bmatrix} \cdot \begin{bmatrix} v_1 \\ v_2 \\ v_3 \\ v_4 \\ v_5 \\ v_6 \\ v_7 \\ v_8 \end{bmatrix}$$

$$= \begin{bmatrix} g_{-1}v_8 + g_0 v_1 + g_1 v_2 + g_2 v_3 + g_3 v_4 \\ g_{-1}v_2 + g_0 v_3 + g_1 v_4 + g_2 v_5 + g_3 v_6 \\ g_{-1}v_4 + g_0 v_5 + g_1 v_6 + g_2 v_7 + g_3 v_8 \\ g_{-1}v_6 + g_0 v_7 + g_1 v_8 + g_2 v_1 + g_3 v_2 \end{bmatrix}$$

$$= \begin{bmatrix} -h_2 v_8 + h_1 v_1 - h_0 v_2 + h_1 v_3 - h_2 v_4 \\ -h_2 v_2 + h_1 v_3 - h_0 v_4 + h_1 v_5 - h_2 v_6 \\ -h_2 v_4 + h_1 v_5 - h_0 v_6 + h_1 v_7 - h_2 v_8 \\ -h_2 v_6 + h_1 v_7 - h_0 v_8 + h_1 v_1 - h_2 v_2 \end{bmatrix} \tag{11.31}$$

We now multiply

$$\tilde{G}_7 = \begin{bmatrix} g_0 & g_1 & g_2 & g_3 & 0 & 0 & 0 & 0 & 0 & 0 & 0 & 0 & 0 & g_{-1} \\ 0 & g_{-1} & g_0 & g_1 & g_2 & g_3 & 0 & 0 & 0 & 0 & 0 & 0 & 0 & 0 \\ 0 & 0 & 0 & g_{-1} & g_0 & g_1 & g_2 & g_3 & 0 & 0 & 0 & 0 & 0 & 0 \\ 0 & 0 & 0 & 0 & 0 & g_{-1} & g_0 & g_1 & g_2 & g_3 & 0 & 0 & 0 & 0 \\ 0 & 0 & 0 & 0 & 0 & 0 & 0 & g_{-1} & g_0 & g_1 & g_2 & g_3 & 0 & 0 \\ 0 & 0 & 0 & 0 & 0 & 0 & 0 & 0 & 0 & g_{-1} & g_0 & g_1 & g_2 & g_3 \\ g_2 & g_3 & 0 & 0 & 0 & 0 & 0 & 0 & 0 & 0 & 0 & g_{-1} & g_0 & g_1 \end{bmatrix}$$

$$= \begin{bmatrix} h_1 & -h_0 & h_1 & -h_2 & 0 & 0 & 0 & 0 & 0 & 0 & 0 & 0 & 0 & -h_2 \\ 0 & -h_2 & h_1 & -h_0 & h_1 & -h_2 & 0 & 0 & 0 & 0 & 0 & 0 & 0 & 0 \\ 0 & 0 & 0 & -h_2 & h_1 & -h_0 & h_1 & -h_2 & 0 & 0 & 0 & 0 & 0 & 0 \\ 0 & 0 & 0 & 0 & 0 & -h_2 & h_1 & -h_0 & h_1 & -h_2 & 0 & 0 & 0 & 0 \\ 0 & 0 & 0 & 0 & 0 & 0 & 0 & -h_2 & h_1 & -h_0 & h_1 & -h_2 & 0 & 0 \\ 0 & 0 & 0 & 0 & 0 & 0 & 0 & 0 & 0 & -h_2 & h_1 & -h_0 & h_1 & -h_2 \\ h_1 & -h_2 & 0 & 0 & 0 & 0 & 0 & 0 & 0 & 0 & 0 & -h_2 & h_1 & -h_0 \end{bmatrix}$$

and the periodized 14-vector $\mathbf{v}^p$ given in (11.28) to obtain

$$\mathbf{z}^p = \tilde{G}_7 \mathbf{v}^p = \begin{bmatrix} g_{-1}v_2 + g_0 v_1 + g_1 v_2 + g_2 v_3 + g_3 v_4 \\ g_{-1}v_2 + g_0 v_3 + g_1 v_4 + g_2 v_5 + g_3 v_6 \\ g_{-1}v_4 + g_0 v_5 + g_1 v_6 + g_2 v_7 + g_3 v_8 \\ g_{-1}v_6 + g_0 v_7 + g_1 v_8 + g_2 v_7 + g_3 v_6 \\ \hline g_{-1}v_8 + g_0 v_7 + g_1 v_6 + g_2 v_5 + g_3 v_4 \\ g_{-1}v_6 + g_0 v_5 + g_1 v_4 + g_2 v_3 + g_3 v_2 \\ g_{-1}v_4 + g_0 v_3 + g_1 v_2 + g_2 v_1 + g_3 v_2 \end{bmatrix}$$

$$= \begin{bmatrix} -h_2 v_2 + h_1 v_1 - h_0 v_2 + h_1 v_3 - h_2 v_4 \\ -h_2 v_2 + h_1 v_3 - h_0 v_4 + h_1 v_5 - h_2 v_6 \\ -h_2 v_4 + h_1 v_5 - h_0 v_6 + h_1 v_7 - h_2 v_8 \\ -2h_2 v_6 + 2h_1 v_7 - h_0 v_8 \\ \hline -h_2 v_8 + h_1 v_7 - h_0 v_6 + h_1 v_5 - h_2 v_4 \\ -h_2 v_6 + h_1 v_5 - h_0 v_4 + h_1 v_3 - h_2 v_2 \\ -h_2 v_4 + h_1 v_3 - h_0 v_2 + h_1 v_1 - h_2 v_2 \end{bmatrix}$$

Note that the elements of $\mathbf{z}^p$ are symmetric as well, although the symmetry is a bit different than that possessed by the elements of $\mathbf{y}^p$. We have $z_5^p = z_3^p$, $z_6^p = z_2^p$, and $z_7^p = z_1^p$, so that $\mathbf{z}^p$ has the form

$$\mathbf{z}^p = \begin{bmatrix} z_1^p, & z_2^p, & z_3^p, & z_4^p & z_3^p, & z_2^p, & z_1^p \end{bmatrix}^T \tag{11.32}$$

We also observe that $z_2^p = z_2$ and $z_3^p = z_3$. Unlike z_1, z_1^p uses only the first four elements of $\mathbf{v}$ and in a similar way, we see that z_4^p uses only the last three elements of $\mathbf{v}$. Thus we will take the first four elements of $\mathbf{z}^p$ as the highpass portion of the wavelet transform. Our modified wavelet transform is

$$[y_1^p, y_2^p, y_3^p, y_4^p \mid z_1^p, z_2^p, z_3^p, z_4^p]^T \qquad (11.33)$$

□

A Modified Biorthogonal Wavelet Transform

The algorithm for Example 11.1 is as follows:

1. Form $\mathbf{v}^p$ using (11.28).

2. Compute $\mathbf{w} = W_{14}\mathbf{v}^p$ using the routine BWT1D1 (Algorithm 11.1).

3. If $\mathbf{y}^p$ and $\mathbf{z}^p$ denote the lowpass and highpass portions of $\mathbf{w}$, respectively, then take as the modified transform the 8-vector given by (11.33).

Assuming this process works for arbitrary N-vectors (N even) and odd-length biorthogonal filter pairs, it is easy to produce a general algorithm.

Algorithm 11.4 (Modified BWT for Odd-Length Filter Pairs). *This algorithm takes as input the N-vector $\mathbf{v}$ (N even) and an odd-length biorthogonal filter pair $\mathbf{h}$ and $\tilde{\mathbf{h}}$ and returns a modified version of the biorthogonal wavelet transform. The vector $\mathbf{v}$ is periodized to form vector $\mathbf{v}^p \in \mathbb{R}^{2N-2}$. Then we use Algorithm 11.1 to compute the biorthogonal wavelet transform of $\mathbf{v}^p$. If $\mathbf{y}^p$ and $\mathbf{z}^p$ are the lowpass and highpass portions, respectively, of the biorthogonal wavelet transform, then the algorithm returns the modified biorthogonal wavelet transform*

$$[y_1, \ldots, y_{N/2} \mid z_1, \ldots, z_{N/2}]^T$$

1. Given $\mathbf{v} \in \mathbb{R}^N$, N even, form $\mathbf{v}^p \in \mathbb{R}^{2N-2}$ by the rule

$$\mathbf{v}^p = [v_1, v_2, \ldots, v_{N-1}, v_N, v_{N-1}, \ldots, v_3, v_2]^T \qquad (11.34)$$

2. Compute the biorthogonal wavelet transform $\mathbf{w} = \tilde{W}_{2N-2}\mathbf{v}^p$ using Algorithm 11.1. Denote the lowpass portion of the transform by $\mathbf{y}^p$ and the highpass portion of the transform by $\mathbf{z}^p$.

3. Return $[y_1^p, \ldots, y_{N/2}^p, z_1^p, \ldots, z_{N/2}^p]^T$.

□

Let's look at an example that illustrates Algorithm 11.4.

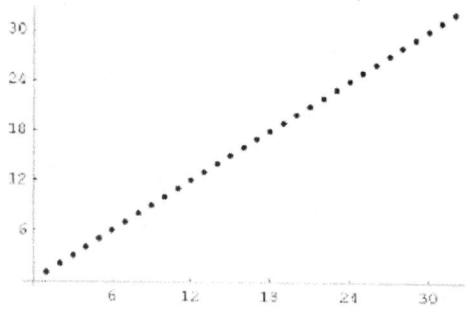

Figure 11.1   The components of vector **v**.

Example 11.2 (Using Algorithm 11.4). *Consider the vector* $\mathbf{v} \in \mathbb{R}^{24}$ *whose components are given by the rule* $v_k = k$, $k = 1, \ldots, 24$. *The components of vector* **v** *are plotted in Figure 11.1.*

We next compute the biorthogonal wavelet transform (Algorithm 11.1) and the modified biorthogonal wavelet transform (Algorithm 11.4) of **v** *using the* $(5, 3)$ *biorthogonal spline filter pair*

$$\mathbf{h} = (h_2, h_1, h_0, h_1, h_2) = \left(-\frac{\sqrt{2}}{8}, \frac{\sqrt{2}}{4}, \frac{3\sqrt{2}}{4}, \frac{\sqrt{2}}{4}, -\frac{\sqrt{2}}{8} \right)$$

and

$$\tilde{\mathbf{h}} = (\tilde{h}_1, \tilde{h}_0, \tilde{h}_1) = \left(\frac{\sqrt{2}}{4}, \frac{\sqrt{2}}{2}, \frac{\sqrt{2}}{4} \right)$$

Denote these transforms by **w** *and* $\mathbf{w}^m$, *respectively. In Figure 11.2 we plot the lowpass portions of* **w** *and* $\mathbf{w}^m$.

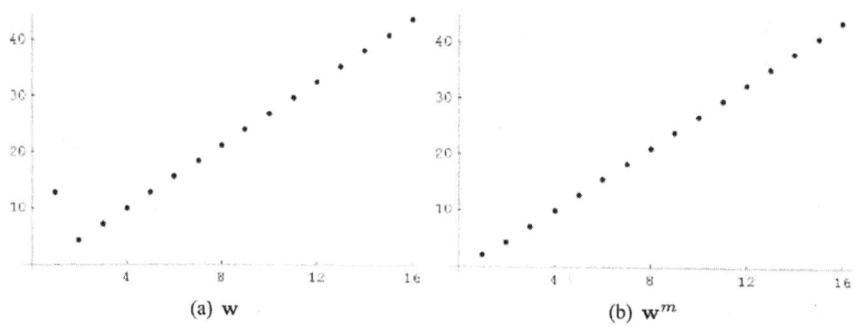

(a) **w**   (b) $\mathbf{w}^m$

Figure 11.2   Plots of the lowpass portions of **w** and $\mathbf{w}^m$.

The lowpass portion of $\mathbf{w}^m$ is a better approximation of $\mathbf{v}$ than is the lowpass portion of $\mathbf{w}$.   □

Why Does Algorithm 11.4 Work?

We have seen that Algorithm 11.4 works for vectors of length 8 and a biorthogonal filter pair whose lengths are 5 and 3. In Computer Lab 11.6 you will use a **CAS** to test the algorithm for other odd-length biorthogonal filter pairs and vectors of even length. When you work through the lab, you will see that the lowpass and highpass portions $\mathbf{y}^p$ and $\mathbf{z}^p$ always exhibit symmetry. For even N, $\mathbf{y}^p$ has the form

$$\mathbf{y}^p = [y_1^p, y_2^p, \ldots, y_{N/2}^p, y_{N/2}^p, y_{N/2-1}^p, \ldots, y_2^p]^T \tag{11.35}$$

and $\mathbf{z}^p$ has the form

$$\mathbf{z}^p = [z_1^p, z_2^p, \ldots, z_{N/2}^p, z_{N/2-1}^p, z_{N/2-2}^p, \ldots, z_1^p]^T \tag{11.36}$$

To prove these facts, we must return to the ideas of convolution and filters from Chapter 5.

Symmetry in the Lowpass Portion $\mathbf{y}^p$

Let's first analyze the lowpass portion of the transform returned by Algorithm 11.4. Recall the entire transform process originated from the idea of convolving two bi–infinite sequences $\tilde{\mathbf{h}}$ and $\mathbf{x}$. We also learned that we could write the convolution process as an infinite-dimensional matrix times $\mathbf{x}$. We first write our symmetric filter as a bi-infinite sequence

$$\tilde{\mathbf{h}} = (\ldots, 0, 0, \tilde{h}_{\tilde{L}}, \tilde{h}_{\tilde{L}-1}, \ldots, \tilde{h}_1, \tilde{h}_0, \tilde{h}_1, \ldots, \tilde{h}_{\tilde{L}}, 0, 0, \ldots)$$

If $\mathbf{x} = (\ldots, x_{-2}, x_{-1}, x_0, x_1, x_2, \ldots)$, we can write $\mathbf{y} = \mathbf{h} * \mathbf{x} = H\mathbf{x}$ using the matrix

$$\tilde{H} = \begin{bmatrix} \ddots & \ddots & \ddots & \ddots & \ddots & & & & & & & \\ \tilde{h}_{\tilde{L}} & \ldots & \tilde{h}_2 & \tilde{h}_1 & \mathbf{\tilde{h}_0} & \tilde{h}_1 & \tilde{h}_2 & \ldots & \tilde{h}_{\tilde{L}} & 0 & 0 & 0 & 0 \\ 0 & \tilde{h}_{\tilde{L}} & \ldots & \tilde{h}_2 & \tilde{h}_1 & \mathbf{\tilde{h}_0} & \tilde{h}_1 & \tilde{h}_2 & \ldots & \tilde{h}_{\tilde{L}} & 0 & 0 & 0 \\ \ldots & 0 & 0 & \tilde{h}_{\tilde{L}} & \ldots & \tilde{h}_2 & \tilde{h}_1 & \mathbf{\tilde{h}_0} & \tilde{h}_1 & \tilde{h}_2 & \ldots & \tilde{h}_{\tilde{L}} & 0 & 0 & \ldots \\ 0 & 0 & 0 & \tilde{h}_{\tilde{L}} & \ldots & \tilde{h}_2 & \tilde{h}_1 & \mathbf{\tilde{h}_0} & \tilde{h}_1 & \tilde{h}_2 & \ldots & \tilde{h}_{\tilde{L}} & 0 \\ 0 & 0 & 0 & 0 & \tilde{h}_{\tilde{L}} & \ldots & \tilde{h}_2 & \tilde{h}_1 & \mathbf{\tilde{h}_0} & \tilde{h}_1 & \tilde{h}_2 & \ldots & \tilde{h}_{\tilde{L}} \\ & & \ddots & \ddots & \ddots & \ddots & \ddots & & & & & \ddots \end{bmatrix}$$

where the boldface elements are on the main diagonal of H. Recall that the elements of $\mathbf{y}$ are given by the formula[2]

$$y_n = \sum_k \tilde{h}_k \, x_{n-k} = \sum_k \tilde{h}_{n-k} \, x_k \tag{11.37}$$

---

[2]By Problem 5.7 in Section 5.1, we know that the convolution product is commutative.

To form the lowpass portion of a wavelet transform, we first *downsample* $H\mathbf{x}$ by discarding every other row and then truncate the output. The truncation process typically involves the introduction of wrapping rows — let's hold off on truncation for the moment. We can represent the downsampling by slightly altering the convolution formula (11.37). We form the bi-infinite sequence a using the rule

$$a_n = y_{2n} = \sum_k \tilde{h}_{2n-k}\, x_k \tag{11.38}$$

Suppose that $\mathbf{v} \in \mathbb{R}^N$ where N is even. Let's load $\mathbf{x}$ with repeated copies of $\mathbf{v}^p$. Using (11.34), we know that $\mathbf{v}^p$ is of length $2N - 2$. We store $\mathbf{v}^p$ as $x_0, \dots, x_{2N-3}$. That is, we take

$$\begin{bmatrix} x_0, & x_1, & \cdots & x_{N-2}, & x_{N-1}, & x_N & \cdots & x_{2N-4}, & x_{2N-3} \end{bmatrix}^T = \begin{bmatrix} v_1, & v_2, & \cdots & v_{N-1}, & v_N, & v_{N-1}, & \cdots & v_3, & v_2 \end{bmatrix}^T$$
$$= \mathbf{v}^p \tag{11.39}$$

Now periodically extend (11.39) to fill $\mathbf{x}$. So

$$\mathbf{x} = (\dots, \mathbf{v}^p, \mathbf{v}^p, \mathbf{v}^p, \dots) \tag{11.40}$$

By construction, $\mathbf{x}$ is a $(2N - 2)$-periodic sequence so that

$$x_k = x_{2N-2+k} \tag{11.41}$$

for all $k \in \mathbb{Z}$. In Problem 11.9, you will show that $\mathbf{x}$ is symmetric about zero.

Now we consider a where a_n is given by (11.38). The following proposition tells us that a is periodic and symmetric.

Proposition 11.1 (Sequence a Is Periodic and Symmetric). *Assume that* $\mathbf{v} \in \mathbb{R}^N$ *where N is even and that* $\mathbf{v}^p$ *is given by (11.34). Further suppose that*

$$\tilde{\mathbf{h}} = (\dots, 0, 0, \tilde{h}_{\tilde{L}}, \tilde{h}_{\tilde{L}-1}, \dots, \tilde{h}_1, \tilde{h}_0, \tilde{h}_1, \dots, \tilde{h}_{\tilde{L}-1}, \tilde{h}_{\tilde{L}}, 0, 0, \dots)$$

and $\mathbf{x}$ is given by (11.40). If a is formed componentwise by the rule (11.38), then a *is $(N - 1)$-periodic and $a_n = a_{-n}$ for all $n \in \mathbb{Z}$.*   □

Proof of Proposition 11.1. Let's first show that a is an $(N - 1)$-periodic sequence. That is, we must show that $a_n = a_{n+N-1}$ for all $n \in \mathbb{Z}$. Using (11.38), we have

$$a_{n+N-1} = \sum_k \tilde{h}_{2(n+N-1)-k}\, x_k = \sum_k \tilde{h}_{2n-(k+2-2N)}\, x_k$$

Now make the substitution $m = k + 2 - 2N$. Then $k = m + 2N - 2$, and since k runs over all integers, so does m. We have

$$a_{n+N-1} = \sum_m \tilde{h}_{2n-m}\, x_{m+2N-2}$$

But x is $(2N - 2)$-periodic, so by (11.41) we can rewrite the previous identity as

$$a_{n+N-1} = \sum_m \tilde{h}_{2n-m} \, x_m$$

But this is exactly a_n so the first part of the proof is complete.

To show that $a_n = a_{-n}$, we use (11.38) to write

$$a_{-n} = \sum_k \tilde{h}_{2(-n)-k} \, x_k = \sum_k \tilde{h}_{-(2n+k)} \, x_k = \sum_k \tilde{h}_{2n+k} \, x_k$$

since we are given that $\tilde{h}$ satisfies $\tilde{h}_m = \tilde{h}_{-m}$. Now by Problem 11.9, we know that $x_k = x_{-k}$, so we can write

$$a_{-n} = \sum_k \tilde{h}_{2n+k} \, x_{-k}$$

Now make the substitution $m = -k$ in the previous identity. Then $k = -m$ and since k runs over all integers, so will m. We have

$$a_{-n} = \sum_m \tilde{h}_{2n-m} x_m = a_n$$

and the proof is complete. □

The following corollary is almost immediate.

Corollary 11.1 (Characterizing the Elements of Sequence a). *Each period of* a *has the form*

$$a_0, a_1, a_2, \ldots, a_{N/2-1}, a_{N/2-1}, \ldots, a_2, a_1$$

□

Proof of Corollary 11.1. We know from Proposition 11.1 that a is $(N - 1)$-periodic, so the elements of each period of a are

$$a_0, a_1, \ldots, a_{N/2-1}, a_{N/2}, \ldots, a_{N-2} \tag{11.42}$$

We must show that for $k = 1, \ldots, \frac{N}{2} - 1$, we have $a_k = a_{N-1-k}$.

Let $k = 1, \ldots, \frac{N}{2} - 1$. Since a is $(N - 1)$-periodic, we know that

$$a_{-k} = a_{N-1-k}$$

But from Proposition 11.1, we know that the elements of a satisfy $a_k = a_{-k}$. So

$$a_k = a_{-k} = a_{N-1-k}$$

and the proof is complete. □

Let's look at an example.

Example 11.3 (Using Proposition 11.1 and Corollary 11.1). *Let's analyze* a *in the case where* $N = 8$. *Then by Proposition 11.1,* a *is 7-periodic sequence, and from Corollary 11.1 we have*

$$[a_0, a_1, a_2, a_3, a_4, a_5, a_6]^T = [a_0, a_1, a_2, a_3, a_3, a_2, a_1]^T$$

This is exactly the structure (modulo shifting the index by 1) of $\mathbf{y}^p$ *in (11.30)! It is left for you in Problem 11.10 to show that* $y_k^p = a_{k-1}$ *for* $k = 1, \ldots, 4$ *when we use the three-term symmetric filter* $\tilde{\mathbf{h}}$ *from Example 11.1.* □

Characterizing the $(N - 1)$-Periodic Values of a

Our next proposition confirms that the elements of the lowpass vector $\mathbf{y}^p$ from Algorithm 11.4 are given by (11.42).

Proposition 11.2 (Connecting $\mathbf{y}^p$ and the Elements of a). *Assume that* $\mathbf{v} \in \mathbb{R}^N$ *where N is even and that* $\mathbf{v}^p$ *is given by (11.34). Further suppose that*

$$\tilde{\mathbf{h}} = (\ldots, 0, 0, \tilde{h}_{\tilde{L}}, \tilde{h}_{\tilde{L}-1}, \ldots, \tilde{h}_1, \tilde{h}_0, \tilde{h}_1, \ldots, \tilde{h}_{\tilde{L}-1}, \tilde{h}_{\tilde{L}}, 0, 0, \ldots)$$

satisfies $\tilde{h}_m = \tilde{h}_{-m}$ *for all* $m \in \mathbb{Z}$ *and* $\mathbf{x}$ *is given by (11.40). If* a *is formed componentwise by the rule (11.38) and* $\mathbf{y}^p = \tilde{H}_{N-1}\mathbf{v}^p$ *is the lowpass portion of the wavelet transform built using the filter* $(\tilde{h}_{\tilde{L}}, \tilde{h}_{L-1}, \ldots, \tilde{h}_1, \tilde{h}_0, \tilde{h}_1, \ldots, \tilde{h}_{L-1}, \tilde{h}_{\tilde{L}})$, *then* $y_k^p = a_{k-1}$ *for* $k = 1, \ldots, \frac{N}{2}$. □

Proof of Proposition 11.2. This proof is technical and is left as Problem 11.11. □

Symmetry in the Highpass Portion $\mathbf{z}^p$

The procedure for verifying the symmetry (11.36) exhibited by $\mathbf{z}^p$ from Algorithm 11.4 is quite similar to the steps to show the symmetry (11.35) for $\mathbf{y}^p$. We use $\mathbf{x}$ defined by (11.39) but now use the bi-infinite sequence

$$\tilde{\mathbf{g}} = (\ldots, 0, 0, \tilde{g}_{-L+1}, \tilde{g}_{-L+2}, \ldots, \tilde{g}_{-1}, \tilde{g}_0, \ldots, \tilde{g}_L, \tilde{g}_{L+1}, 0, 0, \ldots)$$

in our convolution product

$$z_n = \sum_k g_k x_{n-k} = \sum_k g_{n-k} x_k$$

Recall that the filter $\tilde{\mathbf{g}}$ is built by the component-wise rule $\tilde{g}_k = (-1)^k h_{1-k}$, where $\mathbf{h}$ is a symmetric filter of odd length.

We define the bi-infinite sequence b of downsampled terms by

$$b_n = z_{2n} = \sum_k g_{2n-k} x_k \qquad (11.43)$$

The following proposition is analogous to Proposition 11.1.

Proposition 11.3 (Sequence b Is Periodic and Symmetric). *Assume that* $\mathbf{v} \in \mathbb{R}^N$ *where N is even and that* $\mathbf{v}^p$ *is given by (11.34). Further suppose that*

$$\tilde{\mathbf{g}} = (\ldots, 0, 0, \tilde{g}_{-L+1}, \tilde{g}_{-L+2}, \ldots, \tilde{g}_{-1}, \tilde{g}_0, \tilde{g}_1, \ldots, \tilde{g}_L, \tilde{g}_{L+1}, 0, 0, \ldots)$$

is constructed from $\mathbf{h} = (\ldots, 0, 0, h_L, h_{L-1}, \ldots, h_1, h_0, h_1, \ldots, h_{L-1}, h_L, 0, 0, \ldots)$ *using the rule* $g_k = (-1)^k h_{1-k}$. *If* $\mathbf{x}$ *is given by (11.40) and* $\mathbf{b}$ *is formed component-wise by the rule (11.43), then* $\mathbf{b}$ *is* $(N-1)$-*periodic and* $b_n = b_{1-n}$ *for all* $n \in \mathbb{Z}$. □

Proof of Proposition 11.3. The proof that b is $(N-1)$-periodic is quite similar to the proof that a is $(N-1)$-periodic given for Proposition 11.1, so we leave it as Problem 11.12.

To see that $b_n = b_{1-n}$, we use (11.43) and the fact that $x_k = x_{-k}$ to write

$$b_{1-n} = \sum_k g_{2(1-n)-k} x_k = \sum_k g_{2-2n-k} x_k = \sum_k g_{2-2n-k} x_{-k}$$

Now we make the substitution $m = -k$. Since k runs through all the integers, so does m and we have

$$b_{1-n} = \sum_m g_{2-2n+m} x_m = \sum_m g_{2-(2n-m)} x_m$$

In Problem 10.12 of Section 10.1, we showed that $\tilde{g}_k = \tilde{g}_{2-k}$. Thus we can write the last identity as

$$b_{1-n} = \sum_m g_{2n-m} x_m = b_n$$

and the proof is complete. □

The following corollary characterizes the structure of each period of b.

Corollary 11.2 (Characterizing the Elements of Sequence b). *Each period of* b *has the form*

$$b_1, b_2, \ldots, b_{N/2-1}, b_{N/2}, b_{N/2-1}, \ldots, b_2, b_1$$

□

Proof of Corollary 11.2. The proof is left as Problem 11.13. □

Note that the symmetry is a bit different for **b** than it was for **a**. Let's again consider Proposition 11.3 and Corollary 11.2 in the $N = 8$ case.

Example 11.4 (Using Proposition 11.3 and Corollary 11.2). *Let's analyze* **b** *in the case where* $N = 8$. *Then by Proposition 11.3,* **b** *is a 7-periodic sequence and from Corollary 11.2 we have*

$$[b_1, b_2, b_3, b_4, b_5, b_6, b_7]^T = [b_1, b_2, b_3, b_4, b_3, b_2, b_1]^T$$

The structure of the period is identical to that of $\mathbf{z}^p$ *in (11.32). Moreover, if we use the five-term symmetric filter from Example 11.1, we see that* $b_k = z_k^p$, $k = 1, \ldots, 4$ *(see Problem 11.10).*   □

The next proposition is similar to Proposition 11.4. Again, the proof is quite technical and is left as a challenging problem.

Proposition 11.4 (Connecting $\mathbf{z}^p$ **and the Elements of b).** *Assume that* $\mathbf{v} \in \mathbb{R}^N$ *where* N *is even and that* $\mathbf{v}^p$ *is given by (11.34), and suppose that*

$$\tilde{\mathbf{g}} = (\ldots, \tilde{g}_{-L+1}, \tilde{g}_{-L+2}, \ldots, \tilde{g}_{-1}, \tilde{g}_0, \tilde{g}_1, \ldots, \tilde{g}_L, \tilde{g}_{L+1}, \ldots)$$

is constructed from $\mathbf{h} = (\ldots, 0, 0, h_L, h_{L-1}, \ldots, h_1, h_0, h_1, \ldots, h_{L-1}, h_L, 0, 0, \ldots)$ *using the rule* $g_k = (-1)^k h_{1-k}$. *If* $\mathbf{x}$ *is given by (11.40),* **b** *is formed component-wise by the rule (11.43), and* $\mathbf{z}^p = \tilde{G}_{N-1}\mathbf{v}^p$ *is the highpass portion of the wavelet transform built using the filter* $(\tilde{g}_{-L+1}, \ldots, \tilde{g}_0, \ldots, \tilde{g}_L, \tilde{g}_{L+1})$, *then* $z_k^p = b_k$ *for* $k = 1, \ldots, \frac{N}{2}$.   □

Proof of Proposition 11.4. The proof is left as Problem 11.14.   □

Inverting the Modified Transform for Odd-Length Filter Pairs

Now that we have verified Algorithm 11.4 for odd-length biorthogonal filter pairs, we need to consider the inverse process. There are a couple of ways we could proceed. The first is the most computationally efficient way to proceed but is difficult to implement. The second option is simple to implement but slows both the modified biorthogonal transform and its inverse by roughly a factor of 2.

Finding the Matrix for the Inverse Modified Transform

Let's first consider the computationally efficient method. The idea here is that the modified biorthogonal wavelet transform maps the input vector $\mathbf{v} \in \mathbb{R}^N$ to an N-vector $\mathbf{w}$, so we should be able to represent this transformation with an $N \times N$ matrix $\tilde{W}_N^m$. Here the superscript m denotes a modified transform. Once we have identified this matrix, then we compute its inverse and write an algorithm to implement its use. To better understand this method, let's again look at the modified transform of Example 11.1.

Example 11.5 (Inverting the Modified Biorthogonal Transform: Filter Lengths 3 and 5). *Consider the modified biorthogonal wavelet transformation from Example 11.1. Recall that the transform worked on vectors of length 8 and used a biorthogonal filter pair* $\mathbf{h}$ *and* $\tilde{\mathbf{h}}$ *whose lengths are 5 and 3, respectively.*

Using (11.29) and (11.31), we can write down the mapping:

$$
\mathbf{v} = \begin{bmatrix} v_1 \\ v_2 \\ v_3 \\ v_4 \\ v_5 \\ v_6 \\ v_7 \\ v_8 \end{bmatrix} \rightarrow \begin{bmatrix} \tilde{h}_0 v_1 + 2\tilde{h}_1 v_2 \\ \tilde{h}_1 v_2 + \tilde{h}_0 v_3 + \tilde{h}_1 v_4 \\ \tilde{h}_1 v_4 + \tilde{h}_0 v_5 + \tilde{h}_1 v_6 \\ \tilde{h}_1 v_6 + \tilde{h}_0 v_7 + \tilde{h}_1 v_8 \\ \hline h_1 v_1 - (h_0 + h_2) v_2 + h_1 v_3 - h_2 v_4 \\ -h_2 v_2 + h_1 v_3 - h_0 v_4 + h_1 v_5 - h_2 v_6 \\ -h_2 v_4 + h_1 v_5 - h_0 v_6 + h_1 v_7 - h_2 v_8 \\ -2h_2 v_6 + 2h_1 v_7 - h_0 v_8 \end{bmatrix}
$$

It is straightforward to identify the transform matrix in this case. We have

$$
\left[\frac{\mathbf{y}^m}{\mathbf{z}^m} \right] = \tilde{W}_8^m \mathbf{v}
$$

where

$$
\tilde{W}_8^m = \begin{bmatrix}
\tilde{h}_0 & 2\tilde{h}_1 & 0 & 0 & 0 & 0 & 0 & 0 \\
0 & \tilde{h}_1 & \tilde{h}_0 & \tilde{h}_1 & 0 & 0 & 0 & 0 \\
0 & 0 & 0 & \tilde{h}_1 & \tilde{h}_0 & \tilde{h}_1 & 0 & 0 \\
0 & 0 & 0 & 0 & 0 & \tilde{h}_1 & \tilde{h}_0 & \tilde{h}_1 \\
\hline
h_1 & -h_0-h_2 & h_1 & -h_2 & 0 & 0 & 0 & 0 \\
0 & -h_2 & h_1 & -h_0 & h_1 & -h_2 & 0 & 0 \\
0 & 0 & 0 & -h_2 & h_1 & -h_0 & h_1 & -h_2 \\
0 & 0 & 0 & 0 & 0 & -2h_2 & 2h_1 & -h_0
\end{bmatrix} \tag{11.44}
$$

and $\mathbf{y}^m$ *and* $\mathbf{z}^m$ *are nothing more than the first four elements in (11.30) and (11.32), respectively.*

This matrix is similar to $\tilde{W}_8$, *but rows 1, 5, and 8 are different. These changes will undoubtedly alter the inverse of* W_8^m — *we cannot expect the inverse to be the transpose of a wavelet matrix. There are other unanswered questions. How does the inverse change when N changes? We will need* W_8^m *for a two-dimensional transform* — *what does its inverse look like? What happens if we use other odd-length filters?*

As you can see, this method has several disadvantages. It is possible, however, to analyze the inverse for particular filters. For example, if we use the (5,3) *biorthogonal spline filter pair* $\mathbf{h} = (h_2, h_1, h_0, h_1, h_2) = \left(-\frac{\sqrt{2}}{8}, \frac{\sqrt{2}}{4}, \frac{3\sqrt{2}}{4}, \frac{\sqrt{2}}{4}, -\frac{\sqrt{2}}{8} \right)$ *and* $\tilde{\mathbf{h}} =$

$(\tilde{h}_1, \tilde{h}_0, \tilde{h}_1) = \left(\frac{\sqrt{2}}{4}, \frac{\sqrt{2}}{2}, \frac{\sqrt{2}}{4}\right)$ in (11.44), we have

$$\tilde{W}_8^m = \left[\begin{array}{cccccccc} \frac{\sqrt{2}}{2} & \frac{\sqrt{2}}{2} & 0 & 0 & 0 & 0 & 0 & 0 \\ 0 & \frac{\sqrt{2}}{4} & \frac{\sqrt{2}}{2} & \frac{\sqrt{2}}{4} & 0 & 0 & 0 & 0 \\ 0 & 0 & 0 & \frac{\sqrt{2}}{4} & \frac{\sqrt{2}}{2} & \frac{\sqrt{2}}{4} & 0 & 0 \\ 0 & 0 & 0 & 0 & 0 & \frac{\sqrt{2}}{4} & \frac{\sqrt{2}}{2} & \frac{\sqrt{2}}{4} \\ \hline \frac{\sqrt{2}}{4} & -\frac{5\sqrt{2}}{8} & \frac{\sqrt{2}}{4} & \frac{\sqrt{2}}{8} & 0 & 0 & 0 & 0 \\ 0 & \frac{\sqrt{2}}{8} & \frac{\sqrt{2}}{4} & -\frac{3\sqrt{2}}{4} & \frac{\sqrt{2}}{4} & \frac{\sqrt{2}}{8} & 0 & 0 \\ 0 & 0 & 0 & \frac{\sqrt{2}}{8} & \frac{\sqrt{2}}{4} & -\frac{3\sqrt{2}}{4} & \frac{\sqrt{2}}{4} & \frac{\sqrt{2}}{8} \\ 0 & 0 & 0 & 0 & 0 & \frac{\sqrt{2}}{4} & \frac{\sqrt{2}}{2} & \frac{3\sqrt{2}}{4} \end{array}\right] \qquad (11.45)$$

and we can use a **CAS** to compute the inverse

$$(\tilde{W}_8^m)^{-1} = \left[\begin{array}{cccc|cccc} \frac{3\sqrt{2}}{4} & -\frac{\sqrt{2}}{4} & 0 & 0 & \frac{\sqrt{2}}{2} & 0 & 0 & 0 \\ \frac{\sqrt{2}}{4} & \frac{\sqrt{2}}{4} & 0 & 0 & -\frac{\sqrt{2}}{2} & 0 & 0 & 0 \\ -\frac{\sqrt{2}}{8} & \frac{3\sqrt{2}}{4} & -\frac{\sqrt{2}}{8} & 0 & \frac{\sqrt{2}}{4} & \frac{\sqrt{2}}{4} & 0 & 0 \\ 0 & \frac{\sqrt{2}}{4} & \frac{\sqrt{2}}{4} & 0 & 0 & -\frac{\sqrt{2}}{2} & 0 & 0 \\ 0 & -\frac{\sqrt{2}}{8} & \frac{3\sqrt{2}}{4} & -\frac{\sqrt{2}}{8} & 0 & \frac{\sqrt{2}}{4} & \frac{\sqrt{2}}{4} & 0 \\ 0 & 0 & \frac{\sqrt{2}}{4} & \frac{\sqrt{2}}{4} & 0 & 0 & -\frac{\sqrt{2}}{2} & 0 \\ 0 & 0 & -\frac{\sqrt{2}}{8} & \frac{11\sqrt{2}}{8} & 0 & 0 & \frac{\sqrt{2}}{4} & -\frac{\sqrt{2}}{2} \\ 0 & 0 & 0 & -\sqrt{2} & 0 & 0 & 0 & -\sqrt{2} \end{array}\right]$$

There are some similarities here between $(\tilde{W}_8^m)^{-1}$ and $\tilde{W}_8^{-1} = W_8^T$. We have

$$W_8^T = \left[\begin{array}{cccc|cccc} \frac{3\sqrt{2}}{4} & -\frac{\sqrt{2}}{8} & 0 & -\frac{\sqrt{2}}{8} & \frac{\sqrt{2}}{4} & 0 & 0 & \frac{\sqrt{2}}{4} \\ \frac{\sqrt{2}}{4} & \frac{\sqrt{2}}{4} & 0 & 0 & -\frac{\sqrt{2}}{2} & 0 & 0 & 0 \\ -\frac{\sqrt{2}}{8} & \frac{3\sqrt{2}}{4} & -\frac{\sqrt{2}}{8} & 0 & \frac{\sqrt{2}}{4} & \frac{\sqrt{2}}{4} & 0 & 0 \\ 0 & \frac{\sqrt{2}}{4} & \frac{\sqrt{2}}{4} & 0 & 0 & -\frac{\sqrt{2}}{2} & 0 & 0 \\ 0 & -\frac{\sqrt{2}}{8} & \frac{3\sqrt{2}}{4} & -\frac{\sqrt{2}}{8} & 0 & \frac{\sqrt{2}}{4} & \frac{\sqrt{2}}{4} & 0 \\ 0 & 0 & \frac{\sqrt{2}}{4} & \frac{\sqrt{2}}{4} & 0 & 0 & -\frac{\sqrt{2}}{2} & 0 \\ -\frac{\sqrt{2}}{8} & 0 & -\frac{\sqrt{2}}{8} & \frac{3\sqrt{2}}{4} & 0 & 0 & \frac{\sqrt{2}}{4} & \frac{\sqrt{2}}{4} \\ \frac{\sqrt{2}}{4} & 0 & 0 & \frac{\sqrt{2}}{4} & 0 & 0 & 0 & -\frac{\sqrt{2}}{2} \end{array}\right]$$

and we can see that columns 1, 4, and 8 are the only ones that differ between the two matrices.

We can also write down W_8^m and check $(W_8^m)^T$ against $(\tilde{W}_8^m)^{-1}$ (see Problem 11.15). You will see that these two matrices are also similar and differ only in columns 4 and 8.

In Problems 11.16 and 11.17 you will further investigate the structure of $\tilde{W}_N^m$ and W_N^m, and their inverses for general even N. In Computer Lab 11.7 you will *write a set of modules that utilize the results of this problem to create a fast modified biorthogonal wavelet transform and inverse for the* $(5, 3)$ *biorthogonal spline filter pair.*   □

The point of Example 11.5 is to see that it is difficult to analyze the inverse matrix for our modified biorthogonal wavelet transform.

A Simple Algorithm for Inverting the Modified Transform

We use an alternative method to invert the transform.

Recall that Algorithm 11.4 takes an N-vector $\mathbf{v}$, periodizes it using (11.34) to obtain $\mathbf{v}^p$, and then applies Algorithm 11.1 to obtain

$$\begin{bmatrix} \mathbf{y}^p \\ \mathbf{z}^p \end{bmatrix}$$

where $\mathbf{y}^p$ and $\mathbf{z}^p$ are the lowpass and highpass portions of the transform, respectively.

The algorithm then returns

$$\begin{bmatrix} \mathbf{y}^m \\ \mathbf{z}^m \end{bmatrix} = \begin{bmatrix} y_1^p & \cdots & y_{N/2}^p \,\big|\, z_1^p & \cdots & z_{N/2}^p \end{bmatrix}^T \tag{11.46}$$

But from Propositions 11.2 and 11.4, we know that $\mathbf{y}^p$ and $\mathbf{z}^p$ are always of the form

$$\begin{aligned} \mathbf{y}^p &= \begin{bmatrix} y_1^p, & y_2^p, & \cdots, & y_{N/2}^p \,\big|\, y_{N/2}^p, & y_{N/2-1}^p, & \cdots, & y_3^p, & y_2^p \end{bmatrix}^T \\ \mathbf{z}^p &= \begin{bmatrix} z_1^p, & z_2^p, & \cdots, & z_{N/2}^p \,\big|\, z_{N/2-1}^p, & z_{N/2-2}^p, & \cdots, & z_2^p, & z_1^p \end{bmatrix}^T \end{aligned} \tag{11.47}$$

So our process for inverting is straightforward:

Algorithm 11.5 (Inverting the Modified Biorthogonal Transform for Odd-Length Filters). *Given the N-vector $\mathbf{w} = (\mathbf{y}^m \,|\, \mathbf{z}^m)^T$ where $\mathbf{y}^m$ and $\mathbf{z}^m \in \mathbb{R}^{\frac{N}{2}}$ are given by (11.46), and a biorthogonal filter pair whose lengths are odd, perform the following steps:*

1. *Form the* $(2N - 2)$*-vector*

$$\mathbf{w}^p = \begin{bmatrix} y_1^p, & \cdots, & y_{N/2}^p, & y_{N/2}^p, & y_{N/2-1}^p, & \cdots, & y_2^p \,\big|\, z_1^p, & \cdots, & z_{N/2}^p, & z_{N/2-1}^p, & z_{N/2-2}^p, & \cdots, & z_1^p \end{bmatrix}^T$$

2. *Apply Algorithm 11.3 to $\mathbf{w}^p$ to obtain $\mathbf{v}^p$ given by (11.34).*

3. *Return $[v_1, \ldots, v_N]^T$.*

☐

Of course, the problem with Algorithms 11.4 and 11.5 is that we must process vectors of length $2N - 2$ in order to obtain vectors of length N. These algorithms are bound to be slower, but considering the flexibility and ease of programming, they will work fine for our purposes.

The Modified Biorthogonal Wavelet Transform for Even-Length Filter Pairs

We now turn our attention to developing an algorithm for computing the modified biorthogonal transform for even-length filter pairs. The ideas are very similar to those discussed in detail for the odd-length filter pairs. As a result, we motivate the process with an example, state the algorithm, and leave the technical results that verify the process as problems. Let's consider an example.

Example 11.6. *Suppose that* $\mathbf{v}$ *is a vector of length* $N = 12$ *and suppose that we wish to compute the biorthogonal wavelet transform of* $\mathbf{v}$ *using the even-length, symmetric filter pair*[3]

$$\tilde{\mathbf{h}} = (\tilde{h}_{-1}, \tilde{h}_0, \tilde{h}_1, \tilde{h}_2) = (\tilde{h}_2, \tilde{h}_1, \tilde{h}_1, \tilde{h}_2)$$

and

$$\mathbf{h} = (h_{-3}, h_{-2}, h_{-1}, h_0, h_1, h_2, h_3, h_4) = (h_4, h_3, h_2, h_1, h_1, h_2, h_3, h_4)$$

The lowpass portion of the transform is

$$\mathbf{y} = \tilde{H}_6 \mathbf{v} = \begin{bmatrix} \tilde{h}_1 & \tilde{h}_1 & \tilde{h}_2 & 0 & 0 & 0 & 0 & 0 & 0 & 0 & 0 & \tilde{h}_2 \\ 0 & \tilde{h}_2 & \tilde{h}_1 & \tilde{h}_1 & \tilde{h}_2 & 0 & 0 & 0 & 0 & 0 & 0 & 0 \\ 0 & 0 & 0 & \tilde{h}_2 & \tilde{h}_1 & \tilde{h}_1 & \tilde{h}_2 & 0 & 0 & 0 & 0 & 0 \\ 0 & 0 & 0 & 0 & 0 & \tilde{h}_2 & \tilde{h}_1 & \tilde{h}_1 & \tilde{h}_2 & 0 & 0 & 0 \\ 0 & 0 & 0 & 0 & 0 & 0 & 0 & \tilde{h}_2 & \tilde{h}_1 & \tilde{h}_1 & \tilde{h}_2 & 0 \\ \tilde{h}_2 & 0 & 0 & 0 & 0 & 0 & 0 & 0 & 0 & \tilde{h}_2 & \tilde{h}_1 & \tilde{h}_1 \end{bmatrix} \cdot \begin{bmatrix} v_1 \\ v_2 \\ v_3 \\ v_4 \\ v_5 \\ v_6 \\ v_7 \\ v_8 \\ v_9 \\ v_{10} \\ v_{11} \\ v_{12} \end{bmatrix}$$

$$= \begin{bmatrix} \tilde{h}_2 v_{12} + \tilde{h}_1 v_1 + \tilde{h}_1 v_2 + \tilde{h}_2 v_3 \\ \tilde{h}_2 v_2 + \tilde{h}_1 v_3 + \tilde{h}_1 v_4 + \tilde{h}_2 v_5 \\ \tilde{h}_2 v_4 + \tilde{h}_1 v_5 + \tilde{h}_1 v_6 + \tilde{h}_2 v_7 \\ \tilde{h}_2 v_6 + \tilde{h}_1 v_7 + \tilde{h}_1 v_8 + \tilde{h}_2 v_9 \\ \tilde{h}_2 v_8 + \tilde{h}_1 v_9 + \tilde{h}_1 v_{10} + \tilde{h}_2 v_{11} \\ \tilde{h}_2 v_{10} + \tilde{h}_1 v_{11} + \tilde{h}_1 v_{12} + \tilde{h}_2 v_1 \end{bmatrix}$$

The highpass filter is obtained using the rule $\tilde{g}_k = (-1)^k h_k$, $k = -3, \ldots, 4$. *Using the symmetry of* $\mathbf{h}$, *we have*

$$\tilde{\mathbf{g}} = (\tilde{g}_{-3}, \tilde{g}_{-2}, \tilde{g}_{-1}, \tilde{g}_0, \tilde{g}_1, \tilde{g}_2, \tilde{g}_3, \tilde{g}_4)$$
$$= (-h_4, h_3, -h_2, h_1, -h_1, h_2, -h_3, h_4)$$

---

[3] For example, we could use the $(8, 4)$ biorthogonal spline filter pair developed in Section 10.2.

The highpass portion of the transform is

$$
\mathbf{z} = \tilde{G}_6 \mathbf{v} = \begin{bmatrix}
h_1 & -h_1 & h_2 & -h_3 & h_4 & 0 & 0 & 0 & 0 & -h_4 & h_3 & -h_2 \\
h_3 & -h_2 & h_1 & -h_1 & h_2 & -h_3 & h_4 & 0 & 0 & 0 & 0 & -h_4 \\
0 & -h_4 & h_3 & -h_2 & h_1 & -h_1 & h_2 & -h_3 & h_4 & 0 & 0 & 0 \\
0 & 0 & 0 & -h_4 & h_3 & -h_2 & h_1 & -h_1 & h_2 & -h_3 & h_4 & 0 \\
h_4 & 0 & 0 & 0 & 0 & -h_4 & h_3 & -h_2 & h_1 & -h_1 & h_2 & -h_3 \\
h_2 & -h_3 & h_4 & 0 & 0 & 0 & 0 & -h_4 & h_3 & -h_2 & h_1 & -h_1
\end{bmatrix} \cdot \begin{bmatrix} v_1 \\ v_2 \\ v_3 \\ v_4 \\ v_5 \\ v_6 \\ v_7 \\ v_8 \\ v_9 \\ v_{10} \\ v_{11} \\ v_{12} \end{bmatrix}
$$

$$
= \begin{bmatrix}
-h_4v_{10}+h_3v_{11}-h_2v_{12}+h_1v_1-h_1v_2+h_2v_3-h_3v_4+h_4v_5 \\
-h_4v_{12}+h_3v_1-h_2v_2+h_1v_3-h_1v_4+h_2v_5-h_3v_6+h_4v_7 \\
-h_4v_2+h_3v_3-h_2v_4+h_1v_5-h_1v_6+h_2v_7-h_3v_8+h_4v_9 \\
-h_4v_4+h_3v_5-h_2v_6+h_1v_7-h_1v_8+h_2v_9-h_3v_{10}+h_4v_{11} \\
-h_4v_6+h_3v_7-h_2v_8+h_1v_9-h_1v_{10}+h_2v_{11}-h_3v_{12}+h_4v_1 \\
-h_4v_8+h_3v_9-h_2v_{10}+h_1v_{11}-h_1v_{12}+h_2v_1-h_3v_2+h_4v_3
\end{bmatrix}
$$

Note that there is wrapping in both portions of the transform. For example, the first element in $\mathbf{y}$ *is a linear combination of* v_1, v_2, v_3 *and* v_{12}. *We proceed as we did in Example 11.1. We define the 24-vector*

$$
\mathbf{v}^p = \begin{bmatrix} v_1, & v_2, & \dots, & v_{11}, & v_{12} \end{bmatrix} v_{12}, & v_{11}, & \dots, & v_2, & v_1 \end{bmatrix}^T
$$

and compute $\mathbf{y}^p = \tilde{W}_{24}\mathbf{v}^p$. *In Problem 11.18 you will verify that the lowpass portion of the transform is*

$$
\mathbf{y}^p = \begin{bmatrix}
(\tilde{h}_1+\tilde{h}_2)v_1+\tilde{h}_1v_2+\tilde{h}_2v_3 \\
\tilde{h}_2v_2+\tilde{h}_1v_3+\tilde{h}_1v_4+\tilde{h}_2v_5 \\
\tilde{h}_2v_4+\tilde{h}_1v_5+\tilde{h}_1v_6+\tilde{h}_2v_7 \\
\tilde{h}_2v_6+\tilde{h}_1v_7+\tilde{h}_1v_8+\tilde{h}_2v_9 \\
\tilde{h}_2v_8+\tilde{h}_1v_9+\tilde{h}_1v_{10}+\tilde{h}_2v_{11} \\
\tilde{h}_2v_{10}+\tilde{h}_1v_{11}+(\tilde{h}_1+\tilde{h}_2)v_{12} \\
\hline
\tilde{h}_2v_{10}+\tilde{h}_1v_{11}+(\tilde{h}_1+\tilde{h}_2)v_{12} \\
\tilde{h}_2v_8+\tilde{h}_1v_9+\tilde{h}_1v_{10}+\tilde{h}_2v_{11} \\
\tilde{h}_2v_6+\tilde{h}_1v_7+\tilde{h}_1v_8+\tilde{h}_2v_9 \\
\tilde{h}_2v_4+\tilde{h}_1v_5+\tilde{h}_1v_6+\tilde{h}_2v_7 \\
\tilde{h}_2v_2+\tilde{h}_1v_3+\tilde{h}_1v_4+\tilde{h}_2v_5 \\
(\tilde{h}_1+\tilde{h}_2)v_1+\tilde{h}_1v_2+\tilde{h}_2v_3
\end{bmatrix} \tag{11.48}
$$

and the highpass portion of the transform is

$$
\mathbf{z}^p = \begin{bmatrix}
(h_1-h_2)v_1+(h_3-h_1)v_2-(h_4-h_2)v_3-h_3v_4+h_4v_5 \\
(h_3-h_4)v_1-h_2v_2+h_1v_3-h_1v_4+h_2v_5-h_3v_6+h_4v_7 \\
-h_4v_2+h_3v_3-h_2v_4+h_1v_5-h_1v_6+h_2v_7-h_3v_8+h_4v_9 \\
-h_4v_4+h_3v_5-h_2v_6+h_1v_7-h_1v_8+h_2v_9-h_3v_{10}+h_4v_{11} \\
-h_4v_6+h_3v_7-h_2v_8+h_1v_9-h_1v_{10}+h_2v_{11}-(h_3-h_4)v_{12} \\
-h_4v_8+h_3v_9-(h_2-h_4)v_{10}+(h_1-h_3)v_{11}-(h_1-h_2)v_{12} \\
\hline
h_4v_8-h_3v_9+(h_2-h_4)v_{10}-(h_1-h_3)v_{11}+(h_1-h_2)v_{12} \\
h_4v_6-h_3v_7+h_2v_8-h_1v_9+h_1v_{10}-h_2v_{11}+(h_3-h_4)v_{12} \\
h_4v_4-h_3v_5+h_2v_6-h_1v_7+h_1v_8-h_2v_9+h_3v_{10}-h_4v_{11} \\
h_4v_2-h_3v_3+h_2v_4-h_1v_5+h_1v_6-h_2v_7+h_3v_8-h_4v_9 \\
-(h_3-h_4)v_1+h_2v_2-h_1v_3+h_1v_4-h_2v_5+h_3v_6-h_4v_7 \\
-(h_1-h_2)v_1+(h_1-h_3)v_2+(h_4-h_2)v_3+h_3v_4-h_4v_5
\end{bmatrix} \tag{11.49}
$$

There is some symmetry in $\mathbf{y}^p$ and $\mathbf{z}^p$. In fact, the structure of these vectors is

$$\mathbf{y}^p = \left[\, y_1^p,\; y_2^p,\; y_3^p,\; y_4^p,\; y_5^p,\; y_6^p, \middle| y_6^p,\; y_5^p,\; y_4^p,\; y_3^p,\; y_2^p,\; y_1^p \,\right]^T$$

$$\mathbf{z}^p = \left[\, z_1^p,\; z_2^p,\; z_3^p,\; z_4^p,\; z_5^p,\; z_6^p, \middle| -z_6^p,\; -z_5^p,\; -z_4^p,\; -z_3^p,\; -z_2^p,\; -z_1^p \,\right]^T$$

We take as our modified the 12-vector

$$\left[\frac{\mathbf{y}^m}{\mathbf{z}^m}\right] = \left[\, y_1^p,\; \cdots,\; y_6^p \,\middle|\, z_1^p,\; \cdots,\; z_2^p \,\right]^T$$

□

Algorithms for the Modified Biorthogonal Wavelet Transform and its Inverse for Even-Length Filter Pairs

We are now ready to state the algorithms for computing the modified biorthogonal wavelet transform for even-length filter pairs and its inverse. The algorithms are similar to Algorithms 11.4 and 11.5. In Problems 11.19 to 11.21 you will verify the validity of the following algorithm.

Algorithm 11.6 (Modified Biorthogonal Transform for Even-Length Filter Pairs).
This algorithm takes as input the N-vector $\mathbf{v}$ (N even) and an even-length biorthogonal filter pair $\mathbf{h}$ and $\tilde{\mathbf{h}}$ and returns a modified version of the biorthogonal wavelet transform. The vector $\mathbf{v}$ is periodized to form vector $\mathbf{v}^p \in \mathbb{R}^{2N}$. We then use Algorithm 11.1 to compute the biorthogonal wavelet transform of $\mathbf{v}^p$. If $\mathbf{y}^p$ and $\mathbf{z}^p$ are the lowpass and highpass portions, respectively, of the biorthogonal wavelet transform, then the algorithm returns the modified biorthogonal wavelet transformation

$$\left[y_1, \ldots, y_{N/2} \,\middle|\, z_1, \ldots, z_{N/2}\right]^T$$

1. Given $\mathbf{v} \in \mathbb{R}^N$, N even, form $\mathbf{v}^p \in \mathbb{R}^{2N}$ by the rule

$$\mathbf{v}^p = [v_1, \ldots, v_N, v_N, \ldots, v_1]^T \tag{11.50}$$

2. Compute the biorthogonal wavelet transform $\mathbf{w} = \tilde{W}_{2N}\mathbf{v}^p$ using Algorithm 11.1. Denote the lowpass portion of the transform by $\mathbf{y}^p$ and the highpass portion of the transform by $\mathbf{z}^p$.

3. Return $\left[y_1^p, \ldots, y_{N/2}^p, z_1^p, \ldots, z_{N/2}^p\right]^T$.

□

The algorithm for the inverse modified biorthogonal transform is very similar to Algorithm 11.5.

Algorithm 11.7 (Modified Inverse Biorthogonal Transform for Even-Length Filter Pairs). *Given the N-vector* $\mathbf{w} = (\mathbf{y}^m \mid \mathbf{z}^m)^T$, *where* $\mathbf{y}^m$ *and* $\mathbf{z}^m \in \mathbb{R}^{\frac{N}{2}}$ *are the lowpass and highpass portions, respectively, of the modified biorthogonal wavelet transform, and a biorthogonal filter pair whose lengths are even, perform the following steps:*

1. *Form the $2N$-vector*

$$\mathbf{w}^p = \left[y_1^p, \ldots, y_{N/2}^p, y_{N/2}^p, \ldots, y_1^p \mid z_1^p, \ldots, z_{N/2}^p, -z_{N/2}^p, \ldots, -z_1^p \right]^T$$

2. *Apply Algorithm 11.3 to $\mathbf{w}^p$ to obtain*

$$\mathbf{v}^p = \left[v_1, \ldots, v_N, v_N, \ldots, v_1 \right]^T$$

3. *Return* $\left[v_1, \ldots, v_N \right]^T$.

$\square$

PROBLEMS

11.9 Let $\mathbf{v}^p$ be given by (11.34) and $\mathbf{x}$ be given by (11.40). Show that $\mathbf{x}$ is symmetric about zero. In other words, show that for all $m \in \mathbb{Z}$, we have $x_m = x_{-m}$.

11.10 Consider the biorthogonal filter pair from Example 11.1. Suppose that we form

$$\tilde{\mathbf{h}} = (\ldots, 0, 0, \tilde{h}_1, \tilde{h}_0, \tilde{h}_1, 0, 0, \ldots)$$

and

$$\mathbf{h} = (\ldots, 0, 0, h_2, h_1, h_0, h_1, h_2, 0, 0, \ldots)$$

For $N = 8$, use (11.28) to form $\mathbf{v}^p$, and from there, use (11.39) to form $\mathbf{x}$. From Examples 11.3 and 11.4 we know that each period of $\mathbf{a}$ and $\mathbf{b}$ (defined in (11.38) and (11.43)) have the structures $(a_0, a_1, a_2, a_3, a_3, a_2, a_1)$ and $(b_1, b_2, b_3, b_4, b_3, b_2, b_1)$, respectively. Show that $y_k^p = a_{k-1}$ and $z_k^p = b_k$, $k = 1, 2, 3, 4$ where the values y_k^p and z_k^p are given by (11.29) and (11.31), respectively.

11.11 Prove Proposition 11.2.

11.12 Complete the proof of Proposition 11.3 and show that $\mathbf{b}$ is $(N-1)$-periodic.

11.13 Prove Corollary 11.2. (*Hint:* Show $b_n = b_{N-n}$.)

11.14 Prove Proposition 11.4.

11.15   In Example 11.5 we built the matrix $\tilde{W}_8^m$ that computes the modified biorthogonal wavelet transform using the $(5,3)$ biorthogonal spline filter pair. Write down W_8^m for the $(5,3)$ biorthogonal spline filter pair. (*Hint:* It differs from W_8 only in rows 1, 4, and 8. To help you with the construction, use W_{10} given in Example 10.7.) Observe that $(W_8^m)^T$ differs from $(\tilde{W}_8^m)^{-1}$ in columns 4 and 8.

11.16   Suppose that N is even and let $\tilde{W}_N$ and W_N be matrices associated with the biorthogonal transform and its inverse using the $(5,3)$ biorthogonal spline filter pair. If $\tilde{\mathbf{w}}^k$ are the rows of $\tilde{W}_N$ and $\mathbf{w}^k$ are the rows of W_N, $k = 1, \ldots, N$, show that

$$
\tilde{W}_N^m =
\left[
\begin{array}{cccccccc}
\frac{\sqrt{2}}{2} & \frac{\sqrt{2}}{2} & 0 & 0 & \cdots & 0 & 0 & 0 & 0 \\
\multicolumn{9}{c}{\tilde{\mathbf{w}}^2} \\
\multicolumn{9}{c}{\vdots} \\
\multicolumn{9}{c}{\tilde{\mathbf{w}}^{N/2}} \\
\hline
\frac{\sqrt{2}}{4} & -\frac{5\sqrt{2}}{8} & \frac{\sqrt{2}}{4} & \frac{\sqrt{2}}{8} & 0 & \cdots & 0 & 0 & 0 \\
\multicolumn{9}{c}{\tilde{\mathbf{w}}^{N/2+2}} \\
\multicolumn{9}{c}{\vdots} \\
\multicolumn{9}{c}{\tilde{\mathbf{w}}^{N-1}} \\
0 & 0 & 0 & 0 & \cdots & 0 & \frac{\sqrt{2}}{4} & \frac{\sqrt{2}}{2} & \frac{3\sqrt{2}}{4}
\end{array}
\right]
$$

and

$$
W_N^m =
\left[
\begin{array}{cccccccc}
\frac{3\sqrt{2}}{4} & \frac{\sqrt{2}}{2} & -\frac{\sqrt{2}}{4} & 0 & \cdots & 0 & 0 & 0 & 0 \\
\multicolumn{9}{c}{\mathbf{w}^2} \\
\multicolumn{9}{c}{\vdots} \\
\multicolumn{9}{c}{\mathbf{w}^{N/2-1}} \\
0 & 0 & 0 & 0 & \cdots & -\frac{\sqrt{2}}{8} & \frac{\sqrt{2}}{4} & \frac{5\sqrt{2}}{8} & \frac{\sqrt{2}}{4} \\
\multicolumn{9}{c}{\mathbf{w}^{N/2+1}} \\
\hline
\multicolumn{9}{c}{\mathbf{w}^{N/2+2}} \\
\multicolumn{9}{c}{\vdots} \\
\multicolumn{9}{c}{\mathbf{w}^{N-1}} \\
0 & 0 & 0 & 0 & \cdots & 0 & 0 & \frac{\sqrt{2}}{2} & -\frac{\sqrt{2}}{2}
\end{array}
\right]
$$

(*Hint:* Look at the entries of $\tilde{W}_{2N-2}\mathbf{v}^p$ and $W_{2N-2}\mathbf{v}^p$ where $\mathbf{v}^p$ is given by (11.34).)

11.17   Under the assumptions of Problem 11.16, let

$$
U = \begin{bmatrix}
\frac{3\sqrt{2}}{4} & \frac{\sqrt{2}}{4} & -\frac{\sqrt{2}}{8} & 0 & 0 & 0 & \cdots & 0 & 0 \\
-\frac{\sqrt{2}}{4} & \frac{\sqrt{2}}{4} & \frac{3\sqrt{2}}{4} & \frac{\sqrt{2}}{4} & -\frac{\sqrt{2}}{8} & 0 & \cdots & 0 & 0 \\
 & & & & \mathbf{w}^3 & & & & \\
 & & & & \vdots & & & & \\
 & & & & \mathbf{w}^{N/2-1} & & & & \\
0 & 0 & 0 & \cdots & 0 & -\frac{\sqrt{2}}{8} & \frac{\sqrt{2}}{4} & \frac{11\sqrt{2}}{8} & -\sqrt{2} \\
 & & & & \mathbf{w}^{N/2+1} & & & & \\
 & & & & \vdots & & & & \\
 & & & & \mathbf{w}^{N-1} & & & & \\
0 & 0 & 0 & \cdots & 0 & 0 & 0 & -\frac{\sqrt{2}}{2} & \sqrt{2}
\end{bmatrix}
$$

and

$$
V = \begin{bmatrix}
\frac{\sqrt{2}}{2} & \frac{\sqrt{2}}{4} & 0 & 0 & \cdots & 0 & 0 & 0 \\
 & & \tilde{\mathbf{w}}^2 & & & & & \\
 & & \vdots & & & & & \\
 & & \tilde{\mathbf{w}}^{N/2} & & & & & \\
\frac{\sqrt{2}}{2} & -\frac{5\sqrt{2}}{8} & \frac{\sqrt{2}}{4} & \frac{\sqrt{2}}{8} & 0 & \cdots & 0 & 0 \\
 & & \tilde{\mathbf{w}}^{N/2+2} & & & & & \\
 & & \vdots & & & & & \\
 & & \tilde{\mathbf{w}}^{N-1} & & & & & \\
0 & 0 & 0 & \cdots & 0 & \frac{\sqrt{2}}{8} & \frac{\sqrt{2}}{4} & -\frac{3\sqrt{2}}{4}
\end{bmatrix}
$$

Show that U^T is the inverse of $\tilde{W}_N^m$ and V is the inverse of W_N^m. (*Hint:* You know that the inner product

$$
\tilde{\mathbf{w}}^k \cdot \mathbf{w}^\ell = \begin{cases} 1, & k = \ell \\ 0, & k \neq \ell \end{cases}
$$

You need to check the "special rows" — for example, the inner product of the first row $\left(\frac{\sqrt{2}}{2}, \frac{\sqrt{2}}{4}, 0, \ldots, 0\right)$ of $\tilde{W}_N^m$ and the ℓth row of U should be 1 if $\ell = 1$ and 0 otherwise. It is easy to check this condition as well as to verify that the other "special rows" satisfy the necessary orthogonality conditions.)

11.18   Verify the computations (11.48) and (11.49) in Example 11.6.

11.19   Suppose that $\mathbf{v} = [v_1, \ldots, v_N]^T \in \mathbf{R}^N$, N even. Form the vector $\mathbf{v}^p \in \mathbb{R}^{2N}$ as

$$
\mathbf{v}^p = [v_1, \ldots, v_N, v_N, \ldots, v_1]^T
$$

Let $\mathbf{x}$ be the bi-infinite sequence whose first $2N$ terms satisfy

$$
[x_1, x_2, \ldots, x_N, x_{N+1}, \ldots, x_{2N}]^T = \mathbf{v}^p
$$

and the rest are obtained by periodic extension. That is, $\mathbf{x} = (\ldots, \mathbf{v}^p, \mathbf{v}^p, \mathbf{v}^p, \ldots)$. Show that

(a) $\mathbf{x}$ is $2N$-periodic.

(b) $x_k = x_{1-k}$ for all $k \in \mathbb{Z}$.

11.20   Let $\mathbf{v}, \mathbf{v}^p$, and $\mathbf{x}$ be as defined in Problem 11.19 and suppose that $\tilde{\mathbf{h}}$ is the symmetric bi-infinite sequence given by

$$\tilde{\mathbf{h}} = (\ldots, \tilde{h}_{-\tilde{L}+1}, \ldots, \tilde{h}_0, \tilde{h}_1, \ldots, \tilde{h}_{\tilde{L}}, \ldots)$$
$$= (\ldots, \tilde{h}_{\tilde{L}}, \ldots, \tilde{h}_1, \tilde{h}_1, \ldots, \tilde{h}_{\tilde{L}}, \ldots)$$

Define the bi-infinite sequence $\mathbf{a}$ componentwise by

$$a_n = \sum_k \tilde{h}_{2n-k} x_k$$

(a) Show that $\mathbf{a}$ is $2N$-periodic. *Hint:* Recall that convolution is commutative (see Problem 5.7 in Section 5.1) and use Problem 11.19(a).

(b) Use the fact that $\tilde{h}_k = \tilde{h}_{1-k}$, $k \in \mathbb{Z}$ and Problem 11.19(b) to show that for all $n \in \mathbb{Z}$, $a_n = a_{1-n}$.

(c) Use parts (a) and (b) to show that $a_{N+1-n} = a_n$, $n \in \mathbb{Z}$.

(d) Use parts (a) and (c) to show that each period of $\mathbf{a}$ has the form

$$[a_1, a_2, \ldots, a_{N/2}, a_{N/2}, \ldots, a_2, a_1]^T$$

(e) Let $\mathbf{y}^p$ be the lowpass portion of the modified biorthogonal wavelet transform returned by Algorithm 11.6 with lowpass filter $\tilde{\mathbf{h}} = (\tilde{h}_{\tilde{L}}, \ldots, \tilde{h}_1, \tilde{h}_1, \ldots, \tilde{h}_{\tilde{L}})$. Show that $(a_1, \ldots, a_{N/2}) = (y_1^p, \ldots, y_{N/2}^p)$.

11.21   Let $\mathbf{v}, \mathbf{v}^p$, and $\mathbf{x}$ be as defined in Problem 11.19 and suppose that $\tilde{\mathbf{g}}$ is the symmetric bi-infinite sequence

$$\tilde{\mathbf{g}} = (\ldots, \tilde{g}_{-L+1}, \ldots, \tilde{g}_0, \tilde{g}_1, \ldots, \tilde{g}_L, \ldots)$$

constructed from the symmetric bi-infinite sequence

$$\mathbf{h} = (\ldots, 0, h_{-L+1}, \ldots, h_0, h_1, \ldots, h_L, 0, \ldots)$$
$$= (\ldots, 0, h_L, \ldots, h_1, h_1, \ldots, h_L, 0, \ldots)$$

Define the bi-infinite sequence **b** component-wise by

$$b_n = \sum_k \tilde{g}_{2n-k} x_k$$

(a) Show that **b** is a $2N$-periodic sequence. (*Hint:* The proof should be very similar to that given for Problem 11.20(a).)

(b) Use the fact that $\tilde{g}_k = -\tilde{g}_{1-k}$, $k \in \mathbb{Z}$ (see Problem 10.12 in Section 10.1) and Problem 11.19(b) to show that for all $n \in \mathbb{Z}$, $b_n = -b_{1-n}$.

(c) Use parts (a) and (b) to show that $b_{N+1-n} = -b_n$, $n \in \mathbb{Z}$.

(d) Use parts (a) and (c) to show that each period of **a** has the form

$$[b_1, b_2, \ldots, b_{N/2}, -b_{N/2}, \ldots, -b_2, -b_1]^T$$

(e) Let $\mathbf{z}^m$ be the highpass portion of the modified biorthogonal wavelet transform returned by Algorithm 11.6 with highpass filter $\tilde{\mathbf{g}} = (\tilde{g}_{-L+1}, \ldots, \tilde{g}_0, \tilde{g}_1, \ldots, \tilde{g}_L)$. Show that $[b_1, \ldots, b_{N/2}]^T = [z_1^p, \ldots, z_{N/2}^p]^T$.

Computer Labs

11.6   Investigating Algorithm 11.4. From the text Web site, access the lab bwtmod. In this lab you can try various even-length vectors **v** and odd-length biorthogonal filter pairs with Algorithm 11.4 and see the symmetry exhibited by the lowpass and highpass portions of the biorthogonal wavelet transform of $\mathbf{v}^p$.

◆ **11.7   Software Development: Modified Biorthogonal Wavelet Transform for the $(5, 3)$ Biorthogonal Spline Filter Pair.** From the text Web site, access the package fastbiorth53. In this development lab you will you generate code that will produce a fast algorithm for computing the modified biorthogonal wavelet transform and its inverse using the $(5, 3)$ biorthogonal spline filter pair. Problems 11.16 and 11.17 will be useful for this lab. Instructions are provided that allow you to add the code to the software package DiscreteWavelets.

◆ **11.8   Software Development: Modified Biorthogonal Wavelet Transform.** From the text Web site, access the package modifiedbiorth. In this development lab you will write code for the implementation of Algorithms 11.4, 11.5, 11.6, and 11.7 that computes the modified biorthogonal wavelet transform and its inverse. Instructions are provided that allow you to add the code to the software package DiscreteWavelets.

CHAPTER 12

THE JPEG2000 IMAGE COMPRESSION STANDARD

In 1992, JPEG became an international standard for compressing digital still images. JPEG is an acronym for the Joint Photographic Experts Group that was formed in the 1980s by members of the International Organization for Standardization (ISO) and the International Telecommunication Union (ITU). The JPEG compression standard remains very popular. According to Charrier et al. [16], 80% of all images appearing on the World Wide Web[1] are stored using it. Despite the popularity of the JPEG standard, members of JPEG decided that the algorithm could be improved in several areas and that it should be enhanced so as to meet the growing needs of applications using digital images. In 1997, work began on the development of the improved JPEG2000 compression standard. The core system of the JPEG2000 compression method is now a published standard by the ISO (ISO 15444). Unfortunately, due to patent issues involving the encoding method, the complete algorithm is not yet an ISO standard. Thus, JPEG2000 is not supported by Web browsers such as Microsoft's Internet Explorer or Mozilla's Firefox. It is hoped that once the patent issues are

[1] Digital images using the JPEG standard typically have the suffix . jpg attached to their file name.

Discrete Wavelet Transformations: An Elementary Approach With Applications. By P. J. Van Fleet   **459**
Copyright © 2008 John Wiley & Sons, Inc.

resolved, JPEG2000 will replace the current JPEG standard as the most popular tool for image compression.

One of the biggest changes in the new JPEG2000 standard is the fact that the biorthogonal wavelet transform has replaced the discrete cosine transform as the preferred method for organizing data in a way that can best be utilized by encoders. The new JPEG2000 standard also offers lossless and lossy compression. For lossy compression, JPEG2000 uses the CDF97 filter pair (Section 10.3), and for lossless compression, JPEG2000 uses a variant of the $(5,3)$ biorthogonal spline filter pair (Section 10.2). Both forms of compression also utilize the modified biorthogonal wavelet transform developed in Section 11.3 to handle boundary issues in a more effective manner.

We begin this chapter with a brief overview of the JPEG standard. In the second section, we describe the basic JPEG2000 standard. Both standards involve many more options than we present here. Our aim is to show the reader how the biorthogonal wavelet transform is used in the JPEG2000 standard. Towards this end, we do not utilize any of the sophisticated encoders used by either standard. Instead, we use the basic Huffman coding method described in Section 3.4. One of the best features of the JPEG2000 standard is the ability to perform lossless compression. We learn about *lifting* Section 12.3. This simple and clever method introduced by Sweldens in [75] can be used to quickly compute the biorthogonal wavelet transform. In the case of the modified $(5,3)$ biorthogonal spline filter pair, we can slightly alter the lifting method and produce a reversible integer to integer transform! In the final section of the chapter, examples are given to illustrate the implementation and effectiveness of the JPEG2000 standard.

12.1 AN OVERVIEW OF JPEG

Although it is beyond the scope of this book to present the JPEG standard in its complete form, an overview of the basic algorithm will help us understand the need for improvements and enhancements included in the JPEG2000 standard. For the sake of simplicity, let's assume that the grayscale image we wish to compress is of size $N \times M$ where both N and M are divisible by 8.

The first step in the JPEG method is to substract 128 from each of the pixel values. This step has the effect of centering the intensity values about zero. The next step is to partition the image into nonoverlapping 8×8 blocks. The *discrete cosine transform* (DCT) is then applied to each 8×8 matrix. The intent of this step is similar to that of applying a wavelet transform — the DCT will concentrate the energy of each block into just a few elements.

If A is an 8×8 matrix, then the DCT is computed by

$$C = UAU^T$$

where

$$U = \frac{1}{2} \begin{bmatrix} \frac{\sqrt{2}}{2} & \frac{\sqrt{2}}{2} & \frac{\sqrt{2}}{2} & \frac{\sqrt{2}}{2} & \frac{\sqrt{2}}{2} & \frac{\sqrt{2}}{2} & \frac{\sqrt{2}}{2} & \frac{\sqrt{2}}{2} \\ \cos(\frac{\pi}{16}) & \cos(\frac{3\pi}{16}) & \cos(\frac{5\pi}{16}) & \cos(\frac{7\pi}{16}) & \cos(\frac{9\pi}{16}) & \cos(\frac{11\pi}{16}) & \cos(\frac{13\pi}{16}) & \cos(\frac{15\pi}{16}) \\ \cos(\frac{\pi}{8}) & \cos(\frac{3\pi}{8}) & \cos(\frac{5\pi}{8}) & \cos(\frac{7\pi}{8}) & \cos(\frac{9\pi}{8}) & \cos(\frac{11\pi}{8}) & \cos(\frac{13\pi}{8}) & \cos(\frac{15\pi}{8}) \\ \cos(\frac{3\pi}{16}) & \cos(\frac{9\pi}{16}) & \cos(\frac{15\pi}{16}) & \cos(\frac{21\pi}{16}) & \cos(\frac{27\pi}{16}) & \cos(\frac{33\pi}{16}) & \cos(\frac{39\pi}{16}) & \cos(\frac{45\pi}{16}) \\ \cos(\frac{\pi}{4}) & \cos(\frac{3\pi}{4}) & \cos(\frac{5\pi}{4}) & \cos(\frac{7\pi}{4}) & \cos(\frac{9\pi}{4}) & \cos(\frac{11\pi}{4}) & \cos(\frac{13\pi}{4}) & \cos(\frac{15\pi}{4}) \\ \cos(\frac{5\pi}{16}) & \cos(\frac{15\pi}{16}) & \cos(\frac{25\pi}{16}) & \cos(\frac{35\pi}{16}) & \cos(\frac{45\pi}{16}) & \cos(\frac{55\pi}{16}) & \cos(\frac{65\pi}{16}) & \cos(\frac{75\pi}{16}) \\ \cos(\frac{3\pi}{8}) & \cos(\frac{9\pi}{8}) & \cos(\frac{15\pi}{8}) & \cos(\frac{21\pi}{8}) & \cos(\frac{27\pi}{8}) & \cos(\frac{33\pi}{8}) & \cos(\frac{39\pi}{8}) & \cos(\frac{45\pi}{8}) \\ \cos(\frac{7\pi}{16}) & \cos(\frac{21\pi}{16}) & \cos(\frac{35\pi}{16}) & \cos(\frac{49\pi}{16}) & \cos(\frac{63\pi}{16}) & \cos(\frac{77\pi}{16}) & \cos(\frac{91\pi}{16}) & \cos(\frac{105\pi}{16}) \end{bmatrix} \quad (12.1)$$

Element–wise, we have

$$u_{jk} = \begin{cases} \frac{\sqrt{2}}{4}, & j = 1 \\ \frac{1}{2}\cos((j-1)(2k-1)\pi/16), & j = 2, \ldots, 8 \end{cases}$$

It can be shown that U is an orthogonal matrix (a nice proof is given by Strang in [70]). Moreover, fast algorithms exist for computing the DCT (see, e. g., Rao and Yip [62] or Wickerhauser [85]).

The elements in each 8×8 block of the DCT are normalized and the results are rounded to the nearest integer. As stated by Gonzalez and Woods [39], the normalization matrix is

$$Z = \begin{bmatrix} 16 & 11 & 10 & 16 & 24 & 40 & 51 & 61 \\ 12 & 12 & 14 & 19 & 26 & 58 & 60 & 55 \\ 14 & 13 & 16 & 24 & 40 & 57 & 69 & 56 \\ 14 & 17 & 22 & 29 & 51 & 87 & 80 & 62 \\ 18 & 22 & 37 & 56 & 68 & 109 & 103 & 77 \\ 24 & 35 & 55 & 64 & 81 & 104 & 113 & 92 \\ 49 & 64 & 78 & 87 & 103 & 121 & 120 & 101 \\ 72 & 92 & 95 & 98 & 112 & 100 & 103 & 99 \end{bmatrix} \quad (12.2)$$

Thus, the quantization step renders an 8×8 matrix $\hat{C}$ whose elements are given by

$$\hat{c}_{jk} = \text{Round}\left(\frac{c_{jk}}{z_{jk}}\right)$$

The elements of $\hat{C}$ are next loaded into the vector

$$\mathbf{d} = [d_0, \ldots, d_{63}]^T$$

using the following *zigzag* array

$$
\begin{array}{|c|c|c|c|c|c|c|c|}
\hline
0 & 1 & 5 & 6 & 14 & 15 & 27 & 28 \\
\hline
2 & 4 & 7 & 13 & 16 & 26 & 29 & 42 \\
\hline
3 & 8 & 12 & 17 & 25 & 30 & 41 & 43 \\
\hline
9 & 11 & 18 & 24 & 31 & 40 & 44 & 53 \\
\hline
10 & 19 & 23 & 32 & 39 & 45 & 52 & 54 \\
\hline
20 & 22 & 33 & 38 & 46 & 51 & 55 & 60 \\
\hline
21 & 34 & 37 & 47 & 50 & 56 & 59 & 61 \\
\hline
35 & 36 & 48 & 49 & 57 & 58 & 62 & 63 \\
\hline
\end{array}
\tag{12.3}
$$

That is,

$$[d_0, d_1, d_2, d_3, d_4, \ldots, d_{61}, d_{62}, d_{63}]^T = [\hat{c}_{11}, \hat{c}_{12}, \hat{c}_{21}, \hat{c}_{31}, \ldots, \hat{c}_{78}, \hat{c}_{87}, \hat{c}_{88}]^T \tag{12.4}$$

These vectors are concatenated left to right, top to bottom to form a vector of length NM. This vector is then encoded using Huffman coding[2] to complete the process.

The JPEG compression standard can also handle color digital images. The red, green, and blue channels of a color image are first converted to YCbCr space, and the process described above is then applied to each of the Y, Cb, and Cr channels. Let's look at an example.

Example 12.1 (The JPEG Compression Method). *Let's consider the* 200×200 *grayscale image plotted in Figure 12.1. The image is partitioned into* 8×8 *blocks, resulting in a* 25×25 *block matrix representation of the original image.*

We next subtract 128 *from each of the elements in each block. Let's look at the result for the block in the* 3,13 *position. The original intensities in this block are*

$$
A_{3,13} = \begin{bmatrix}
114 & 117 & 121 & 102 & 0 & 101 & 203 & 219 \\
118 & 115 & 121 & 141 & 36 & 153 & 194 & 216 \\
120 & 120 & 126 & 79 & 84 & 203 & 196 & 216 \\
117 & 139 & 75 & 65 & 154 & 188 & 221 & 216 \\
132 & 107 & 32 & 150 & 195 & 209 & 192 & 220 \\
141 & 43 & 85 & 127 & 180 & 179 & 208 & 210 \\
26 & 15 & 95 & 161 & 184 & 192 & 205 & 217 \\
5 & 87 & 107 & 173 & 164 & 195 & 206 & 209
\end{bmatrix}
$$

---

[2]The JPG standard provides default *AC and DC Huffman tables*, but we can also use the Huffman coding scheme developed in Section 3.4.

(a) Grayscale image A

(b) Partitioned image

Figure 12.1   An image A and a partition of the image into a 25×25 block matrix where each block is 8×8. We have highlighted the 3,13 block. We use it to illustrate the JPEG algorithm throughout this example.

and the shift by -128 *gives*

$$
\tilde{A}_{3,13} =
\begin{bmatrix}
-14 & -11 & -7 & -26 & -128 & -27 & 75 & 91 \\
-10 & -13 & -7 & 13 & -92 & 25 & 66 & 88 \\
-8 & -8 & -2 & -49 & -44 & 75 & 68 & 88 \\
-11 & 11 & -53 & -63 & 26 & 60 & 93 & 88 \\
4 & -21 & -96 & 22 & 67 & 81 & 64 & 92 \\
13 & -85 & -43 & -1 & 52 & 51 & 80 & 82 \\
-102 & -113 & -33 & 33 & 56 & 64 & 77 & 89 \\
-123 & -41 & -21 & 45 & 36 & 67 & 78 & 81
\end{bmatrix}
$$

We next compute the DCT of each block. In the case of the $\tilde{A}_{3,13}$, we obtain (rounded to three decimal places)

$$T_{3,13} = U\tilde{A}_{3,13}U^T$$

$$= \begin{bmatrix}
106.125 & -358.94 & 94.406 & 4.768 & -7.375 & 37.074 & 3.515 & -43.029 \\
-34.435 & 110.662 & 170.188 & -50.391 & -69.658 & 32.725 & 17.088 & -31.676 \\
-55.859 & -4.425 & -23.146 & -124.169 & -37.895 & 57.964 & 7.534 & -24.109 \\
-13.689 & 11.344 & 17.895 & 13.94 & 68.516 & 41.226 & -5.414 & 14.157 \\
3.625 & 5.685 & 13.407 & 3.739 & -8.875 & -36.96 & -52.232 & -15.378 \\
-26.538 & -24.704 & 12.334 & -5.977 & -23.14 & -8.142 & 7.319 & 43.106 \\
1.354 & 23.671 & 31.534 & 7.182 & -17.993 & 1.504 & 18.396 & 20.056 \\
0.757 & -14.879 & -23.833 & 1.512 & 9.906 & -6.844 & 10.164 & -24.961
\end{bmatrix}$$

where U is given by (12.1). We next normalize the elements of each block T_{jk}, $j, k = 1, \ldots, 25$ using (12.2). That is, if t_{mn}, $m, n = 1, \ldots, 8$, are the elements of block T_{jk}, we compute $\tilde{t}_{mn} = t_{mn}/z_{mn}$. We will call the resulting block $\tilde{T}_{jk}$. In the case of the 3,13 block, we have (rounded to three decimal places)

$$\tilde{T}_{3,13} = \begin{bmatrix}
6.633 & -32.631 & 9.441 & 0.298 & -0.307 & 0.927 & 0.069 & -0.705 \\
-2.87 & 9.222 & 12.156 & -2.652 & -2.679 & 0.564 & 0.285 & -0.576 \\
-3.99 & -0.34 & -1.447 & -5.174 & -0.947 & 1.017 & 0.109 & -0.431 \\
-0.978 & 0.667 & 0.813 & 0.481 & 1.343 & 0.474 & -0.068 & 0.228 \\
0.201 & 0.258 & 0.362 & 0.067 & -0.131 & -0.339 & -0.507 & -0.2 \\
-1.106 & -0.706 & 0.224 & -0.093 & -0.286 & -0.078 & 0.065 & 0.469 \\
0.028 & 0.37 & 0.404 & 0.083 & -0.175 & 0.012 & 0.153 & 0.199 \\
0.011 & -0.162 & -0.251 & 0.015 & 0.088 & -0.062 & 0.099 & -0.252
\end{bmatrix}$$

We quantize the elements of each block by rounding them to the nearest integer. We will call the resulting blocks $\hat{C}_{jk}$, $j, k = 1, \ldots, 25$. In the case of the 3,13 block, we have

$$\hat{C}_{3,13} = \begin{bmatrix}
7 & -33 & 9 & 0 & 0 & 1 & 0 & -1 \\
-3 & 9 & 12 & -3 & -3 & 1 & 0 & -1 \\
-4 & 0 & -1 & -5 & -1 & 1 & 0 & 0 \\
-1 & 1 & 1 & 0 & 1 & 0 & 0 & 0 \\
0 & 0 & 0 & 0 & 0 & 0 & -1 & 0 \\
-1 & -1 & 0 & 0 & 0 & 0 & 0 & 0 \\
0 & 0 & 0 & 0 & 0 & 0 & 0 & 0 \\
0 & 0 & 0 & 0 & 0 & 0 & 0 & 0
\end{bmatrix}$$

In $\hat{C}_{3,13}$, we can see the effects of the DCT. This transform "pushes" the large elements of the block toward the upper left-hand corner. As a result, the energy of each block is contained in only a few of the elements. We have plotted the cumulative energies of $A_{3,13}$ and $\hat{C}_{3,13}$ in Figure 12.2. In Figure 12.3 we have plotted the quantized DCT for the image stored in A.

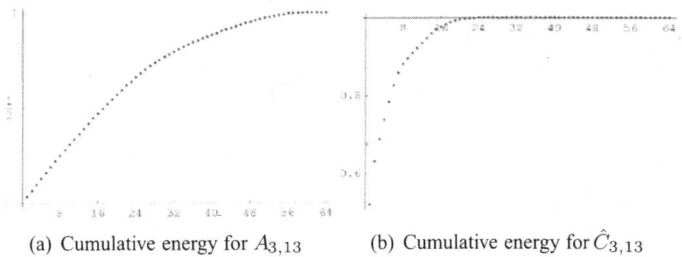

(a) Cumulative energy for $A_{3,13}$ (b) Cumulative energy for $\hat{C}_{3,13}$

Figure 12.2   The cumulative energy vectors for the block $A_{3,13}$ and the quantized DCT block $\hat{C}_{3,13}$.

Figure 12.3   The quantized DCT for image A.

We now load the elements of $\hat{C}_{jk}$ into vector $\mathbf{d}_{jk}$ using (12.3). In the case of $\hat{C}_{3,13}$, we have

$$\mathbf{d}_{3,13} = [7, -3, -33, 9, 9, -4, -1, 0, 12, 0, 0, -3, -1, 1, 0, -1, 0, 1, -5, -3, 1,$$
$$0, 1, -1, 0, 0, -1, 0, 0, 0, 0, 0, 1, 1, 0, -1, -1, 0, \ldots, 0, -1, 0, \ldots, 0]^T$$

The last step is to perform the Huffman coding. We create the 40,000-vector $\mathbf{v}$ by chaining the $\mathbf{d}_{jk}$ together left to right, top to bottom. Huffman coding is then performed on $\mathbf{v}$. Using Algorithm 3.1 from Section 3.4, we find that the image can be encoded using a total of 599,241 bits or at a rate of 1.4981 bpp.

The encoded version of $\mathbf{v}$ can now be stored or placed on a server. To view the compressed image, we must reverse the process. That is, we decode the Huffman codes and then rebuild the blocks $\hat{C}_{jk}$. We multiply each of the elements $\hat{c}_{mn}$ of $\hat{C}_{jk}$

by z_{mn} to obtain an approximate $\tilde{T}_{jk}$ of T_{jk}. The fact that we have rounded data in our quantization step means that we will never be able to exactly recover the original image. The inversion process is completed by applying the inverse DCT to each block $\tilde{T}_{jk}$. That is, we obtain approximations $\tilde{A}_{jk}$ to A_{jk} by computing

$$\tilde{A}_{jk} = U^T \tilde{T}_{jk} U$$

where $j, k = 1, \ldots, 25$. In Figure 12.4, we plot both $A_{3,13}$ and $\tilde{A}_{3,13}$.

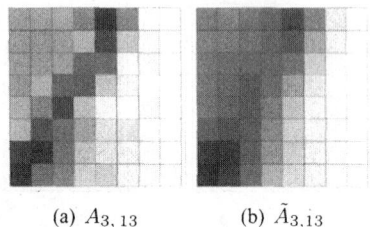

(a) $A_{3,13}$ (b) $\tilde{A}_{3,13}$

Figure 12.4   The block $A_{3,13}$ from the original image in Figure 12.1(a) and the JPEG-compressed $\tilde{A}_{3,13}$.

The approximate $\tilde{A}$ of the original image A is formed by constructing the 25×25 block matrix whose entries are $\tilde{A}_{jk}$ and adding 128 to each element in $\tilde{A}_{jk}$. We have plotted the image in JPEG format in Figure 12.5. The PSNR is computed to be 31.9113.   □

Figure 12.5   The JPEG-compressed version of the original image.

PROBLEMS

12.1   For each 8-vector $\mathbf{v}$, compute the one-dimensional DCT $\mathbf{y} = U\mathbf{v}$, where U is given by (12.1).

(a) $v_k = c, k = 1, \ldots, 8$ where c is any nonzero number.

(b) $v_k = k, k = 1, \ldots, 8$.

(c) $v_k = (-1)^k, k = 1, \ldots, 8$.

(d) $v_k = mk + b, k = 1, \ldots, 8$, where m and b are real numbers with $m \neq 0$.
   (*Hint:* Use parts (a) and (b).)

12.2   Suppose that $\mathbf{v} = [v_1, v_2, v_3, v_4, v_4, v_3, v_2, v_1]^T$ and let $\mathbf{y} = U\mathbf{v}$ where U is given by (12.1). Show that the even elements of $\mathbf{y}$ are zero. (*Hint:* Use the trigonometric identity $\cos\theta = -\cos(\pi - \theta)$.)

12.3   Let $\mathbf{u}^1$ be the first row of U, where U is given by (12.1). Show that $\| \mathbf{u}^1 \| = 1$ and explain why $\mathbf{u}^1$ is orthogonal to the other seven rows of U.

12.4   Suppose that A is an 8×8 matrix whose entries are each $a_{jk} = c$, where $k = 1, \ldots, 8$ and c is any real nonzero number. Compute UAU^T where U is given by (12.1). (*Hint:* Use part (a) from Problem 12.1.)

12.5   Suppose that, for a given 8×8 matrix A, we compute $B = UAU^T$ and then create the matrix C whose elements are $b_{jk}/z_{jk}, j, k = 1, \ldots, 8$, where Z is given by (12.2). The elements of C are rounded to give

$$
\hat{C} = \begin{bmatrix}
3 & 0 & 2 & -1 & -4 & -1 & 0 & -2 \\
7 & -1 & - & 1 & 2 & 1 & 1 & 0 \\
0 & 10 & -1 & 2 & 0 & -3 & 1 & -1 \\
0 & 1 & -2 & -3 & 1 & 0 & 1 & 0 \\
3 & -4 & 0 & -4 & 0 & -1 & 1 & -1 \\
-4 & -1 & 0 & -3 & -1 & 1 & -1 & 0 \\
0 & 0 & -1 & 1 & 0 & 1 & 1 & 0 \\
-1 & -1 & 0 & 0 & 0 & 0 & 0 & 0
\end{bmatrix}
$$

Use the zigzag array (12.3) to create the zigzag vector $\mathbf{d}$ (12.4) that will be sent to the Huffman coder.

Computer Lab

12.1   **JPEG Compression.** From the text Web site, access the lab jpeg. In this lab you will use the ideas of this section to write a module that performs JPEG compression on a digital image.

12.2 THE BASIC JPEG2000 ALGORITHM

Example 12.1 illustrates a couple of problems with the current JPEG standard. The first problem is that the compression standard is *lossy*. That is, due to the rounding that occurs in the quantization step, it is generally impossible to exactly recover the original image from the compressed version.

In Figure 12.6 we have plotted an enlarged version of the bottom left-hand corner (50×50 submatrix) of both the original image and the JPEG compressed image.

(a) Original image. (b) JPEG compressed version.

Figure 12.6 A plot of the bottom left corner of the original image A and the same region of the JPEG compressed version.

If you look closely at the image in Figure 12.6(b), you will notice that some pixels appear in "blocks." Partitioning the image into 8×8 blocks and processing these blocks independently leads to the artifacts in the compressed image.

JPEG2000 - Improving on the JPEG Standard

The JPEG2000 standard aspires to resolve many of the issues faced by the current standard. Christopoulos et al. [17] identify several features of JPEG2000. Some of these features are

1. **Lossy and lossless compression.** The JPEG2000 standard allows the user to perform lossless compression as well as lossy compression.

2. **Better compression.** The JEPG2000 standard produces higher-quality images for very low bit rates (below 0.25 bpp according to JPEG [49]).

3. **Progressive signal transmission.** JPEG2000 can reconstruct the digital image (at increasingly higher resolution) as it is received via the internet.

4. **Tiling.** JPEG2000 allows the user to *tile* the image. If desired, the image can be divided into nonoverlapping tiles, and each tile can be compressed independently. This is a more general approach to the 8×8 block partitioning of JPEG. If tiling is used, the resulting uncompressed image is much smoother than that produced by the current JPEG standard.

5. **Regions of Interest.** The JPEG2000 standard allows users to define Regions of Interest (ROIs) and encode these ROIs so that they can be uncompressed at a higher resolution than the rest of the image.

6. **Larger image size.** The current JPEG standard can handle images of size $64,000 \times 64,000$ or smaller. JPEG2000 can handle images of size $2^{32} - 1 \times 2^{32} - 1$ or smaller (see JPEG [49]).

7. **Multiple channels.** The current JPEG standard can only support three channels (these channels are used for color images). The new JPEG2000 standard can support up to 256 channels. JPEG2000 can also handle color models other than YCbCr.

The Basic JPEG2000 Algorithm

We are now ready to present a basic JPEG2000 algorithm. This algorithm appears in Gonzalez and Woods [39]. There are many features we omit at this time. The interested reader is encouraged to consult Taubman [78] for a more detailed description of the compression standard.

Algorithm 12.1 (Basic JPEG2000 Compression Standard). *This algorithm takes a grayscale image A whose dimensions are even and whose intensities are integers ranging from 0 to 255 and returns a compressed image using the JPEG2000 compression format.*

1. *Subtract 128 from each element of A.*

2. *Apply i iterations of the biorthogonal wavelet transform. If the dimensions of A are $N \times M$ with $N = 2^p$ and $M = 2^q$, then the largest i can be is the minimum of p and q. For lossless compression, use a modified version of the $(5, 3)$ biorthogonal spline filter pair and for lossy compression, use the CDF97 biorthogonal filter pair.*

3. *For lossy compression, apply a quantizer similar to the shrinkage function (9.4) defined in Section 9.1.*

4. *Use an arithmetic coding algorithm known as Embedded Block Coding with Optimized Truncation (EBCOT) to encode the quantized transform data.*

☐

To compress color images using the JPEG2000 standard, we first convert from RGB space to YCbCr space (see Section 3.2) and apply Algorithm 12.1 to each of the Y, Cb, and Cr channels. Next we look at step 2 of Algorithm 12.1 in a bit more detail.

The Biorthogonal Wavelet Transform for Lossy Compression

Perhaps the biggest change in the JPEG2000 compression standard is the use of the biorthogonal wavelet transform instead of the discrete cosine transform. The biorthogonal wavelet transform is one reason the compression rates for JPEG2000 are superior to those of JPEG. The iterative process of the transform makes it easy to see how it can be implemented to perform progressive image transmission. In a more advanced implementation of JPEG2000, the user can opt to reconstruct only a portion of the compressed image. The local nature of the biorthogonal wavelet transform can be exploited to develop fast routines to perform this task.

For lossy compression, the CDF97 filter pair developed in Section 10.3 is used. The transform is performed using the modified biorthogonal wavelet transformation (Algorithm 11.4) from Section 11.3. In this way, the transform is better suited to handle pixels at the right and bottom boundaries of the image.

In our derivation of the CDF97 filter pair, $\tilde{\mathbf{h}}$ is the seven-term filter and it is used to construct the lowpass portion of the transform matrix $\tilde{W}$. The nine-term filter is $\mathbf{h}$ and it is used to construct the highpass portion of $\tilde{W}$. We then use $\mathbf{h}$ and $\tilde{\mathbf{h}}$ to form the lowpass and highpass portions of W, respectively. Thus our two-dimensional transform is $B = \tilde{W}AW^T$.

JPEG2000 reverses the roles of the filters (see Christopoulos [17]). That is, the biorthogonal transform is implemented as $B = WA\tilde{W}^T$. The inverse transform is then $A = \tilde{W}^TBW$. It is easy to implement this change in Algorithm 11.1 — we simply call the algorithm with the filters reversed.

The Biorthogonal Wavelet Transform for Lossless Compression

For lossless compression, the members of the JPEG settled on the $(5,3)$ biorthogonal spline filter pair. These filters are symmetric and close in length. The best feature of this filter is that it can easily be adapted to map integers to rational numbers. We can then use these modified filters in conjunction with a technique called *lifting* to produce a fast algorithm that maps integers to integers. This technique is reversible, so no information is lost.

The LeGall Filter Pair

Recall from Example 10.6 of Section 10.2, that the $(5, 3)$ biorthogonal spline filter pair is

$$\tilde{\mathbf{h}} = (\tilde{h}_{-1}, \tilde{h}_0, \tilde{h}_1) = \left(\frac{\sqrt{2}}{4}, \frac{\sqrt{2}}{2}, \frac{\sqrt{2}}{4} \right)$$

$$\mathbf{h} = (h_{-2}, h_{-1}, h_0, h_1, h_2) = \left(-\frac{\sqrt{2}}{8}, \frac{\sqrt{2}}{4}, \frac{3\sqrt{2}}{4}, \frac{\sqrt{2}}{4}, -\frac{\sqrt{2}}{8} \right)$$

As is the case in lossy compression, the JPEG2000 reverses the roles of these filters. We will use the five-term filter in the lowpass portion of the transform and the three-term filter in the highpass portion of the transform.

The filter pair is further modified by multiplying the five-term filter by $\frac{\sqrt{2}}{2}$ and multiplying the three-term filter by $\sqrt{2}$. The resulting filter is often called the *LeGall filter pair* (see LeGall [54]).

Definition 12.1 (The LeGall Filter Pair). *We define the* LeGall filter pair *as*

$$\boxed{\begin{aligned} \tilde{\mathbf{h}} &= (\tilde{h}_{-2}, \tilde{h}_{-1}, \tilde{h}_0, \tilde{h}_1, \tilde{h}_2) = (-\tfrac{1}{8}, \tfrac{1}{4}, \tfrac{3}{4}, \tfrac{1}{4}, -\tfrac{1}{8}) \\ \mathbf{h} &= (h_{-1}, h_0, h_1) = (\tfrac{1}{2}, 1, \tfrac{1}{2}) \end{aligned}} \tag{12.5}$$

$$\square$$

The JPEG2000 standard uses $b = 0$ and $n = 1$ in Corollary 10.1 to obtain the highpass filters. Inserting $b = 0$ into (10.30) gives

$$\begin{aligned} \tilde{g}_k &= (-1) \cdot (-1)^k h_{1-k} = (-1)^{k+1} h_{1-k} \\ g_k &= (-1) \cdot (-1)^k \tilde{h}_{1-k} = (-1)^{k+1} \tilde{h}_{1-k} \end{aligned} \tag{12.6}$$

Using (12.5) and (12.6), we see the highpass filters associated with the LeGall filters are

$$\boxed{\begin{aligned} \tilde{\mathbf{g}} &= (\tilde{g}_0, \tilde{g}_1, \tilde{g}_2) = (-\tfrac{1}{2}, 1, -\tfrac{1}{2}) \\ \mathbf{g} &= (g_{-1}, g_0, g_1, g_2, g_3) = (-\tfrac{1}{8}, -\tfrac{1}{4}, \tfrac{3}{4}, -\tfrac{1}{4}, -\tfrac{1}{8}) \end{aligned}} \tag{12.7}$$

The Quantization Process

Suppose that we have computed i iterations of the biorthogonal wavelet transform. We denote by $\mathcal{V}^n$, $\mathcal{H}^n$, $\mathcal{D}^n$ the vertical, horizontal, and diagonal portions, respectively,

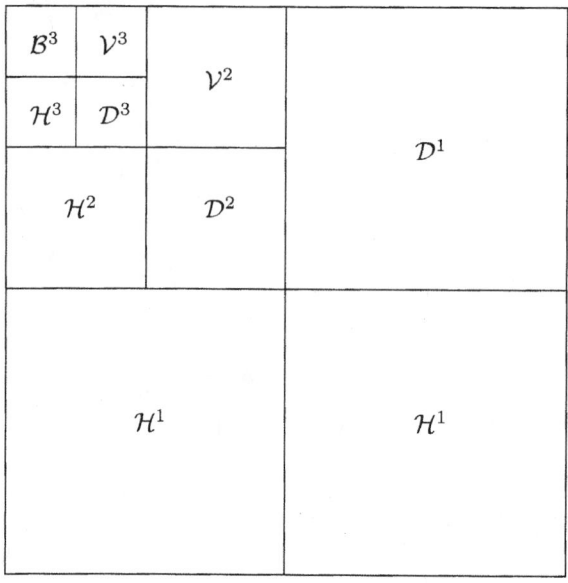

Figure 12.7   Three iterations of the biorthogonal wavelet transformation.

of the nth iteration of the wavelet transform. Here $n = 1, \ldots, i$. We denote by $\mathcal{B}^i$ the blur portion of the transform. An example of this notation is given in Figure 12.7.

We quantize each of $\mathcal{V}^n, \mathcal{H}^n, \mathcal{D}^n$, and $\mathcal{B}^i, n = 1, \ldots, i$ separately. Suppose that t is any element in $\mathcal{V}^n, \mathcal{H}^n, \mathcal{D}^n$, or $\mathcal{B}^i, i = 1, \ldots, n$. Then the quantization function is given by

$$q(t) = \text{sgn}(t) \lfloor |t|/d \rfloor \qquad (12.8)$$

Here $\text{sgn}(t)$ is the sign function given by (9.1) in Section 9.1 and $\lfloor \cdot \rfloor$ is the floor function defined in Problem 6.26(c) in Section 6.4. The value d is called the *quantization step size*. We discuss the computation of d in the next subsection. A quantization step size must be provided for each of $\mathcal{V}^n, \mathcal{H}^n, \mathcal{D}^n$, and $\mathcal{B}^i, n = 1, \ldots, i$.

It is easy to understand how the quantization process operates. We first compute the absolute value of t. We divide this number by the quantization step size. The floor function truncates the result to an integer and the sgn function reattaches the original sign of t.

Note that if $|t|$ is smaller than the quantization step size, the floor function converts the result to zero. In this regard, (12.8) acts much like the shrinkage function (9.4) used in denoising in Chapter 9. If $d \le |t| < 2d$, the floor function returns 1 and the sgn function reattaches the original sign of t. In general, if $\ell d \le |t| < (\ell + 1)d$, then the floor function returns ℓ. In Problem 12.9 you are asked to plot $q(t)$ for various values of d.

Creating the Quantization Step Sizes

To create the quantization step sizes, we form 3-vectors $\mathbf{d}^n = [d_V^n, d_H^n, d_D^n]^T$, for $n = 1, \ldots, i-1$ and a 4-vector $\mathbf{d}^i = [d_V^i, d_H^i, d_D^i, d_B^i]^T$. These vectors contain step sizes for the quantization. For $n = 1, \ldots, i-1$, d_V^n, d_H^n, and d_D^n are used to quantize $\mathcal{V}^n$, $\mathcal{H}^n$, and $\mathcal{D}^n$. The same pattern holds for $\mathbf{d}^i$ except that d_B^i will be used to quantize the blur portion of the transform.

To form the vectors $\mathbf{d}^n$, $n = 1, \ldots, i$, we first need a *base step size*. We take

$$\tau = 2^{R-c+i}\left(1 + \frac{f}{2^{11}}\right)$$

where R is the number of bits needed to represent the values of the original image, and c and f are the number of bits used to represent the exponent and mantissa, respectively, of the elements in the blur portion of the transform. Since our pixels range in intensity from 0 to $255 = 2^8 - 1$, we will take $R = 8$. Typical values for c and f are 8.5 and 8, respectively (see [39] for more details). Thus

$$\tau = 2^{8-8.5+i}\left(1 + \frac{8}{2^{11}}\right) = 2^{i-1/2}1.00390625 \tag{12.9}$$

We use τ to write

$$\mathbf{d}^k = [d_V^k, d_H^k, d_D^k]^T = \frac{1}{2^{k-1}}[\tau, \tau, 2\tau]^T \tag{12.10}$$

for $k = 1, \ldots, i-1$ and for $\mathbf{d}^i$, we have

$$\mathbf{d}^i = [d_V^i, d_H^i, d_D^i, d_B^i]^T = \frac{1}{2^{i-1}}[\tau, \tau, 2\tau, \frac{\tau}{2}]^T \tag{12.11}$$

Although we do not discuss the motivation for the derivation of $\mathbf{d}^k$ at this time, the formulation depends on the number of *analysis gain bits* for each level of the transformation. For more information, see Gonzalez and Woods [39] or Christopoulos [17]. This method for determining the step size is used in *implicit quantization*. In the case of *explicit quantization*, the user would supply all vectors $\mathbf{d}^n$, $n = 1, \ldots, i$. Let's look at an example.

Example 12.2 (Computing Quantization Step Sizes). *Suppose that we performed* $i = 3$ *iterations of the biorthogonal wavelet transformation. Using (12.9), we see that*

$$\tau = 2^{3-1/2}1.00390625 \approx 5.678951$$

The step size vectors for the first and second iterations are

$$\mathbf{d}^1 = [d_V^1, d_H^1, d_D^1]^T = [\tau, \tau, 2\tau]^T = [5.678951, 5.678951, 11.357903]^T$$

$$\mathbf{d}^2 = [d_V^2, d_H^2, d_D^2]^T = \left[\frac{1}{2}\tau, \frac{1}{2}\tau, \tau\right]^T = [2.839476, 2.839476, 5.678951]^T$$

and the step size vector for the third iteration is

$$\mathbf{d}^3 = [d_V^3, d_H^3, d_D^3, d_B^3]^T$$
$$= \left[\frac{1}{4}\tau, \frac{1}{4}\tau, \frac{1}{2}\tau, \frac{1}{8}\tau\right]^T$$
$$= [1.419738, 1.419738, 2.839476, 0.709869]^T$$

Note that the step sizes get smaller as the iteration number increases.   □

Encoding in JPEG2000

Rather than Huffman coding, the JPEG2000 standard uses a form of *arithmetic coding* to encode quantized transform values. Arithmetic coding is different than Huffman coding in that it takes a long string of integer values and stores them as a single rational number between 0 and 1. Naturally, this number has a binary representation, and the primary difference between arithmetic coding and Huffman coding is that Huffman coding assigns a single character to a list of bits, whereas arithmetic coding might use some combination of characters to define a list of bits. According to Wayner [84], it is possible for arithmetic coding methods to be 10% more efficient than Huffman coding.

The JPEG2000 uses a method of arithmetic coding called *Embedded Block Coding with Optimized Truncation* (EBCOT). This form of arithmetic coding was introduced by Taubman [77]. This coding method utilizes the wavelet transform and facilitates many of the desired features of the JPEG2000 standard. The development of arithmetic coding and EBCOT are outside the scope of this book. The interested reader is referred to Sayood [65] or Taubman [78].

Inverting the Process

To invert the process, we begin by decoding the encoded quantized transform. Next, we multiply the lowpass and the highpass portions by the associated quantization step size. We then apply the inverse biorthogonal wavelet transform using the CDF97 filter pair. Recall that Algorithm 12.1 stipulates that the transform is computed using $B = WA\tilde{W}^T$ so the inverse transform uses $\tilde{W}^TBW$. We then add 128 to the result and obtain a compressed version of the original image.

For color images, we decode the quantized transforms of the Y, Cb, and Cr channels and then apply the inversion process to each channel. As a last step, we convert from YCbCr space to RGB space.

PROBLEMS

12.6 Let $v = [3, 1, 0, 5, 10, 20, 0, 17]^T$. Using the LeGall filter pair in conjunction with Algorithm 11.4, compute the modified biorthogonal wavelet transformation of v by hand.

12.7 Suppose that h and $\tilde{h}$ are the $(5, 3)$ biorthogonal spline filter whose lengths are 5 and 3 respectively. Suppose that you have an algorithm BWT that takes as input an even length vector v and the filters $\tilde{h}$ and h and returns the modified biorthogonal wavelet transform via Algorithm 11.4. The exact call is $y = \text{BWT}(v, \tilde{h}, h)$. The lowpass portion of the transform is constructed from $\tilde{h}$, and the highpass portion of the transform is constructed from h.

(a) What arguments would you use in BWT to have the routine return the modified biorthogonal transform using the LeGall filter pair? Write your arguments in terms of $\tilde{h}$ and h.

(b) Assuming that $\text{IBWT}(y, \tilde{h}, h)$ returns the inverse biorthogonal wavelet transform, what arguments would you give IBWT to have the routine return the inverse modified biorthogonal transform using the LeGall filter pair? Write your arguments in terms of $\tilde{h}$ and h.

12.8 Suppose that v is an even-length vector whose entries are integers from the set $\{-128, \ldots, 127\}$. Suppose that we compute the modified biorthogonal wavelet transform of v using the LeGall filter.

(a) What are the largest and smallest possible values of the lowpass portion of the transform?

(b) Describe the types of values returned by the lowpass portion of the transform.

(c) What are the largest and smallest possible values of the highpass portion of the transform?

(d) Describe the types of values returned by the highpass portion of the transform.

12.9 Plot $q(t)$ given by (12.8) for $d = 1,2,5,10$.

12.10 Suppose that

$$c = [5, 103/4, 43, 21/2, 33/4, 11/4, -29, -121/4, 11/4, 25/2, -17/2, -1]^T$$

Apply the quantization function $q(t)$ given by (12.8) to c using

(a) $d = 1$.

(b) $d = 5$.

(c) $d = 10$.

(d) $d = 15$.

12.11   Suppose that we compute i iterations of the modified biorthogonal wavelet transformation and for $k = 1, \ldots, i - 1$. Let $\mathbf{d}^k$ be given by (12.10).

(a) Show that for $k = 1, \ldots, i - 2$, $\mathbf{d}^{k+1} = \frac{1}{2}\mathbf{d}^k$.

(b) Why do you think the step sizes decrease by a factor of 2 for each successive iteration?

(c) Why do you think the step size for the diagonal block is twice as large as that used for the vertical and horizontal blocks?

12.3   LIFTING AND LOSSLESS COMPRESSION

The lifting method for implementing the biorthogonal wavelet transform was introduced by Wim Sweldens[3] in [75]. For a quick introduction to the topic, see Sweldens tutorial paper [74]. Another good reference on lifting is the book by Jensen and la Cour-Harbo [50]. The lifting scheme is a fast implementation of the modified biorthogonal wavelet transform and uses less memory than the method outlined in Algorithm 11.4. If you worked through Problem 6.26 in Section 6.4, then you have had a brief introduction to lifting and how to use it in lossless compression.

The Lifting Method and the LeGall Filter Pair

The lifting method produces the same results as those we obtain by using the modified biorthogonal wavelet transformation (Algorithm 11.4) from Section 11.3. Let's look at an example to see how the process works. If we apply Algorithm 11.4 with the LeGall filter pair to an arbitrary 10-vector,

$$\mathbf{v} = [v_1, v_2, \ldots, v_{10}]^T$$

the lowpass portion of the transform is

$$\mathbf{s} = \begin{bmatrix} -\frac{1}{8}v_3 + \frac{1}{4}v_2 + \frac{3}{4}v_1 + \frac{1}{4}v_2 - \frac{1}{8}v_3 \\ -\frac{1}{8}v_1 + \frac{1}{4}v_2 + \frac{3}{4}v_3 + \frac{1}{4}v_4 - \frac{1}{8}v_5 \\ -\frac{1}{8}v_3 + \frac{1}{4}v_4 + \frac{3}{4}v_5 + \frac{1}{4}v_6 - \frac{1}{8}v_7 \\ -\frac{1}{8}v_5 + \frac{1}{4}v_6 + \frac{3}{4}v_7 + \frac{1}{4}v_8 - \frac{1}{8}v_9 \\ -\frac{1}{8}v_7 + \frac{1}{4}v_8 + \frac{3}{4}v_9 + \frac{1}{4}v_{10} - \frac{1}{8}v_9 \end{bmatrix} \qquad (12.12)$$

---

[3] The term *lifting* was introduced by Sweldens to name the method he devised for increasing the number of derivatives m and n for which the Fourier series $\tilde{H}(\omega)$ and $H(\omega)$ associated with filters $\tilde{\mathbf{h}}$, $\mathbf{h}$, respectively, satisfy $\tilde{H}^{(m)}(\pi) = H^{(n)}(\pi) = 0$.

while the highpass portion of the transform is

$$\mathbf{d} = \begin{bmatrix} -\frac{1}{2}v_1 + v_2 - \frac{1}{2}v_3 \\ -\frac{1}{2}v_3 + v_4 - \frac{1}{2}v_5 \\ -\frac{1}{2}v_5 + v_6 - \frac{1}{2}v_7 \\ -\frac{1}{2}v_7 + v_8 - \frac{1}{2}v_9 \\ -\frac{1}{2}v_9 + v_{10} - \frac{1}{2}v_9 \end{bmatrix} \tag{12.13}$$

We could certainly simplify s_1 and s_5 in (12.12) and d_5 in (12.13). To better illustrate the lifting method, we choose not to do so. We now show how the same results can be obtained via the lifting scheme.

Note that the center terms in (12.12) are built from the odd elements of $\mathbf{v}$, and the center terms in (12.13) are built from the even elements of $\mathbf{v}$. Thus, the first step in lifting is to partition $\mathbf{v}$ into odd and even vectors:

$$\mathbf{o} = [o_1, o_2, o_3, o_4, o_5]^T = [v_1, v_3, v_5, v_7, v_9]^T$$

and

$$\mathbf{e} = [e_1, e_2, e_3, e_4, e_5]^T = [v_2, v_4, v_6, v_8, v_{10}]^T$$

We can easily compute $\mathbf{d}$ from $\mathbf{e}$ and $\mathbf{o}$. We have

$$d_1 = e_1 - \frac{1}{2}(o_1 + o_2)$$

$$d_2 = e_2 - \frac{1}{2}(o_2 + o_3)$$

$$d_3 = e_3 - \frac{1}{2}(o_3 + o_4)$$

$$d_4 = e_4 - \frac{1}{2}(o_4 + o_5)$$

$$d_5 = e_5 - \frac{1}{2}(o_5 + o_5)$$

If we introduce $o_6 = o_5$, we can write the elements of $\mathbf{d}$ as

$$d_k = e_k - \frac{1}{2}(o_k + o_{k+1}), \qquad k = 1, \ldots, 5 \tag{12.14}$$

The so-called *lifting step* is evident when we make the observation that the sum of d_1 and d_2 is related to s_2. For example, if we compute $d_1 + d_2$, we have

$$d_1 + d_2 = e_1 - \frac{1}{2}(o_1 + o_2) + e_2 - \frac{1}{2}(o_2 + o_3)$$

$$= v_2 - \frac{1}{2}(v_1 + v_3) + v_4 - \frac{1}{2}(v_3 + v_5)$$

$$= -\frac{1}{2}v_1 + v_2 - v_3 + v_4 - \frac{1}{2}v_5$$

The result is "close" to

$$s_2 = -\frac{1}{8}v_1 + \frac{1}{4}v_2 + \frac{3}{4}v_3 + \frac{1}{4}v_4 - \frac{1}{8}v_5$$

Indeed, if we divide the sum by 4 and manipulate the term containing v_3, we see that

$$\begin{aligned}
\frac{d_1 + d_2}{4} &= -\frac{1}{8}v_1 + \frac{1}{4}v_2 - \frac{1}{4}v_3 + \frac{1}{4}v_4 - \frac{1}{8}v_5 \\
&= \left(-\frac{1}{8}v_1 + \frac{1}{4}v_2 + \frac{3}{4}v_3 + \frac{1}{4}v_4 - \frac{1}{8}v_5\right) - v_3 \\
&= s_2 - v_3 \\
&= s_2 - o_2
\end{aligned}$$

Solving for s_2 gives

$$s_2 = o_2 + \frac{1}{4}(d_1 + d_2)$$

In a similar manner, we can show that

$$s_k = o_k + \frac{1}{4}(d_k + d_{k-1}), \qquad k = 2,3,4,5 \qquad\qquad (12.15)$$

Some special consideration is needed for s_1. We can write

$$\begin{aligned}
s_1 &= -\frac{1}{8}v_3 + \frac{1}{4}v_2 + \frac{3}{4}v_1 + \frac{1}{4}v_2 - \frac{1}{8}v_3 \\
&= -\frac{1}{4}v_3 + \frac{1}{2}v_2 + \frac{3}{4}v_1 \\
&= \left(-\frac{1}{4}v_3 + \frac{1}{2}v_2 - \frac{1}{4}v_1\right) + v_1 \\
&= \frac{1}{4}(-v_3 + 2v_2 - v_1) + v_1 \\
&= \frac{1}{4}(d_1 + d_1) + v_1
\end{aligned}$$

Thus, we have constructed the lowpass portion of the transform from the highpass portion of the transform!

We can easily generalize (12.14) and (12.15) to create a lifting process for the modified biorthogonal transform using the LeGall filter pair.[4] For an N-vector $\mathbf{v}$, where N is even, we first partition $\mathbf{v}$ into even and odd vectors $\mathbf{e}$ and $\mathbf{o}$ each of length

---

[4]Actually, the JPEG2000 standard stipulates that the user can choose to compute the modified biorthogonal wavelet transform with the using either lifting process or Algorithm 11.4. The development of the lifting process for the CDF97 filter pair requires a bit more work than that needed for the LeGall filter pair so we do not cover it in this book. The interested reader is referred to work by Daubechies and Sweldens [25].

$N/2$. We then compute the highpass portion **d** of the transform by first introducing the value $o_{N/2+1} = o_{N/2}$ and using

$$d_k = e_k - \frac{1}{2}(o_k + o_{k+1}), \qquad k = 1, \ldots, \frac{N}{2} \qquad (12.16)$$

Once we have **d**, we can compute the lowpass portion **s** of the transformation. We set $d_0 = d_1$ and then use

$$s_k = o_k + \frac{1}{4}(d_k + d_{k-1}), \qquad k = 1, \ldots, \frac{N}{2} \qquad (12.17)$$

An Algorithm for Performing Lifting with the LeGall Filter

We can easily generalize the ideas employed to transform a 10-vector via lifting to develop a lifting algorithm for arbitrary even-length vectors.

Algorithm 12.2 (Lifting with the LeGall Filter Pair). *This algorithm takes a vector* **v** *of even length N and, using the LeGall filter (12.5), returns the same modified biorthogonal wavelet transform that is returned by Algorithm 11.4 from Section 11.3.*

The algorithm starts by partitioning **v** *into vectors* **e** *and* **o***. These vectors contain the even and odd components of* **v***, respectively. We next compute the highpass portion* **d** *of the transformation using (12.16), and then use* **d** *and* **o** *in (12.17) to construct the lowpass portion* **s** *of the transformation.*

Algorithm: LWT1D1

Input: Vector **v** of even length N.

Output: Modified biorthogonal wavelet transformation using LeGall filters.

// Partition **v** into even and odd vectors.

For $[k = 1, \, k \le \frac{N}{2}, \, k++,$
    $e_k = v_{2k}$
    $o_k = v_{2k-1}$
]

// Add an extra component to **o** for use computing $d_{N/2}$.

$o_{N/2+1} = o_{N/2}$

// Compute the highpass portion **d**.

For $[k = 1, k \leq \frac{N}{2}, k++,$
$$d_k = e_k - \frac{1}{2}(o_k + o_{k+1})$$
$]$

```
// Add an extra component to d for use computing s_1.
```

$$d_0 = d_1$$

```
// Compute the lowpass portion s.
```

For $[k = 1, k \leq \frac{N}{2}, k++,$
$$s_k = o_k + \frac{1}{4}(d_k + d_{k-1})$$
$]$

Return [**Join** [s,d]]

$\square$

Let's look at an example.

Example 12.3 (Using the Lifting Scheme). *Let* $\mathbf{v} = [-5, 5, 0, 1, -3, 4, -1, -2]^T$. *Compute the modified biorthogonal transform with the LeGall filter pair using the lifting method.*

Solution. *We begin by parititioning* $\mathbf{v}$ *into even and odd vectors each of length* 4. *We have*

$$\mathbf{o} = [-5, 0, -3, -1]^T \qquad and \qquad \mathbf{e} = [5, 1, 4, -2]^T$$

We next set $o_5 = o_4 = -1$ *and use (12.16) to compute the highpass portion* $\mathbf{d}$. *We have*

$$
\begin{aligned}
d_1 &= e_1 - \tfrac{1}{2}(o_1 + o_2) = & 5 - \tfrac{1}{2}(-5 + 0) = & \quad \tfrac{15}{2} \\
d_2 &= e_2 - \tfrac{1}{2}(o_2 + o_3) = & 1 - \tfrac{1}{2}(0 - 3) = & \quad \tfrac{5}{2} \\
d_3 &= e_3 - \tfrac{1}{2}(o_3 + o_4) = & 4 - \tfrac{1}{2}(-3 - 1) = & \quad 6 \\
d_4 &= e_4 - \tfrac{1}{2}(o_4 + o_5) = & -2 - \tfrac{1}{2}(-1 - 1) = & \quad -1
\end{aligned}
$$

Next we set $d_0 = d_1 = \frac{15}{2}$ *and use (12.17) to compute the lowpass portion* $\mathbf{s}$ *of the transform.*

$$
\begin{aligned}
s_1 &= o_1 + \tfrac{1}{4}(d_1 + d_0) = -5 + \tfrac{1}{4}\left(\tfrac{15}{2} + \tfrac{15}{2}\right) = -\tfrac{5}{4} \\
s_2 &= o_2 + \tfrac{1}{4}(d_2 + d_1) = 0 + \tfrac{1}{4}\left(\tfrac{5}{2} + \tfrac{15}{2}\right) = \tfrac{5}{2} \\
s_3 &= o_3 + \tfrac{1}{4}(d_3 + d_2) = -3 + \tfrac{1}{4}\left(6 + \tfrac{5}{2}\right) = -\tfrac{7}{8} \\
s_4 &= o_4 + \tfrac{1}{4}(d_4 + d_3) = -1 + \tfrac{1}{4}(-1 + 6) = \tfrac{1}{4}
\end{aligned}
$$

$\square$

Inverting the Lifting Process

It is a straightforward process to invert the lifting process. We assume that s and d are given and follow the notation in Algorithm 12.2 to set $d_0 = d_1$. Then solving for s_k in (12.17) gives

$$o_k = s_k - \frac{1}{4}(d_k + d_{k-1}), \qquad k = 1, \ldots, \frac{N}{2} \qquad (12.18)$$

Now we have the odd components **o** of **v**! If we set $o_{N/2+1} = o_{N/2}$ we can solve (12.16) for e_k:

$$e_k = d_k + \frac{1}{2}(o_k + o_{k+1}), \qquad k = 1, \ldots, \frac{N}{2} \qquad (12.19)$$

Thus, to obtain **v**, we simply intertwine the elements of **o** and **e**! In Problem 12.13 you are asked to write an algorithm for performing inverse lifting.

Lifting and Integer to Integer Transforms

Suppose that we are given a vector **v** of even length whose elements are integers. If we apply the modified biorthogonal transform via Algorithm 11.4 using the LeGall filter pair, integers or rational numbers with denominators 2, 4, or 8 are returned (see Problem 12.15 from Section 12.2). Rounding or truncating these values to the nearest integer, applying the inverse transform, and rounding or truncating the result does not necessarily return the original vector **v** (see Problem 12.14). As you will see in Problem 12.16, it is possible to modify the LeGall filter pair and then use Algorithm 11.4 to map integers to integers. The process is reversible, but it is not desirable because of its adverse effect on the quantization step.

The big advantage to the lifting process is that it can easily be adjusted to map integers to integers in a such a way as to make the inverse transform exact. Let **v** be a vector of even length N and suppose that it has been partitioned into even and odd parts **e** and **o**. Recall that for $o_{N/2+1} = o_{N/2}$, we built the highpass portion **d** of the transform via the formula (12.16)

$$d_k = e_k - \frac{1}{2}(o_k + o_{k+1}), \qquad k = 1, \ldots, \frac{N}{2}$$

and then constructed the highpass portion by setting $d_0 = d_1$ and using (12.17) to compute

$$s_k = o_k + \frac{1}{4}(d_k + d_{k-1}), \qquad k = 1, \ldots, \frac{N}{2}$$

To alter these formulas so that they map integers to integers, we simply take

$$\boxed{d_k^* = e_k - \left\lfloor \frac{1}{2}(o_k + o_{k+1}) \right\rfloor, \qquad k = 1, \ldots, \frac{N}{2}} \qquad (12.20)$$

where $o_{N/2+1} = o_{N/2}$ to obtain the integer-valued highpass portion $\mathbf{d}^*$ of the transform. Here, $\lfloor \cdot \rfloor$ is the floor function introduced in Problem 7.1(c) in Section 6.4. Once the integer-valued d_k^* are known, we set $d_0^* = d_1^*$ and compute

$$s_k^* = o_k + \left\lfloor \frac{1}{4}\left(d_k^* + d_{k-1}^*\right) + \frac{1}{2}\right\rfloor, \qquad k = 1, \ldots, \frac{N}{2} \tag{12.21}$$

to obtain the integer-valued lowpass portion $\mathbf{s}^*$ of the transform.

As the authors point out in [13], $\frac{1}{2}$ is added to the floored part of s_k^* to eliminate bias. In Problem 12.18 you will further explore this step.

Let's look at an example.

Example 12.4 (Mapping Integers to Integers Using Lifting). *We return to Example 12.3. If we apply (12.20) and (12.21) to* $\mathbf{v}$, *we obtain*

$$
\begin{aligned}
d_1^* &= e_1 - \left\lfloor \tfrac{1}{2}\left(o_1 + o_2\right)\right\rfloor = 5 - \left\lfloor \tfrac{1}{2}(-5 + 0)\right\rfloor = 8 \\
d_2^* &= e_2 - \left\lfloor \tfrac{1}{2}\left(o_2 + o_3\right)\right\rfloor = 1 - \left\lfloor \tfrac{1}{2}(0 - 3)\right\rfloor = 3 \\
d_3^* &= e_3 - \left\lfloor \tfrac{1}{2}\left(o_3 + o_4\right)\right\rfloor = 4 - \left\lfloor \tfrac{1}{2}(-3 - 1)\right\rfloor = 6 \\
d_4^* &= e_4 - \left\lfloor \tfrac{1}{2}\left(o_4 + o_5\right)\right\rfloor = -2 - \left\lfloor \tfrac{1}{2}(-1 - 1)\right\rfloor = -1
\end{aligned}
$$

and

$$
\begin{aligned}
s_1^* &= o_1 + \left\lfloor \tfrac{1}{4}\left(d_1^* + d_0^*\right) + \tfrac{1}{2}\right\rfloor = -5 + \left\lfloor \tfrac{1}{4}(8 + 8) + \tfrac{1}{2}\right\rfloor = -1 \\
s_2^* &= o_2 + \left\lfloor \tfrac{1}{4}\left(d_2^* + d_1^*\right) + \tfrac{1}{2}\right\rfloor = 0 + \left\lfloor \tfrac{1}{4}(3 + 8) + \tfrac{1}{2}\right\rfloor = 3 \\
s_3^* &= o_3 + \left\lfloor \tfrac{1}{4}\left(d_3^* + d_2^*\right) + \tfrac{1}{2}\right\rfloor = -3 + \left\lfloor \tfrac{1}{4}(6 + 3) + \tfrac{1}{2}\right\rfloor = -1 \\
s_4^* &= o_4 + \left\lfloor \tfrac{1}{4}\left(d_4^* + d_3^*\right) + \tfrac{1}{2}\right\rfloor = -1 + \left\lfloor \tfrac{1}{4}(-1 + 6) + \tfrac{1}{2}\right\rfloor = 0
\end{aligned}
$$

$\square$

Inverting the Integer Lifting Process

To invert this process, we simply first solve (12.21) for o_k to obtain the identity

$$o_k = s_k^* - \left\lfloor \frac{1}{4}\left(d_k^* + d_{k-1}^*\right) + \frac{1}{2}\right\rfloor, \qquad k = 1, \ldots, \frac{N}{2} \tag{12.22}$$

where $d_0^* = d_1^*$. Note that this part of the inversion process is exact — s_k^* was built by adding some number to o_k, so o_k can be recovered by subtracting that value from s_k^*.

Once we have the odd vector $\mathbf{o}$, we solve (12.20) for e_k to write

$$e_k^* = d_k^* + \left\lfloor \frac{1}{2} \left(o_k + o_{k+1} \right) \right\rfloor, \qquad k = 1, \ldots, \frac{N}{2} \tag{12.23}$$

where $o_{N/2+1} = o_{N/2}$. For now we have denoted the even vector in (12.23) by $\mathbf{e}^*$. The reason we do this is that we have not yet verified that (12.23) returns $\mathbf{e}$.

Why does this process work? It turns out that the answer relies on a simple proposition.

Proposition 12.1 (Flooring Half-Integers). *Suppose that a and b are integers. Then*

$$\left\lfloor \frac{1}{2} \left(a + b \right) \right\rfloor = \begin{cases} \frac{1}{2} \left(a + b \right), & a + b \ even \\ \frac{1}{2} \left(a + b \right) - \frac{1}{2}, & a + b \ odd \end{cases}$$

$\square$

Proof of Proposition 12.1. The proof is left as Problem 12.20. $\square$

If we apply Proposition 12.1 to (12.20), simplify, and compare the result to (12.16), we see that

$$d_k^* = e_k - \left\lfloor \frac{1}{2} \left(o_k + o_{k+1} \right) \right\rfloor$$

$$= e_k - \begin{cases} \frac{1}{2} \left(o_k + o_{k+1} \right), & \text{if } o_k + o_{k+1} \text{ is even} \\ \frac{1}{2} \left(o_k + o_{k+1} \right) - \frac{1}{2}, & \text{if } o_k + o_{k+1} \text{ is odd} \end{cases}$$

$$= \begin{cases} e_k - \frac{1}{2} \left(o_k + o_{k+1} \right), & \text{if } o_k + o_{k+1} \text{ is even} \\ e_k - \frac{1}{2} \left(o_k + o_{k+1} \right) + \frac{1}{2}, & \text{if } o_k + o_{k+1} \text{ is odd} \end{cases}$$

$$= \begin{cases} d_k, & \text{if } o_k + o_{k+1} \text{ is even} \\ d_k + \frac{1}{2}, & \text{if } o_k + o_{k+1} \text{ is odd} \end{cases} \tag{12.24}$$

Let's consider (12.23) in the case where $o_k + o_{k+1}$ is even. By (12.24) we know that $d_k^* = d_k$ and Proposition 12.1 gives $\left\lfloor \frac{1}{2} \left(o_k + o_{k+1} \right) \right\rfloor = \frac{1}{2} \left(o_k + o_{k+1} \right)$. Inserting these results into (12.23) and comparing to (12.19) gives

$$e_k^* = d_k^* + \left\lfloor \frac{1}{2} \left(o_k + o_{k+1} \right) \right\rfloor$$

$$= d_k + \frac{1}{2} \left(o_k + o_{k+1} \right)$$

$$= e_k$$

Now if $o_k + o_{k+1}$ is odd, (12.24) tells us that $d_k^* = d_k + \frac{1}{2}$ and Proposition 12.1 gives $\lfloor \frac{1}{2}(o_k + o_{k+1}) \rfloor = \frac{1}{2}(o_k + o_{k+1}) - \frac{1}{2}$. Inserting these results into (12.23) and comparing to (12.19) gives

$$
e_k^* = d_k^* + \lfloor \frac{1}{2}(o_k + o_{k+1}) \rfloor
$$

$$
= d_k + \frac{1}{2} + \frac{1}{2}(o_k + o_{k+1}) - \frac{1}{2}
$$

$$
= e_k
$$

In both cases we see that $e_k^* = e_k$, so (12.23) exactly returns the even vector e!

PROBLEMS

12.12   Let $v = [3, 1, 0, 5, 10, 20, 0, 17]^T$. Compute the modified biorthogonal transform of v by hand using the lifting method (Algorithm 12.2). Compare your result to that of Problem 11.6 in Section 12.2.

12.13   Write an algorithm that implements (12.18) and (12.19) and inverts the lifting scheme.

12.14   For v from Problem 12.12, compute the modified biorthogonal transform (Algorithm 11.4) using the LeGall filter pair and use $\lfloor \cdot \rfloor$ to truncate the elements of the transform to integers. Next, compute the inverse biorthogonal transform using Algorithm 11.5. Do you recover v?

12.15   Suppose that v is a vector of even length whose values are integers from the set $\{-128, \ldots, 0, \ldots, 127\}$. If we apply Algorithm 11.4 to v using the LeGall filter pair, explain why elements of the lowpass portion of the data are integers or half-integers from the set $\{-256, \ldots, 254\}$ and the elements of the highpass portion of the data are rational numbers with denominators 1, 2, 4, or 8 from the set $\{-765/4, \ldots, 765/4\}$.

12.16   How could you modify the modified biorthogonal transform that uses the LeGall filter pair to map integers to integers? How would this modification affect the encoding process?

12.17   Let v be as given in Problem 12.12. Use (12.20) and (12.21) to compute the integer-valued biorthogonal wavelet transform of v. Now use (12.22) and (12.23) to exactly recover v.

12.18   In this problem we will investigate further the reason for adding $\frac{1}{2}$ to the floored part of (12.21). The following steps will help you organize your work.

(a) Suppose that a is an integer. Explain why, for some integer n, a can written as exactly one of $4n$, $4n + 1$, $4n + 2$, or $4n + 3$.

(b) Now suppose that a and b are integers. Then from part (a), a is either $4n$, $4n + 1$, $4n + 2$, or $4n + 3$, and b is either $4m$, $4m + 1$, $4m + 2$, or $4m + 3$. Now consider the value $(a + b)/4$. How many ways can this term be expressed in terms of n and m?

(c) How many of the expressions that you obtained in part (b) are mapped to $n + m$ when we compute $\lfloor (a + b)/4 \rfloor$?

(d) How many of the expressions that you obtained in part (b) are mapped to $n + m$ when we compute $\lfloor (a + b)/4 + \frac{1}{2} \rfloor$?

12.19   Consider $\mathbf{v}$ as given in Example 12.3. In Example 12.4 we used (12.20) and (12.21) to create an integer-valued transform. Use (12.22) and then (12.22) on the vectors $\mathbf{d}^*$ and $\mathbf{s}^*$ from Example 12.4 to show that we recover $\mathbf{v}$ exactly.

12.20   Prove Proposition 12.1.

12.21   Suppose that we replace the floor function $\lfloor \cdot \rfloor$ by the *ceiling function* $\lceil \cdot \rceil$ in (12.20), (12.20), (12.22), and (12.22). Is the integer-to-integer transformation process still reversible? If your answer is yes, provide a proof. Here, the ceiling function $\lceil t \rceil$ returns the smallest integer greater than or equal to t.

12.22   Repeat Problem 12.21, but in this case use normal rounding instead of the ceiling function.

12.23   A floating-point operation (FLOP) is defined to be an arithmetic operation (usually either addition or multiplication) performed on two real numbers using floating-point arithmetic (see Burden and Faires [11] for more information on floating-point arithmetic). FLOPs are often used to measure the speed and efficiency of an algorithm.

Suppose that our task is to compute one iteration of the modified biorthogonal wavelet transformation (stored in $\mathbf{y}$) of N-vector $\mathbf{v}$ (N even) using the LeGall filter pair. In this problem you will calculate how many FLOPs are necessary to compute $\mathbf{y}$ using Algorithm 11.4 and the number of FLOPs needed to compute $\mathbf{y}$ using lifting (Algorithm 12.2).

(a) Algorithm 11.4, modulo increasing the length of $\mathbf{v}$ and then dropping some terms, computes an arbitrary element of the lowpass portion of the transform as

$$-\frac{1}{8}v_{j-2} + \frac{1}{4}v_{j-1} + \frac{3}{4}v_j + \frac{1}{4}v_{j+1} - \frac{1}{8}v_{j+2}$$

Assuming that we do not count the FLOPs necessary to form $\frac{1}{8}$, $\frac{1}{4}$, and $\frac{3}{4}$, how many additions are performed in this computation? How many multiplications are performed in this computation?

(b) Using part (a) determine how many FLOPs are needed to compute the lowpass portion of the transform.

(c) An arbitrary element of the highpass portion of the transform is given by

$$-\frac{1}{2}v_{j-1} + v_j - \frac{1}{2}v_{j+1}$$

How many FLOPs are needed to compute the highpass portion of the transform? Combine your answer with that obtained in part (b) to determine the total number of FLOPs needed to compute the transform of $\mathbf{v}$ using Algorithm 11.4.

(d) Using (12.16), verify that the number of FLOPs needed to compute the highpass portion of the transform using lifting is the same as that needed for Algorithm 11.4.

(e) Using (12.17), find the number of FLOPs needed to compute the lowpass portion of the transform via lifting.

(f) Combine your answers in parts (d) and (e) to determine the number of FLOPs needed to compute the transform using Algorithm 12.2. Which algorithm requires fewer FLOPs?

Computer Lab

♦ **12.2 Software Development: The Lifting Scheme.** From the text Web site, access the package `lifting`. In this development lab you will generate code that will implement Algorithm 12.2 and use the lifting scheme to produce the modified wavelet transform of an even-length vector $\mathbf{v}$ using the LeGall filter pair. You will also write a module to perform the inverse transform via lifting. Modules for iterative one- and two-dimensional transforms and their inverses will be created as well. A flag will be passed to each module to indicate whether or not the output should be integer valued. Instructions are provided that allow you to add the code to the software package `DiscreteWavelets`.

12.4   EXAMPLES

We conclude this chapter with some examples of how the naive JPEG2000 algorithm works. We perform lossy compression and lossless compression of the image in Figure 12.1(a) and then compress a color image.

Example 12.5 (Lossy JPEG2000 Compression). *We first subtract 128 from each element of the image matrix and then perform three iterations of the modified biorthogonal wavelet transform (Algorithm 11.4) using the CDF97 filter pair. The result is plotted in Figure 12.8(a).*

The next step is to quantize the wavelet transform. The quantization step size vectors are given in Example 12.2. We use these step sizes with the quantization

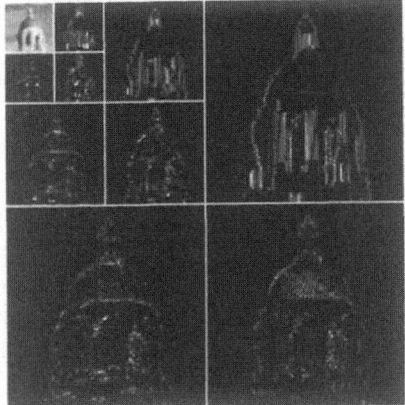

(a) Biorthogonal wavelet transformation

(b) Quantized biorthogonal wavelet transformation

Figure 12.8   Three iterations of the biorthogonal wavelet transform using the CDF97 filter pair and the quantized version of the transform.

function (12.8) and quantize each portion of the wavelet transform. The quantized transform is plotted in Figure 12.8(b).

At this point, the quantized transform would be encoded. Since the development of arithmetic coding (not to mention the EBCOT coding method) is beyond the scope of this course, we are left only with Huffman coding as a way to encode the quantized transform. Using Algorithm 3.1 on the quantized transform reduces the data from $320,000$ bits (the original image is 200×200) to $89,418$ bits. Thus, we would need *about* 2.23545 *bpp to encode the data. Rest assured that the EBCOT method would produce much better results.*

The encoded transform would be submitted, and to view the image, the steps are reversed. The data are decoded, and then each portion of the transform is multiplied by the appropriate quantization step size. The inverse biorthogonal wavelet transform is applied and we add 128 *to the result. The compressed image is plotted in Figure 12.9.*

Figure 12.9   The JPEG2000 compressed version of the original image.

The PSNR is computed to be 41.6303 *and this value is higher than the PSNR value* 31.9113 *we achieved using JPEG. Note also that the block artifacts that appear using the JPEG standard (see Figure 12.5) are no longer present.*

In Figure 12.10 we have plotted the bottom left corner of Figure 12.9. We have also plotted the same region for the JPEG compressed version. Note that the block artifacts that are present in the JPEG version no longer appear in the JPEG2000 version. You can also compare the results with the plot of the bottom left corner of the original image in Figure 12.6(a).   □

Example 12.6 (Lossless JPEG2000 Compression). *In this example we apply lossless compression to the image plotted in Figure 12.1(a). Let A be the* 200×200 *matrix that holds the grayscale intensity values of the image.*

We begin by subtracting 128 *from each element in A. We now apply three iterations of the modified biorthogonal wavelet transformation with the LeGall filter using the lifting method (Algorithm 12.2). We utilize (12.21) and (12.20) so that the transform values are integer valued. The biorthogonal wavelet transform is plotted in Figure 12.11.*

At this point we would use the EBCOT method to encode the data. Since we do not develop EBCOT in this text, we will instead use the basic Huffman coding developed

(a) JPEG2000 compressed version. (b) JPEG compressed version.

Figure 12.10   A plot of the bottom left corner of the JPEG2000 and the JPEG compressed versions of the original image A.

in Section 3.4. Using Huffman encoding, the transform can be stored using 163937 *bits and since the original image is comprised of* $200 \times 200 = 40,000$ *pixels, the storage rate is* 4.098425 *bpp.*

To invert the process and recover image A *exactly, we decode and then lift in conjunction with (12.22) and (12.23) to compute the inverse biorthogonal wavelet transform.*   □

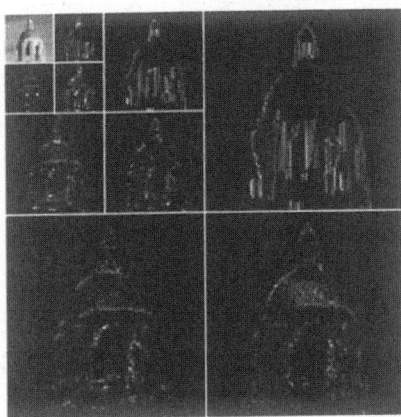

Figure 12.11   The biorthogonal wavelet transform of the image.

Example 12.7 (Color JPEG2000 Compression). *In this example we use Algorithm 12.1 to compress the image plotted in Figure 12.12. The dimensions of this image are 256×384.*

Figure 12.12 The original color image. See the color plates for a color version of this image.

The original image is given in RGB space. The first step is to subtract 128 from each of the R, G, and B matrices and then convert the result to YCbCr space. These channels are plotted in Figure 12.13.

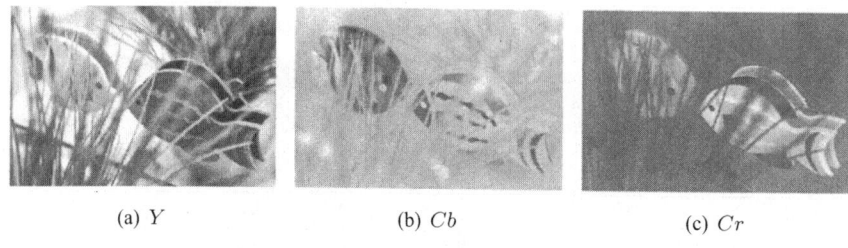

(a) Y (b) Cb (c) Cr

Figure 12.13 The Y, Cb, and Cr channels for the image in Figure 12.12.

We next compute three iterations of the biorthogonal wavelet transform using the CDF97 filter on each of Y, Cb, and Cr. We then use vectors $\mathbf{d}^1, \mathbf{d}^2$, and $\mathbf{d}^3$ in conjunction with the quantization function (12.8) to quantize each transform. The process here is exactly the same as in Example 12.5 except that here we must perform it on three different matrices. The wavelet transforms and their quantized versions are plotted in Figure 12.14.

At this point we would encode using EBCOT. As in Example 12.5, we instead use basic Huffman encoding to obtain a basic understanding of the savings. Table 12.1

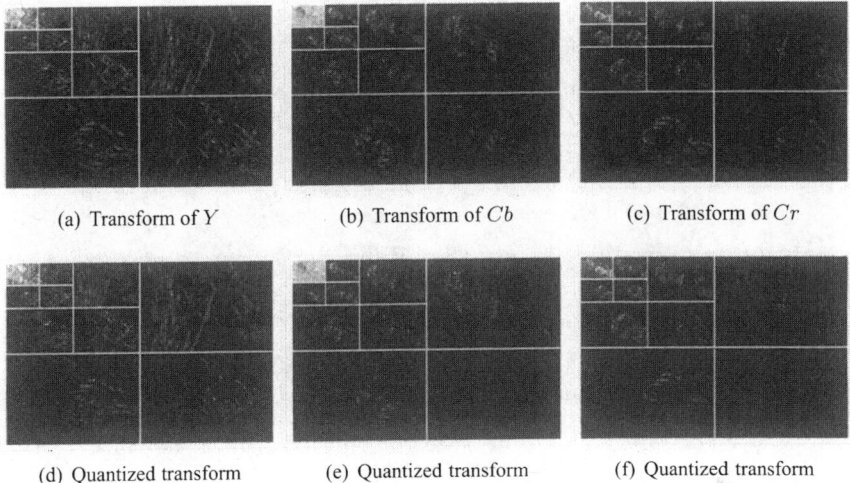

(a) Transform of Y (b) Transform of Cb (c) Transform of Cr

(d) Quantized transform (e) Quantized transform (f) Quantized transform

Figure 12.14 Three iterations of the biorthogonal wavelet transform using the CDF97 filter on each of the Y, Cb, and Cr channels of the original image. The quantized transforms appear below each transform.

gives the results. Recall that each of the R, G, and B channels of the original image consists of $256 \times 384 \times 8 = 786{,}432$ bits.

Table 12.1 Bits needed to encode the Y, Cb, and Cr channels using basic Huffman encoding.

Channel	Bits Needed	bpp
Y	224,836	2.287150
Cb	143,138	1.456075
Cr	136,070	1.384176

To invert the process, we multiply each portion of the transform of the quantized transform of Y by the appropriate quantization step size and then apply three iterations of the inverse biorthogonal wavelet transform. We repeat this process for the quantized transforms of the Cb and Cr channels. We now have compressed versions of Y, Cb, and Cr. These compressed versions are converted back to RGB space, and we then add 128 to each of the R, G, and B channels. The result is plotted in Figure 12.15(a). For comparative purposes we have plotted a compressed version of the original image using the JPEG standard in Figure 12.15(b).

To measure the PSNR for color images, we use only the Y channels. The PSNR for the JPEG2000 compressed image is 41.3876 while the PSNR for the JPEG compressed

(a) JPEG2000   (b) JPEG

Figure 12.15 The compressed image using JPEG2000 and JPEG. See the color plates for color versions of these images.

image is 34.6919. *In Figure 12.16, we plot the upper left quarter of each compressed image. Note that the JPEG version has block artifacts that are not present in the JPEG2000 image.* ☐

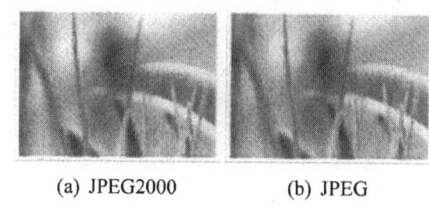

(a) JPEG2000   (b) JPEG

Figure 12.16 The bottom left quarters of the images from Figure 12.15. See the color plates for color versions of these images.

Computer Lab

12.3 JPEG2000 Compression. From the text Web site, access the lab jpeg2000. In this lab you will use the ideas of this section to write a module that performs JPEG2000 compression on a digital image.

APPENDIX A

BASIC STATISTICS

The material in Chapter 9 makes use of several ideas from basic statistics. We also use some simple descriptive statistics in the edge detection application in Section 6.4. In this appendix we review some basic statistical concepts.

A.1 DESCRIPTIVE STATISTICS

In Chapter 6 we make use of some basic descriptive statistics. We utilize the mean absolute deviation in Chapter 9. We start first with the definition of the mean.

Definition A.1 (Mean). *Let* $\mathbf{x} = (x_1, x_2, \ldots, x_n)$ *be a list of numbers. Then the mean* $\overline{x}$ *is defined as*

$$\overline{x} = \frac{x_1 + x_2 + \cdots + x_n}{n} = \frac{1}{n} \sum_{k=1}^{n} x_k$$

$\square$

A second statistic that is used to measure the center of the data is the median. The median is a number x_{med} such that the number of data below x_{med} is equal to the number of data above x_{med}.

Definition A.2 (Median). *Let* $\mathbf{x} = (x_1, x_2, \ldots, x_n)$ *be a list of numbers. The median* x_{med} *is constructed by first sorting the values* $x_1, x_2, \ldots, x_n$ *from smallest to largest. Call this new list* $\mathbf{y} = (y_1, y_2, \ldots, y_n)$. *If* n *is odd, we set* $x_{med} = y_{\frac{n+1}{2}}$. *If* n *is even, we take the median to be the average of the two middle components. That is,* $x_{med} = \frac{1}{2}\left(y_{n/2} + y_{n/2+1}\right)$. $\square$

The mean and the median are two descriptive statistics that tell us something about the "center" of the data. Unfortunately, neither tell us about the spread of the data. We use the ideas of variance and standard deviation to assist us in analyzing the spread of a data set.

Definition A.3 (Variance and Standard Deviation). *Suppose that the list of numbers* $\mathbf{x} = (x_1, x_2, \ldots, x_n)$ *has mean* $\bar{x}$. *The* variance *of this data set is defined by*

$$s^2 = \frac{1}{n-1}\sum_{k=1}^{n}(x_k - \bar{x})^2 \tag{A.1}$$

The standard deviation *is defined to be the nonnegative square root of the variance.* $\square$

It is interesting to note that the divisor in (A.1) is $n - 1$ rather than n. You will sometimes see the sample variance defined with an n in the divisor, but this produces a *biased estimator*. If you are interested in exploring this topic further, consult an introductory text on probability and statistics such as DeGroot [27]. The following example illustrates the descriptive statistics defined thus far.

Example A.1 (Computing Descriptive Statistics). *Find the mean, median, variance, and standard deviation of* $\mathbf{x} = (6, -3, 17, 0, 5)$, $\mathbf{y} = (12, 19, -6, 0, -2, -11)$. *Solution. For the list* $\mathbf{x}$, *we have* $\bar{x} = \frac{1}{5}(6 - 3 + 17 + 0 + 5) = 5$. *If we sort this list, we have* $(-3, 0, 5, 6, 17)$ *so that* $x_{med} = 5$. *Finally, we compute the variance as*

$$s^2 = \frac{1}{4}\left((6-5)^2 + (-3-5)^2 + (17-5)^2 + (0-5)^2 + (5-5)^2\right) = 58.5$$

Thus, the standard deviation is $s = \sqrt{58.5} \approx 7.64853$.

For the list $\mathbf{y}$, *we have* $\bar{y} = \frac{1}{6}(12 + 19 - 6 + 0 - 2 - 11) = 2$. *If we sort the list, we obtain* $(-11, -6, -2, 0, 12, 19)$, *and since* $n = 6$ *is even, we take* $y_{med} = \frac{-2+0}{2} = -1$. *For the variance, we compute*

$$s^2 = \frac{1}{5}\left((12-2)^2 + (19-2)^2 + (-6-2)^2 + (-2-2)^2 + (-11-2)^2 + (0-2)^2\right)$$

$$= 128.4$$

Thus the standard deviation is $s = \sqrt{127.6} \approx 11.296$. □

The *median absolute deviation* turns out to be a useful descriptive statistic for estimating the noise level in noisy signals.

Definition A.4 (Median Absolute Deviation). *Let $\mathbf{v} = (v_1, \ldots, v_N)$ and let v_{med} denote the median of $\mathbf{v}$. Form the vector $\mathbf{w} = (|v_1 - v_{med}|, \ldots, |v_N - v_{med}|)$. We define the* median absolute deviation *of $\mathbf{v}$ as:*

$$MAD(\mathbf{v}) = w_{med} \tag{A.2}$$

□

Example A.2 (Computing the MAD). *Consider the list $\mathbf{y}$ from Example A.1. To compute $MAD(\mathbf{y})$, we recall that $y_{med} = -1$ and then form the new vector*

$$\mathbf{w} = [\,|12 - (-1)|, |19 - (-1)|, |-6 - (-1)|,$$
$$|0 - (-1)|, |-2 - (-1)|, |-11 - (-1)|\,]$$
$$= (13, 20, 5, 1, 1, 10)$$

We next sort this list to obtain $[\,1, 1, 5, 10, 13, 20\,]$. Since $n = 6$ is even, we form the median of this list by averaging 5 and 10. Hence, $MAD(\mathbf{y}) = 7.5$. □

PROBLEMS

A.1 Let $\mathbf{x} = [\,-3, 5, 12, -9, 5, 0, 2\,]$. Find $\bar{x}$, x_{med}, $MAD(\mathbf{x})$, the variance s^2, and the standard deviation s for $\mathbf{x}$.

A.2 The mean of $x_1, x_2, \ldots, x_9$ is $\bar{x} = 7$. Suppose that $x_{10} = 12$. Find $\bar{x}$ for $x_1, x_2, \ldots, x_9, x_{10}$.

A.3 Let $\mathbf{x} = (x_1, \ldots, x_n)$ and let a be any number. Form $\mathbf{y}$ by adding a to each element of $\mathbf{x}$. That is, $\mathbf{y} = (x_1 + a, x_2 + a, \ldots, x_n + a)$. Show that $\bar{y} = \bar{x} + a$ and that the variance of $\mathbf{y}$ is the same as the variance of $\mathbf{x}$.

A.4 Let $\mathbf{x} = (1, 2, \ldots, n)$. Find $\bar{x}$ and the variance of $\mathbf{x}$. (*Hint:* See the hint given in Problems 2.1 and 2.3 in Section 2.1.)

A.5 Let $\mathbf{x} = [\,1, 2, \ldots, n\,]$. Show that $\bar{x} = x_{med}$.

A.6 Let $\mathbf{x} = [\,x_1, \ldots, x_n\,]$ with $a \le x_k \le b$ for each $k = 1, 2, \ldots, n$. Show that $a \le \bar{x} \le b$.

A.7 Let $\mathbf{x} = (x_1, x_2, \ldots, x_n)$. Show that the variance s^2 can be written as

$$s^2 = \frac{1}{n-1} \left(\sum_{k=1}^{n} x_k^2 - n\bar{x}^2 \right)$$

A.2   SAMPLE SPACES, PROBABILITY, AND RANDOM VARIABLES

In probability and statistics, we often analyze an experiment or repeated experiments. As a basis for this analysis, we need to identify a *sample space*.

Sample Spaces

Definition A.5 (Sample Space). *The sample space for an experiment is the set of all possible outcomes of the experiment.*   □

Example A.3 (Identifying Sample Spaces). *Here are some examples of sample spaces.*

(a) *The experiment is to toss a penny, nickel, and dime. If we let H denote heads and T denote tails, then the sample space is*

$$S = \{HHH, HHT, HTH, HTT, THH, THT, TTH, TTT\}$$

(b) *The experiment is to toss a standard red die and a standard green die and record the numbers that appear on top of the dice. The sample space is the set of ordered pairs*

$$S = \{(j, k) \mid j, k = 1, \dots, 6\}$$

The previous two sample spaces are examples of finite sample spaces. Many experiments have sample spaces that are not finite.

(c) *The experiment of measuring the height of the woman from a certain population. Although there are finitely many women in the population, their heights (in feet) can assume any value in the interval* [0,9].

(d) *The experiment of measuring the difference of the true arrival time of a city bus versus the expected arrival time at a certain stop results in an infinite number of possible outcomes.*

□

Probability

Let A be any subset of a sample space S. The set A is also called an *event*. If $A_1, A_2, \dots$ is a sequence of events, we say that $A_1, A_2, \dots$ is a *sequence of disjoint events* if $A_j \cap A_k = \varnothing$ whenever $j \neq k$.

We assign a *probability* to an event A. This value, denoted by $Pr(A)$, indicates the likeliness that A will occur. The range of the numbers is 0 (impossible that A occurs) to 1 (A will certainly occur). Mathematically, we can think of Pr as a function that maps an event to some number in $[0,1]$. The function Pr must satisfy the following three axioms:

(Pr 1) For any event A in sample space S, we have $Pr(A) \geq 0$.

(Pr 2) If S is the sample space, then $Pr(S) = 1$.

(Pr 3) Let $A_1, A_2, \ldots$ be a sequence of disjoint events. Then

$$Pr(A_1 \cup A_2 \cup \ldots) = \sum_{k=1}^{\infty} Pr(A_k)$$

Given a function Pr that satisfies the probability axioms Pr1 to Pr3, we can prove several properties obeyed by Pr.

Proposition A.1. *Let S be a sample space and Pr a probability function that satisfies (Pr 1) to (Pr 3). Then*

(a) $Pr(\varnothing) = 0$.

(b) If $A_1, \ldots, A_n$ is a finite list of disjoint events, then

$$Pr(A_1 \cup A_2 \cup \ldots \cup A_n) = \sum_{k=1}^{n} Pr(A_k)$$

(c) If A^c denotes all the elements in S but not in A, then $Pr(A^c) = 1 - Pr(A)$.

(d) If A and B are events in S with $A \subseteq B$, then $Pr(A) \leq Pr(B)$.

(e) If A and B are any events in S, then

$$Pr(A \cup B) = Pr(A) + Pr(B) - Pr(A \cap B)$$

$\square$

Proof of Proposition A.1. We will prove part (a) and leave the remaining proofs as problems. For part (a), let $A_k = \varnothing$ for $k = 1, 2, \ldots$. Then by (Pr 3), we have

$$Pr(A_1 \cup A_2 \cup \cdots) = \sum_{k=1}^{\infty} Pr(A_k)$$

Now $A_1 \cup A_2 \cup \ldots = \varnothing$, so the identity above can be written as

$$Pr(\varnothing) = \sum_{k=1}^{\infty} Pr(\varnothing)$$

The only way the infinite series on the right converges is if each term is zero.

$\square$

Let's look at an example illustrating these properties.

Example A.4 (Using Probability Properties). *A small remote town has two grocery stores only and the citizens only shop for groceries at these two stores. The probability that a citizen shops at store 1 is 0.65 and the probability that a citizen shops at store 2 is 0.43.*

(a) *What is the probability that a citizen does not shop at store 2?*

(b) *What is the probability that a citizen shops at both stores?*

Solution. *Let A be the event that a citizen shops at Store 1 and B be the event that a citizen shops at Store 2.*

For part (a) we can use part (c) of Proposition A.1. We have $Pr(B^c) = 1 - Pr(B) = 1 - 0.43 = 0.57$.

For part (b), we need to compute $Pr(A \cap B)$. We use the last property of Proposition A.1. Since $A \cup B = S$, we have

$$1 = Pr(S) = Pr(A \cup B) = Pr(A) + Pr(B) - Pr(A \cap B)$$
$$= 0.65 + 0.43 - Pr(A \cap B)$$
$$= 1.08 - Pr(A \cap B)$$

We can solve for $Pr(A \cap B)$ in the above equation to find that $Pr(A \cap B) = 0.08$.

$\square$

Random Variables

We are now ready to define a *random variable*. Random variables are very important in statistical modeling.

Definition A.6 (Random Variable). *Let S be a sample space. A real-valued function $X : S \rightarrow \mathbb{R}$ is called a* random variable. $\square$

Example A.5 (Identifying Random Variables). *Here are some examples of random variables.*

(a) *Consider the sample space S from part (a) of Example A.3. Let's define the random variable X to be number of tails recorded when the three coins are tossed. So if the three coins were tossed and the outcome was HTH, then $X = 1$.*

(b) *Consider the sample space S from part (b) of Example A.3. We can define a random variable X to be the sum of the numbers on the dice. For example, if* the roll was $(2, 5)$, then $X = 7$. We could also define a random variable X to be the number of even numbers that appear on the roll. So if the roll was $(2,5)$, then $X = 1$.

(c) For the sample space S in part (c) of Example A.3, the height of a woman chosen from the population is a random variable.

☐

PROBLEMS

A.8   Suppose that the experiment we conduct is to record various two-card hands dealt from a deck of cards. Describe the sample space.

A.9   Suppose that the experiment we conduct is to measure the outcome of drawing three balls from a container containing two red ball and five blue balls. Describe the sample space.

A.10   Prove parts (b) – (e) of Proposition A.1.

A.11   Children at a summer camp have the opportunity to attend either a music class or a science class. All children must attend at least one class. The probability that a student attends music class is 0.39 and the probability that a student attends science class is 0.67. What is the probability a student attends both classes?

A.12   Let S be the sample space that consists of the possible sums that occur when two fair dice are rolled. Describe three possible random variables for S.

A.13   Let A and B be two events. Let C be the event that exactly one of the events will occur. Write $Pr(C)$ in terms of probabilities involving A and B.

A.14   Let S be the sample space that consists of the possible temperature measurements at an airport. Describe three possible random variables for S.

A.3   CONTINUOUS DISTRIBUTIONS

The random variables in parts (a) and (b) of Example A.5 are examples of *discrete random variables* — i.e., random variables can only assume a finite number of values. A *continuous random variable* can assume every value in a given interval. In part (c) of Example A.5, X can theoretically assume any positive value, say between 0 and 9 feet.

Distributions and Probability Density Functions

In this appendix we are primarily concerned with continuous random variables. We have need to study the *distribution* of a continuous random variable. The following definition makes precise what we mean by the distribution of a continuous random variable.

Definition A.7 (Probability Density Function). *Let X be a random variable. We say that X has a* continuous distribution *if there exists a nonnegative real-valued function $f(t)$ so that for every interval (or finite union of intervals) $A \subseteq \mathbb{R}$, the probability that X assumes a value in A can be written as the integral of f over A. If we use the notation Pr to denote probability, we have*

$$Pr(X \in A) = \int_A f(t)\, dt \tag{A.3}$$

The function $f(t)$ is called a probability density function *and in addition to being nonnegative for $t \in \mathbb{R}$, we have*

$$\int_{-\infty}^{\infty} f(t)\, dt = 1 \tag{A.4}$$

□

The relation (A.4) follows immediately if we take $A = \mathbb{R}$. The probability that X assumes a real value is 1, so we have

$$1 = Pr(X \in \mathbb{R}) = \int_{-\infty}^{\infty} f(t)\, dt$$

Note that if $A = [a, b]$, then (A.4) becomes

$$Pr(a \le X \le b) = \int_a^b f(t)\, dt$$

A simple example of a distribution is the *uniform distribution*. Any random variable X whose probability density function $f(t)$ is constant on the set of nonzero values of X is called a uniform distribution.

Example A.6 (Uniform Distribution). *Political polling is a complicated process, and the results are often rounded to the nearest percent when they are presented to the public. Suppose it is reported that support for a candidate is $p\%$ of the voters, $p = 0, 1, 2, \ldots, 100$. Then the actual percent X of those who support the candidate could be thought of as following a uniform distribution where the probability density function is*

$$f(t) = \begin{cases} 1, & t \in [p - \frac{1}{2}, p + \frac{1}{2}) \\ 0, & t \notin [p - \frac{1}{2}, p + \frac{1}{2}) \end{cases} \tag{A.5}$$

This probability density function is plotted in Figure A.1.
Certainly, $f(t) \ge 0$ and it is easy to check that $\int_{\mathbb{R}} f(t)\, dt = 1$.   □

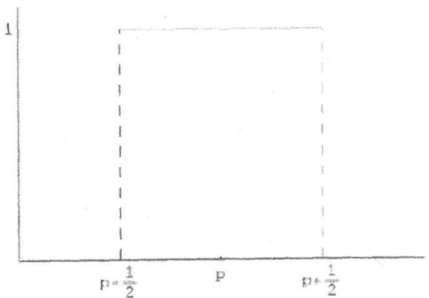

Figure A.1 The uniform distribution for X for Example A.6.

Example A.7 (Computing Probabilities). *Suppose X is the random variable that represents the percentage of tax returns that have errors. It is known (see [36, page 119]) that X follows the distribution*

$$f(t) = \begin{cases} 90t(1-t)^8, & 0 < t < 1 \\ 0, & \text{otherwise} \end{cases}$$

It can easily be verified that $f(t) \geq 0$ and $\int_{\mathbb{R}} f(t)\, dt = 1$.

(a) *Find the probability that there will be fewer than 10% erroneous tax forms in a given year.*

(b) *Find the probability that between 10% and 25% of the tax forms for a given year will have errors.*

Solution. *For part (a) we compute*

$$Pr(X < .1) = \int_0^{.1} 90t(1-t)^8\, dt \approx 26.39\%$$

and for part (b) we compute

$$Pr(.1 < X < .25) = \int_{.1}^{.25} 90t(1-t)^8\, dt \approx 49.21\%$$

$\square$

Distributions Involving Several Random Variables

We can extend the notion of the distribution of a random variable X to higher dimensions. Much of what follows utilizes ideas of multiple integrals from multivariate calculus.

Definition A.8 (Joint Distribution). *If $X_1, \ldots, X_N$ are random variables, we say that $X_1, \ldots, X_N$ have a* continuous joint distribution *if there exists a nonnegative function $f(t_1, \ldots, t_N)$ defined for all points $(t_1, \ldots, t_N) \in \mathbb{R}^N$ so that for every N-dimensional interval (or finite union of intervals) $A \subseteq \mathbb{R}^N$, the probability that $(X_1, \ldots, X_N)$ assumes a value in A can be written as the integral of f over A. That is,*

$$Pr((X_1, \ldots, X_N) \in A) = \int_A f(t_1, \ldots, t_N) \, dt_1 \cdots dt_N \qquad (A.6)$$

The function $f(t_1, \ldots, t_N)$ is called a joint probability density function. *This function also satisfies*

$$1 = Pr((X_1, \ldots, X_N) \in \mathbb{R}^N) = \int_{\mathbb{R}^N} f(t_1, \ldots, t_N) \, dt_1 \ldots dt_N = 1 \qquad (A.7)$$

$\square$

Here is an example of a joint probability function in $\mathbb{R}^2$.

Example A.8 (Bivariate Distribution). *Consider the function*

$$f(x, y) = \begin{cases} c x^2 y & 0 \le y \le 4 - x^2 \\ 0 & \text{otherwise} \end{cases}$$

Find the value for c that makes $f(x, y)$ a joint probability density function on $\mathbb{R}^2$ and then find $P(Y \ge -2X + 4)$.
Solution. *Let $A = \{(x, y) \mid 0 \le y \le 4 - x^2\}$. The set A is plotted in Figure A.2(a). For $f(x, y)$ to be a joint probability density function on $\mathbb{R}^2$, we need $f(x, y) \ge 0$ on $\mathbb{R}^2$ and*

$$\int_{\mathbb{R}^2} f(x, y) \, dy \, dx = 1$$

Certainly, for $c > 0$, we have $f(x, y) \ge 0$ for all $(x, y) \in \mathbb{R}^2$. Since $f(x, y) = 0$ for $(x, y) \notin A$, we have

$$\int_{\mathbb{R}^2} f(x, y) \, dy \, dx = \int_A f(x, y) \, dy \, dx = \int_{x=-2}^{2} \int_{y=0}^{4-x^2} c x^2 y \, dy \, dx$$

$$= c \int_{-2}^{2} x^2 \left(\int_0^{4-x^2} y \, dy \right) dx = \frac{c}{2} \int_{-2}^{2} x^2 (4 - x^2)^2 \, dx$$

$$= c \int_0^2 x^2 (4 - x^2)^2 \, dx = \frac{1024}{105} c$$

Since we want the integral equal to 1, we choose $c = \frac{105}{1024}$.
   To find the desired probability, we must integrate over the region where A intersects the set $B = \{(x, y) \mid y \ge -2x + 4\}$. This region is plotted in Figure A.2(b).

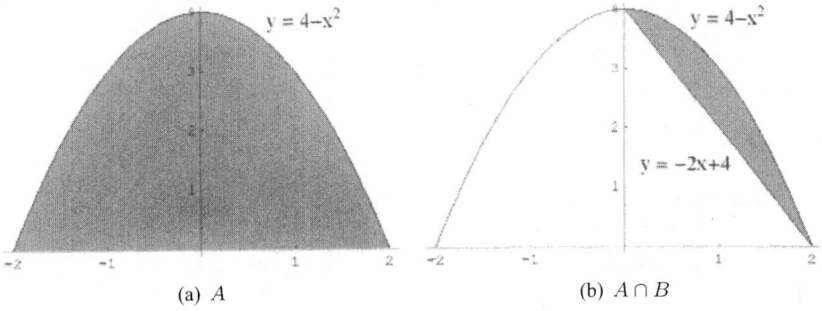

Figure A.2   The sets A and $A \cap B$ from Example A.8.

We can thus compute

$$Pr(Y \geq -2X + 4) = \int_{x=0}^{2} \int_{y=-2x+4}^{4-x^2} \frac{105}{1024} x^2 y \, dy \, dx = \frac{9}{32}$$

□

In many applications we are interested in identifying the distribution function of a particular random variable when we have started with a joint density distribution.

Definition A.9 (Marginal Distribution). *Suppose that f is the joint probability distribution function of random variables $X_1, X_2, \ldots, X_N$. Then the distribution function f_j of random variable X_j is called the* marginal distribution function *of X_j. In the case where $X_1, X_2, \ldots, X_N$ are continuous random variables, we can write $f_j(x)$ as*

$$f_j(x) = \int_{\mathbb{R}^{N-1}} f(x_1, x_2, \ldots, x_N) \, dx_1 \cdots dx_{j-1} \, dx_{j+1} \cdots dx_N \qquad (A.8)$$

□

Thus, we see from (A.8) that we form the marginal distribution function $f_j(x)$ by "integrating out" the remaining variables in the joint distribution $f(x_1, \ldots, x_N)$.

Example A.9 (Marginal Distribution). *Let $f(x, y)$ be the joint probability density function from Example A.8. Compute the marginal distributions $f_1(x)$ and $f_2(y)$.*
Solution. *We first note from Figure A.2 that $f_1(x) = 0$ for $x < -2$ and $x > 2$. For $-2 \leq x \leq 2$, we need to integrate $f_1(x)$ over all possible values of y. But for a fixed $x \in [-2, 2]$, y ranges from 0 to $4 - x^2$, so we can compute*

$$f_1(x) = \int_{-\infty}^{\infty} f(x, y) \, dy = \int_{0}^{4-x^2} \frac{105}{1024} x^2 y \, dy = \frac{105}{2048} x^2 (4 - x^2)^2$$

For $f_2(y)$, we again use Figure A.2 to see that $f_2(y) = 0$ for $y < 0$ and $y > 4$. Moreover, when $0 \leq y \leq 4$, we can solve $y = 4 - x^2$ for x and deduce that $f(x, y) = 0$ unless $-\sqrt{4 - y} \leq x \leq \sqrt{4 - y}$. So the marginal distribution is

$$f_2(y) = \int_{-\infty}^{\infty} f(x, y)\, dx = \frac{105}{1024}\, y \int_{-\sqrt{4-y}}^{\sqrt{4-y}} x^2\, dx = \frac{35}{512}\, y\, (4 - y)^{3/2}$$

The marginal distributions $f_1(x)$ and $f_2(y)$ are plotted in Figure A.3.

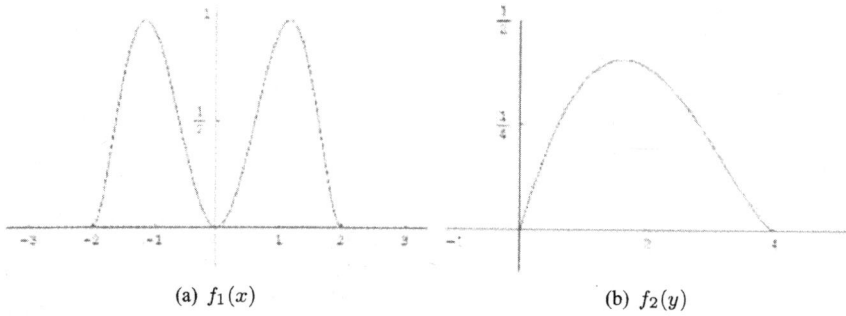

(a) $f_1(x)$ (b) $f_2(y)$

Figure A.3   Marginal distributions.

□

Independent Random Variables

We will often be interested in random variables that are *independent*. For example, suppose that X and Y are random variables. Then X and Y are independent if the value of X has no affect on the value of Y, and vice versa. We can write this mathematically as follows:

Definition A.10 (Independent Random Variables). *Let X and Y be random variables. Then for any two sets of real numbers A and B, we say X and Y are independent if*

$$Pr(X \in A \text{ and } Y \in B) = Pr(X \in A) \cdot Pr(Y \in B)$$

We can easily extend this definition for N random variables. We say that the random variables $X_1, \ldots, X_N$ are independent if for any N sets of real numbers $A_1, \ldots, A_N$, we have

$$Pr(X_1 \in A_1, X_2 \in A_2, \ldots, X_N \in A_N) = Pr(X_1 \in A_1) \cdots Pr(X_N \in A_N)$$

□

The next result follows without proof. The proof uses Definition A.10 along with marginal density functions (see [27], pp. 150-151).

Proposition A.2. *Suppose that $X_1, X_2, \ldots, X_N$ are continuous random variables with joint probability density function $f(x_1, \ldots, x_N)$. Further, let $f_j(x_j)$ denote the marginal probability density function of X_j, $j = 1, \ldots, N$. Then $X_1, X_2, \ldots, X_N$ are independent random variables if and only if*

$$f(x_1, x_2, \ldots, x_N) = f_1(x_1)f_2(x_2) \cdots f_N(x_N) \tag{A.9}$$

$\square$

PROBLEMS

A.15   Suppose that $f(t)$ is the uniform distribution on $[a, b]$ with $a < b$. Find $f(t)$.

A.16   Let $r \in \mathbb{R}$ and set $S = \{r\}$. Using (A.3), compute $Pr(X \in S)$. Why does your answer makes sense for a continuous random variable X?

A.17   Consider the function

$$f(t) = \begin{cases} ce^{-2t}, & t \geq 0 \\ 0, & t < 0 \end{cases}$$

(a) Find the value for c so that $f(t)$ is a probability density function.

(b) Find $Pr(2 < X < 3)$.

(c) Find $Pr(X > 1)$.

A.18   Let c be a constant and suppose that X and Y have a continuous joint distribution

$$f(x, y) = \begin{cases} c, & 0 \leq x \leq 2, 0 \leq y \leq 3, \\ 0, & \text{otherwise} \end{cases}$$

(a) What value must c assume?

(b) Find the marginal densities $f_1(x)$ and $f_2(y)$.

A.19   Suppose that a point (X, Y) is selected at random from the semicircle $0 \leq y \leq \sqrt{4 - x^2}$, $-2 \leq x \leq 2$.

(a) Find the joint probability density function for X and Y. (Hint: Since the point is chosen at random, the joint probability density function must be constant inside the semicircle and zero outside the semicircle.)

(b) Find $Pr(X \geq Y)$.

A.20   Let X and Y be random variables with joint probability density function

$$f(x, y) = \begin{cases} 9xy, & 0 \le x \le 2, \, 0 \le y \le \frac{1}{3} \\ 0, & \text{otherwise} \end{cases}$$

Are X and Y independent random variables?

A.21   For the joint density function in Problem A.19, find the marginal densities $f_1(x)$ and $f_2(y)$.

A.4 EXPECTATION

If you have had a calculus class, you may remember covering a section on moments and centers of mass. In particular, the x-coordinate of the center of mass of a plate is given by

$$\bar{x} = \frac{1}{A} \int_a^b x f(x) \, dx \tag{A.10}$$

where A is the area of the plate. This concept can serve as motivation for the statistical concept of *expectation* or *expected value*.

Definition A.11 (Expected Value). *The* expected value *of a random variable X with probability density function $f(t)$ is*

$$\boxed{E(X) = \int_{-\infty}^{\infty} t f(t) \, dt} \tag{A.11}$$

□

Suppose that the probability density function $f(t)$ is zero outside an interval $[a, b]$. Then

$$E(X) = \int_a^b t f(t) \, dt$$

looks a lot like (A.10). As a matter of fact, it is the same since the "plate" we consider is the region bounded by $f(t)$ and the t-axis, and we know from (A.4) that this area is 1.

The expected value of X is also called the *mean* of X. We can think of the expected value as an average value of the random variable X.

Example A.10 (Expected Value). *Consider the probability density function given by (A.5). We compute the expected value*

$$E(X) = \int_{-\infty}^{\infty} t f(t) \, dt = \int_{p-1/2}^{p+1/2} t \, dt = p$$

What this computation tells us is that p is the average value for X. □

We will have the occasion to compute the expected value of a function $g(x_1, \ldots, x_N)$.

Definition A.12 (Expected Value of a Function). *Suppose that $X_1, \ldots, X_N$ are random variables with joint probability density function $f(x_1, \ldots, x_N)$. Let $g : \mathbb{R}^N \mapsto \mathbb{R}$ be some function. Then we define the expected value of function $g(x_1, \ldots, x_N)$ as*

$$E(g(x_1, \ldots, x_N)) = \int_{\mathbb{R}^N} g(x_1, \ldots, x_N) f(x_1, \ldots, x_N) \, dx_1 \cdots dx_N$$

□

Properties of Expected Values

There are several useful properties for expected values. We have the following result regarding the expected value of linear combinations of random variables.

Proposition A.3 (Addition Properties of Expected Value). *Let a and b be any real numbers and suppose that X and Y are random variables. Then*

(a) $E(a) = a$

(b) $E(aX + b) = aE(X) + b$

(c) $E(X + Y) = E(X) + E(Y)$

More generally, if $X_1, \ldots, X_N$ are random variables, then:

(d) $E\left(\sum_{k=1}^{N} X_k\right) = \sum_{k=1}^{N} E(X_k)$

□

Proof of Proposition A.3. The proof is left as Problem A.24. □

A final property of expectation that is quite useful has to do with expected values of products of independent random variables.

Proposition A.4 (Products of Expected Values). *Suppose that $X_1, X_2, \ldots, X_N$ are independent random variables such that $E(X_j)$ exists for $j = 1, 2, \ldots, N$. Suppose that f is the joint probability density function for $X_1, \ldots, X_N$ and f_j is the marginal density function for $X_j, j = 1, \ldots, N$. Then*

$$E(X_1 \cdot X_2 \cdots X_N) = E(X_1) \cdot E(X_2) \cdots E(X_N) \tag{A.12}$$

□

Proof of Proposition A.4. We start by writing

$$E(X_1 \cdot X_2 \cdots X_N) = \int_{\mathbb{R}^N} x_1 \cdot x_2 \cdots x_N f(x_1, \ldots, x_N) \, dx_1 \cdots dx_N$$

Since $X_1, \ldots, X_N$ are independent we can write this integral as

$$
\begin{aligned}
E(X_1 \cdot X_2 \cdots X_N) &= \int_{\mathbb{R}^N} x_1 \cdot x_2 \cdots x_N f(x_1, \ldots, x_N) \, dx_1 \cdots dx_N \\
&= \int_{\mathbb{R}^N} x_1 \cdots x_N f_1(x_1) \cdots f_N(x_N) \, dx_1 \cdots dx_N \\
&= \int_{\mathbb{R}} \cdots \int_{\mathbb{R}} x_N f_N(x_N) \cdots x_2 f_2(x_2) x_1 f_1(x_1) \, dx_1 \cdots dx_N
\end{aligned}
$$

Consider the innermost integral. We can factor out $x_N f_N(x_N) \cdots x_2 f_2(x_2)$ so that the innermost integral is simply

$$\int_{\mathbb{R}} x_1 f_1(x_1) \, dx_1 = E(X_1)$$

Now $E(X_1)$ is a constant and it can be factored out of the integral to leave

$$E(X_1 \cdot X_2 \cdots X_N) = E(X_1) \int_{\mathbb{R}} \cdots \int_{\mathbb{R}} x_N f_N(x_N) \cdots x_2 f_2(x_2) \, dx_2 \cdots dx_N$$

Again, we can factor out all but $x_2 f_2(x_2)$ from the innermost integral. Thus the innermost integral is

$$\int_{\mathbb{R}} x_2 f_2(x_2) \, dx_2 = E(X_2)$$

Continuing this process gives the desired result.

$\square$

Variance

While the expected value is useful for analyzing a distribution, it is restricted in what it can tell us. Consider the probability density functions

$$f_1(x) = \begin{cases} 1, & -\frac{1}{2} \le x \le \frac{1}{2} \\ 0, & \text{otherwise} \end{cases}$$

and

$$f_2(x) = \frac{1}{2} e^{-|x|}$$

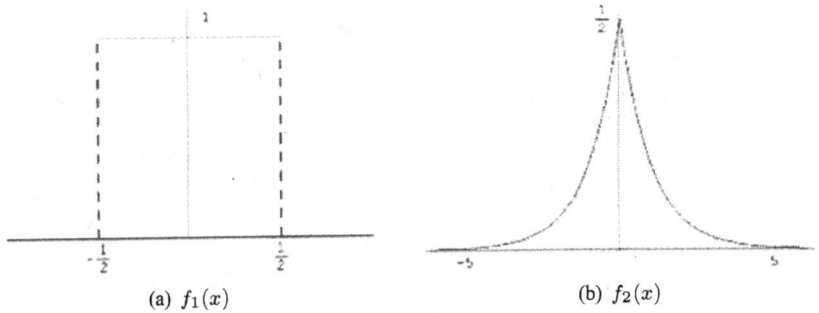

(a) $f_1(x)$ (b) $f_2(x)$

Figure A.4   The probability density functions $f_1(x)$ and $f_2(x)$.

Both have expected value zero but as you can see in Figure A.4, the distributions are quite different.

Another measure is needed to help us better analyze distributions. The tool that is often used in this endeavor is the *variance* of a distribution. The variance is a measure that indicates how much the distribution is spread out.

Definition A.13 (Variance). *Let* X *be a random variable with* $\mu = E(X)$. *Then the variance of* X *is given by*

$$\boxed{Var(X) = E((X - \mu)^2)}$$

The standard deviation *of* X *is given by the square root of* $Var(X)$.   □

Example A.11 (Computing Variance). *Find the variance for the distributions* $f_1(x)$ *and* $f_2(x)$ *plotted in Figure A.4.*
Solution. *For* $f_1(x)$, $\mu = 0$, *so*

$$Var(X) = E(X^2) = \int_{\mathbb{R}} x^2 f_1(x)\, dx = \int_{-1/2}^{1/2} x^2\, dx = \frac{1}{12}$$

Now $\mu = 0$ *for* $f_2(x)$ *as well so*

$$Var(X) = E(X^2) = \frac{1}{2}\int_{\mathbb{R}} x^2 e^{-|x|}\, dx = \int_0^\infty x^2 e^{-x}\, dx$$

We can use integration by parts twice to see that

$$\int x^2 e^{-x}\, dx = -(x^2 + 2x + 2)e^{-x}$$

so that

$$\int_0^\infty x^2 e^{-x}\, dx = \lim_{L\to\infty} -(x^2 + 2x + 2)e^{-x}\Big|_0^L = 2$$

□

Properties of Variance

Just as in the case of expectation, we can write down some basic properties obeyed by the variance of a distribution.

Proposition A.5 (Properties of Variance). *Let X be a random variable with a and b any constants and $\mu = E(X)$.*

(a) *$Var(aX + b) = a^2 \, Var(X)$.*

(b) *$Var(X) = E(X^2) - \mu^2$.*

$\square$

Proof of Proposition A.5. For part (a), we first use (c) from Proposition A.3 to write $E(aX + b) = aE(X) + b = a\mu + b$. We can thus compute

$$
\begin{aligned}
Var(aX + b) &= E\left((aX + b - a\mu - b)^2\right) \\
&= E((aX - a\mu)^2) \\
&= a^2 E((X - \mu)^2) \\
&= a^2 \, Var(X)
\end{aligned}
$$

Note that we have also utilized (b) from Proposition A.3 (with $b = 0$) in the argument. For part (b), we again use Proposition A.3 to write

$$
\begin{aligned}
Var(X) &= E((X - \mu)^2) \\
&= E(X^2 - 2X\mu + \mu^2) \\
&= E(X^2) - 2\mu E(X) + E(\mu^2) \\
&= E(X^2) - 2\mu^2 + \mu^2 \\
&= E(X^2) - \mu^2
\end{aligned}
$$

$\square$

Proposition A.6 (Sums of Variances). *Let $X_1, X_2, \ldots, X_N$ be independent random variables. Then*

$$
Var(X_1 + X_2 + \cdots + X_N) = Var(X_1) + Var(X_2) + \cdots + Var(X_N)
$$

$\square$

Proof of Proposition A.6. This proof is by induction. Let $N = 2$ and set $\mu_1 = E(X_1)$ and $\mu_2 = E(X_2)$. From (c) of Proposition A.3 we have

$$E(X_1 + X_2) = E(X_1) + E(X_2) = \mu_1 + \mu_2$$

We use this to compute

$$
\begin{aligned}
Var(X_1 + X_2) &= E\left((X_1 + X_2 - \mu_1 - \mu_2)^2\right) \\
&= E\left((X_1 - \mu_1)^2 + (X_2 - \mu_2)^2 + 2(X_1 - \mu_1)(X_2 - \mu_2)\right) \\
&= E((X_1 - \mu_1)^2) + E((X_2 - \mu_2)^2) + 2E((X_1 - \mu_1)(X_2 - \mu_2)) \\
&= Var(X_1) + Var(X_2) + 2E((X_1 - \mu_1)(X_2 - \mu_2))
\end{aligned}
$$

Since X_1 and X_2 are independent, we can use Proposition A.4 to infer that

$$E((X_1 - \mu_1)(X_2 - \mu_2)) = E(X_1 - \mu_1)E(X_2 - \mu_2)$$

But $E(X_1 - \mu_1) = E(X_1) - E(\mu_1) = \mu_1 - \mu_1 = 0$. In a similar way, $E(X_2 - \mu_2) = 0$, so the result holds for $N = 2$. You are asked to prove the induction step in Problem A.30. $\qquad\square$

PROBLEMS

A.22 Let X be a random variable whose distribution is

$$f(t) = \begin{cases} 3t^2, & 0 < t < 1 \\ 0, & \text{otherwise} \end{cases}$$

Find $E(X)$.

A.23 Let X be a random variable with probability density function f and let g be a function. Consider the statement $E(g(x)) = g(E(x))$. Either prove this statement or find a counterexample that renders it false.

A.24 Prove Proposition A.3.

A.25 Suppose that X is a random variable with probability distribution function $f(t)$.

(a) Suppose that there exists a constant a such that $Pr(X \geq a) = 1$. Show that $E(X) \geq a$.

(b) Suppose that there exists a constant b such that $Pr(X \leq b) = 1$. Show that $E(X) \leq b$.

(c) Use parts (a) and parts (b) to show that if $Pr(a \leq X \leq b) = 1$, then $a \leq E(X) \leq b$.

A.26 Let a be some constant and assume that $Pr(X \geq a) = 1$ and $E(X) = a$. Show that

(a) $Pr(X > a) = 0$.

(b) $Pr(X = a) = 1$.

(*Hint:* Problem A.25 will be useful.)

A.27 Use Proposition A.3 to show that if $X_1, X_2, \ldots, X_N$ are random variables such that each $E(X_j)$ exists for $j = 1, \ldots, N$ and $a_1, a_2, \ldots, a_N$ and b are any constants, then

$$E(a_1 X_1 + a_2 X_2 + \cdots + a_N X_N + b) = \sum_{j=1}^{N} a_k E(X_k) + b$$

A.28 Suppose that $f(t)$ is the uniform distribution on $[0,3]$. That is,

$$f(t) = \begin{cases} \frac{1}{3}, & 0 \leq t \leq 3 \\ 0, & \text{otherwise} \end{cases}$$

Assume that X_1 and X_2 are random variables. Find $E((X_1 + X_2)(2X_1 - X_2))$. You may assume that $X_1 + X_2$ and $2X_1 - X_2$ are independent random variables.

A.29 Find $Var(X)$ for the distribution in Problem A.22.

A.30 Complete the proof of Proposition A.6. That is, assume $Var(X_1 + \cdots + X_N) = Var(X_1) + \cdots + Var(X_N)$ and show that

$$Var(X_1 + \cdots + X_N + X_{N+1}) = Var(X_1) + \cdots + Var(X_N) + Var(X_{N+1})$$

A.31 Use Proposition A.6 in conjunction with Proposition A.5 to show that if $X_1, X_2, \ldots, X_N$ are independent random variables and $b, c_1, c_2, \ldots, c_N$ are constants, then

$$Var(c_1 X_1 + \cdots c_N X_N + b) = c_1^2 Var(X_1) + \cdots c_N^2 Var(X_N)$$

A.32 Suppose X_1 and X_2 are independent random variables with $E(X_1) = E(X_2)$. Assume also that $Var(X_1)$ and $Var(X_2)$ exist. Show that

$$E\big((X_1 - X_2)^2\big) = Var(X_1) + Var(X_2)$$

A.5 TWO SPECIAL DISTRIBUTIONS

We will make use of two special distributions in the text. The first is the well-known normal distribution and the second is the χ^2 distribution.

The Normal Distribution and the χ^2 Distribution

Definition A.14 (Normal Distribution). *A random variable X has a* normal dis-tribution *if for real numbers μ and σ with $\sigma > 0$, the probability density function is*

$$f(t) = \frac{1}{\sqrt{2\pi}\,\sigma}\, e^{-\frac{(t-\mu)^2}{2\sigma^2}} \qquad (A.13)$$

$\mu = E(X)$ *is the mean and σ^2 is the variance.* □

Since $\sigma > 0$, it is clear that $f(t) > 0$. In Problem A.36, you will show that $\int_{\mathbb{R}} f(t)\,dt = 1$.

The probability density function $f(t)$ from (A.13) with mean $\mu = 0$ is plotted in Figure A.5.

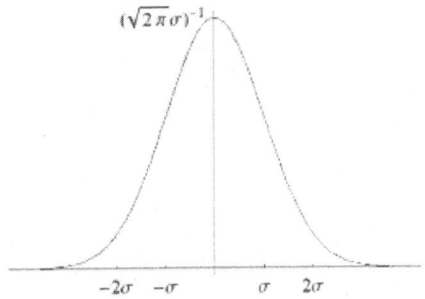

Figure A.5 The normal distribution with mean zero and standard deviation σ.

Definition A.15 (χ^2 Distribution). *Let n be a positive integer and X be a random variable. Then X has a χ^2 distribution with n degrees of freedom if the probability density function is*

$$f(t) = \begin{cases} \frac{1}{2^{n/2}\Gamma(n/2)}\, t^{(n/2)-1}\, e^{-t/2}, & t > 0 \\ 0, & \text{otherwise} \end{cases} \qquad (A.14)$$

□

The *gamma function* $\Gamma(t)$ is utilized in the definition of the χ^2 distribution. You can think of $\Gamma(t)$ as a generalized factorial function. Indeed, $\Gamma(k) = (k-1)!$ whenever k is a positive integer. In the case where k is an odd positive integer, formulas exist for computing $\Gamma(k/2)$. For more information on the gamma function, see Carlson [15].

It can be shown that $\Gamma(n/2) > 0$ whenever n is a positive integer, so that $f(t) \geq 0$. In Problem A.41 you will show that $\int_{\mathbb{R}} f(t)\, dt = 1$.

In Figure A.6, χ^2 distributions are plotted for $n = 3, 4, 5, 6$. In Problem A.40, you are asked to plot the χ^2 distributions for $n = 1, 2, 7$.

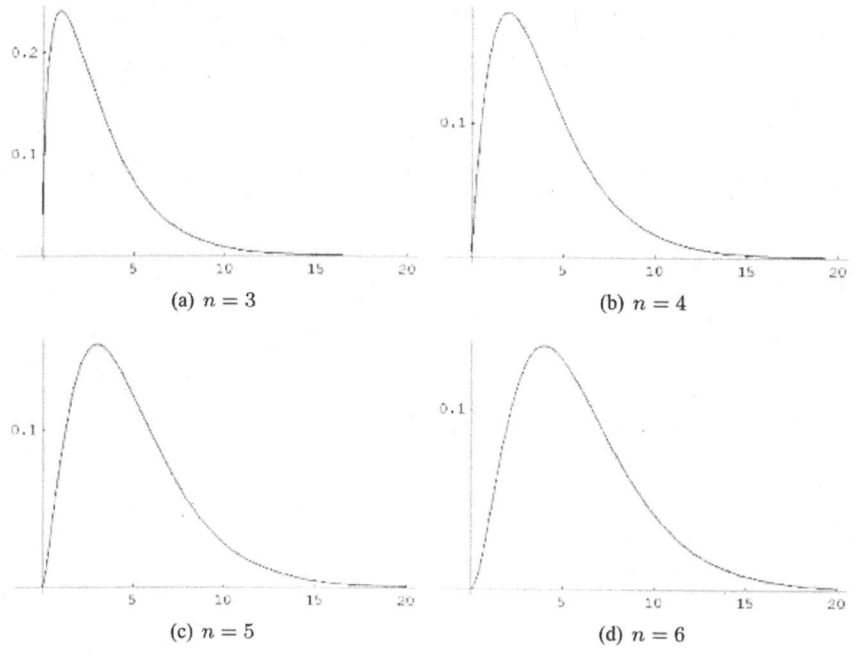

(a) $n = 3$

(b) $n = 4$

(c) $n = 5$

(d) $n = 6$

Figure A.6   The χ^2 probability density functions with $n = 3, 4, 5, 6$.

The expected value for X when X has the χ^2 distribution is easy to compute.

Proposition A.7 (Expected Value of χ^2 Distribution). *If X is a χ^2 random variable with n degrees of freedom, then*

$$E(X) = n$$

☐

Proof of Proposition A.7. The proof is left as Problem A.42.   ☐

The following result appears in [82]. We need it in Chapter 9.

Theorem A.1. *Suppose that X is a normally distributed random variable with mean $E(X) = \mu$ and standard deviation σ. Define the random variable*

$$Z = \frac{X - \mu}{\sigma}$$

Then Z^2 has a χ^2 distribution with 1 degree of freedom.   □

PROBLEMS

A.33 Use a **CAS** to plot the following normal distributions on the same set of axes:

(a) $\mu = 1, \sigma = 2$.

(b) $\mu = 1, \sigma = \frac{1}{4}$.

(c) $\mu = 1, \sigma = 5$.

A.34 Suppose that X is normally distributed with mean zero and $\sigma = 1$. Use a **CAS** (if necessary) to compute the following probabilities:

(a) $Pr(X \geq 0)$.

(b) $Pr(X < 1.5)$.

(c) $Pr(|X| < 2.5)$.

(d) $Pr(0 \leq -3X + 2 \leq 4)$.

A.35 Use what you know about the normal distribution with mean 0 to compute the following integrals:

(a) $\int_{\mathbb{R}} e^{-t^2/2} \, dt$.

(b) $\int_0^\infty e^{-t^2/2} \, dt$.

(c) $\int_0^\infty e^{-2t^2} \, dt$.

A.36 In this problem you will show that the pdf (A.13) has a unit integral. This problem requires some knowledge of multivariable calculus. The following steps will help you organize your work.

(a) First consider the integral $I = \int_{\mathbb{R}} e^{-x^2} \, dx$. Then

$$I^2 = \left(\int_{\mathbb{R}} e^{-x^2} \, dx \right)^2 = \int_{\mathbb{R}} e^{-x^2} \, dx \int_{\mathbb{R}} e^{-y^2} \, dy$$

Convert the integral above to polar coordinates and show that

$$I^2 = 2\pi \int_0^\infty r e^{-r^2} \, dr$$

(b) Use u-substitution to compute the integral above and thereby show that $I^2 = \pi$, or equivalently,

$$I = \int_{\mathbb{R}} e^{-x^2} \, dx = \sqrt{\pi} \qquad\qquad (A.15)$$

(c) Now consider the pdf (A.13). Make the substitution $u = \frac{t-\mu}{\sqrt{2}\,\sigma}$ in conjunction with (A.15) to show that the normal pdf has a unit integral.

A.37   In this problem we formally define the gamma function and prove a fundamental property about it. For $a > 0$, define the *gamma function* as

$$\Gamma(a) = \int_0^\infty t^{a-1} e^{-t} \, dt \qquad\qquad (A.16)$$

(a) Show that $\Gamma(1) = 1$.

(b) Use integration by parts to show that for $a > 1$, $\Gamma(a) = (a-1)\Gamma(a-1)$.

(c) Use parts (a) and (b) to show that if n is a positive integer, then $\Gamma(n) = (n-1)!$.

A.38   In this problem you will show that $\Gamma(1/2) = \sqrt{\pi}$. *Hint:* Use (A.16) with $a = \frac{1}{2}$ to write down the integral form of $\Gamma(1/2)$ and then make the substitution $u = \sqrt{t}$. Use (A.15).

A.39   Use Problem A.38 along with Problem A.37(b) to show that $\Gamma(7/2) = \frac{15\sqrt{\pi}}{8}$.

A.40   Use a **CAS** to sketch the χ^2 distribution for $n = 1, 2, 7$. You will find Problems A.38 and A.39 helpful.

A.41   In this problem, you will show that the pdf (A.14) has unit integral. *Hint:* Make the substitution $u = t/2$ in (A.14), simplify, and then use (A.16).

A.42   In this problem you will prove Proposition A.7. You will need the definition of the gamma function (A.16) to do so. The following steps will help you organize your work:

(a) Show that $E(X) = \frac{1}{2^{n/2}\Gamma(n/2)} \int\limits_0^\infty t^{n/2} e^{-t/2} \, dt$

(b) Make an appropriate u-substitution to show that

$$E(X) = \frac{2}{\Gamma(n/2)} \Gamma\left(\frac{n}{2} + 1\right)$$

(c) Use part (b) of Problem A.37 on part (b) to conclude that $E(X) = n$.

A.43   Let $k = 1, 2, \ldots$ and suppose that X is a random variable that has a χ^2 distribution with n degrees of freedom. In this problem you will compute $E(X^k)$. You will need the definition of the gamma function (A.16) to do so. The following steps will help you organize your work:

(a) Write down the integral that represents $E(X^k)$ and then make an appropriate u-substitution to write

$$E(X^k) = \frac{2^k}{\Gamma(n/2)} \int_0^\infty u^{n/2+k-1} e^{-u} \, du = \frac{2^k}{\Gamma(n/2)} \Gamma\left(\frac{n}{2} + k\right)$$

(b) Repeatedly use (b) from Problem A.37 to show that

$$\Gamma\left(\frac{n}{2} + k\right) = \frac{n}{2}\left(\frac{n}{2} + 1\right) \cdots \left(\frac{n}{2} + k - 1\right) \Gamma(n/2)$$

and thus show that

$$E(X^k) = 2^k \frac{n}{2}\left(\frac{n}{2} + 1\right) \cdots \left(\frac{n}{2} + k - 1\right)$$

A.44   Use Problem A.43 and Proposition A.7 to show that $Var(X) = 2n$ when X has the χ^2 distribution with n degrees of freedom.

REFERENCES

1. Edward Aboufadel, Julia Olsen, and Jesse Windle. Breaking the Holiday Inn Priority Club CAPTCHA. *The Coll. Math. J.*, 36(2), March 2005.

2. Robert G. Bartle and Donald R. Sherbert. *Introduction to Real Analysis, 3rd ed.* Wiley, Hoboken, NJ, 1999.

3. George Benke, Maribeth Bozek-Kuzmicki, David Colella, Garry M. Jacyna, and John J. Benedetto. Wavelet-based analysis of electroencephalogram (EEG) signals for detection and localization of epileptic seizures. In Harold H. Szu, editor, *Proceedings of SPIE: Wavelet Applications II*, volume 2491, pages 760–769, April 1995.

4. J. Berger, R. R. Coifman, and M. J. Goldberg. Removing noise from music using local trigonometric bases and wavelet packets. *J. Audio Eng. Soc.*, 42(10):808–818, 1994.

5. Jonathan Berger. Brahms at the piano. CCRMA, Stanford University, Stanford, CA, 1999. http://www-ccrma.stanford.edu/~brg/brahms2.html.

6. A. Bijaoui, E. Slezak, F. Rue, and E. Lega. Wavelets and the study of the distant universe. *Proc. of the IEEE*, 84(4):670–679, 1996.

7. Albert Boggess and Francis J. Narcowich. *A First Course in Wavelets with Fourier Analysis.* Prentice Hall, Upper Saddle River, NJ, 2001.

8. C. M. Brislawn. Fingerprints go digital. *Notices Amer. Math. Soc.*, 42(11):1278–1283, November 1995.

9. Richard A. Brualdi. *Introductory Combinatorics, 4th ed.* Prentice Hall, Upper Saddle River, NJ, 2004.

10. A. G. Bruce and H.-Y. Gao. Understanding waveshrink: Variance and bias estimation. *Biometrika*, 83:727–745, 1996.

11. Richard L. Burden and J. Douglas Faires. *Numerical Analysis, 8th ed.* Brooks/Cole, Pacific Grove, CA, 2005.

12. P. J. Burt and E. H. Adelson. The Laplacian pyramid as a compact image code. *IEEE Trans. Comm.*, 31(4):532–540, April 1983.

13. A. Calderbank, I. Daubechies, W. Sweldens, and B.-L. Yeo. Wavelet transforms that map integers to integers. *Appl. Comp. Harm. Anal.*, 5(3):332–369, 1998.

14. J. Canny. A computational approach to edge detection. *IEEE Trans. Pattern Anal. and Mach. Intell.*, 8:679–714, 1986.

15. Bille C. Carlson. *Special Functions of Applied Mathematics.* Academic Press, San Diego, CA, 1977.

16. Maryline Charrier, Diego Santa Cruz, and Mathias Larsson. JPEG2000, the next millenium compression standard for still images. In *Proceedings of the IEEE International Conference on Multimedia Computing and Systems (ICMCS)*, volume 1, pages 131–132, June 1999.

17. C. Christopoulos, A. Skodras, and T. Ebrahimi. The JPEG2000 still image coding system: An overview. *IEEE Trans. on Consumer Electronics*, 46(4):1103–1127, November 2000.

18. Charles K. Chui. *An Introduction to Wavelets, Volume 1, (Wavelet Analysis and Its Applications).* Academic Press, San Diego, CA, 1992.

19. Barry Cipra. Wavelet applications come to the fore. *SIAM News*, 26(7), November 1993.

20. A. Cohen, I. Daubechies, and J.-C. Feauveau. Biorthogonal bases of compactly supported wavelets. *Comm. Pure Appl. Math.*, 45:485–560, 1992.

21. Keith Conrad. Probability distributions and maximum entropy. University of Connecticut, 2005. http://www.math.uconn.edu/~kconrad/blurbs/entropypost.pdf.

22. Wolfgang Dahmen, Andrew Kurdila, and Peter Oswald. *Multiscale Wavelet Methods for Partial Differential Equations (Wavelet Analysis and Its Applications).* Academic Press, San Diego, CA, 1997.

23. I. Daubechies. Orthogonal bases of compactly supported wavelets. *Comm. Pure Appl. Math.*, 41:909–996, 1988.

24. I. Daubechies. Orthonormal bases of compactly supported wavelets: II. Variations on a theme. *SIAM J. Math. Anal.*, 24(2):499–519, 1993.

25. I. Daubechies and W. Sweldens. Factoring wavelet transforms into lifting steps. *J. Fourier Anal. Appl.*, 4(3):245–267, 1998.

26. Ingrid Daubechies. *Ten Lectures on Wavelets.* Society for Industrial and Applied Mathematics, Philadelphia, PA, 1992.

27. Morris H. DeGroot and Mark J. Schervish. *Probability and Statistics, 3rd ed.* Addison Wesley, Reading, MA, 2002.

28. D. Donoho. Wavelet shrinkage and W.V.D.: A 10-minute tour. In Y. Meyer and S. Rogues, editors, *Progress in Wavelet Analysis and Applications*, Editions Frontiers, pages 109–128, Toulouse, France, 1992.

29. D. Donoho and I. Johnstone. Ideal spatial adaptation via wavelet shrinkage. *Biometrika*, 81:425–455, 1994.

30. D. Donoho and I. Johnstone. Adapting to unknown smoothness via wavelet shrinkage. *J. Amer. Stat. Assoc.*, 90(432):1200–1224, December 1995.

31. David Donoho, Mark Reynold Duncan, Xiaoming Huo, and Ofer Levi-Tsabari. Wavelab 802, August 1999. http://www-stat.stanford.edu/~wavelab/.

32. Rakesh Dugad, Krishna Rataconda, and Narendra Ahuja. A new wavelet-based scheme for watermarking images. In *Proceedings of the International Conference on Image Processing 1998*, volume 2, pages 419–423, Chicago, IL, October 1998.

33. Paul M. Embree and Bruce Kimble. *C Language Algorithms for Digital Signal Processing*. Prentice Hall, Upper Saddle River, NJ, 1991.

34. J. Q. Fan, P. Hall, M. A. Martin, and P. Patil. On local smoothing of nonparametric curve estimates. *J. Amer. Stat. Assoc.*, 91(433):258–266, 1996.

35. Michael W. Frazier. *An Introduction to Wavelets Through Linear Algebra*. Undergraduate Texts in Mathematics. Springer-Verlag, New York, NY, 1999.

36. John E. Freund. *Mathematical Statistics, 2nd ed.* Prentice-Hall, Upper Saddle River, NJ, Second edition, 1971.

37. H.-Y. Gao and A. Bruce. Waveshrink with firm shrinkage. *Statistica Sinica*, 7:855–874, 1997.

38. Ramazan Gençay, Faruk Selçuk, and Brandon Whitcher. *An Introduction to Wavelets and Other Filtering Methods in Finance and Economics*. Academic Press, San Diego, CA, 2002.

39. Rafael C. Gonzalez and Richard E. Woods. *Digital Image Processing, 2nd ed.* Pearson Prentice Hall, Upper Saddle River, NJ, 2002.

40. Rafael C. Gonzalez, Richard E. Woods, and Steven L. Eddins. *Digital Image Processing Using Matlab*. Pearson Prentice Hall., Upper Saddle River, NJ, 2004.

41. Alexander Grossmann and Jean Morlet. Decomposition of Hardy functions into square integrable wavelets of constant shape. *SIAM J. Math. Anal.*, 15(4):723–736, July 1984.

42. F. Hampel. The influence curve and its role in robust estimation. *J. Amer. Stat. Assoc.*, 69(346):383–393, June 1974.

43. Taosong He, Sydney Wang, and Arie Kaufman. Wavelet-based volume morphing. In *Proceedings of Visualization 94*, pages 85–92, Washington, D. C., October 1994.

44. M. Hennessey, J. Jalkio, C. Greene, and C. Sullivan. Optimal routing of a sailboat in steady winds. Preprint, May 2007.

45. Stephen G. Henry. Catch the (seismic) wavelet. *AAPG Explorer*, pages 36–38, March 1997.

46. Stephen G. Henry. Zero phase can aid interpretation. *AAPG Explorer*, pages 66–69, April 1997.

47. Barbara Burke Hubbard. *The World According to Wavelets: The Story of a Mathematical Technique in the Making.* A. K. Peters, 2nd ed., Wellesley, MA, 1998.

48. David A. Huffman. A method for the construction of minimum-redundancy codes. *Proc. Inst. Radio Eng.*, 40:1098–1101, September 1952.

49. ISO/IEC JTC1/SC29/WG1 N505. New work item: JPEG2000 image coding system, March 1997. http://www.jpeg.org/public/wg1n505.pdf.

50. Arne Jensen and Anders la Cour-Harbo. *Ripples in Mathematics: The Discrete Wavelet Transform.* Springer-Verlag, New York, NY, 2001.

51. David W. Kammler. *A First Course in Fourier Analysis.* Prentice Hall, Upper Saddle River, NJ, 2000.

52. Fritz Keinert. *Wavelets and Multiwavelets.* Chapman & Hall/CRC, Boca Raton, FL, 2004.

53. Thomas W. Körner. *Fourier Analysis.* Cambridge University Press, New York, NY, 1989.

54. D. LeGall and A. Tabatabai. Subband coding of digital images using the wavelet transform. *IEEE Trans. Image Proc.*, 1:205–220, April 1992.

55. Jun Li. A wavelet approach to edge detection. Master's thesis, Sam Houston State University, Huntsville, TX, 2003.

56. S. Mallat and W. L. Hwang. Singularity detection and processing with wavelets. *IEEE Trans. Inf. Th.*, 38(2):617–643, March 1992.

57. Stéphane Mallat. Multiresolution approximations and wavelet orthonormal bases of $l^2(\mathbb{R})$. *Trans. Amer. Math. Soc.*, 315:69–87, September 1989.

58. D. Marr and E. Hildreth. Theory of edge detection. *Proc. R. Soc. London*, 207, 1980.

59. Yves Meyer. *Wavelets and Operators.* Advanced Mathematics. Cambridge University Press, New York, NY, 1992.

60. Steve Oualline. *Practical C++ Programming.* O'Reilly, Sebastopol, CA, 2003.

61. Charles A. Poynton. *A Technical Introduction to Digital Video.* Wiley, Hoboken, NJ, 1996.

62. K. Ramamohan Rao and P. Yip. *Discrete Cosine Transform: Algorithms, Advantages, Applications.* Academic Press, San Diego, CA, 1990.

63. Walter Rudin. *Principles of Mathematical Analysis, 3rd ed.* McGraw-Hill, New York, NY, 1976.

64. John C. Russ. *The Image Processing Handbook, 4th ed.* CRC Press, Boca Raton, FL, 2002.

65. Khalid Sayood. *Introduction to Data Compression, 2nd ed.* Morgan Kaufmann, San Francisco, CA, 2000.

66. Claude E. Shannon. A mathematical theory of communication. *Bell Syst. Tech. J.*, 27(3):379–423, July 1948. Continued in 27(4): 623-656, October 1948.

67. Steven Smith. *Digital Signal Processing: A Practical Guide for Engineers and Scientists.* Newnes, Elsevier Science, Amsterdam, 2002.

68. C. M. Stein. Estimation of the mean of a multivariate normal distribution. *Ann. Stat.*, 9(6):1135–1151, November 1981.

69. James Stewart. *Calculus: Early Transcendentals, 6th ed.* Brooks/Cole, Pacific Grove, CA, 2007.

70. Gilbert Strang. The discrete cosine transform. *SIAM Rev.*, 41(1):135–147, 1999.

71. Gilbert Strang. *Introduction to Linear Algebra, 3rd ed.* Wellesley Cambridge Press, Wellesley, MA, 2003.

72. Gilbert Strang. *Linear Algebra and Its Applications, 4th ed.* Brooks/Cole, Pacific Grove, CA, 2006.

73. Gilbert Strang and Truong Nguyen. *Wavelets and Filter Banks.* Wellesley Cambridge Press, Wellesley, MA, 1996.

74. W. Sweldens. Wavelets and the lifting scheme: A 5 minute tour. *Z. Angew. Math. Mech.*, 76 (Suppl. 2):41–44, 1996.

75. W. Sweldens. The lifting scheme: A construction of second generation wavelets. *SIAM J. Math. Anal.*, 29(2):511–546, 1997.

76. C. Taswell. The what, how, and why of wavelet shrinkage denoising. Technical Report CT-1998-09, Computational Toolsmiths, Stanford, CA, January 1999.

77. David Taubman. High performance scalable image compression with EBCOT. *IEEE Trans. Image Proc.*, 9:1158–1170, July 2000.

78. David Taubman and Michael Marcellin. *JPEG2000: Image Compression Fundamentals, Standards and Practice.* The International Series in Engineering and Computer Science. Kluwer Academic Publishers, Norwell, MA, 2002.

79. M. Unser and T. Blu. Mathematical properties of the JPEG2000 wavelet filters. *IEEE Trans. Image Proc.*, 12(9):1080–1090, 2003.

80. Krishnaraj Varma and Amy Bell. JPG2000 - choices and tradeoffs for encoders. *IEEE Signal Proc. Mag.*, pages 70–75, November 2004.

81. Brani Vidakovic. *Statistical Modeling by Wavelets.* Wiley, Hoboken, NJ, 1999.

82. Dennis D. Wackerly, William Mendenhall III, and Richard L. Scheaffer. *Mathematical Statistics with Applications, 6th Ed.* Duxbury, Belmont, CA, 2002.

83. David F. Walnut. *An Introduction to Wavelet Analysis.* Birkhäuser, Cambridge, MA, 2002.

84. Peter Wayner. *Compression Algorithms for Real Programmers.* Morgan Kaufmann, San Francisco, CA, 2000.

85. Mladen Victor Wickerhauser. *Adapted Wavelet Analysis from Theory to Software.* A. K. Peters, Wellesley, MA, 1994.

86. Xiang-Gen Xia, C. G.Boncelet, and G. R. Arce. A multiresolution watermark for digital images. In *International Conference on Image Processing 1997*, volume 1, pages 548–551, Santa Barbara, CA, October 1997.

87. B. L. Yoon and P. P. Vaidyanathan. Wavelet-based denoising by customized thresholding. In *Proceedings of the 29th International Conference on Acoustics, Speech, and Signal Processing*, Montreal, Canada, May 2004.

INDEX

$(5, 3)$ biorthogonal filter pair
   applied to an image, 382
$(5, 3)$ biorthogonal spline filter pair, **378**–380,
     434, 436, 440, 447, 454, 460, 471, 475
   in JPEG2000, 469–470
   in lifting, 460
   signal compression, 390
$(8, 4)$ biorthogonal filter pair, 382
$(8, 4)$ biorthogonal spline filter pair, **383**
$(9, 7)$ biorthogonal spline filter pair, 394, 403
   derivative conditions, 394
   Fourier series, 394
     derivative conditions, 403

A

Adelson, Edward, 370
algorithms
   BWT1D1, 414
   HWA, 191
   HWT1D, 181
   HWT1D1, 169
   HWT2D, 195

   HWT2D1, 192
   HWTB, 192
   IBWT1D1, 433
   IBWTht, 430
   IHT1D, 182
   IHWT1D1, 170
   IHWT2D1, 193
   IWHT2D, 196
   IWT1D1, 277
   IWTht, 276
   LWT1D1, 479
   WT1D1, 270
   modified biorthogonal transformation, 452
   modified biorthogonal wavelet transformation,
     439
   modified inverse biorthogonal transformation,
     449
   modified inverse biorthogonal wavelet
     transformation, 453
American Standard Code for Information
     Interchange (ASCII), 50, 52, 70–71, 88
analysis gain bits, 473
antisymmetric filter, 370

ach?

applications first approach, xiv
applications of wavelets, 4
arithmetic coding, 474
averages block, 164

B

Bartle, Robert, 111
base step size, **473**
basis, 110
Bell, Amy, 74
Berger, Jonathan, 5
Bezout's theorem, **405**
    proof, 406
biased estimator, 494
bi-infinite sequences, 127–128, 135, 140
binary form, 50
binomial coefficient, 125–126, 264, 299
binomial theorem, 125, 404
biorthogonal filter pair, 282, 352–353, **358**,
        370–371, 378, 393, 407
    $(5, 3)$
        associated wavelet filter pair, 379
    $(8, 4)$, 382
    $(9, 7)$, 394
    associated wavelet filters, 360
    building $\mathbf{g}$ and $\tilde{\mathbf{g}}$ from $\tilde{\mathbf{h}}$ and $\mathbf{h}$, 360
    building G and $\tilde{G}$ from $\tilde{H}$ and H, 359
    example, 361, 367, 391
    symmetric, 352, 371
    wavelet filter
        example, 362
biorthogonal spline filter pair, 352, **389**, 393–394
    $(5, 3)$, 378–380
    $(8, 4)$, **383**
    Daubechies formulation, 386
    example, 389
biorthogonal wavelet transform matrix, 379, 391
    $(5, 3)$ biorthogonal spline filter pair, 380–381
    $(8, 4)$ biorthogonal spline filter pair, 384–385
    example, 361–363
biorthogonal wavelet transformation, 407
    algorithm, 407
    algorithm for 1D transformation, 414
        modified using symmetry, 439, 450
    algorithm for inverse, 407, 417
    algorithm
        modified for symmetry, 408
        modified using symmetry, 452
    highpass portion, 412, 437
    highpass portion and symmetry, 444
    highpass portion
        modified using symmetry, 438
        odd-length filters, 413

odd length filters, 413
in JPEG2000, 460, 469–470
inverse, 475
inverse algorithm, 433
    modified using symmetry, 453
inverse in block matrix form, 417
inverse
    $G^T \mathbf{t}$ for even-length filters, 428
    $G^T \mathbf{t}$ for odd-length filters, 428
    $H^T \mathbf{s}$ for even-length filters, 423–426, 428
    $H^T \mathbf{s}$ for odd-length filters, 417–420, 422
    algorithm, 433
    algorithm for $H^T \mathbf{s}$ and $G^T \mathbf{t}$, 429
    algorithm modified using symmetry, 449
        modified using symmetry, 446–447, 449
lossless compression, 470
lowpass portion, 408–412
lowpass portion and symmetry, 441
lowpass portion
    $(5, 3)$, 436
    modified using symmetry, 437
    modified using symmetry, 439
biorthogonality condition, **358**–359, 395
    example, 361
    Fourier series
        existence of solution, 398
bit, 50
bit stream, 88
bit stream in image compression, 204, 209–210
bits per pixel (bpp), **94**
block $\mathcal{B}$
    in JPEG2000, 472
block $\mathcal{D}$
    in JPEG2000, 471–473
block $\mathcal{H}$
    in JPEG2000, 471–473
block $\mathcal{V}$
    in JPEG2000, 471–473
Blu, Thierry, 394
Boggess, Albert, 11, xxi, 111, 124, 377
box function, 375
bpp, 202
    image compression, 204, 209–210
    image compression example, 240
Brualdi, Richard, 312
Bruce, Andrew, 348
Burden, Richard, 485
Burt, Peter, 370
byte, 50

C

C12 filter, **308**
C6 filter, **305**

in VisuShrink denoising, 331
signal compression example, 311
solving for, 303, 306–307
Calderbank, Robert, 251
Canny, John, 210
CAPTCHA, 221
Carlson, Bille, 513
causal filter, **139**
CDF97 filter pair, 394–395, 400–**402**, 402, 478
  associated wavelet filters, **402**
  derivative conditions, 394
  Fourier series, 401
    derivative conditions, 406
    plots, 402
  in inverse JPEG2000 process, 474
  in JPEG2000, 460, 469–470
  values for h, 402
  values for h̄, 402
ceiling function, 64, 485
Charrier, Maryline, 459
chi-square distribution, 512–**513**
  degrees of freedom, 513
  expected value, 514
Christopoulos, Charilaos, 468, 470, 473
chrominance, 73
  YCbCr color space, 74
Chui, Charles, 376
Cohen–Daubechies–Feauveau 9/7 filter pair, 353
Cohen, Albert, 9, 394–395, 398–400
Coiflet filter, **305**, 400
  $K = 2$, 308
  $K = 1$, 303
  associated wavelet filter, 309
    Fourier series, 309
  finite-length limits, 307
  length 12, 308
  length 6, 305
  solving for, 307
  wavelet transform matrix, 310
Coiflet filters, 282, 435
  Fourier series, 301
Coifman, Ronald, 4, 298
color image
  Haar wavelet transform
    compression, 207
  HWT, 197
  in JPEG, 462
  JPEG2000, 474
color space, 70
color space conversions
  RGB to YCbCr, 74
  RGB to YIQ, 77
  YCbCr to RGB, 76
  YUV to YCbCr, 77

color spaces
  CYM, 76
  HSI, 70
  HSI space, 73
  RGB, 71
  YCbCr, 70, 72
  YIQ color space, 77
  YUV space, 77
complex arithmetic, 99
  addition, 100
  multiplication, 100
  subtraction, 100
complex exponential function, 104
complex exponential functions, **107**
  orthogonality relation, 108
complex number, **98**
  conjugate, 100
  imaginary part, 98
  modulus, **102**
  real part, 98
computer lab
  color image denoising, 350
computer labs
  complex arithmetic, 104
  Fourier series, 126
  1D Haar wavelet transform and inverse, 183
  2D Haar wavelet transforms, 199
  block matrix arithmetic, 48
  CAPTCHA, 221
  color histogram equalization, 78
  color space transformations, 78
  color to grayscale and pseudocolor maps, 78
  convolution, 137
  cumulative energy and data compression, 88
  data compression, 280, 316, 435
  Daubechies filters, 265
  elementary image processing, 70
  entropy and PSNR, 88
  filters, 150
  general wavelet transform matrix, 316
  Haar wavelet transform, 173
  image compression, 252
  image edges, 70
  image processing and block matrices, 70
  incremental image resolution, 199
  inner products and norms, 22
  intensity transformations, 70
  JPEG, 467
  JPEG2000 compression, 492
  matrix arithmetic, 38
  matrix completion, 280
  modified biorthogonal wavelet transformation,
    457
  partial image inversion, 199

progressive image reconstruction, 280
software development
   color space transformations, 78
   1D biorthogonal wavelet transformation, 416
   1D Daubechies wavelet transforms, 279
   1D Haar wavelet transform, 173
   1D Haar wavelet transform and inverse, 183
   2D biorthogonal wavelet transformation, 416
   2D Daubechies wavelet transforms, 280
   2D Haar wavelet transforms, 199
   biorthogonal spline filter pairs, 393
   CDF97 filter pair, 406
   Coiflet filters, 316
   constructing λ^{sure}, 350
   cumulative energy, entropy, PSNR, 88
   Daubechies filters, 252
   fast algorithm using (5**3**) biorthogonal spline
      filter pair, 457
   generating Daubechies filters, 265
   Huffman coding, 95
   intensity transformations, 70
   inverse 1D biorthogonal wavelet transform,
      435
   inverse 2D biorthogonal wavelet transform,
      435
   lifting, 486
   modified biorthogonal wavelet
      transformation, 457
   wavelet shrinkage, 324, 334, 350
  sparseness and λ^{sure}, 350
  SureShrink, 350
  VisuShrink, 335
  wavelet shrinkage, 325
Conrad, Keith, 85
convex combination, 350
convolution, 127–**128**, 140, 441
convolution matrix, 155
  averaging filter, 158
convolution theorem, **133**, 152
  example, 134
convolution
  as a shifting inner product, 130
    examples, 131
  as matrix product, 150, 152
  commutative process, 136
  examples, 129
  fast multiplication, 135
  of functions, 375
cumulative distribution function, 65
cumulative energy, **81**
  color image compression, 207
  example computations, 81
  image compression, 202
  signal compression example, 239

customized threshold function, 349
CYM color space, 76
CYMK color space, 76

D

D10 filter, **262**
D4 filter, **233**
  $|H(\omega)|$, 238, 283
  building **g** from **h**, 227
  derivative condition, 231
  Fourier series $H(\omega)$, 288
  highpass (wavelet) filter, 234
  image compression example, 239
  in VisuShrink denoising, 329
  lowpass conditions, 227
  orthogonality conditions, 226
  signal compression, 238
  solving system algebraically, 245
  solving the system, 231
  system of equations, 231
  wavelet filter, 234
D6 filter, **237**
  $|H(\omega)|$, 238
  derivative conditions, 236
  image compression example, 239
  in SureShrink denoising, 344
  in VisuShrink denoising, 328, 331
  lowpass conditions, 236
  orthogonality conditions, 235
  signal compression, 238
  solving system algebraically, 248
  system to solve, 237
  wavelet matrix, 235
data compression
  naive with HWT, 164
Daubechies filters, 224, 435
Daubechies four-tap scaling function, 8
Daubechies system, 258–**259**
  characterizing solutions, 261
  derivative conditions, 257
  length 10 filter, 262
  lowpass conditions, 256–257, 263
  number of solutions, 259
  orthogonality conditions, 253–254, 256, 263
  reflections are solutions, 260
  wrapping rows, 253
Daubechies wavelet transform
  algorithm for $H^T\mathbf{y}$, 276
  algorithm for 1D inverse, 277
  algorithm for 1D transform, 270
  block matrix form for inverse, 271
  pseudocode, 265
  pseudocode for inverse, 271

wrapping rows, 266–267
  inverse, 273–274
Daubechies, Ingrid, 9, 224, 259, 261, 282, 298,
  301–302, 308, 351, 353, 370–371, 377,
  386, 406, 478
DCT, 460–461, 464, 467
  matrix defined, 460
  matrix elements, 461
  quantized, 464–465
deconvolution, 127, 150, 152, 155
  example, 153
  Fourier series, 152
DeGroot, Morris, 494
DeMoivre's theorem, 110, 403
denoising, 4, 317–318, 320–321
denoising example, 328–329
denoising
  block $\mathcal{D}$, 343
  block $\mathcal{H}$, 343
  block $\mathcal{V}$, 343
  Brahms recording, 4
density estimation, 5
derivative, **218**
derivative conditions
  C12 filter, 308
descriptive statistics, 493
  example, 494
differences block, 164
digital watermarking, 4
dilation equation, 9, 376
  for spline functions, 376
discrete cosine transform, **460**
  matrix defined, 461
  matrix elements, 461
discrete Fourier transform, 7
discrete Haar wavelet transformation, 13
DiscreteWavelets software package, xix
disjoint events, **496**–497
distribution, 499
  chi-square, **513**
  continuous, **500**
  joint, **502**
  marginal distribution, **503**
    computing, 503
  normal, **513**
  uniform, 500
Donoho, David, 317–318, 325–326, 328, 332,
  335, 342
downsampling, 161

E

EBCOT, 469, 474
Eddins, Steven, 49, 210–211

edge detection, 4, 56, 210
  Canny method, 211
  Haar wavelet transform, 212
  Marr – Hildreth method, 212
  Sobel method, 211
  thresholding, 214
    automated selection, 214
embedded block coding with optimized truncation,
  469
Embedded Block Coding with Optimized
  Truncation, 474
Embree, Paul, 150
entropy, 79
  example computations, 80
  image compression, 202, 204, 209–210
  image compression example, 240
error function, **83**
estimator function, 338–339, 341
  example, 340
  minimizing on a computer, 341
  recursive formula, 347
Euclidean algorithm, 406
Euler's formula, 104–**105**
event, 496, 499
expectation, 506
expected value, 321, **506**
expected value of a function, **507**
expected value
  χ^2 distribution, 514
  computing, 506
  products, 507
  properties, 507
explicit quantization, 473

F

Faires, J. Douglas, 485
Fan, Jianqing, 332
father wavelet function, 8
Feauveau, Jean-Christophe, 9
filter, 127, 138
filter bank, 155
filter elements g_k from $G(\omega)$, 295
filter
  antisymmetric, 370
  averaging filter, 138, 158
  causal filter, **139**, 148
  finite-length, 368, 371
  finite-length filter, **282**, 284, 286, 288
  FIR filter, 139–**140**, 147
  Haar filter, 166
  highpass filter, 140, 145, **147**–148
    examples, 145
  highpass

from lowpass, 149
how to apply, 138
ideal lowpass filter, 147
lowpass, 369, 371
lowpass filter, 140, 142–**143**, 148
   examples, 144
odd-length filter, 416
symmetric, 351, 353, **366**, 368, 371
symmetric and lowpass, 353, 371, 416
filters, 13
   biorthogonal, **358**
finite-length filter, 368–369, 371
   condition for orthogonality, 288
   orthogonality, 363, 366
FIR filter, 139–**140**, 140, 147, 153, 155, 158, 282
Firefox, 459
firm threshold function, 348
floating-point operation, 485
floor function, 218, 259, 422, 426, 472, 485
FLOP, 485–486
Fourier series, 110–**111**, 140, 147
   absolute convergence, 111, 121
   coefficients, **112**
   convergence, 111
   even function, 121
   finite length, 119
   modulation rule, 118, 123
   odd function, 121
   translation rule, 117
   uniform convergence, 121
$(9, 7)$ biorthogonal spline filter pair, 394
$\cos^{\tilde{N}}\left(\frac{\omega}{2}\right)$, 372–374
$\cos^{N}\left(\frac{\omega}{2}\right)$, 372–373
biorthogonal filter, 393
biorthogonal filter pair
   example, 361, 377
biorthogonal filter
   Daubechies formulation, 386
biorthogonal spline filter pair, 389
   Daubechies formulation, 386
biorthogonality condition
   existence of solution, 398
building G and $\tilde{G}$ from $\tilde{H}$ and H, 359
CDF97 filter pair, 394, 401
   plots, 402
Coiflet filter, 301–303, 400
   $K = 1$, 303–304
   associated wavelet filter, 309
convolution theorem, 133
deconvolution, 152
derivative condition
   Daubechies filters, 298
lowpass filters, 141

odd-length symmetric filter, 395–396
   example, 397–398
   in terms of cos, 399
   lowpass conditions, 396–397
orthogonality condition for finite-length filters, 286
   example, 286
orthogonality conditions, 357–358
   two filters, 356
spline filter, 374, 382, 389
Fourier, Joseph, 111
Frazier, Michael, 11, xxi, 377

G

gamma correction, 62
gamma function, 513, **516**
Gao, Hong-Ye, 348
Gaussian white noise, **318**, 323, 327, 333–334, 342, 348
   distribution, 318
Gibbs phenomenon, 114
Gonzalez, Rafael, 49, 92, 150, 210–211, 214, 461, 469, 473
gradient, 211
grayscale image, 200
greatest integer function, 218
Green, Bo, 312
Grossman, Alexander, 9

H

Haar filter, 166, 223–224, 283, 371
$|H(\omega)|$, 234, 283
   image compression example, 239
   signal compression, 238
   VisuShrink denoising, 334
Haar wavelet filter, 166, 223, 225
Haar wavelet transform, **166**, 200, 223, 353
   algorithm for 1D transform, 169
   algorithm for inverse 1D transform, 170
   algorithm for iteration, 181
   application to vector, 167
   application
      color image compression, 207
      edge detection, 210, 212
      image compression, 200–201
   as a vector-valued function, 345–347
   block $\mathcal{B}$, 187
   block $\mathcal{D}$, 187
      in edge detection, 212
   block $\mathcal{H}$, 187
      in edge detection, 212
   block $\mathcal{V}$, 187
      in edge detection, 212

block matrix form, 178
color images, 197
diagonal changes, 188
highpass portion, 200
horizontal changes, 188
inverting, 162
iterating, 173
iterating inverse, 179
iterating
    example, 174
    product of matrices, 174–175, 177
lifting, 217
modified, 201
nondecimated, 220
orthogonalizing, 165
processing columns of a matrix, 183–184
processing rows of a matrix, 184, 186
pseudocode for one-dimensional inverse, 170
pseudocode for one-dimensional transform, 168
pseudocodes
    iteration, 180
two-dimensional block form, 186
two-dimensional
    algorithm, 191, 195
    algorithm for inverse, 192
    applied to an image, 189
    block form, 189
    inverse algorithm, 196
    inverting, 190
    iterating, 193
    iterating inverse, 194
    pseudocode, 191
    pseudocodes, 195
    vertical changes, 188
Haar, Alfréd, 166
half-angle formulas, 110
Hampel, Frank, 328
hard threshold, 321
hard threshold rule, 321, 324
heavisine function, 318–319
highpass conditions
    in terms of Fourier series, 297
highpass filter, 145, **147**–149, 155, 371
    condition on Fourier series, 146
    example, 145
Hildreth, Ellen, 210
histogram, 63, 65
histogram equalization, 63–65
    color images, 78
    example, 66
HSI color space, 70, 73
hue, 73
Huffman coding, 88, 94, 200, 460, 462, 465, 467, 474

AC and DC tables, 462
algorithm, 91
decoding, 93
example, 88
image compression example, 241
Huffman, David, 88
Hwang, Wen Liang, 212
HWT, **166**, 173, 223

I

ideal lowpass filter, 142
ideal mean squared error, 326
image compression
    comparing filters, 239
    first example, 1
    Haar wavelet transform, 200
        color, 207
        example, 201
        quantization, 200
image denoising
    example, 343
image morphing, 4
image negation, 54
    color image, 72
image processing
    color to grayscale, 75
    gamma correction, 62
    histogram equalization, 63–64
    image negation, 54
    masking, 59
    naive edge detection, 56
imaginary number, **98**
implicit quantization, 473
independent random variable, **504**
integer to integer biorthogonal wavelet
        transformation, 460, 470
intermodulation, 77
International Organization for Standardization, 459
International Telecommunication Union, 459
Internet Explorer, 459
ISO, 459
ITU, 459

J

Jensen, Arne, 476
Johnstone, Iain, 318, 325–326, 328, 332, 335, 342
joint distribution, **502**
Joint Photographic Experts Group, 459
joint probability density function, 503
JPEG, xviii, 459–460, 462–463, 466–468, 470
    example, 462
    problems with standard, 468

JPEG2000, 3, 9, 13, xviii, 200, 207, 282, 352, 394, 459–460, 468–471, 474
JPEG2000 Image Compression Standard, 282
JPEG2000
  algorithm, 469
  better compression than JPEG, 468
  biorthogonal wavelet transformation, 470
  coding, 474
  color image compression
    example, 490
  color images, 470
  examples, 460, 486
  inverting the process, 474
  lifting, 478
  lossless compression
    example, 488
  lossy and lossless compression, 468
  lossy compression
    example, 486
  new features, 468–469
  progressive signal transmission, 468
  quantization function, 472
  regions of interest, 469
  tiling, 469

K

Kammler, David, 7, 111, 114, 128
Keinert, Fritz, 251
Kimble, Bruce, 150
Körner, Thomas, 111

L

la Cour-Harbo, Anders, 476
Laplacian, 212
LeGall filter pair, **471**, 475–476, 478, 481, 484–486
LeGall, Didier, 471
Leibniz generalized product rule, 264
Li, Jun, 212
lifting, xviii, 460, 470, 476–477, 484, 486
  Haar wavelet transform, 476
  algorithm for 1D biorthogonal wavelet transformation, 479
  even and odd vectors, 477
  example, 480
  Haar wavelet transform, 217
  highpass portion, 478
  integer to integer map, 481
    bias, 482, 484
    example, 482
    highpass portion, 481
    inverse, 482
    lowpass portion, 482
    recovering **e**, 483–484
    recovering **o**, 482
    inverse, 481
    lifting step, 477
    lowpass portion, 476–478
lossless compression, 79, 200, 460
  in JPEG2000, 469
lossy compression, 200, 352, 460
  in JPEG2000, 469
lowpass conditions
  in terms of Fourier series, 297
lowpass filter, 140, 142–**143**, 148–149, 155, 158, 370–371, 384
  condition for Fourier series, 143
  examples, 144
  Fourier series, 141
  ideal, 142, 147
luminance, 73, 77
  YCbCr color space, 74

M

m-bit image, 52
Maclaurin series, 104
MacLaurin series, 405
Maclaurin series
  $\cos t$, 105
  $\sin t$, 105
  e^t, 105
MAD, 328, **495**
  estimating noise level, 328
  median absolute deviation, 495
Mallat, Stéphane, 9, 212
marginal distribution, **503**
Marr, David, 210
matrix completion, 251
matrix
  addition and subtraction defined, **23**
  arithmetic examples, 24
  banded, 151
  block matrix, 38
    addition and substraction, 39
    inverting, 48
    multiplication, 40
    transpose, 44
  color image, 72
  determinant, 36
  diagonal matrix, 30
  dimension defined, **23**
  examples, 22
  gray scale image, 53
  identity matrix I_n defined, **29**
  image masking, 59
  inverting, 29

lower triangular, 151
multiplication defined, **27**
multiplication described, 25
multiplication examples, 27
multiplication
    interpretation, 42
nonsingular matrix
    defined, **29**
nonsingular
    properties, 31
orthogonal matrix, **34**
    preserves distance, 34
    rotation, 35
scalar product defined, **24**
submatrix, 38
symmetric, **33**
transpose defined, **32**
transpose examples, 33
max norm, 21
mean, **493**–494, 506
mean squared error, 320–321
    ideal, 326
mean
    normal distribution, 513
median, **494**–495
median absolute deviation, 328, **495**
Meyer, Yves, 9
Microsoft, 459
modeling of galaxies, 5
modified biorthogonal transformation
    inverse, 449
modified biorthogonal wavelet transformation,
        439, 447, 450, 452, 470, 475–476, 478,
        481, 484
    in JPEG2000, 460
    why it works, 441
modulus of a complex number, 102
Morlet, Jean, 9
moving inner product, 140
Mozilla, 459
multiresolution analysis, 9

N

Narcowich, Francis, 11, xxi, 111, 124, 377
National Television System Committee (NTSC),
    75
Nguyen, Truong, xiv, 251
noise, 318
noise level, 318, 327, 331, 333
    estimating using MAD, 328
noise
    file transmission, 318
    in photography, 318

recording, 318
scanning image, 318
norm, **21**
    max norm, 21
    taxicab norm, 21
    vector norm, 19
normal distribution, 512–**513**
normalization matrix, **461**
NTSC, 75, 77
numerical PDEs, 5

O

orthogonal matrix, 10
orthogonality conditions
    biorthogonal filters, 355
    building $G(\omega)$ from $H(\omega)$, 292, 295
    Daubechies filter, 298
    Fourier series, 357–358
    in terms of finite-length Fourier series, 292
    in terms of Fourier series, 290, 296
    symmetric finite-length filters, 367
    two filters, 356
Oualline, Steve, 50, 88

P

parity, **286**, 288, 389, 392
partial derivative, **336**
Pascal's identity, 126
Pascal's triangle, 374
peak signal-to-noise ratio (PSNR), 82–**83**
peak signal to noise ratio (PSNR)
    example computations, 83
phase angle, 149
pixel, 52
portable gray map (PGM), 52
portable pixel map (PPM), 71
Poynton, Charles, 73–75
probability, 496, 499
probability density function, **500**
probability function, 497
probability properties, 497, 499
probability
    computing, 501
pseudoinverse matrix, 220
PSNR, **83**
    image compression, 205–206, 210

Q

quadrature channel, 77
quantization function, 472
quantization process
    in JPEG2000, 472
    JPEG2000, 472

quantization step size, **472**–473
quantization step
   JPEG, 461

R

random variable, **498**–499
   continuous, 499, 501
   discrete, 499
   examples, 498
   independent, 504
Rao, K. Ramamohan, 461
resolution, 52
RGB color cube, 71
RGB color space, 71–72, 77
RGB space, 470, 474
RGB triple, 71
ROI, 469
rounding function, 485
Rudin, Walter, 133, 285
Russ, John, 49

S

sample space, **496**, 498–499
   examples, 496
saturation, 73
sawtooth function, 113
Sayood, Khalid, 92, 474
scaling function, 377, 390
seismic data analysis, 5
seizure detection, 5
sgn function, 319, 324
Shannon, Claude, 79
Sherbert, Donald, 111
shrink, 321
sign function, 319, 472
signal compression
   (5, 3) biorthogonal spline filter pair, 389
   C6 filter, 311
   D4 filter, 238
   D6 filter, 238
   example, 238
   Haar filter, 238
singular value decomposition, 250
Smith, Steven, 142, 146
Sobel, Irwin, 210
soft threshold function, 322
soft threshold rule, 321, 327, 331–332
   as a vector-valued function, 335–336
   example, 322
software labs
   Huffman coding, 95
spline filter, **374**, 386, 391–392
   which to use, 390

spline filters, 352
spline from architecture, 375
spline function, 375
   constant, 375
   linear, 375
square integrable function, 111
standard deviation, **494**
   continuous, **509**
Stein's unbiased risk estimator, 335, 337
Stein, Charles, 335, 337
Stewart, James, 20, 104, 111, 211–212, 336
Strang, Gilbert, xiv, 30, 151, 220, 250–251, 461
SURE, 335
SureShrink, 318, 332, 335, 337, 342–343
   tolerance, 335
   2D example, 343
   choosing tolerance, 342
   tolerance, 335, 342, 344, 348
Sweldens, Wim, xviii, 217, 460, 476, 478
symmetric filter, 351, 353, **366**, 368–369, 371, 416
   exploiting in biorthogonal wavelet
      transformation, 436
   starting and stopping indices, 366

T

Taswell, Carl, 317
Taubman, David, 469, 474
taxicab norm, 21
text Web site, xx
threshold rule, 319–320
   edge detection, 214
tolerance
   for denoising, 321
triangle function, 375

U

uniform distribution, 500
   example, 500
universal threshold, 325, **327**–328, 334–335, 342, 344
   1D denoising example, 331
Unser, Michael, 394

V

Vaidyanathan, Palghat, 324, 348–349
variance, **494**
   computing, 509
   continuous, **509**
   normal distribution, 513
   properties, 510
   sums of, 510
Varman, Krishnaraj, 74

vectors
  described, 16
  entropy, 79
  inner product, **17**
  norm, **19**
  orthogonal set, 18
  orthogonal vectors, **18**
  standard basis vector, 29
  sum and difference, 16
  transpose, **17**
Vidakovic, Brani, 317, 327
VisuShrink, 318, 325, 343
  1D denoising example, 328–329
  2D denoising example, 331
  accuracy result, 326
  universal threshold, 327

W

Wackerly, Dennis, 317
Walnut, David, 11, xxi, 329, 377, 436
WaveLab software package, 342, 348
wavelet matrix
  block form, 226
  orthogonality condition, 226
wavelet shrinkage, 317–318
  accuracy result, 326
  algorithm, 319

why it works, 323
wavelet transform in wavelet shrinkage, 319
wavelet transform matrix
  block form, 353
  Coiflet filter, 310
  lowpass portion, 354
wavelet transform wrapping rows, 435
Wavelet/Scalar Quantization Specification, 4
Wayner, Peter, 49, 474
weakly differentiable, 337
Wickerhauser, M. Victor, 461
Woods, Richard, 49, 92, 150, 210–211, 214, 461, 469, 473
wrapping rows, 225, 436

Y

Yang, Yongzhi, 299
YCbCr color space, 70, 72, 74, 77, 197
YCbCr space, 207, 462, 470, 474
Yip, Patrick, 461
YIQ color space, 77
Yoon, Byung-Jun, 324, 348–349
YUV color space, 77

Z

zigzag array, **462**, 467